Statistics

Concepts and Applications

Amir D. Aczel

Bentley College

IRWIN

Chicago • Bogotá • Boston • Buenos Aires • Caracas
London • Madrid • Mexico City • Sydney • Toronto

Senior sponsoring editor:	Richard T. Hercher, Jr.
Associate editor:	Colleen Tuscher
Development:	Burrston House
Project editor:	Stephanie M. Britt
Production supervisor:	Laurie Kersch
Interior designer:	Stuart Paterson; Imagehouse, Inc.
Cover designer:	Stuart Paterson; Imagehouse, Inc.
Cover photographer:	Art Resource
Art studio:	Electronic Publishing Services
Graphics supervisor:	Eurnice Harris
Compositor:	Carlisle Communications, Ltd.
Typeface:	10/12 Times Roman
Printer:	Von Hoffmann Press, Inc.

Library of Congress Cataloging-in-Publication Data

Aczel, Amir D.
    Statistics : concepts and applications / Amir D. Aczel
      p.   cm.
    Includes index.
    ISBN 0-256-11935-X
    1. Statistics. I. Title.
    QA276. 12.A23   1995
    519.5—dc20
                                                      94–41399
                                                      CP

Printed in the United States of America
2 3 4 5 6 7 8 9 0 VH 1 0 9 8 7 6 5

For Debra

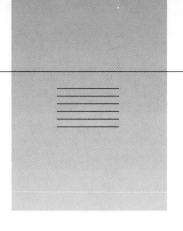

Preface

We live in the information age. The half-life of human knowledge is currently ten years overall, and only two years in science. This means that new knowledge accumulates so fast that in ten years (and a mere two years in science), half of what we now know will no longer be valid and will be replaced with new knowledge. The methods for gathering, assessing, understanding and interpreting, and drawing inferences from data constitute the field of *statistics.*

Yet, despite the great importance of the field of statistics, it seems that we have not been able to teach this discipline well. In almost two decades of dealing with students of different levels, backgrounds, and prior statistical education, I have rarely met a student who—a semester or more after completing an introductory statistics course—could demonstrate a true understanding of basic statistical concepts. I believe that one of the culprits is poorly written statistics textbooks.

Most statistics textbooks have fallen into one of two categories. In the first group are books that are theoretically correct, but abstract, mathematical, and intimidating to the student. They have no direct relevance to, or bearing on, the real world. These books perpetuate the well-known "Oh no! Not statistics!" mentality among beginning students. In the second category are books designed to be very user friendly. These books aim at what their authors perceive as the level of understanding of today's student—they focus on simple "cookbook" procedures and calculations. In my opinion, such books miss the boat completely. They tend to avoid the important concepts, their explanations tend to be superficial, and ultimately they do not really motivate

or teach statistics. Students who use such books come out of the introductory course with no lasting understanding of this important discipline or appreciation of its applications. Days after taking the final exam, students cannot remember what a p-value is, and in the future they are likely to misuse statistical methods and misinterpret results.

WHAT'S DIFFERENT, AND WHY?

The key features of this book that distinguish it from the competition are

- Superior treatment of the topics that usually cause the most difficulty for students, such as sampling, confidence intervals, and hypothesis testing.

- Optimal, research-based choice of topics to be covered.

- Careful selection of stimulating, relevant, and instructive examples that motivate the student and clarify key concepts.

- Wide variety of problems covering all levels and application areas.

- A flexible, modern introduction to statistical use of the computer, which stresses application rather than "syntax."

- A unique cross-referencing system for key concepts.

- A friendly writing style and an approach that makes statistics understandable and interesting.

In writing this book, I felt that I could make statistics both interesting and useful to students—I have been doing this for many years, teaching students of very different backgrounds and interests in many locations. The variety of students and their interests and abilities helped me develop methods for teaching statistics in a meaningful way to *anyone*. I have been aided as well by my academic research and publications, which have kept me on the leading edge of statistical knowledge. This book goes the extra mile in the sense that rather than shy away from the important and beautiful theory of statistics, the explanations of statistical ideas are thorough and relevant. I have tried to explain every concept in a clear way, often with a picture, to show the student what these important elements of statistics actually mean. Overall, reviewers have been uniformly and genuinely enthusiastic in their assessment of how well I was able to accomplish this task.

HOW THE BOOK WAS DEVELOPED

More than any other statistics book, this text was designed specifically for the needs of the student and the instructor in an introductory course. Richard D. Irwin collaborated with Burrston House, Ltd., a publishing house specializing in developing the highest-quality textbooks, for the purpose of researching the market to define the optimal content for a text designed to meet the objectives of this course.

In the first phase of this exhaustive research effort, a survey was sent to a large random sample of instructors of statistics. Respondents were asked a variety of questions about their courses, the composition of the class in terms of majors and interests, and which topics students found to be most difficult or confusing. The information and insights gleaned from the survey became the cornerstone of this book.

In the next phase, a prospectus, table of contents and detailed questionnaire were sent out for review by a total of 72 experienced instructors. Their detailed feedback was used in writing the initial draft. Next, a group of eight instructors carefully reviewed and critiqued the entire manuscript. This process was followed by a conference with first-draft reviewers at the Joint Statistical Meetings in San Francisco in 1993. Based on this input, the second draft was composed. A panel of 15 instructors evaluated each chapter in the revised manuscript. Finally, in the third draft, reviewers' comments were used in yet another revision of the text.

Here is a summary of some of what we learned from this research and reviewing process.

- The three topics causing the most difficulty for students in this course are probability, hypothesis testing (including *p*-values), and sampling distributions. All three topics as covered in this book received extremely high marks from reviewers. Most said that the treatment of these critical topics was superior to that in any other book they have used.

- Very few instructors (less than 15%) teach the Poisson distribution at this level; the same holds true for multiple regression, two-way ANOVA, and nonparametrics. Congruent with this virtual consensus, we do not cover these topics.

- Reviewers responded favorably to the integrated inclusion of the topics of sampling distributions and confidence intervals within the same chapter, and found the combination especially successful in teaching both topics more effectively.

- Reviewers were uniformly impressed with and enthusiastic about the novel use of contemporary news articles and other excerpts in teaching statistics in the examples and problems.

This book contains the following additional features, all endorsed by the majority of reviewers, and very important for effective teaching of modern statistics.

Chapter Opening Stories. Statistics is more than a science, and more than an art. One new approach to teaching statistics is to view it as a liberal art. Every chapter in the book opens with what I believe is a fascinating story. I have collected these stories over years of research, often going to the exact geographical location—an island in the Mediterranean, Westminster Abbey, or the Bancroft Library—in search of original material with statistical relevance and interest. Based on their use in my own teaching, I am confident that these stories will make the students more interested in the subject matter. Virtually all the reviewers of this book were complimentary about the chapter openers, describing them as a great motivating tool for the student.

Extensive Illustrations. Nowhere is the saying "a picture is worth a thousand words" truer than here, in trying to explain statistical concepts. This text has a number of photos as well as graphs that are unique. In the regression chapter, for example, I use a human eye looking at a data set from different angles to demonstrate the idea of the variance of Y as compared with the variance of the data around the regression line. The figures aim to enhance the student's intuition for, and comfort with, data and their display and analysis.

Everyday-Life Examples from Current Newspapers and Magazines.   Statistics surround us. One cannot open a newspaper without seeing the results of some poll showing that 70 percent of Americans do not exercise, or that 47 percent approve of the manner the president handled this or that issue; and what causes cancer? or heart attacks? is a Mediterranean diet healthier than others? Statistical issues—in fact, statistical *inference* in the essence—are everywhere. By reprinting interesting and relevant articles from *The New York Times* and other leading newspapers as well as leading national magazines, I was able to bring these issues from everyday life to the students' attention. I often ask the student to analyze the statistical issues raised in the articles. This feature, too, was highly praised as a key element in differentiating this book.

Cross-Referencing System.   For students to understand and use statistics, they need certain key concepts, or building blocks, of the subject (e.g. standard deviation, central limit theorem, or a random variable). To assist the students in this process I have included a unique cross-referencing system. Throughout this text, when students encounter an icon symbolizing links in a chain following a term, they should recognize it as a key concept. In later chapters, when this building block is used with other concepts, the icon will appear again, with the page reference to where the concept was originally developed directly underneath the icon. If the students need a quick review of the earlier topic to enhance their understanding of the current concept, they can easily do so. This study aid was ranked first (out of twelve) in utility by students in recent research conducted by the publisher.

Interviews.   I was fortunate enough to have a group of leading experts and real-world practitioners in a number of fields agree to be interviewed for the book. The interviewees talk to the student about their professions and how statistics is important in their areas of expertise. For example, celebrated defense attorney and Harvard Professor Alan M. Dershowitz discusses how the concepts of probability and statistical inference are used—and how, in his opinion, they *should* be used—by defense attorneys in criminal trials. One of this century's leading statisticians, Professor Bradley Efron of Stanford University, tells us how he used statistics in verifying Hubble's Law, as well as his view of the future of statistics and the role computing will have. Other interviews span a variety of fields and subjects, including television, the NFL, government, quality control, and medicine. Each interview is placed in a stand-alone module between chapters.

A Global Perspective.   In today's world, a text that does not address global issues—population, economics, or the environment—does a disservice to the student. A highlight of this book is the international setting and character of many of the problems and examples. A graphical comparison of life in India and China, the risk of malaria in various countries, problems on exchange rates at Orly airport in Paris, the number of miles driven by a tourist in Europe, and using statistics in scientific efforts to save the Collosseum in Rome from collapse are just a few examples.

Comprehensive, Thoroughly Explained Examples.
There are few things in an introductory textbook that are as important as the examples. I have chosen examples carefully from a large variety of fields. The situations are all realistic and often include the actual data. The student is taken through each example step by step, de-emphasizing formulas and concentrating on concepts and their application. The examples do not end with a numerical answer—interpretations are offered to show the student how to use and understand the results obtained from statistics.

Ample Real-World Problems.   The problem sets in this book were highly praised during the review process. They were chosen from many fields—medicine, education, archaeology, social science, psychology, engineering, nursing, business, economics, and more. The problems are leveled: Starting with simple exercises in each section to build the student's confidence, and leading to more involved problems that make the student think not only about statistics, but about the problem area and its relevance. I have also included some problems that refer to large data sets at the end of the book, and some that ask for a written summary or report.

Integrated Computer Use. Since statistics is most often applied with the aid of computers, the computer applications are developed throughout the text, starting in Chapter 1. The package MINITAB was chosen as the example package for its ease of use. The commands are explained throughout the book as needed, so that the student does not need to consult a manual and can use the computer with ease early on. Certain problems are designated as Computer Problems and marked with an icon. These problems may also be solved by hand, but their preferred solution is by computer.

The Entire Package. The book is accompanied by a complete array of teaching and learning aids. These include:

- An Instructor's Solutions Manual.
- A Student's Solutions Guide.
- An Instructor Resource Guide.
- An Instructor's Transparency Master.
- A Data Disk.
- A Test Bank.
- A Computerized Test Bank.

In addition, there are several packaging options available from the publisher.

ACKNOWLEDGMENTS

I am deeply indebted to the following statistics instructors who have evaluated all ten chapters and with their comments, ideas, and advice have influenced every page of this text. Without their encouragement, willingness, and dedication to assist in producing a better text it would not have happened. Each of them will recognize their contributions and I'm hopeful they are pleased with the result. Special gratitude is extended to Patricia Buchanan of Pennsylvania State University, Chris Franklin of the University of Georgia, and Joseph Ibrahim of Harvard University, all of whom labored through two entire drafts. Chris and Joe also participated in the reviewer conference at the ASA meetings.

First Draft Reviewers

Patricia Buchanan, Pennsylvania State University

Shyam Chadha, University of Wisconsin-Eau Claire

Chris Franklin, University of Georgia

Brenda Gunderson, University of Michigan

Joseph Ibrahim, Harvard University

Y. Leon Maksoudian, California State Polytechnic University-San Louis Obispo

E. D. McCune, Stephen F. Austin State University

Mack Shelley, Iowa State University

Nola Tracey, McNeese State University

Second Draft Reviewers

William Beyer, University of Akron

Patricia Buchanan, Pennsylvania State University

Michael Doviak, Old Dominion University

Eugene Enneking, Portland State University

Chris Franklin, University of Georgia

Gavin Gregory, University of Texas-El Paso

Bernard Harris, University of Wisconsin-Madison

Jason Hsu, Ohio State University

Joseph Ibrahim, Harvard University

J. David Mason, University of Utah

Ronald McCuiston, Pensacola Junior College

Linda McDonald, University of Texas-Arlington

Wendy McGuire, Santa Fe Community College

John Unbehaun, University of Wisconsin-LaCrosse

John Whitesitt, Southern Oregon State College

As mentioned previously, the following instructors helped shape the scope, sequence, and depth of coverage in this text by completing an in-depth questionnaire, evaluating a preliminary table of contents, and critiquing a sample chapter. Their collective judgment was instrumental in putting this project on the right track from the outset.

Madeline A. Bradley, University of North Carolina-Greensboro

Shyam Chadha, University of Wisconsin-Eau Claire

Harold Jacobs, East Stroudsburg University

Jim Jacobs, Valencia Community College

Phillip E. Johnson, University of North Carolina-Charlotte

Don Loftsgarden, University of Montana

Gary Klingler, Grand Valley State University

Linda McDonald, University of Texas-Arlington

David K., Neal, Western Kentucky University

John Whitesitt, Southern Oregon State College

Shelemyahu Zacks, State University of New York-Binghampton

I also wish to thank the following instructors and students who read the interviews and gave me advice regarding the types of individuals and subjects they think students taking an introductory statistics course would find valuable and interesting:

E. Jacquelin Dietz, North Carolina State University

Michael Doviak, Old Dominion University

Chris Franklin, University of Georgia

Susan K. Herring, Sonoma State University

Joseph Ibrahim, Harvard University

Barry Katz, St. Louis University

J. David Mason, University of Utah

University of Georgia Students: Heather Adams, Matthew Freeman, Wood Pope, Trent Schueneman, Lisa Spitz. Thank you to Carl Mueller and Gene Enneking, who double checked all the calculations and solutions in the text. They were both extremely thorough and this text is much cleaner and more correct as a result. I take full responsibility for any errors that remain.

Thank you also to the interviewees for agreeing to provide their candid and detailed comments and information about their fields and the role statistics play in them. When we set out to do the interviews, we wanted to get the real, detailed, dramatic, inside scoop, and all of the interviewees were overwhelmingly supportive of that effort. In fact, they all graciously provided much more than we ended up using in the text. There were several goals we had in mind, that I believe are accomplished in the final versions included in the text. First, we wanted to provide the student with a personal insight into the people who participate in the field of statistics. Many students leave their course with the impression statistics is strictly numbers and mathematics. We tried therefore to include some biographical background, real personality, and some authority.

Second, we wanted the interviews to be 'deep' enough to be of real value to read, and honest. The interviewees didn't hesitate, and we hope that you and your students will appreciate the candor. Finally, we wanted to provide a broad range of application areas and show students how useful and important

statistics are in the world. A quick summary of who the interviewees are and why they are included is listed below.

Melissa Anderson was interviewed about statistics in education as well as issues in research and ethics. Her academic interest is in the study of education, and she has worked with the Acadia Institute.

Alan Dershowitz was interviewed about how statistics, especially probability, is used and not used in Law, particularly in criminal defense cases.

Arnold Zellner is included to provide background on the field of statistics, the Bayesian approach, and the interplay of economics and statistics.

Bud Goode talks about a popular field, sports, particularly the National Football League, and how he uses various statistics to analyze action on the gridiron.

Bradley Efron provides a research perspective on trends in the field of Statistics and on advances in the field based on improving uses of the computer.

Lee Wilkinson describes how his dual interests in statistics and computers led him into the role of entrepreneur, and founding Systat, Inc.

Toni Falvo provided an overview of the use of data and market research in programming and advertising decisions at WLS, a Chicago television station.

Nancy Kirkendall describes the information gathering and analysis used by the federal government, in this case, to monitor and plan for U.S. energy consumption.

Mary Guinan is active in the battle against AIDS and her interview describes some of the statistical methods and issues which are important to the CDC.

Phil Crosby is interviewed about the relationship between traditional statistics and the current trend toward Total Quality Management.

My warmest thanks also to the following friends and colleagues: Ray Ledoux, Art O'Leary, Professors Scott Callan, Richard Frese, Richard Fristensky, Steven Grubaugh, David Gulley, Charles Hadlock, Dominique Haughton, Jack Hegarty, Erl Sorensen, Andrew Stollar, Nicholas Teebagy, James Zeitler. I am also indebted to my friend Mario Demanuele of Malta for his kind help with pictures of the Mosta bomb.

Warm thanks to my editor, Richard T. Hercher Jr of Irwin, for his encouragement and direction throughout this entire project, and for his many excellent ideas and suggestions. I thank Colleen Tuscher, the associate editor, for her enthusiasm and for her many contributions to the book. Many thanks to Stephanie Britt, the project editor, for her superb editing of the manuscript. Thanks also go to Stuart Paterson, Imagehouse, Inc., for the interior and cover design.

I want to express my deep gratitude to Glenn Turner of Burrston House for his wise suggestions, direction, and encouragement, and for his countless insights in developing this book. Thanks also to Meg Turner for her extensive help on this project. I thank the staff at Burrston House for their efforts.

Finally, I thank my wife, Debra, for all her help and for some of the photographs.

Amir D. Aczel

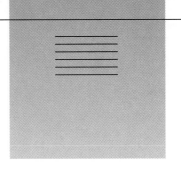

Contents in Brief

CHAPTER 1   Information Everywhere   2

Interview with Melissa Anderson   67

CHAPTER 2   What Are the Chances?   70

Interview with Alan Dershowitz   107

CHAPTER 3   Chance Quantities   112

Interview with Arnold Zellner   151

CHAPTER 4   The Bell-Shaped Curve   154

Interview with Bud Goode   185

CHAPTER 5   Let's Take a Sample   190

Interview with Bradley Efron   239

CHAPTER 6   Trial by Probability   244

Interview with Lee Wilkinson   309

CHAPTER 7   Making Comparisons   314

Interview with Toni Falvo   351

CHAPTER 8   Making Extended
Comparisons   356

Interview with Nancy Kirkendall   401

CHAPTER 9   Is There a Relationship?   406

Interview with Mary Guinan   461

CHAPTER 10   The Quest for Quality   464

Interview with Phil Crosby   481

Contents

CHAPTER 1 Information Everywhere 2

1.1 Introduction 2
    Samples and Populations 5
    Data and Data Collection 5
    Data Distributions 7
1.2 Methods of Displaying Data 9
    Pie Charts 9
    Bar Graphs 12
    The Histogram 13
    Stemplots (Stem-and-Leaf Displays) 15
    Frequency Polygons and Ogives 19
    A Caution about Graphs 21
1.3 Percentiles, Quartiles, and the Median 35
    Percentiles 35
    Quartiles 35
    The Median 36
1.4 Box Plots 42
1.5 Measures of Centrality 45
1.6 Measures of Variability 50
    The Empirical Rule 55
1.7 Caution, Care with Numbers, and Ethics 56
    Ethics 57

Interview with Melissa Anderson 67

CHAPTER 2 What Are the Chances? 70

2.1 Introduction 72
2.2 Basic Definitions: Events, Sample Space, and Probabilities 75
2.3 Basic Rules for Probability 81
    The Range of Values 81
    The Rule of Complements 82
    The Rule of Unions 83
    Mutually Exclusive Events 83

2.4 Conditional Probability 87
2.5 Independence of Events 92
    Product Rules for Independent Events 93
2.6 Bayes' Theorem 97

Interview with Alan Dershowitz 107

CHAPTER 3 Chance Quantities 112

3.1 Random Variables 115
    Discrete and Continuous Random Variables 120
3.2 Expected Values of Discrete Random Variables 124
    The Variance and the Standard Deviation of a Random Variable 127
    Some Properties of Means and Variances 129
3.3 The Binomial Distribution 130
    The Binomial Table 132
    Mean, Variance, and Shape of the Binomial Distribution 135
    Sampling with or without Replacement 137
3.4 Continuous Random Variables 140

Interview with Arnold Zellner 151

CHAPTER 4 The Bell-Shaped Curve 154

4.1 The Normal Probability Distribution 156
    The Shape of the Normal Distribution 158
    Notation 159
4.2 The Standard Normal Distribution 159
    Finding Probabilities of the Standard Normal Distribution 160
    Finding Values of Z Given a Probability 164

4.3 The Transformation of Normal Random
     Variables 167
         Using the Normal Transformation 169
4.4 The Relationship between X and Z and the Use
     of the Inverse Transformation 173
4.5 The Normal Distribution as an Approximation
     to Other Probability Distributions 176
4.6 Normal Data 178

Interview with Bud Goode 185

CHAPTER 5   Let's Take a Sample 190

5.1 Statistical Inference 192
         Sampling Methods and Statistical Inference 196
5.2 Sampling Distributions 199
5.3 The Sampling Distribution of the Sample
     Mean 200
5.4 Confidence Intervals 205
         Confidence Interval for the Population Mean
             When the Population Standard Deviation Is
             Known 205
5.5 Confidence Intervals for μ When σ Is
     Unknown— the t Distribution 214
         The t Distribution 214
5.6 The Sampling Distribution of the Sample
     Proportion 222
         Large-Sample Confidence Intervals for the
             Population Proportion 223
5.7 Sample Size Determination 228
5.8 Epilogue: The Polls 234

Interview with Bradley Efron 239

CHAPTER 6   Trial by Probability 244

6.1 Introduction 247
6.2 Statistical Hypothesis Testing 257
6.3 Tests about the Population Mean, μ, When
     the Population Standard Deviation, σ, Is
     Known 257
         Standardizing the Test 257
         One-Tailed Tests 260
         The p-Value 265
6.4 Tests about the Population mean, μ, When
     the Population Standard Deviation, σ, Is
     Unknown 274
6.5 Large-Sample Tests for the Population
     Proportion 284

6.6 How Hypothesis Testing Works 291
         Some Comments and Caveats 294
6.7 The Probability of a Type II Error and the
     Power of the Test 297

Interview with Lee Wilkinson 309

CHAPTER 7   Making Comparisons 314

7.1 Introduction 316
7.2 Paired-Observations Comparisons 316
         Experimental Design 316
         Confidence Intervals 318
7.3 A Test for the Difference between Two
     Population Means, Using Independent Random
     Samples 322
         Confidence Intervals 325
7.4 A Test for the Difference between Two
     Population Means, Assuming Equal Population
     Variances 330
         Confidence Intervals 333
7.5 A Large-Sample Test for the Difference
     between Two Population Proportions 336
         Confidence Intervals 338

Interview with Toni Falvo 351

CHAPTER 8   Making Extended
              Comparisons 356

8.1 Introduction 358
8.2 The Hypothesis Test of Analysis of Variance
     (ANOVA) 359
         The Test Statistic and the F Distribution 361
8.3 The Theory and the Computations
     of ANOVA 365
         The Sum of Squares Principle 369
         The Degrees of Freedom 372
         The Mean Squares 373
         The Expected Values of the Statistics MSTR and
             MSE under the Null Hypothesis 373
         The F Statistic 374
         The ANOVA Table 374
         More on Experimental Design 379
         Randomized Complete Block Design 379
8.4 Contingency Table Analysis: A Chi-Square Test
     for Independence 384
         The Chi-Square Distribution 386
         A Chi-Square Test for Equality of Proportions 389

Interview with Nancy Kirkendall 401

CHAPTER 9   Is There a Relationship? 406

9.1 Introduction 408
9.2 Statistical Models 409
9.3 The Simple Linear Regression Model 410
9.4 Estimation: The Method of Least Squares 414
9.5 Error Variance and the Standard Errors of
     Regression Estimators 421
        Confidence Intervals for the Regression
           Parameters 424
        A Hypothesis Test for the Existence of a Linear
           Regression Relationship 425
9.6 How Good Is the Regression? 431
        Analysis of Variance Table and an F Test of the
           Regression Model 435
9.7 Residual Analysis and Checking for Model
     Inadequacies 437
        A Check for the Equality of Variance of the
           Errors 438
        Testing for Missing Variables 438
        Detecting a Curvilinear Relationship between Y
           and X 439
        Detecting Deviations from the Normal
           Distribution Assumption 439
9.8 Use of the Regression Model
     for Prediction 444
        Point Predictions 444
        Prediction Intervals 445
        A Confidence Interval for the Average Y, Given a
           Particular Value of X 446
9.9 Correlation 448

Interview with Mary Guinan 461

CHAPTER 10   The Quest for Quality 464

10.1 Control Charts 466
10.2 The $\bar{x}$ Chart 468
10.3 The R Chart and the s Chart 472
        The R Chart 472
        The s Chart 473
10.4 The p Chart 474
10.4 The c Chart 476

Interview with Phil Crosby 481

GLOSSARY   483

BIBLIOGRAPHY   487

APPENDIX A: DATA TABLES   489

**APPENDIX B: STATISTICAL
TABLES   493**

**APPENDIX C: ANSWERS
TO MOST ODD-NUMBERED
PROBLEMS   519**

INDEX   531

Statistics

Concepts and Applications

Information Everywhere

The Phaestos Disk

1.1 Introduction 4

1.2 Methods of Displaying Data 9

1.3 Percentiles, Quartiles, and the Median 35

1.4 Box Plots 42

1.5 Measures of Centrality 45

1.6 Measures of Variability 50

1.7 Caution, Care with Numbers, and Ethics 52

O f the several archaeological studies begun in Crete in the 1800s, the most famous is Arthur Evans's excavation of King Minos's palace at Knossos. But farther south on Crete lies the site of the 3,500-year-old palace of Phaestos. Many artifacts were recovered from the site, and archaeologists were able to understand the purpose of most of the objects they discovered. Then they discovered the Phaestos disk. The disk is like no other object ever seen. It is a flat, light round disk of clay about 7 inches in diameter. Hieroglyphic-like characters, similar to the ones seen in Egypt, are impressed on it in a spiral. The disk was a compact and unusual device for storing and displaying large amounts of information in the ancient Minoan civilization. Yet the Phaestos disk remains undeciphered.

In this chapter we discuss information: its collection, display, and summarization. These are the first steps of any statistical analysis.

1.1 INTRODUCTION

Information may be *qualitative* or *quantitative*. To illustrate the difference between these two types of information, let's consider an example related to the chapter opener. The Minoan civilization, which produced the Phaestos disk, worshiped the bull, and bulls roamed freely in Minoan palaces.[1] A qualitative characteristic of a bull may be his color. A quantitative characteristic may be his weight, height, or age. When the Phaestos disk is finally deciphered, we may find that it is a record of both qualitative and quantitative information.

A piece of information that may be expressed as a number, changing from item to item, is called a **variable.** The weight of a Minoan bull, which changes from bull to bull, is a quantitative variable. A qualitative variable (also called a categorical variable) is harder, conceptually, to express as a number. If the Minoans had only white or black bulls, they might have decided to code a black bull's color by assigning it a 1 and a white one's by assigning it a 0. (Or the reverse could have been used, 0 for black and 1 for white.) The use of 0 and 1 to code qualitative variables may be arbitrary, but it is the usual practice. With more colors than just black or white, 2, 3, 4, and so on could be used.

> A **quantitative variable** can be described by a number for which arithmetic operations such as averaging make sense. A **qualitative (or categorical) variable** simply records a quality. If a number is used for distinguishing members of different categories of a qualitative variable, the number assignment is arbitrary.

The field of statistics deals with **measurements**, some quantitative and some qualitative. The measurements are the actual numerical values of a variable. To describe two white bulls and three black ones, our color measurements could be 0, 0, 1, 1, 1. The weight measurements for the same five bulls, in kilograms, might be 651.5, 702, 599.1, 683, 707.4.

There are four generally used **scales of measurement,** listed here from weakest to strongest.

Nominal Scale.   In the **nominal scale** of measurement, numbers are used simply as labels for groups or classes. If our data set consists of blue, green, and red items, we may designate blue as 1, green as 2, and red as 3. In this case, the numbers 1, 2, and 3 stand only for the category to which a data point belongs. "Nominal" stands for "name" of category. The nominal scale of measurement is used for qualitative rather than quantitative data: blue, green, red; male, female; professional classification; geographic classification; and so on.

Ordinal Scale.   In the **ordinal scale** of measurement, data elements may be ordered according to their relative size or quality. Four products ranked by a consumer may be ranked as 1, 2, 3, and 4, where 4 is the best and 1 is the worst. In this scale of

---

[1]The legend of the Minotaur derives from this practice.

measurement we do not know how much better one product is than others, only that it is better.

Interval Scale.   In the **interval scale** of measurement, we can assign a meaning to distances between any two observations. The data are on an *interval* of numbers, and the distances between elements can be measured in units. In January 1992, the Dow Jones average was 3108; in January 1994, it was 3914. These numbers are on an interval scale.

Ratio Scale.   The **ratio scale** is the strongest scale of measurement. Here not only do distances between paired observations have a meaning, but there is also a meaning to ratios of distances. Salaries are measured on a ratio scale: A salary of $50,000 is twice as large as a salary of $25,000. Such a comparison is not possible with temperatures, which are on an interval scale but not on a ratio scale (we can't say that 50°F is twice as warm as 25°F). The ratio scale contains a *meaningful zero* (0°F in temperature is not meaningful in this respect). The distinction between the interval and ratio scales, however, is not always immediately clear.

Samples and Populations

In statistics we make a distinction between two concepts: a population and a sample.

Courtesy of Grant Heilman
Photography, Inc.

The **population** consists of the set of all measurements in which the investigator is interested. The population is also called the **universe**.

A **sample** is a subset of measurements selected from the population. Sampling from the population is often done randomly, such that every possible sample of *n* elements will have an equal chance of being selected. A sample selected in this way is called a **simple random sample**, or just a **random sample**. A random sample allows chance to determine its elements.

For example, Farmer Jane owns 1,264 sheep. These sheep constitute her entire *population* of sheep. If 15 sheep are selected to be sheared, then these 15 represent a *sample* from Jane's population of sheep. Further, if the 15 sheep were selected at *random* from Jane's population of 1,264 sheep, then they would constitute a *random sample* of sheep.

The definitions of sample and population are relative to what we want to consider. If Jane's sheep are all we care about, then they constitute a population. If, however, we are interested in all the sheep in the county, then all of Jane's 1,264 sheep are a sample of that larger population (although this sample would *not* be random).

The distinction between a sample and a population is very important in statistics.

Data and Data Collection

A set of measurements obtained on some variable is called a **data set.** For example, heart rate measurements for 10 patients may constitute a data set. The variable we're interested in is heart rate, and the scale of measurement here is a ratio scale. (A heart that beats 80 times a minute is twice as fast as a heart that beats 40 times a minute.) Our actual observations of the patients' heart rates, the data set, might be: 60, 70, 64, 55, 70, 80, 70, 74, 51, 80.

Data are collected by various methods. Sometimes our data set consists of the entire population we're interested in. If we have the actual point spread for five football games, and if we are interested only in these five games, then our data set of five measurements is the entire population of interest. (In this case, our data are on a ratio scale. Why? Suppose the data set for the five games tells only whether the home or visiting team won. What would be our measurement scale in this case?)

In other situations data may constitute a sample from some population. If the data are to be used to draw some conclusions about the larger population they were drawn from, then we must collect the data with great care. A conclusion drawn about a population based on the information in a sample from the population is called a **statistical inference.** Statistical inference is an important topic of this book. To ensure the accuracy of statistical inference, data must be drawn randomly from the population of interest, and we must make sure that every segment of the population is adequately and proportionally represented in the sample.

Statistical inference may be based on data collected in surveys or experiments, which must be carefully constructed. For example, when we want to obtain information from people, we may use a mailed questionnaire or a telephone interview as a convenient instrument. In such surveys, however, we want to minimize any **nonresponse bias.** This is the biasing of the results that occurs when we disregard the fact that some people will simply not respond to the survey. The bias distorts the findings, because the people who do not respond may belong more to one segment of the population than to another. In social research, some questions may be sensitive, for example, "Have you ever been arrested?" This may easily result in a nonresponse bias, because people who have indeed been arrested will be less likely to answer the question (unless they can be perfectly certain of remaining anonymous). Surveys conducted by popular magazines often suffer from nonresponse bias, especially when their questions are provocative. What makes good magazine reading often makes bad statistics.

Suppose we want to measure the speed performance or gas mileage of an automobile. Here the data will come from experimentation. In this case we want to make sure that a variety of road conditions, weather conditions, and other factors are represented. Pharmaceutical testing is also an example where data may come from experimentation. Drugs are usually tested against a placebo as well as against no treatment at all. When designing an experiment to test the effectiveness of a sleeping pill, the variable of interest may be the time, in minutes, that elapses between taking the pill and falling asleep.

In experiments, as in surveys, it is important to **randomize** if inferences are indeed to be drawn. People should be randomly chosen as subjects for the experiment if an inference is to be drawn for the entire population. Randomization should also be used in assigning people to the three groups: pill, no pill, or placebo. Such a design will minimize potential biasing of the results.

In other situations data may come from published sources, such as statistical abstracts of various kinds or government publications. The published unemployment rate over a number of months is one example. Here, data are "given" to us without our having any control over how they are obtained. Again, caution must be exercised. The unemployment rate over a given period is not a random sample of any *future* unemployment rates, and making statistical inferences in such cases may be involved and difficult. If, however, we are interested only in the period we have data for, then our data do constitute an entire population, which may be described. In any case, however, we must also be careful to note any missing data or incomplete observations.

Data Distributions

One way to describe any data set we may have is to construct the data distribution.

> The **distribution** of a data set is a listing of the frequencies of occurrence of the measurements in the data set. The distribution can give us a picture of the data set. It gives us information about the variable being measured.

Computers, along with statistical software, are very useful in analyzing and graphing data and data distributions. In this book we will make use of a statistical computing package called MINITAB, which is described briefly in the accompanying table. Further explanations of MINITAB are given throughout the book as needed so that you can use the package as an aid in solving the problems at the end of each section.

An Introduction to the Computer Package MINITAB

MINITAB is a widely used, easy-to-learn computer package that can save time in the simple statistical computations described in the early chapters of this book. Using MINITAB or another statistical computing package is essential for carrying out the more advanced statistical methods in the latter part of the book. MINITAB is available for both large computers and personal computers, and there are versions that run in DOS Windows and on a Macintosh. The version you have may be menu driven (thus even easier to use than described here), but we will explain the basic commands, which may be used regardless of special features. Further MINITAB commands will be explained where needed throughout this book.

What makes MINITAB so easy to learn is its very simple structure. We enter data into *columns.* The columns are named successively as C1, C2, C3, and so on. No "syntax" is required. You simply type the following at the MINITAB prompt (MTB>):

    MTB> SET into C1

Note that we will capitalize necessary words in each command. Words shown in lower case are not required and are given only to help us remember the commands and make them more like everyday English. The computer will then put you in a DATA mode (new prompt), at which you may enter the data values. Simply type the numbers, separated by spaces:

    DATA> 5 17.3 204.157 12 42.8 (and so on).

We may also use the READ command to enter data into several columns simultaneously or for reading an existing file into one or more columns. To read a simple (ASCII) data file—which contains only numbers (no headings or other information) in five columns separated by blanks—called MYDATA and residing on drive A of your computer, type:

    MTB> READ 'A:MYDATA' into C1-C5

Returning to the heart rate measurement example, what is the data distribution here? We see that the value 80 appears twice (2 of our 10 measurements are 80); the value 70 appears three times (3 measurements are 70); and the rest of the data values appear only once. We can plot this data distribution, using the MINITAB command HISTOGRAM, as follows:

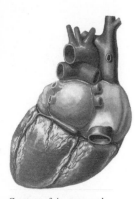

```
MTB > SET into C1
DATA> 60 70 64 55 70 80 70 74 51 80
DATA> END
MTB > HISTOGRAM of data in C1
Histogram of C1        N = 10
Midpoint Count
        50       1  *
        55       1  *
        60       1  *
        65       1  *
        70       3  ***
        75       1  *
        80       2  **
```

Courtesy of *Anatomy and Physiology,* by Seeley, Stephens, and Tate, published by Mosby-Wolfe.

The data here have a simple structure. In general, the HISTOGRAM program in MINITAB will group our observations into intervals of measurements and will then plot the frequency of each interval of numbers. (The midpoints of these intervals are in the left column in the display. This is why the measurements 64 and 74 are plotted at the interval "midpoints" of 65 and 75, respectively.)

Because looking at data is often the first stage of any statistical analysis, graphical methods of displaying data are very important. Looking at the distributions of data may reveal important information about variables and their interactions. The graphs will reveal information about the central tendency of the data (i.e., Where is the distribution **centered**?), as well as information about the **spread** of the data set. Graphical methods will also reveal unusual measurements, called **outliers.** Outliers are larger or smaller than most other measurements in a data set.

Graphical methods for analyzing data (including the histogram used above in the heart rate example) are described in section 1.2, Methods of Displaying Data.

PROBLEMS

1.   A survey by an electric company contains questions on the following:

   1.   Age of household head.
   2.   Sex of household head.
   3.   Number of people in household.
   4.   Use of electric heating (yes or no).
   5.   Number of large appliances used daily.
   6.   Thermostat setting in winter.
   7.   Average number of hours heating is on.
   8.   Average number of heating days.
   9.   Household income.
   10.   Average monthly electric bill.
   11.   Ranking of this electric company as compared with two previous electricity suppliers.

   Describe the variables implicit in these 11 items as quantitative or qualitative, and describe the scales of measurement.

2.   Discuss the various data collection methods described in this section.
3.   Discuss and compare the various scales of measurement.
4.   What is the distribution of each of the variables in the following table of statistics for New Orleans Saints receivers? Create a histogram for each variable.

Saints Statistics

	NO	Yds	Avg	LG	TD
E. Martin	54	735	13.6	34	2
Early	40	565	14.1	63	6
Hilliard	29	185	6.4	19	0
Brown	17	142	8.4	19	1
Small	16	164	10.3	17	1
Smith	14	156	11.1	23	1
Muster	14	142	10.1	31	0
Brenner	11	171	15.5	27	1
Turner	11	111	10.1	16	1
Ned	9	54	6.0	14	0
Newman	8	121	15.1	32	1
Dowdell	6	46	7.7	11	1
McAfee	1	3	3.0	3	0

Source: Reproduced by permission of *The New York Times,* December 20, 1993, p. C7.

5. Five ice cream flavors are rank-ordered by preference. What is the scale of measurement?
6. What is the difference between a qualitative and a quantitative variable?
7. A town has 15 neighborhoods. If you interviewed everyone living in one particular neighborhood, would you be interviewing a population or a sample from the town? Would this be a random sample?

   If you had a list of everyone living in the town, called a **frame,** and you randomly selected 100 people from all the neighborhoods, would this be a random sample?
8. What is the difference between a sample and a population?
9. What is a random sample?
10. For each tourist entering the United States, the U.S. Immigration and Naturalization Service computer is fed the tourist's nationality and length of intended stay. Characterize each variable as quantitative or qualitative.
11. What is the scale of measurement for the color of a karate belt?
12. What is the scale of measurement for the amount of snow deposited by a storm?
13. These are the numbers of pottery remains in sections of an archaeological site:

    14, 17, 5, 12, 16, 8, 9, 7, 14, 16, 9, 10, 5, 18, 17, 17, 13, 8, 2, 15, 14, 6, 14, 9, 4, 16, 8, 17, 11, 8, 14, 9, 15, 14, 16, 5, 13, 12, 17, 17, 6, 15, 10.

    Use a computer to draw a histogram of the distribution of number of pottery remains.

1.2 METHODS OF DISPLAYING DATA

Pie Charts

A **pie chart** is a simple descriptive display of data that sum to a given total. A pie chart is probably the most illustrative way of displaying quantities as percentages of a given total. The total area of the pie represents 100 percent of the quantity of interest (the sum of the variable values in all categories), and the size of each slice is the percentage of the total represented by the category the slice denotes. Pie charts are used to present frequencies for categorical data. The scale of measurement may be nominal or ordinal. Figure 1.1 and Figure 1.2 are examples of pie charts. Note the different categories in each figure and how their frequencies are displayed in the pie.

FIGURE 1.1

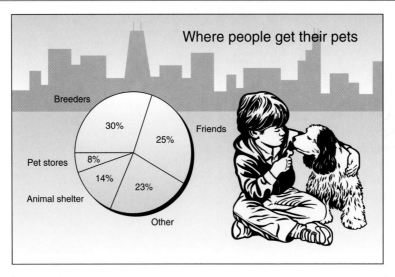

Reprinted by permission from *USA Today,* September 6, 1991, p. 1, from data provided by the American Animal Hospital Association.

FIGURE 1.2

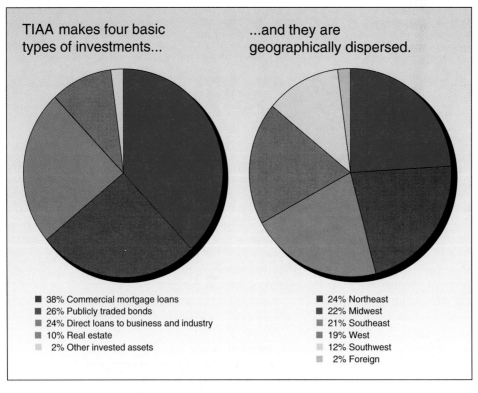

Reprinted by permission from *TIAA/CREF Annual Report,* 1991, p. 13.

The pie chart in Figure 1.1 was constructed as follows. Out of 2,000 pet owners surveyed, 500 said they got their pets from friends. Since 500 out of 2,000 is 25 percent, the slice of the pie corresponding to "friends" must comprise 25 percent of the area of the entire pie. Similarly, 600 out of 2,000 people said they got their pets from breeders. Since this corresponds to 30 percent, the "breeders" slice must comprise 30 percent of the area of the pie. Similar computations lead to 8 percent for "pet stores," 14 percent for "animal shelters," and 23 percent for "other." The total area of the pie is 100 percent.

The data of TIAA/CREF indicated that 38 percent of their investments were in commercial mortgage loans, 26 percent were in publicly traded bonds, and so on. This information, and the information on the geographical distribution of investments, was used in creating the pie charts of Figure 1.2.

The construction of the pie can be done by hand, but since measurements on a circle are difficult, using a computer is more convenient.

The Richest Get Even Richer

FIGURE 1.3

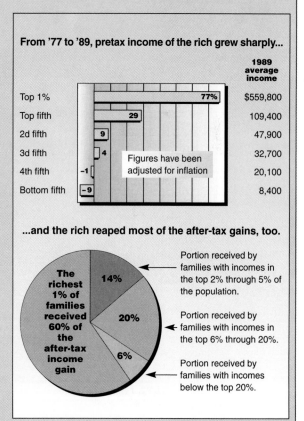

New York, Thursday, March 5, 1992

Even Among the Well-Off, the Richest Get Richer

Data Show the Top 1% Got 60% of the Gain in the 80's Boom

By SYLVIA NASAR

Populist politicians, economists and ordinary citizens have long suspected that the rich have been getting richer. What is making people sit up now is recent evidence that the richest 1 percent of American families appears to have reaped most of the gains from the prosperity of the last decade and a half.

An outsized 60 percent of the growth in after-tax income of all American families between 1977 and 1989—and an even heftier three-fourths of the gain in pretax income—went to the wealthiest 660,000 families, each of which had an annual income of at least $310,000 a year, for a household of four.

From '77 to '89, pretax income of the rich grew sharply...

		1989 average income
Top 1%	77%	$559,800
Top fifth	29	109,400
2d fifth	9	47,900
3d fifth	4	32,700
4th fifth	-1	20,100
Bottom fifth	-9	8,400

Figures have been adjusted for inflation

...and the rich reaped most of the after-tax gains, too.

The richest 1% of families received 60% of the after-tax income gain

14% — Portion received by families with incomes in the top 2% through 5% of the population.

20% — Portion received by families with incomes in the top 6% through 20%.

6% — Portion received by families with incomes below the top 20%.

Bar Graphs

Bar graphs (which use horizontal or vertical rectangles) are often used to display categorical data where there is no emphasis on the percentage of a total represented by each category. The scale of measurement is nominal or ordinal.

Graphs using horizontal bars and those using vertical bars are essentially the same. In some cases, it may be more convenient for the purpose at hand to use one versus the other. For example, if we want to write the name of each category inside the rectangle that represents that category, then a horizontal bar graph may be more convenient. If we want to stress the height of the different columns as measures of the quantity of interest, we would use a vertical bar graph.

Figure 1.3 shows both a bar graph and a pie chart. Figure 1.4 is a collection of descriptive bar graphs. Note the ordinal scale of measurement (year).

FIGURE 1.4   Percentages of Revenues for High-Tech Firms

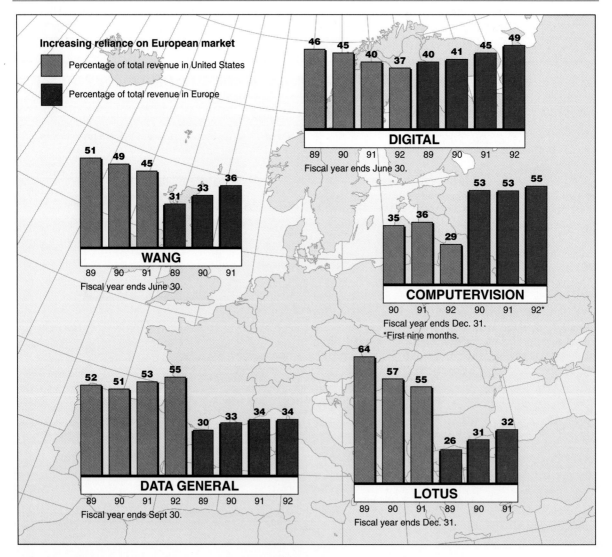

Reprinted by permission from P. Hemp, "Continental Chill," *Boston Globe,* November 15, 1992, p. A1.

The Histogram

A **histogram** is a graphical display of the distribution of a data set. It is used for data
on an interval or ratio scale of measurement. A histogram looks like a vertical bar
graph, except that the columns touch each other (although MINITAB-generated plots
cannot effectively show this without the High-Resolution Graphics option). Since a
histogram displays the distribution of an entire (possibly large) data set with a few
touching columns, there is some aggregation of data into convenient groups, and each
column measures the relative frequency of its particular group of data measurements.
Let's look at an example.

The management of an appliance store recorded the amounts spent at the store by the
184 customers who came in during the last day of the big sale. The data, amounts
spent, were grouped into categories as follows: $0 to less than $100, $100 to less than
$200, and so on up to $600, a bound higher than the amount spent by any single buyer.
The classes and the frequency of each class are shown in Table 1.1. The frequencies,
denoted by $f(x)$, are shown in a histogram in Figure 1.5.

EXAMPLE 1.1

Classes and Frequencies, Example 1.1   **TABLE 1.1**

x Spending Class ($)	$f(x)$ Frequency (number of customers)
0 to less than 100	30
100 to less than 200	38
200 to less than 300	50
300 to less than 400	31
400 to less than 500	22
500 to less than 600	13
	184

A Histogram of the Data in Example 1.1   **FIGURE 1.5**

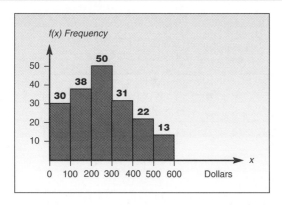

A histogram gives us an idea about the *shape* of the data distribution. It can indicate to us, graphically, where the *center* of the data distribution lies. We'll define measures of centrality of a distribution more precisely later in this chapter. Here, we'll mention informally three measures of centrality: mean, median, and mode. The *mean* is the "center of mass" of the distribution; it is an imaginary fulcrum on which the weight of all the columns together would balance. The *median* is the point that has half the area of the histogram lying to its left and the other half lying to its right. The *mode* is a point under the tallest column (there may be more than one mode).

The histogram will also reveal whether the data distribution is *symmetric* or *skewed*. A symmetric distribution is one that can be "folded" about its median (its center) so that both halves will look identical. Each half of the distribution is a mirror image of the other, the mirror being placed in the center of the distribution. If the distribution has a single highest point in the middle, then the mean, median, and mode are all equal. If it is not symmetric a distribution is skewed. A right-skewed distribution spreads more to the right than to the left (it has a longer "tail" to the right than to the left) and vice versa for a left-skewed distribution. Histograms of symmetric, left-skewed, and right-skewed distributions and their measures of centrality are shown in Figure 1.6.

In Example 1.2 we use the computer to generate histograms (as is most commonly done). Here, the computer will group the data automatically and create the histogram.

FIGURE 1.6   Skewness of Distributions

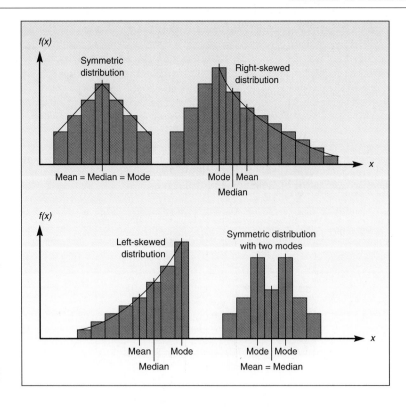

EXAMPLE 1.2

Table 1.2 on page 16 reports percentages of on-time arrivals and on-time departures at major U.S. airports for the second half of 1990 and all of 1991. Using all airports, let's construct histograms of the data for the last (fourth) quarter of 1991. The entire MINITAB program and the resulting histograms are shown in Figure 1.7(a) on page 17.

These histograms provide a great deal of information about the distributions of on-time arrivals and departures. The arrivals distribution is skewed to the right: fewer and fewer airports have higher and higher percentages of on-time arrivals; there is a bigger "lump" at low on-time arrival percentages. The distribution of on-time departure percentages seems more uniformly spread out, over the range of 81 to 93 percent. Look at the distributions carefully and form your own interpretations. (Note that the MINITAB subcommand "increment" would allow a change in interval definition to make the departures histogram tighter.)

We may also construct a histogram by hand. The first step is to divide our data points into **classes** of equal width. This division is not unique. In fact, we will now divide the departure data into classes of different width from those MINITAB produced automatically. We will use the following classes:

Class 1: 80≤Value<83 (80 or above, but less than 83)

Class 2: 83≤Value<86 (83 or above, but less than 86)

Class 3: 86≤Value<89 (86 or above, but less than 89)

Class 4: 89≤Value<92 (89 or above, but less than 92)

Class 5: 92≤Value<95 (92 or above, but less than 95)

Each of our data values falls into one and only one class. The next step is to count the number of data values that fall into each class. We find the following counts:

Class 1: 3 data points

Class 2: 8 data points

Class 3: 11 data points

Class 4: 5 data points

Class 5: 2 data points

The last step is to construct the graph, the histogram. We draw class boundaries on the horizontal axis and counts on the vertical axis. This is shown in Figure 1.7(b) on page 17.

Stemplots (Stem-and-Leaf Displays)

Another useful way to look at data distributions is to use the stem-and-leaf display, or stemplot for short. A **stem-and-leaf display** allows us to look quickly at a data set. It contains some of the features of a histogram but avoids the loss of information in a histogram that results from aggregating the data into intervals. The stem-and-leaf display is based on the tallying principle, | || ||| |||| ||||, but it also uses the decimal base of our number system. In a stem-and-leaf display, the **stem** is the number without its

TABLE 1.2   On-Time Flight Arrivals and Departures at Major U.S. Airports: 1990 and 1991

| AIRPORT | ON-TIME ARRIVALS | | | | | | ON-TIME DEPARTURES | | | | | |
| | 1990 | | 1991 | | | | 1990 | | 1991 | | | |
	3d qtr.	4th qtr.	1st qtr.	2d qtr.	3d qtr.	4th qtr.	3d qtr.	4th qtr.	1st qtr.	2d qtr.	3d qtr.	4th qtr.
Atlanta, Hartsfield International	78.8	77.5	78.0	80.0	80.7	81.1	85.3	82.8	82.4	84.9	88.4	88.1
Baltimore/Washington International	81.1	83.4	84.6	86.6	85.8	84.1	87.4	88.5	88.6	91.2	90.6	89.4
Boston, Logan International	74.9	77.7	76.8	81.2	79.6	75.4	83.6	85.3	84.7	88.3	85.6	83.3
Charlotte, Douglas	84.3	82.9	85.0	85.7	83.7	86.4	87.1	86.3	88.2	89.9	87.9	87.6
Chicago, O'Hare	80.0	78.1	76.0	82.9	85.6	77.0	85.0	82.5	80.3	86.6	89.0	80.8
Dallas/Ft. Worth International	83.5	75.4	78.7	84.1	83.7	77.1	89.0	81.2	84.4	87.7	90.7	84.3
Denver, Stapleton International	82.1	76.4	75.9	78.1	82.4	77.0	87.0	82.2	82.8	84.3	87.5	83.1
Detroit, Metro Wayne	81.7	83.4	85.1	88.6	90.8	87.9	86.9	86.1	87.2	90.6	91.9	89.8
Dulles International	83.5	85.3	83.8	84.6	85.4	85.8	88.4	88.9	87.2	88.4	90.1	88.8
Houston Intercontinental	81.9	79.5	78.3	78.0	85.3	81.3	89.1	86.2	85.2	83.7	89.4	87.1
Las Vegas, McCarran International	86.4	77.7	74.2	84.9	87.0	86.0	90.0	83.5	79.2	88.7	88.9	89.0
Los Angeles International	80.7	75.7	70.0	79.1	78.5	76.0	87.4	84.1	78.0	86.3	85.9	83.9
Miami International	81.7	83.4	82.5	84.1	78.2	77.8	86.8	89.7	87.3	90.7	89.2	88.6
Minneapolis/St. Paul International	81.0	84.1	82.3	88.1	88.6	81.1	85.5	85.8	84.9	90.7	91.0	83.8
Newark International	70.2	72.8	74.4	80.1	79.4	80.1	80.1	82.1	82.3	87.2	85.3	87.0
New York, Kennedy International	69.9	76.5	76.9	81.2	78.5	78.1	74.5	81.4	82.1	85.5	80.2	81.2
New York, LaGuardia	73.2	78.5	81.1	81.6	82.5	83.1	84.4	87.2	86.4	89.1	88.8	88.9
Orlando International	78.8	81.6	79.8	84.0	83.8	84.1	88.7	90.3	87.6	90.0	90.6	91.3
Philadelphia International	70.7	78.6	80.2	82.9	82.3	81.5	80.4	84.8	84.2	88.9	88.4	86.1
Phoenix, Sky Harbor International	83.0	75.5	72.7	86.7	88.8	85.0	86.5	77.6	74.1	87.0	88.6	85.7
Pittsburgh, Greater International	78.2	82.7	81.7	85.6	85.1	80.9	81.9	83.0	83.6	88.1	86.9	81.0
Raleigh/Durham	87.4	85.8	88.9	91.5	88.9	90.3	89.9	88.5	90.1	93.4	92.5	92.6
St. Louis, Lambert	84.0	75.8	69.6	83.4	86.7	81.2	86.7	80.2	75.7	89.5	91.2	85.7
Salt Lake City International	85.3	79.7	77.8	87.5	87.2	80.7	90.8	85.2	83.7	91.8	90.9	87.1
San Diego International, Lindbergh	83.7	76.4	72.2	83.2	85.2	82.5	87.5	83.6	78.7	88.2	88.3	86.5
San Francisco International	78.2	75.4	62.3	73.4	75.1	79.6	87.2	84.7	74.5	82.9	83.3	85.7
Seattle-Tacoma International	78.0	65.7	74.7	81.1	84.9	75.6	84.7	79.9	83.7	88.6	88.1	86.6
Tampa International	80.2	82.1	80.6	84.9	83.2	83.6	89.2	90.3	88.7	92.1	92.9	92.8
Washington International	81.1	84.3	84.9	86.0	86.3	84.9	88.2	88.7	89.8	90.9	91.3	89.9

1992 *Statistical Abstract of the United States*, No. 1045. From data provided by U.S. Department of Transportation, Office of Consumer Affairs, *Air Travel Consumer Report*, monthly.

[In percent. Quarterly, based on gate arrival and departure times for domestic scheduled operations in the 48 contiguous states of major U.S. airlines, per DOT reporting rule effective September 1987. All U.S. airlines with one percent or more of total U.S. domestic scheduled airline passenger revenues are required to report on-time data. A flight is considered on time if it operated less than 15 minutes after the scheduled time shown in the carrier's computerized reservation system. Canceled and diverted flights are considered late. Excludes flight operations delayed/canceled due to aircraft mechanical problems reported on FAA maintenance records (4–5 percent of the reporting airlines' scheduled operations). See source for data on individual airlines.]

MINITAB-Produced Results for Example 1.2   **FIGURE 1.7(a)**

```
MTB > NAME C1 'ARRIVALS'
MTB > NAME C2 'DEPARTS'
MTB > SET C1
DATA> 81.1 84.1 75.4 86.4 77 77.1 77 87.9 85.8 81.3 86 76 77.8 81.1 80.1 78.1
DATA> 83.1 84.1 81.5 85 80.9 90.3 81.2 80.7 82.5 79.6 75.6 83.6 84.9
DATA> END
MTB > SET C2
DATA> 88.1 89.4 83.3 87.6 80.8 84.3 83.1 89.8 88.8 87.1 89 83.9 88.6 83.8 87
DATA> 81.2 88.9 91.3 86.1 85.7 81 92.6 85.7 87.1 86.5 85.7 86.6 92.8 89.9
DATA> END
```

MTB > HISTO C1   MTB > HISTO C2

Histogram of arrivals	N = 29
Midpoint	Count
76	3 ***
78	5 *****
80	4 ****
82	6 ******
84	5 *****
86	4 ****
88	1 *
90	1 *

Histogram of departs	N = 29
Midpoint	Count
81	3 ***
82	0
83	2 **
84	3 ***
85	0
86	4 ****
87	5 *****
88	2 **
89	5 *****
90	2 **
91	1 *
92	0
93	2 **

Another Histogram for the Departure Data in Example 1.2   **FIGURE 1.7(b)**

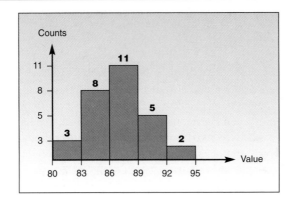

rightmost digit (the **leaf**). The stem is written to the left of a vertical line separating the stem from the leaf. For example, suppose we have the numbers 105, 106, 107, 107, 109. We would display them as follows:

10 | 56779

FIGURE 1.8   MINITAB-Produced Stem-and-Leaf Displays for Example 1.2

```
MTB > STEM C1

Stem-and-leaf of arrivals N = 29
Leaf Unit = 1.0

    2      7 55
    7      7 67777
    9      7 89
   (8)     8 00011111
   12      8 233
    9      8 44455
    4      8 667
    1      8
    1      9 0

MTB > STEM C2

Stem-and-leaf of departs N = 29
Leaf Unit = 0.10

    1      80 8
    3      81 02
    3      82
    7      83 1389
    8      84 3
   11      85 777
   14      86 156
   (4)     87 0116
   11      88 1689
    7      89 0489
    3      90
    3      91 3
    2      92 68
```

Left column: Parentheses signify that the *median* is in this stem. (Median is defined later in this chapter.) The number in the parentheses is the count at the median.

Count of leaves at this stem and all other stems in this direction further from median.

With a more complete data set with different stem values, the last digit of each number is displayed at the appropriate place to the right of its stem digit(s). Stem-and-leaf displays help us identify, at a glance, numbers in our data set that have high frequency.

Figure 1.8 shows the MINITAB commands and output for producing stem-and-leaf displays for the arrivals and departures data in Example 1.2.

The following example shows how a stemplot may be drawn without a computer.

EXAMPLE 1.3

Virtual reality is the name given to a new system of simulating real situations on a computer in a way that gives people the feeling that what they see on the computer screen is a real situation. Flight simulators were the forerunners of virtual reality programs. A particular virtual reality program has been designed to give production engineers experience in real processes. Engineers are supposed to complete certain

Stem-and-Leaf Display of the Task Performance Times of Example 1.3   **FIGURE 1.9**

(a)	1	122355567	(b)	84	1	122355567
	2	0111222346777899		531	2	0111222346777899
	3	012457		988421	3	012457
	4	11257		85200	4	11257
	5	0236		976540	5	0236
	6	02		97655210	6	02

tasks as responses to what they see on the screen. The following data represent the time, in seconds, it took a group of 42 engineers to perform a given task.

11, 12, 12, 13, 15, 15, 15, 16, 17, 20, 21, 21, 21, 22, 22, 22, 23, 24, 26, 27, 27, 27, 28, 29, 29, 30, 31, 32, 34, 35, 37, 41, 41, 42, 45, 47, 50, 52, 53, 56, 60, 62

The data are already arranged in increasing order. We see that the data are in the 10s, 20s, 30s, 40s, 50s, and 60s. We will use the first digit as the stem and the second digit of each number as the leaf. The stem-and-leaf display of our data is shown in Figure 1.9(a).

As you can see, the stem-and-leaf display is a very quick way of arranging the data in a kind of a histogram (turned sideways) that allows us to see what the data look like. Here, we note that the data do not seem to be symmetrically distributed; rather, they are skewed to the right.

Suppose, further, that data are available on another group, this one of 30 engineers. Time measurements, in seconds, to perform a task are

34, 40, 42, 38, 61, 45, 39, 60, 62, 65, 55, 54, 38, 25, 40, 31, 50, 14, 18, 57, 67, 48, 23, 65, 56, 59, 32, 21, 66, 69.

A back-to-back stemplot of these two data sets is shown in Figure 1.9(b). This plot gives a nice comparison of the distributions of performance times for the two groups.

It is possible to increase the number of stems by splitting them in two. The first part gets leaves 0–4, and the second gets leaves 5–9. Try such a display for the data in Example 1.3. Which display is more informative?

Frequency Polygons and Ogives

A **frequency polygon** is similar to a histogram except that there are no rectangles, only a point in the midpoint of each interval at a height proportional to the frequency (or relative frequency, in a relative-frequency polygon) for the category the interval represents. The points are connected. The rightmost and leftmost points are zero; that is, the graph comes down to zero outside the upper and lower limits of the data. Table 1.3 gives the relative frequency of sales volume, in thousands of dollars per week, for pizza at a local establishment.

A relative-frequency polygon for these data is shown in Figure 1.10. Note that the frequency is located in the middle of the interval as a point with height equal to the relative frequency of the interval. Note also that the point zero is added at the left boundary and the right boundary of the data set: the polygon starts at zero and ends at zero even if zero is not part of the data.

TABLE 1.3

Pizza Sales

Sales ($000)	Relative Frequency
6–14	0.20
15–22	0.30
23–30	0.25
31–38	0.15
39–46	0.07
47–54	0.03

FIGURE 1.10   Relative-Frequency Polygon for Pizza Sales

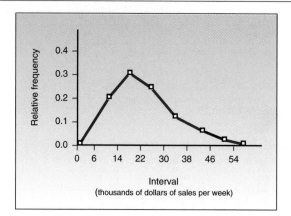

FIGURE 1.11   Ogive of Pizza Sales

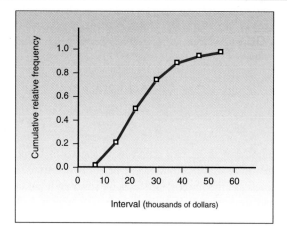

An **ogive** is a cumulative-frequency (or cumulative relative-frequency) graph. An ogive starts at 0 and goes to 1.00 (for a relative-frequency ogive) or to the maximum cumulative frequency. The point with height corresponding to the cumulative frequency is located at the right end point of each interval. An ogive for the data in Table 1.3 is shown in Figure 1.11. While the ogive shown is for the cumulative *relative* frequency, an ogive can also be used for the cumulative absolute frequency.

Space and Time Plots

In recent years, aided by the developments in computer technology, new and greatly improved graphics called **space and time plots** have emerged. These pictures truly aim to be worth a thousand words, and rather than providing all 1,000, we will show only some examples in this section. Figure 1.12 is a space plot. Figure 1.13 is a space and time plot showing poll proportions of the vote for the three candidates in the 1992 presidential race and a map of where they won the actual election. Figure 1.14 is an

The Ring of Fire   **FIGURE 1.12**

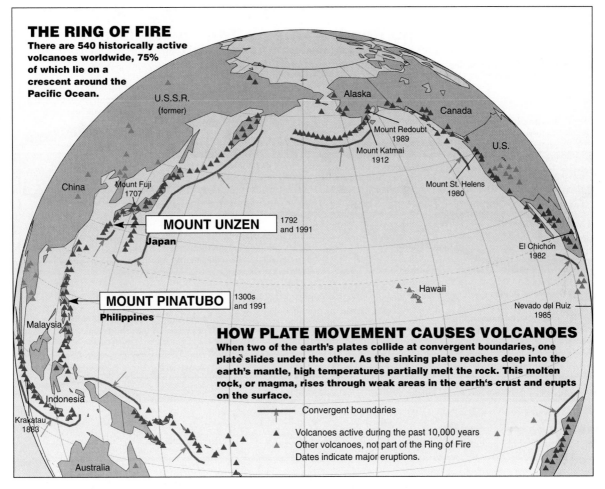

THE RING OF FIRE
There are 540 historically active volcanoes worldwide, 75% of which lie on a crescent around the Pacific Ocean.

HOW PLATE MOVEMENT CAUSES VOLCANOES
When two of the earth's plates collide at convergent boundaries, one plate slides under the other. As the sinking plate reaches deep into the earth's mantle, high temperatures partially melt the rock. This molten rock, or magma, rises through weak areas in the earth's crust and erupts on the surface.

Reprinted by permission from *Time,* June 24, 1991, p. 43, with data from Smithsonian Institution—Global Volcanism Program.

interesting spiral time plot of world trade. Here, world trade is measured away from the center of the graph, and months are measured like time on a clock.

A time plot such as the one in Figure 1.13, but drawn *without* joining the points, is called a *scatter plot.* Such plots will be demonstrated in Chapter 9, where we discuss the relation between two variables.

As you can see from these examples, graphical display of information is a very rich, expanding field.

A Caution about Graphs

A picture is indeed worth a thousand words, but pictures can sometimes be deceiving. This is one example of where the phrase "lying with statistics" comes from: presenting data graphically on a stretched or compressed scale of numbers with the aim of making the data show whatever you want to show. This is one important argument against a merely descriptive approach to data analysis and an argument for

FIGURE 1.13

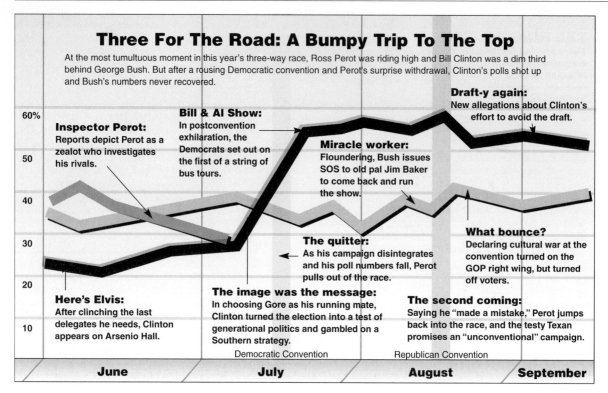

Three For The Road: A Bumpy Trip To The Top

At the most tumultuous moment in this year's three-way race, Ross Perot was riding high and Bill Clinton was a dim third behind George Bush. But after a rousing Democratic convention and Perot's surprise withdrawal, Clinton's polls shot up and Bush's numbers never recovered.

Draft-y again:
New allegations about Clinton's effort to avoid the draft.

Inspector Perot:
Reports depict Perot as a zealot who investigates his rivals.

Bill & Al Show:
In postconvention exhilaration, the Democrats set out on the first of a string of bus tours.

Miracle worker:
Floundering, Bush issues SOS to old pal Jim Baker to come back and run the show.

The quitter:
As his campaign disintegrates and his poll numbers fall, Perot pulls out of the race.

What bounce?
Declaring cultural war at the convention turned on the GOP right wing, but turned off voters.

Here's Elvis:
After clinching the last delegates he needs, Clinton appears on Arsenio Hall.

The image was the message:
In choosing Gore as his running mate, Clinton turned the election into a test of generational politics and gambled on a Southern strategy.

The second coming:
Saying he "made a mistake," Perot jumps back into the race, and the testy Texan promises an "unconventional" campaign.

Democratic Convention   Republican Convention

| June | July | August | September |

Robert Maass/SIPA Press. Phil Huber/Black Star.

FIGURE 1.14

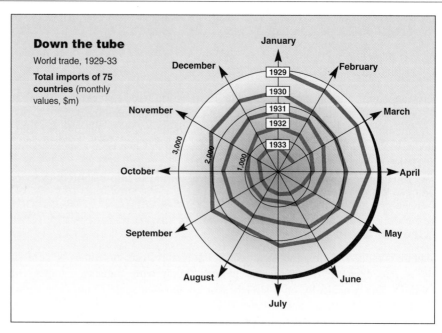

Down the tube

World trade, 1929-33

Total imports of 75 countries (monthly values, $m)

Reprinted by permission from *The Economist,* October 12, 1991, p. 7.

(continued)   Space and Time Plot

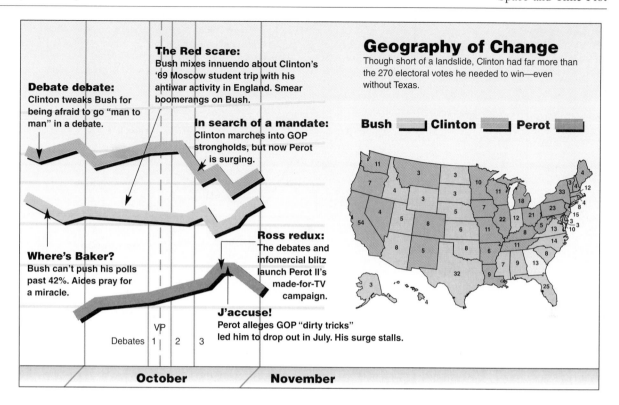

statistical inference. Statistical tests tend to be more objective than our eyes and are less prone to deception as long as our assumptions (primarily that of random sampling) hold. As we will see, statistical inference gives us tools that allow us to evaluate the data objectively.

Pictures are sometimes deceptive even when there is no intention to deceive. When someone shows you a graph of a set of numbers, there may really be no particular scale of numbers that is "right" for the data.

We can see how deceptive graphs can be—even when there is no intention to deceive—by looking at the Crash of '87. On "Bloody Monday," October 19, 1987, the Dow Jones Industrial Average fell 508 points. This much is certain. The question is: How can we best present this decline on a graph? What is the correct distance to use for each day on the horizontal scale (the *x*-axis) of the graph? By the end of that eventful week, almost every newspaper and magazine in the nation had a graph of the Crash of '87 in its pages. Figure 1.15 shows the fluctuations in the Dow over the period October 16 to October 23 as portrayed by *The New York Times* and as portrayed by *Investor's Daily*. The data are the same, yet note the difference in the two graphs. There certainly was no intention to deceive anyone: both publications aimed at conveying the same information to their readers. Yet the graphs are very different because the scales are different. Which scale is correct? There is no clear answer here, no objective way of determining the scale. Notice also that the bottom of the *y*-axis is not zero (even though the Dow is on a ratio scale), but 1,700 in both graphs. This lets the graphs use their space more efficiently, but it also distorts the Crash's effect on the market's total value.

FIGURE 1.15   Two Graphic Displays of the Crash of '87

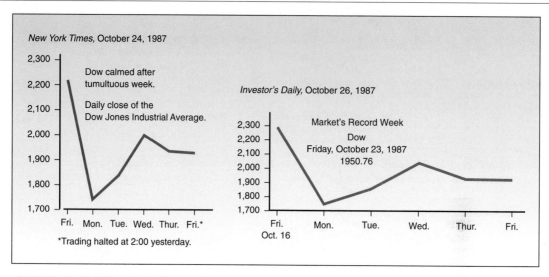

Sources: © 1987 *The New York Times.* Reprinted by permission. © 1987 *Investor's Daily.* Reprinted by permission.

FIGURE 1.16   A Third Graphic Portrayal of the Crash of '87

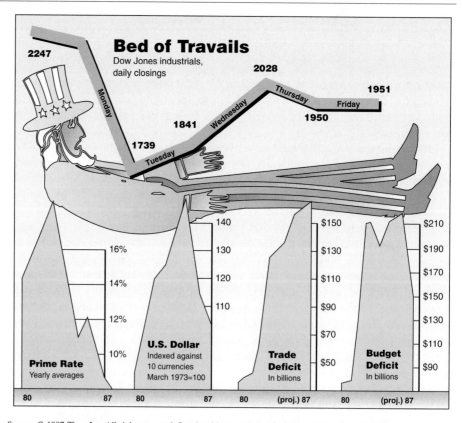

Source: © 1987 Time Inc. All rights reserved. Reprinted by permission from *Time,* November 4, 1987.

If you think the graphs in Figure 1.15 seem illusory (or at least inconsistent), consider the graph in Figure 1.16 from *Time* magazine. The superposition of the graph of the Dow's movements with plots of the prime rate, dollar index, trade deficit, and budget deficit—not to mention Uncle Sam lying on the spikes formed by these plots—might lead the viewer to conclude that there is a relationship between the Dow's movements and the other four variables. While such relationships *may* exist, they are neither proved nor even effectively demonstrated by this rather casual superposition.

---

14. What is displayed in the figure titled "Kings of the Road"? Is the number of cars proportional to the widths of the cars or to their area?

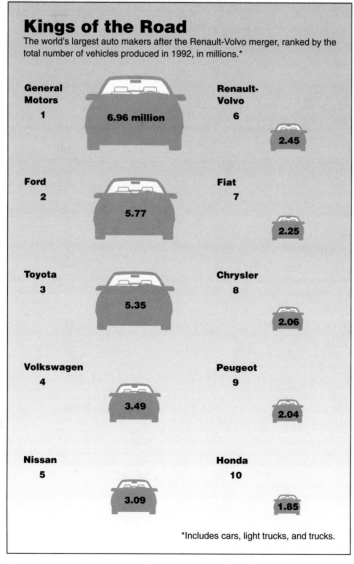

Kings of the Road
The world's largest auto makers after the Renault-Volvo merger, ranked by the total number of vehicles produced in 1992, in millions.*

General Motors
1
6.96 million

Renault-Volvo
6
2.45

Ford
2
5.77

Fiat
7
2.25

Toyota
3
5.35

Chrysler
8
2.06

Volkswagen
4
3.49

Peugeot
9
2.04

Nissan
5
3.09

Honda
10
1.85

*Includes cars, light trucks, and trucks.

Reprinted by permission from A. Riding, "Renault-Volvo Marriage Is On," *The New York Times,* September 7, 1993, p. D1, with data from Renault-Volvo.

15. Interpret the display showing the leading causes of death in the United States.

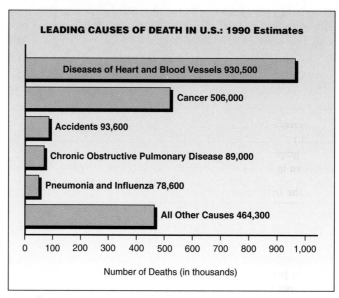

LEADING CAUSES OF DEATH IN U.S.: 1990 Estimates

Diseases of Heart and Blood Vessels 930,500

Cancer 506,000

Accidents 93,600

Chronic Obstructive Pulmonary Disease 89,000

Pneumonia and Influenza 78,600

All Other Causes 464,300

0 100 200 300 400 500 600 700 800 900 1,000

Number of Deaths (in thousands)

Source: National Center for Health Statistics and American Heart Association.

16. What conclusions can be drawn from the graph in the "Peasants of China" article? Does the graph alone provide incriminating evidence?

Peasants of China Discover New Way to Weed Out Girls

By NICHOLAS D. KRISTOF
Special to The New York Times

XIAMEN, China—Half a dozen barefoot peasants took a break during a fierce storm the other day, and as the rain pounded on the rice paddies outside they lounged in a crude stone cottage and marveled at 20th-century technology.

"Last year we had only one girl born in the village—everybody else had boys," Y. H. Chen said in a tone of awe, as the others nodded agreement. He explained that for a bribe of $35 to $50, a doctor will tell whether a woman is pregnant with a boy or girl.

"Then if it's a girl, you get an abortion," he said.

In the China of the 1990's, the modern machine that is having the most far-reaching effect on society is probably not the personal computer, the fax or even the car. It is the ultrasound scanner.

A Skewed Ratio

Partly because of ultrasound scans to check the sex of fetuses, followed by abortions of females, the sex ratio of newborn children in China last year reached 118.5 boys for every 100 girls. That statistic, based on an official survey of 385,000 people conducted last September and October, is a preliminary one, but it so shocked the authorities here that they ordered that it be kept secret.

Normally, women of all races give birth to about 105 or 106 boys for every 100 girls. China's ratio last year was about 13 points off this international norm, meaning that more than 12 percent of all female fetuses were aborted or otherwise unaccounted for.

Because China's population is so huge—1.17 billion—that adds up to more than 1.7 million missing girls each year. In 5 of China's 30 provinces, the sex ratio is already more than 120 boys for every 100 girls.

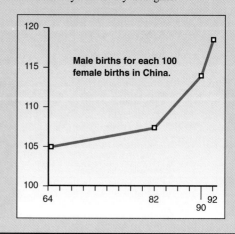

Male births for each 100 female births in China.

Reprinted by permission from *The New York Times,* January 21, 1993, front page, with data from the Chinese census and unpublished data.

17. Using the data in Table 1.2, Example 1.2, construct a histogram for percentages of on-time arrivals for the first, second, and third quarters of 1991. Does it seem that there is an improvement in on-time arrival rates? Compare the histograms with one another.

18. Using the data in Table 1.2, Example 1.2, construct histograms for on-time departures for all the quarters in the table. Is there an improvement over time in on-time departure rates?

19. Using the data in Table 1.2, Example 1.2, draw a stem-and-leaf display for the on-time arrival and the on-time departure rates for the second quarter of 1991 and arrange them back-to-back. (You may do so by cutting and pasting the two displays from the computer, or by hand.) Describe the difference between the two distributions. Why, in your opinion, is there a difference between the distribution of on-time arrivals and on-time departures? (Think airport.)

20. Generate bar graphs and pie charts for number of lines per 100 people for the three countries shown in the table titled "The U.S. Lead in the Information Age."

The U.S. Lead in the Information Age

	U.S.	Japan	Europe
Telephones			
Lines per 100 people	48.9	42.2	42.2
Calls per person per month	43.4	46.1	48.7
Cellular telephones per 100 people	2.6	1.2	1.2
Television			
Households with cable (percent)	55.4	13.3	14.5
VCR-related expenditures per household per year in dollars	44.6	35.3	14.1
Computers			
Personal computers per 100 people	28.1	7.8	9.6
Database production (percent of world)	56.0	2.0	32.0

Reprinted by permission of Gary Stix, "Domesticating Cyberspace," *Scientific American,* August 1993, p. 104, from data provided by the Consumer Federation of America.

21. Using the data in the table in problem 20, create pie charts for cellular phones per 100 people in the United States, Japan, and Europe, and for households with cable and personal computers per 100 people. Use these graphs for comparisons, draw conclusions, and write descriptive statements.

22. Interpret the graph about the pope's travels. Is there an obvious underlying reason for the differences seen in the graph?

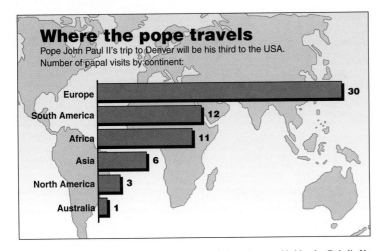

Reproduced by permission of *USA Today,* August 9, 1993, p. 1, from data provided by the Catholic News Service.

23. Draw a pie chart that shows the proportion of each market for Rolls-Royce shown in the following table, and draw a similar chart for Jaguar.

Top Markets in 1992

	Number of Cars Sold	
Market	Jaguar	Rolls-Royce
United States	8,681	392
Britain	5,607	382
Europe (Continental)	5,025	294
Japan	1,501	91
Asia and Africa	740	134
Middle East	250	67

Reprinted by permission of *Fortune,* September 6, 1993, p. 68.

24. For the data in problem 23, draw bar charts for Jaguar and for Rolls-Royce and superimpose them. How do the markets compare with each other? Use your displays in this problem and the previous one to prepare a presentation to the board of directors of Jaguar describing the best and the worst markets for their cars and comparing these markets with those for Rolls-Royce. Where should Jaguar concentrate its marketing efforts? Why?

25. Interpret the pie charts on the growing number of male-headed single-parent families.

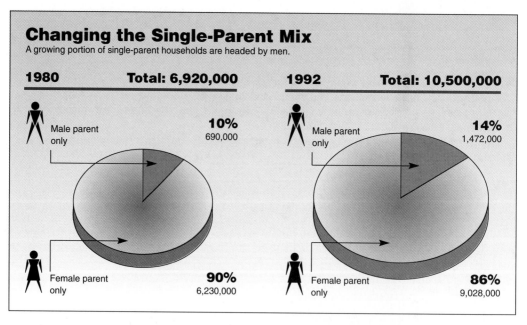

Reproduced by permission of *The New York Times,* August 31, 1993, p. A15, from data provided by the U.S. Census Bureau.

26. Draw a stem-and-leaf display for the data on tax collection. Mark the stem containing the data for the United States. Form conclusions based on the location of this stem as well as the shape of the distribution. What can you say about tax collection in the United States as compared with that in the other industrialized countries in this study?

Government Tax Collections as Fraction of GDP in 1990

Country	Tax Rate (in percent)	Country	Tax Rate (in percent)
United States	30	Germany	37
Australia	30	Finland	37
Turkey	30	New Zealand	40
Japan	30	Italy	40
Switzerland	31	Austria	41
Greece	32	Luxembourg	41
Iceland	32	France	43
Spain	33	Belgium	44
Portugal	33	Netherlands	45
Canada	34	Norway	46
Britain	35	Denmark	47
Ireland	35	Sweden	58

Source: Reproduced with permission from Lester Thurow, *Head to Head* (New York: Warner, 1992), p. 231.

27. Look at the set of nine unlabeled histograms. In each case, describe the distribution as symmetric or skewed; if skewed, state to which side. Describe the distribution as either concentrated in a "center" or spread out. Point out the location of the center. Point out the modes of the distribution—its most frequent values. Finally, look for special patterns and for outliers—data points that are far away from the rest of the data.

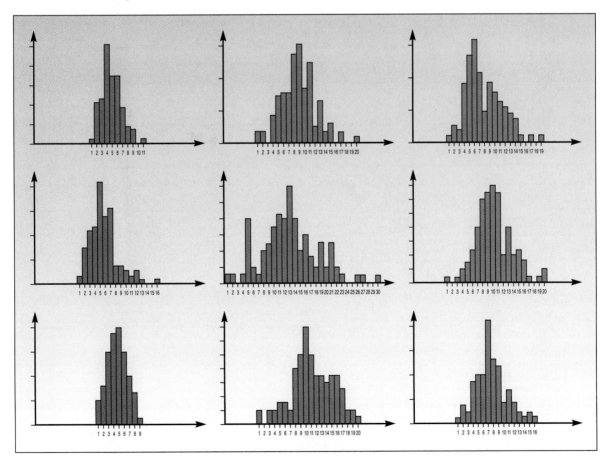

28. The following table gives information on the merchant fleets of the world. Draw a histogram of the number of vessels for all countries listed (Total, Number column). Also construct a stemplot of the average age of ships (Total). Use a computer.

Merchant Fleets of the World: 1980 to 1989 (Vessels of 1,000 gross tons and over. As of Jan. 1 of the following year. Specified countries have 100 or more ships.)

Year and Country of Registry, 1989	Total		Passenger/Cargo Comb.	
	Number	Average age (yr.)	Number	Average age (yr.)
1980, world total	24,867	13	468	24
United States	864	23	65	34
Foreign	24,003	13	403	22
1985, world total	25,555	14	375	25
United States	737	23	37	38
Foreign	24,818	14	338	23
1988, world total	23,468	14	368	23
United States	675	22	21	41
Foreign	22,793	14	347	22
1989, world total[2]	22,983	14	320	23
United States	655	22	19	41
Privately owned	407	16	4	36
Government-owned	248	33	15	43
Foreign	22,328	14	301	22
Argentina	135	18	—	—
Brazil	293	14	2	28
British Colonies	545	15	6	16
Bulgaria	114	15	2	16
China: Mainland	1,281	18	13	25
Taiwan	226	12	1	18
Cyprus	1,054	15	3	32
Denmark (DIS)[3]	199	7	—	—
East Germany	147	14	1	8
Egypt	131	17	4	27
France	134	12	4	10
Greece	914	15	31	28
Honduras	139	25	1	34
India	296	13	3	24
Indonesia	323	19	7	13
Iran	123	13	—	—
Italy	479	16	9	19
Japan	1,007	8	14	10
Liberia	1,409	12	13	16
Malaysia	158	15	—	—
Malta	293	18	2	56
Netherlands	326	9	4	9
Norway (NIS)[3]	587	12	11	8
Panama	3,189	12	38	29
Philippines	558	13	10	28
Poland	245	13	3	9
Romania	304	12	—	—
Singapore	407	12	—	—
South Korea	429	14	—	—
Soviet Union	2,428	17	32	25
Spain	301	13	—	1
Sweden	162	13	4	17
Thailand	124	21	1	—
Turkey	317	14	4	35
United Kingdom	198	14	12	16
West Germany	310	6	6	11
Yugoslavia	272	14	3	26
All others	2,771	(NA)	57	(NA)

— Represents zero. NA Not available. X Not applicable. [1] Includes bulk/oil, ore/oil, and ore/bulk/oil carriers. [2] Average age for the previous year.
[3] International Shipping Registry, which is an open registry, under which the ship flies the flag of the specified nation but is exempt from certain taxation and other regulations.

Source: 1992 *Statistical Abstract of the United States.* U.S. Maritime Administration, *Merchant Fleets of the World,* summary report, annual.

29. Using the information in the table in Problem 28, draw histograms and stemplots for numbers of freighters, bulk carriers, and tankers, as well as their average ages.

Freighters		Bulk Carriers[1]		Tankers	
Number	Average age (yr.)	Number	Average age (yr.)	Number	Average age (yr.)
14,242	14	4,798	10	5,359	12
471	23	20	22	308	20
13,771	13	4,778	10	5,051	11
13,937	15	5,787	11	5,456	13
417	25	25	9	258	19
13,520	15	5,762	11	5,198	13
12,518	15	5,332	12	5,250	14
381	24	26	12	247	19
12,137	15	5,306	12	5,003	13
12,195	15	5,335	12	5,133	14
371	24	26	12	239	19
171	14	26	12	206	18
200	32	—	(X)	33	31
11,824	15	5,309	12	4,894	13
69	18	17	16	49	17
114	19	94	9	83	13
252	17	176	12	111	15
52	16	46	14	14	17
860	19	240	15	168	17
147	13	63	10	15	13
537	15	410	15	104	15
136	8	9	8	54	7
125	15	17	13	4	10
100	18	16	7	11	20
64	11	15	8	51	14
238	18	442	13	203	16
107	26	11	24	20	23
112	15	118	11	63	10
217	21	13	8	86	16
38	17	50	10	35	15
181	16	67	14	222	17
411	7	287	9	295	8
311	12	511	12	574	11
103	17	21	10	34	12
144	18	79	18	68	18
256	9	13	8	53	8
138	12	177	12	261	12
1,707	13	854	10	590	11
225	15	285	8	38	25
143	13	92	12	7	14
220	12	70	12	14	10
216	14	70	9	121	11
211	15	154	12	64	14
1,732	17	244	12	420	15
178	13	43	12	80	15
80	12	17	16	61	14
88	22	2	19	33	21
202	12	58	16	53	16
89	13	29	10	68	15
244	6	10	5	50	8
167	14	90	13	12	13
1,610	(NA)	399	(NA)	705	(NA)

30. Interpret the displayed information on illegal aliens. Recall that an outlier is an observation that is far from the rest (either larger or smaller than the rest of the data). Discuss the outlier in this data set.

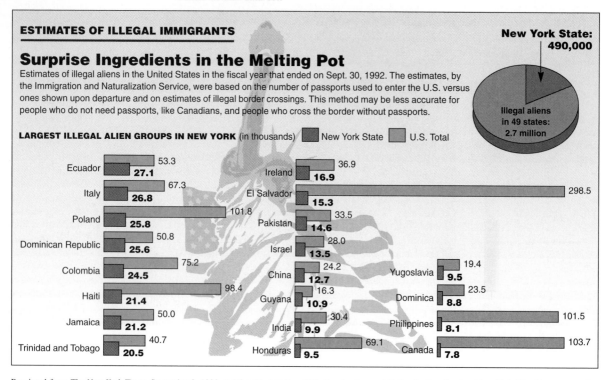

ESTIMATES OF ILLEGAL IMMIGRANTS

Surprise Ingredients in the Melting Pot

Estimates of illegal aliens in the United States in the fiscal year that ended on Sept. 30, 1992. The estimates, by the Immigration and Naturalization Service, were based on the number of passports used to enter the U.S. versus ones shown upon departure and on estimates of illegal border crossings. This method may be less accurate for people who do not need passports, like Canadians, and people who cross the border without passports.

New York State: 490,000

Illegal aliens in 49 states: 2.7 million

LARGEST ILLEGAL ALIEN GROUPS IN NEW YORK (in thousands) ■ New York State ■ U.S. Total

	New York State	U.S. Total
Ecuador	27.1	53.3
Italy	26.8	67.3
Poland	25.8	101.8
Dominican Republic	25.6	50.8
Colombia	24.5	75.2
Haiti	21.4	98.4
Jamaica	21.2	50.0
Trinidad and Tobago	20.5	40.7
Ireland	16.9	36.9
El Salvador	15.3	298.5
Pakistan	14.6	33.5
Israel	13.5	28.0
China	12.7	24.2
Guyana	10.9	16.3
India	9.9	30.4
Honduras	9.5	69.1
Yugoslavia	9.5	19.4
Dominica	8.8	23.5
Philippines	8.1	101.5
Canada	7.8	103.7

Reprinted from *The New York Times,* September 2, 1993, p. B8, with data from U.S. Immigration and Naturalization Service, New York City Department of City Planning.

31. Based on the following display, how does President Clinton's tax increase compare with previous ones this century?

How the Clinton Tax Increase Compares

The tax increase approved by Congress last week has been tagged the biggest "in the history of the world" by opponents. In fact, it does not even rank as the biggest in U.S. history. Here is a look at major tax changes since World War II.

	Impact (in billions)*	As % of GDP
1993: Clinton deficit-reduction plan	$54	0.7%
1990: Bush deficit-reduction plan	30	0.4
1983: Social Security and 1984 budget bill	69	0.9
1982: Reagan tax increase and gas tax	80	1.0
1981: Reagan tax cut	–325	–4.1
1977: Social Security changes	71	0.9
1968–69: Vietnam War tax increase	165	2.1
1950–52: Korean War tax increase	362	4.6
1940–44: World War II tax increase	1,165	14.8

*Expressed in fiscal 1998 dollars, based on government projections.

Reprinted by permission from the *San Francisco Chronicle,* August 9, 1993, p. D1, with data from *Citizens for Tax Justice.*

32. Comment on and critique the following graphs.

(a)   (b)

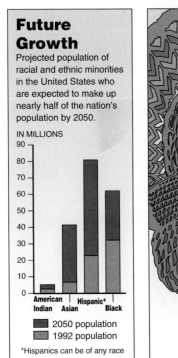

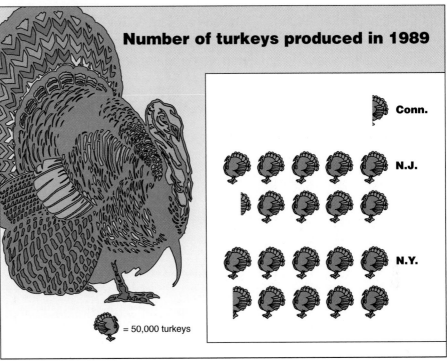

(a) Reprinted by permission from *Boston Globe,* December 4, 1992, p. 3, with data from the U.S. Census Bureau; (b) Reprinted by permission from *The New York Times,* November 23, 1992, p. B1.

(c)

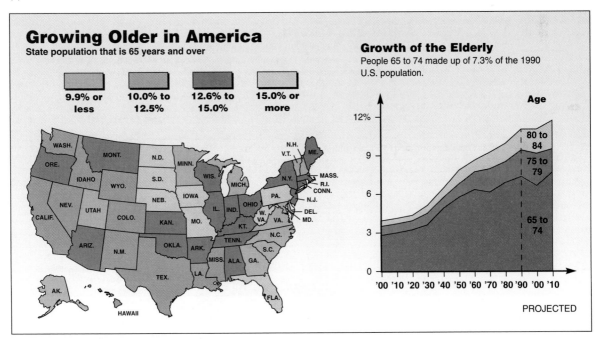

Reprinted by permission from *The New York Times,* November 10, 1992, p. A21, with data from the U.S. Census Bureau.

(d)

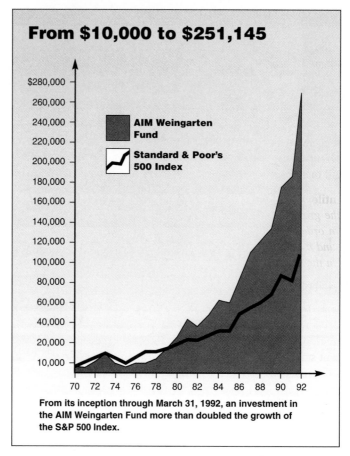

An ad piece for Citibank Investments, 1992.

(e)

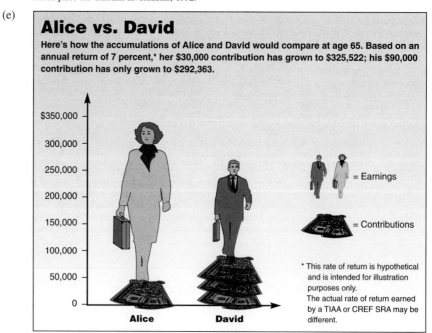

The Participant, November 1992, p. 5.

1.3 PERCENTILES, QUARTILES, AND THE MEDIAN

Percentiles

Given a set of numerical observations, we may order them according to magnitude. Once we have done this, it is possible to define the boundaries of the set. Any student who has taken a nationally administered test, such as the Scholastic Aptitude Test (SAT), is familiar with percentiles. Your score on such a test is compared with the scores of all the people who took the test at the same time, and your position within this group is defined in terms of a percentile. If you are at the 90th percentile, this means that 90 percent of the people who took the test with you obtained either the same score you did or a lower one. In general,

> The p*th percentile of a group of numbers is the value at which* p *percent of the values in the group are less than or equal to this value. When the data are arranged in order, from smallest to largest values, the position of the* pth *percentile is found by using the formula* (n + 1)p/100, *where* n *is the number of data points in the group.*

Let's look at an example.

A large department store collects data on sales made by each of its salespeople. The data, number of sales made on a given day by each of 20 salespeople, are as follows:

EXAMPLE **1.4**

$$9, 6, 12, 10, 13, 15, 16, 14, 14, 16, 17, 16, 24, 21, 22, 18, 19, 18, 20, 17$$

Find the 50th, 80th, and 90th percentiles of this data set. First, let's order the data from smallest to largest:

$$6, 9, 10, 12, 13, 14, 14, 15, 16, 16, 16, 17, 17, 18, 18, 19, 20, 21, 22, 24$$

To find the 50th percentile, we need to determine the data point in position $(n + 1)p/100 = (20 + 1)(50/100) = (21)(0.5) = 10.5$. Thus, we need the data point in position 10.5. Counting the observations from smallest to largest, we find that the 10th observation is 16, and so is the 11th. Therefore, the observation that would lie in position 10.5 (halfway between the 10th and 11th observations) is 16. Thus, the 50th percentile is 16.

SOLUTION

Similarly, we find the 80th percentile of the data set as the observation lying in position $(n + 1)p/100 = (21)(80/100) = 16.8$. The 16th observation is 19, and the 17th is 20; therefore, the 80th percentile is a point lying 0.8 of the way from 19 to 20, that is, 19.8.

The 90th percentile is found as the observation in position $(n + 1)p/100 = (20 + 1)(90/100) = (21)(0.9) = 18.9$, which is 21.9.

Quartiles

A quartile is a certain kind of percentile. Quartiles are more commonly used than other kinds of percentiles, because they break the data distribution into meaningful groups of a quarter of the data set each.

The first or lower **quartile** (Q_L) is the 25th percentile of the data distribution. It is the point at which or below which lie a quarter of our observations. When the data are ordered from smallest to largest, Q_L is the observation in position $(n + 1)/4$.

Similarly,

The second quartile is the 50th percentile of the data distribution. It is the point at which 50 percent of the data are less than or equal to the given value. The second quartile, the 50th percentile, is a very important measure of the center of the data distribution. It is called the **median**, M. When the data are placed in order from smallest to largest, if n is odd, then the median is the center observation in the ordered list. If n is even, the median is the average of the two center observations. The median, in either case, is the observation in position $(n + 1)/2$.

Finally,

The third or upper quartile (Q_U) is the 75th percentile of the data distribution. It is the point at which 75 percent of our data points are less than or equal to the given value. Q_U is the observation in position $3(n + 1)/4$ in the ordered data list.

Example 1.5 will give you some practice with determining quartiles.

EXAMPLE 1.5

Find the lower, middle, and upper quartiles of the data set in Example 1.4.

SOLUTION

Based on the procedure we used in computing the 80th and 90th percentiles, we find that the lower quartile is the observation in position $(21)(0.25) = 5.25$, which is 13.25. The middle quartile was already computed (it is the 50th percentile, the median, which is 16). The upper quartile is the observation in position $(21)(75/100) = 15.75$, which is 18.75.

We define the **interquartile range (IQR)** as the difference between the first and third quartiles.

The interquartile range is a measure of the spread of the data. It gives us an idea of how far away the data points stretch from the center. In Example 1.4 the interquartile range is equal to: Third quartile − First quartile = $18.75 − 13.25 = 5.5$.

The Median

As mentioned earlier, the median (the 50th percentile) is a very important measure of the center of the distribution of the data. If our data distribution is displayed in a histogram, the median is that point on the horizontal axis (the x-axis) such that the area of the histogram to its left is equal to the area of the histogram to its right. The median splits the distribution into two halves: 50 percent of the data to the left and 50 percent to the right. The median is demonstrated using the histograms in Figure 1.17.

The median is a **resistant** measure of the center of the data distribution. It is resistant in the sense that it is not affected by outliers (observations that lie away from

The Median of Data Distributions in Histograms   **FIGURE 1.17**

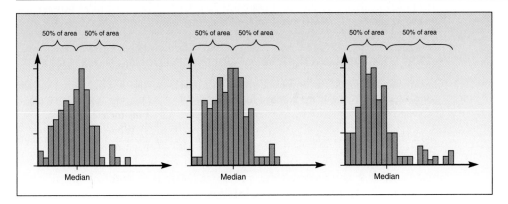

the rest of the data). To see this, look at Figure 1.17 and imagine moving one of the bars lying on the right side of the median to a point farther away to the right. Now ask yourself: does the median change in response to this move? The answer is no. If a point on one side of the median moves farther to the same side, half the data will still be above the former median and the other half at or below it, so the median itself does not change. A resistant measure such as the median does not respond strongly to changes in a few data points, even if these changes are large. The interquartile range is, like the median, resistant. This is not true for another measure of the center of the distribution, the *mean,* which we discuss in a later section. The mean responds to any change in an observation.

For example, find the median of the following set of points: 5, 3, 9, 8, 1, 12, 6. Arranging the seven data points from smallest to largest, we get 1, 3, 5, 6, 8, 9, 12. That is, the first data point is 1, the second is 3, the third is 5, and so on; the seventh is 12. We have $n = 7$, an odd number. So the median is the center observation, the observation in position $(7 + 1)/2 = 4$. The fourth observation is 6. Hence, the median is 6.

As another example, consider the data set 4, 60, 92, 1, 70, 100. Ordering from small to large, we get 1, 4, 60, 70, 92, 100. Our number of observations is $n = 6$, an even number. The median is thus the point that would lie in position $(6+1)/2 = 3.5$; that is, the average between the third and the fourth data points. Hence the median is $(60+70)/2 = 65$.

MINITAB can be used very simply to find the median and the quartiles of any distribution. Just state the command DESCRIBE, followed by the column number (for example, C1). Let's see how this is done, in Figure 1.18.

The MINITAB Command DESCRIBE   **FIGURE 1.18**

```
MTB > SET C1
DATA > 42 33 39 41 40 37 35 29 48 52 23 36 38 50 31
DATA > END
MTB > DESCRIBE C1
```

C1	N	MEAN	MEDIAN	TRMEAN	STDEV	SEMEAN
	15	38.27	38.00	38.38	7.85	2.03
C1	MIN	MAX	Q1	Q3		
	23.00	52.00	33.00	42.00		

PROBLEMS

33. The following data are numbers of passengers on flights of Delta Air Lines between San Francisco and Seattle over 33 days in April and early May.

128, 121, 134, 136, 136, 118, 123, 109, 120, 116, 125, 128, 121, 129, 130, 131, 127, 119, 114, 134, 110, 136, 134, 125, 128, 123, 128, 133, 132, 136, 134, 129, 132

Find the lower, middle, and upper quartiles of this data set. Also find the 10th, 15th, and 65th percentiles. What is the interquartile range?

34. The data in the graph labeled "Sizing Up Airline Fleets" show aircraft numbers for 1990 and 1995 (estimated) for the 12 largest U.S. airlines. Find the quartiles for the 1990 and the 1995 data.

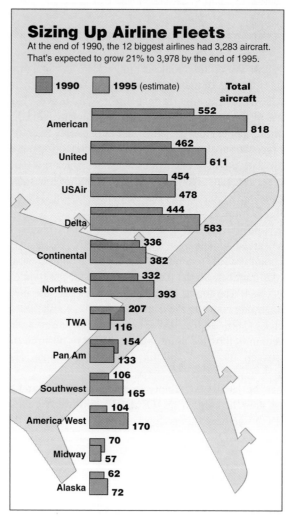

Sizing Up Airline Fleets
At the end of 1990, the 12 biggest airlines had 3,283 aircraft. That's expected to grow 21% to 3,978 by the end of 1995.

| | 1990 | | 1995 (estimate) | | Total aircraft |

Airline	1990	1995
American	552	818
United	462	611
USAir	454	478
Delta	444	583
Continental	336	382
Northwest	332	393
TWA	207	116
Pan Am	154	133
Southwest	106	165
America West	104	170
Midway	70	57
Alaska	62	72

Reprinted by permission from *USA Today*, May 14, 1991, with data from Salomon Bros. Inc.

35. The figure titled "How Old Were They?" shows the ages of U.S. presidents and vice presidents since 1900. Find the quartiles of the age distribution of ages of the presidents and vice presidents, as well as the 10th and 90th percentiles. At what percentile of the distributions lie the present president and the present vice president? Comment on your findings.

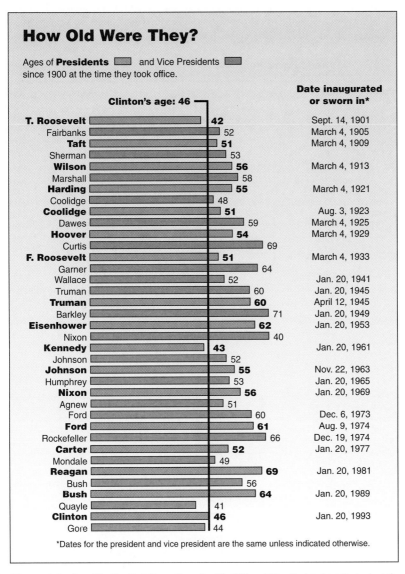

How Old Were They?

Ages of **Presidents** [ ] and Vice Presidents [ ] since 1900 at the time they took office.

		Date inaugurated or sworn in*
Clinton's age: 46		
T. Roosevelt	42	Sept. 14, 1901
Fairbanks	52	March 4, 1905
Taft	51	March 4, 1909
Sherman	53	
Wilson	56	March 4, 1913
Marshall	58	
Harding	55	March 4, 1921
Coolidge	48	
Coolidge	51	Aug. 3, 1923
Dawes	59	March 4, 1925
Hoover	54	March 4, 1929
Curtis	69	
F. Roosevelt	51	March 4, 1933
Garner	64	
Wallace	52	Jan. 20, 1941
Truman	60	Jan. 20, 1945
Truman	60	April 12, 1945
Barkley	71	Jan. 20, 1949
Eisenhower	62	Jan. 20, 1953
Nixon	40	
Kennedy	43	Jan. 20, 1961
Johnson	52	
Johnson	55	Nov. 22, 1963
Humphrey	53	Jan. 20, 1965
Nixon	56	Jan. 20, 1969
Agnew	51	
Ford	60	Dec. 6, 1973
Ford	61	Aug. 9, 1974
Rockefeller	66	Dec. 19, 1974
Carter	52	Jan. 20, 1977
Mondale	49	
Reagan	69	Jan. 20, 1981
Bush	56	
Bush	64	Jan. 20, 1989
Quayle	41	
Clinton	46	Jan. 20, 1993
Gore	44	

*Dates for the president and vice president are the same unless indicated otherwise.

Reprinted by permission from *The New York Times,* November 17, 1992, p. A19, with data from *Facts about the Presidents,* by Joseph Nathan Kane.

36. The following data are the numbers of bacteria colonies that emerge as part of a medical experiment:

    43, 21, 55, 39, 58, 73, 62, 89, 81, 11, 37, 66, 45, 57, 20, 37, 69, 88, 51, 68, 93, 30.

    Find the median and the quartiles of the distribution.

37. The following data are scores on a management examination taken by a group of 22 people.

    88, 56, 64, 45, 52, 76, 54, 79, 38, 98, 69, 77, 71, 45, 60, 78, 90, 81, 87, 44, 80, 41

    Find the median.

38. Find the median and the quartiles of fourth-quarter 1991 on-time arrival percentages using the data in Table 1.2, Example 1.2.

39. Find the *median* country (or dependency or sovereign area) population in the world using the data in the following world population table, for 1991. Note the very useful column "Population Rank, 1991." This makes finding the median of this large data set very easy.

World Population (1,000)

Country	Population 1991	Rank 1991	Country	Population 1991	Rank 1991
World total	5,422,908	(X)			
Afghanistan	16,450	53	El Salvador	5,419	97
Albania	3,335	124	Equatorial Guinea	379	167
Algeria	26,022	36	Estonia	1,584	141
Andorra	53	204	Ethiopia	53,191	23
Angola	8,668	77	Fiji	744	154
Antigua and Barbuda	64	200	Finland	4,991	100
Argentina	32,664	31	France	56,596	19
Armenia	3,357	119	Gabon	1,080	149
Aruba	64	199	Gambia, The	875	151
Australia	17,288	50	Georgia	5,479	95
Austria	7,666	83	Germany	79,548	12
Azerbaijan	7,212	87	West Germany (former)	64,039	(X)
Bahamas, The	252	175	Ghana	15,617	54
Bahrain	537	159	Greece	10,043	68
Bangladesh	116,601	10	Grenada	84	193
Barbados	255	174	Guatemala	9,266	74
Belgium	9,922	69	Guinea	7,456	85
Belize	228	176	Guinea-Bissau	1,024	150
Benin	4,832	103	Guyana	750	153
Bhutan	1,598	142	Haiti	6,287	92
Bolivia	7,157	89	Honduras	4,949	101
Bosnia and Herzegovina	4,517	105	Hungary	10,558	63
Botswana	1,258	146	Iceland	260	173
Brazil	155,356	5	India	869,515	2
Brunei	398	166	Indonesia	193,560	4
Bulgaria	8,911	75	Iran	59,051	18
Burkina	9,360	73	Iraq	19,525	45
Burma	42,112	25	Ireland	3,489	118
Burundi	5,831	94	Israel	4,558	106
Byelarus	10,257	67	Italy	57,772	15
Cambodia	7,146	88	Jamaica	2,489	131
Cameroon	11,390	61	Japan	124,017	7
Canada	26,835	32	Jordan	3,413	123
Cape Verde	387	165	Kazakhstan	16,757	51
Central African Republic	2,952	127	Kenya	25,242	37
Chad	5,122	99	Kiribati	71	196
Chile	13,287	57	Kuwait	2,204	137
China: Mainland	1,151,487	1	Kyrgyzstan	4,394	107
Taiwan	20,659	42	Laos	4,113	112
Colombia	33,778	30	Latvia	2,695	129
Comoros	477	161	Lebanon	3,385	120
Congo	2,309	134	Lesotho	1,801	140
Costa Rica	3,111	12	Liberia	2,730	130
Cote d'Ivoire	12,978	59	Libya	4,353	110
Croatia	4,686	102	Liechtenstein	28	211
Cuba	10,732	62	Lithuania	3,726	114
Cyprus	709	155	Luxembourg	388	164
Czechoslovakia	15,725	52	Macedonia	2,132	136
Denmark	5,133	98	Madagascar	12,185	60
Djibouti	346	171	Malawi	9,438	72
Dominica	86	192	Malaysia	17,982	47
Dominican Republic	7,385	86	Maldives	226	177
Ecuador	10,752	64	Mali	8,339	79
Egypt	54,452	21	Malta	356	168

Source: Reprinted from the 1992 *Statistical Abstract of the United States.*

World Population (1,000)

Country	Population 1991	Rank 1991	Country	Population 1991	Rank 1991
Mauritania	1,996	139	Tonga	102	189
Mauritius	1,081	148	Trinidad and Tobago	1,285	145
Mexico	90,007	11	Tunisia	8,276	80
Moldova	4,393	108	Turkey	58,581	17
Monaco	30	210	Turkmenistan	3,658	116
Mongolia	2,247	135	Tuvalu	9	222
Montenegro	645	156	Uganda	18,690	46
Morocco	26,182	35	Ukraine	51,711	22
Mozambique	15,113	56	United Arab Emirates	2,390	133
Namibia	1,521	144	United Kingdom	57,515	16
Nauru	9	221	United States	252,502	3
Nepal	19,612	44	Uruguay	3,121	125
Netherlands	15,022	55	Uzbekistan	20,569	41
New Zealand	3,309	21	Vanuatu	170	183
Nicaragua	3,752	117	Venezuala	20,189	43
Niger	8,154	81	Vietnam	67,568	13
Nigeria	122,471	8	Western Samoa	190	180
North Korea	21,815	40	Yemen	10,063	71
Norway	4,273	109	Zaire	37,832	29
Oman	1,534	143	Zambia	8,446	78
Pakistan	117,490	9	Zimbabwe	10,720	65
Panama	2,476	132	AREAS OF SPECIAL SOVEREIGNTY AND DEPENDENCIES		
Papua New Guinea	3,913	113	American Samoa	43	207
Paraguay	4,799	104	Anguilla	7	224
Peru	22,362	39	Bermuda	58	201
Philippines	65,759	14	British Virgin Islands	12	218
Poland	37,800	28	Cayman Islands	27	212
Portugal	10,388	66	Cook Islands	18	215
Qatar	518	160	Faroe Islands	48	205
Romania	23,397	38	Federated States of Micronesia	108	188
Russia	148,254	6	French Guiana	102	191
Rwanda	7,903	84	French Polynesia	195	179
Saint Kitts and Nevis	40	208	Gaza Strip	642	157
Saint Lucia	153	184	Gibraltar	30	209
Saint Vincent and the Grenadines	114	187	Greenland	57	203
San Marino	23	214	Guadeloupe	345	169
Sao Tome and Principe	128	186	Guam	145	185
Saudi Arabia	17,870	49	Guernsey	58	202
Senegal	7,953	82	Hong Kong	5,856	93
Serbia	9,883	70	Isle of Man	64	198
Seychelles	69	197	Jersey	84	194
Sierra Leone	4,275	111	Macau	446	162
Singapore	2,756	128	Marshall Islands	48	206
Slovenia	1,954	138	Martinique	345	170
Solomon Islands	347	172	Mayotte	75	195
Somalia	6,709	91	Montserrat	13	219
South Africa	40,601	26	Netherlands Antilles	184	181
South Korea	43,134	24	New Caledonia	172	182
Spain	39,385	27	Northern Mariana Islands	23	213
Sri Lanka	17,424	48	Pacific Islands, Trust Territory of the	14	217
Sudan	27,220	33	Puerto Rico	3,295	122
Suriname	402	163	Reunion	607	158
Swaziland	859	152	Saint Helena	7	223
Sweden	8,564	76	Saint Pierre and Miquelon	6	225
Switzerland	6,784	90	Turkes and Caicos Islands	10	220
Syria	12,966	58	Virgin Islands	99	190
Tajikistan	5,342	96	Wallis and Futuna	17	216
Tanzania	26,869	34	West Bank	1,105	147
Thailand	56,814	20	Western Sahara	197	179
Togo	3,811	115			

1.4 BOX PLOTS

A box plot (also called a *box-and-whisker plot*) is another way of looking at a data set in an effort to determine its central tendency, spread, skewness, and the existence of outliers. We use box plots for data on an interval or ratio scale.

> A **box plot** is a graph of a set of five summary measures of the distribution of the data:
> 1. The median of the data.
> 2. The lower quartile, Q_L.
> 3. The upper quartile, Q_U.
> 4. The smallest observation.
> 5. The largest observation.

Let's make this definition clearer by using a picture. Figure 1.19 shows the parts of a box plot and how they are defined. The median is marked as a vertical line across the box. The **hinges** of the box are the upper and lower quartiles (the rightmost and leftmost sides of the box). The interquartile range (IQR) is the distance from the upper quartile to the lower quartile (the length of the box from hinge to hinge): $\text{IQR} = Q_U - Q_L$.

We define the **upper inner fence** as a point at a distance of 1.5(IQR) above the upper quartile; similarly, the lower inner fence is $Q_L - 1.5(\text{IQR})$. The **outer fences** are defined similarly but are at a distance of 3(IQR) above or below the appropriate hinge. The **whiskers** of the box plot are made by extending a line from the upper quartile to the largest observation that is at or below the upper inner fence, and from the lower quartile to the smallest observation that is at or above the lower inner fence. If one or more observations are outside the inner fence but not outside the corresponding outer fence, they are marked as suspected outliers. If these observations are outside the outer fence, they are marked as outliers.

Figure 1.20 shows the fences (these are not shown on the actual box plot; they are only guidelines for defining the whiskers, suspected outliers, and outliers) and demonstrates how to mark outliers.

FIGURE 1.19   The Box Plot

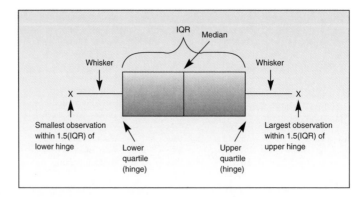

The Elements of a Box Plot   **FIGURE 1.20**

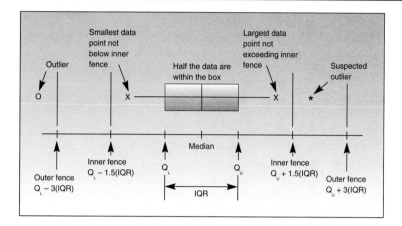

Box plots are very useful for the following purposes:

1. To identify the location of a data set based on the median.
2. To identify the spread of the data based on the length of the box, hinge to hinge (the interquartile range), and the length of the whiskers (the range of the data without extreme observations: outliers or suspected outliers).
3. To identify possible skewness of the distribution of the data set. If the portion of the box to the right of the median is longer than the portion to the left of the median, and/or the right whisker is longer than the left whisker, the data are right-skewed. Similarly, a longer left side of the box and/or left whisker implies a left-skewed data set. If the box and whiskers are symmetric, the data are symmetrically distributed with no skewness.
4. To identify suspected outliers (observations beyond the inner fences but within the outer fences) and outliers (points beyond the outer fences).
5. To compare two or more data sets. By drawing a box plot for each data set and displaying the box plots on the same scale, we can compare several data sets.

Let us now construct a box plot for the data for the first group of engineers in Example 1.3. For this data set, the median is 27, and we find that the lower quartile is 20.75 and the upper quartile is 41. The interquartile range (IQR) is $41 - 20.75 = 20.25$. One and one-half times this distance is 30.38; hence, the inner fences are -9.63 and 71.38. Since no observation lies beyond either point, there are no suspected outliers and no outliers, so the whiskers extend to the extreme values in the data: 11 on the left side and 62 on the right side. The box plot for the data of Example 1.3 is shown in Figure 1.21a. As can be seen from the two parts of Figure 1.21 on page 44, a box plot may also be used to compare two groups of numbers. Notice the difference in location (median) and spread for the two groups of engineers.

---

40. The following data are monthly steel production figures, in millions of tons.   ***PROBLEMS***

    7.0, 6.9, 8.2, 7.8, 7.7, 7.3, 6.8, 6.7, 8.2, 8.4, 7.0, 6.7, 7.5, 7.2, 7.9, 7.6, 6.7, 6.6, 6.3,
    5.6, 7.8, 5.5, 6.2, 5.8, 5.8, 6.1, 6.0, 7.3, 7.3, 7.5, 7.2, 7.2, 7.4, 7.6.

    Draw a box plot of these data.

FIGURE 1.21   MINITAB-Produced Box Plot for the Data of Example 1.3

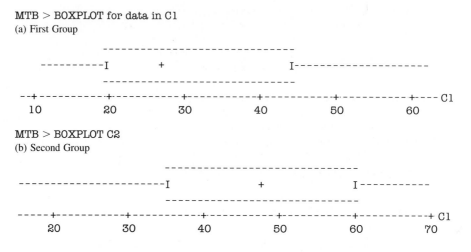

MTB > BOXPLOT for data in C1
(a) First Group

MTB > BOXPLOT C2
(b) Second Group

41. What are the uses of a box plot?

42. *Worker participation in management* is a new concept that involves employees in corporate decision making. The following data are the percentages of employees involved in worker participation programs in a sample of firms. Draw a box plot of these data.

    5, 32, 33, 35, 42, 43, 42, 45, 46, 44, 47, 48, 48, 48, 49, 49, 50, 37, 38, 34, 51, 52, 52, 47, 53, 55, 56, 57, 58, 63, 78

43. Consider the four box plots in the following output, and draw your conclusions about the data sets.

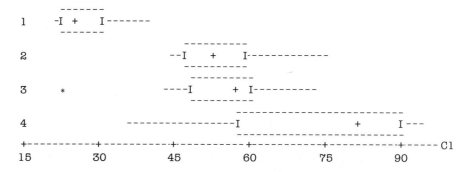

44. The following data (from J. Purba et al., "Decreased number of oxytocin neurons in the hypothalamus in AIDS," *Brain,* Vol. 116, 1993, p. 799) are the weight, in grams, of human brains. One data point is for a person who died from AIDS; the rest are not. Draw a box plot of these data. Identify any outliers. Since AIDS causes loss of brain cells, what conclusions may be drawn here?

    Weight (g): 1090, 1310, 1560, 1330, 1370, 1260, 1350, 1420, 1280, 1400, 1370.

45. Using the data in Table 1.2, Example 1.2, draw a box plot for on-time arrivals in the fourth quarter of 1991.

46. Using the data in Table 1.2, Example 1.2, draw a box plot for the on-time departure rates in the third quarter of 1991.

47. Draw a box plot of the data on numbers of vehicles produced in Problem 14. What is the median?

48. Draw a box plot of government tax collection as percent of GDP, using the data in Problem 26. What is the median tax collection percentage? Describe the distribution.
49. Draw a box plot of the average age of ships registered to different countries of the world, using the data in Problem 28. What is the median age? Are there any outliers? Describe the distribution.
50. Using the display of information on U.S. illegal alien groups in Problem 30, draw a box plot for the number of illegal aliens per country of origin. What is the clear outlier? What is the median?
51. Using the information in Problem 35, draw a box plot of the ages of the president and the vice president. Compare the distributions of these two variables using their box plots. Do the two distributions appear different from each other? Explain.
52. Draw a box plot for peripheral blood and bone marrow cells. Identify any outliers. (Data from S. Giralt et al., "Preliminary results of treatment with Filgrastim for relapse of leukemia after bone marrow transplantation," *New England Journal of Medicine* 329, no. 11, 1993, p. 760.)

> Peripheral blood (%): 58, 88, 56, 71, 81, 80, 34, 86, 46, 75, 82, 11
> Bone marrow cells (%): 25, 60, 45, 30, 32, 40, 62, 65, 54, 89, 41

1.5 MEASURES OF CENTRALITY

Percentiles, and in particular quartiles, are measures of the relative positions of points within a data set or a population (when our data set constitutes the entire population). The median is a special point, since it lies in the center of the data in the sense that half the data lie below it and half above it. The median is thus a measure of the *centrality* of the observations.

In addition to the median, there are two other commonly used measures of centrality. One is the *mode* (or modes—there may be several of them), and the other is the *arithmetic mean,* or just the *mean.*

A **mode** of the data set is a value that occurs most frequently.

Look at the frequencies of occurrence of the data values in Example 1.4, shown in Table 1.4. We see that the value 16 occurs most frequently. There are three data points with this value—more points than for any other value in the data set. Therefore, the mode is 16.

Frequencies of Occurrence of Data Values in Example 1.4   **TABLE 1.4**

Value	Frequency	Value	Frequency
6	1	17	2
9	1	18	2
10	1	19	1
12	1	20	1
13	1	21	1
14	2	22	1
15	1	24	1
16	3		

The most commonly used measure of central tendency of a set of observations is the mean of the observations.

> The **mean** of a set of observations is their **average.** It is equal to the sum of all observations, divided by the number of observations in the set.

Let us denote the observations by $x_1, x_2, \ldots, x_n$. That is, the first observation is denoted by x_1, the second by x_2, and so on to the nth observation, x_n. (In Example 1.4, $x_1 = 9$, $x_2 = 6$, $x_3 = 12$, and $x_n = x_{20} = 17$.) The sample mean is denoted by $\bar{x}$.

> The mean of a sample is defined as follows:
>
> $$\bar{x} = \frac{\text{Sum of } x}{n} = (x_1 + x_2 \ldots + x_n)/n$$

The mean of the observations in Example 1.4 is found as

$$\bar{x} = (x_1 + x_2 + \cdots + x_{20})/20 = (9 + 6 + 12 + 10 + 13 + 15 + 16$$
$$+ 14 + 14 + 16 + 17 + 16 + 24 + 21 + 22 + 18 + 19 + 18 + 20$$
$$+ 17)/20 = 317/20 = 15.85$$

The mean of the observations of Example 1.4, their average, is 15.85.

Figure 1.22 shows the data of Example 1.4 drawn on the number line along with the mean, median, and mode of the observations. If you think of the data points as little balls of equal weight located at the appropriate places on the number line, the mean is that point where all the weights would balance. It is the *fulcrum* of the point-weights, as shown in Figure 1.22.

Our observation set may constitute an entire population.

> A number that describes a *population* is called a population **parameter.** A number that is computed from a *sample* is called a sample **statistic.**

The sample mean is a statistic; the population mean is a parameter. Instead of $\bar{x}$, we denote the population mean by the symbol μ (the Greek letter mu). For a population,

FIGURE 1.22   Mean, Median, and Mode for Example 1.4

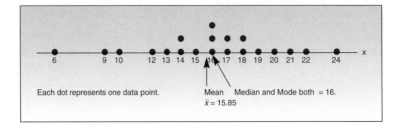

Each dot represents one data point.

Mean   Median and Mode both = 16.
$\bar{x} = 15.85$

we use N as the number of elements instead of n. The mean of a population is defined as follows:

$$\mu = \frac{\text{Sum of } x}{N}$$

The sample mean and the population mean are computed in the same way. This will not be true for other statistics and parameters, as we will see in the next section of this chapter.

What characterizes the three measures of centrality, and what are the relative merits of each? The mean summarizes all of the information in the data. It is the average of all observations. The mean is a single point that can be viewed as the point where all the mass—the weight—of the observations is concentrated. It is the center of mass of the data. If all the observations in our data set were the same size, then (assuming the total is the same) each would be equal to the mean.

The median, on the other hand, is an observation (or a point between two observations) in the center of the data set. One-half of the data lie above this observation, and one-half of the data lie below it. When we compute the median, we do not consider the exact location of each data point on the number line; we only consider whether or not it falls in the half lying above the median or the half lying below the median. If you look at the picture of the data set of Example 1.4, Figure 1.22, you will note that the observation $x_{13} = 24$ lies to the far right. If we shift this particular observation (or any other observation to the right of 16) to the right (say, move it from 24 to 100), what would happen to the median? The answer is: absolutely nothing (prove this to yourself by calculating the new median). The exact location of any data point is not considered in the computation of the median, only its position relative to the central observation. *The median is resistant to extreme observations.*

The mean, on the other hand, is sensitive to extreme observations. Let us see what happens to the mean if we change x_{13} from 24 to 100. The new mean is

$$\bar{x} = (9 + 6 + 12 + 10 + 13 + 15 + 16 + 14 + 14 + 16 + 17$$
$$+ 16 + 100 + 21 + 22 + 18 + 19 + 18 + 20 + 17)/20 = 19.65$$

We see that the mean has shifted almost four units to the right to accommodate the change in the single data point x_{13}.

The mean, however, does have strong advantages as a measure of centrality. *The mean is based on information contained in all of the observations in the data set,* rather than merely lying "in the middle" of the set. The mean also has some desirable mathematical properties that make it useful in many contexts of statistical inference. In cases where we want to guard against the influence of a few outlying observations (outliers), however, we may prefer to use the median.

The mode is less useful than either the mean or the median. In Example 1.4 our data would possess two modes if we had another data point equal to 18, for example. Of the three measures of centrality, we are most interested in the mean.

If a data set or population is **symmetric** (that is, if one side of the distribution of the observations is a mirror image of the other) and if the distribution of the observations has only one mode, then the mode, the median, and the mean are all equal. Such a situation is demonstrated in Figure 1.23. Generally, when the data distribution is not symmetric, then the mean, median, and mode will not all be equal.

FIGURE 1.23   A Symmetrically Distributed Data Set

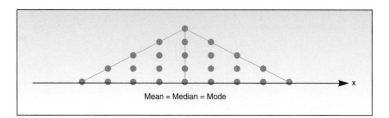

Mean = Median = Mode

For a right-skewed distribution, the mean is to the right of the median, which in turn lies to the right of the mode (assuming a single mode). The opposite is true for left-skewed distributions. This is shown in Figure 1.24.

PROBLEMS

53. Discuss the differences among the three measures of centrality.
54. Find the mean, median, and mode(s) of the observations in problem 40.
55. Find the mean peripheral blood percentage using the data in Problem 52.
56. Find the mean worker participation percentage using the data in Problem 42.
57. The following table lists the exchange rates of several foreign currencies (the value of one U.S. dollar in each) for January 1993. Find the mean and the median for these data.

Date	British Pound	French Franc	German Mark	Italian Lira	Singapore Dollar
01/04/93	0.6666	5.5935	1.6383	1506.50	1.6460
01/05/93	0.6464	5.5460	1.6260	1526.00	1.6540
01/06/93	0.6520	5.5790	1.6456	1529.00	1.6555
01/07/93	0.6522	5.5600	1.6360	1512.00	1.6600
01/08/93	0.6510	5.5875	1.6440	1501.00	1.6640
01/11/93	0.6437	5.5435	1.6399	1477.00	1.6588
01/12/93	0.6477	5.5180	1.6265	1492.00	1.6557
01/13/93	0.6473	5.5110	1.6265	1508.00	1.6590
01/14/93	0.6507	5.5135	1.6220	1490.75	1.6565
01/15/93	0.6536	5.5265	1.6350	1499.00	1.6575
01/18/93	0.6532	5.4535	1.6350	1499.00	1.6575
01/19/93	0.6454	5.4345	1.6070	1476.00	1.6562
01/20/93	0.6486	5.4225	1.6052	1466.00	1.6548
01/21/93	0.6559	5.4520	1.6125	1476.00	1.6558
01/22/93	0.6536	5.3815	1.5912	1463.91	1.6440
01/25/93	0.6427	5.3290	1.5770	1452.43	1.6430
01/26/93	0.6509	5.3370	1.5795	1449.50	1.6418
01/27/93	0.6614	5.3670	1.5866	1487.10	1.6382
01/28/93	0.6604	5.3630	1.5914	1477.00	1.6358
01/29/93	0.6725	5.4475	1.6112	1493.75	1.6370

58. On January 26, 1993, the Bank of England cut interest rates by 1 point. Using the data in the previous problem, check to see whether the mean exchange rate of the British pound (the table values are pounds per dollar) is higher (more pounds per dollar) in the period January 26 to January 29 than it is for the earlier part of the month. If so, why?

Measures of Centrality and Skewness

FIGURE 1.24

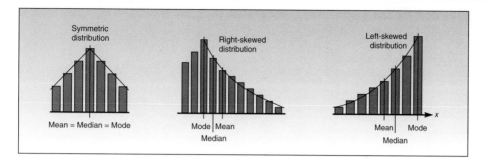

59. The following data are numbers of irrigated farms in the United States, by state. Compare the *average* number of irrigated farms in 1982 with that in 1987. Draw a conclusion.

| | Number (1,000) | |
State	1982	1987
Arizona	4.4	4.2
California	58.4	58.9
Colorado	15.2	14.9
Florida	10.6	12.0
Idaho	17.4	16.6
Kansas	7.3	7.4
Louisiana	3.7	3.9
Montana	9.2	9.5
Nebraska	22.2	22.6
Nevada	2.2	2.2
New Mexico	6.2	7.0
Oregon	15.3	14.4
Texas	19.8	19.8
Utah	11.2	11.1
Washington	16.3	15.4
Wyoming	5.3	5.2

Source: Reprinted from the 1992 *Statistical Abstract of the United States.*

60. Use MINITAB (or another computer package) to analyze the data in Problem 57. Enter the data into five columns: C1, C2, C3, C4, and C5. Then use the command DESCRIBE, followed by each column number separately. The outcome will include the mean, median, and other information for each variable. Also use the command HISTOGRAM and compare the distributions of the currencies' exchange rates for the period of interest.

61. Using the data in Problem 35, find the mean age of a president and the mean age of a vice president. Compare the two, and compare each mean with the corresponding median. Comment on the reasons for the results you found.

62. Using the data in Problem 28, find the mean of the "average age" of vessels in the world's fleets. Compare this mean with the reported "1989 World Total" average age. Why are the two numbers different? (Hint: How is the average computed?)

63. For the data from which the histogram on the right was computed, which is greatest: mean, median, or mode? What is the likely order of these three measures? Explain.

64. The following table shows the twelve operas most frequently produced. Find the mean, median, and mode(s) for the number of productions and for the number of performances of the operas in this group.

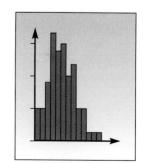

Opera	Productions	Performances
Madama Butterfly	22	102
Le Nozze di Figaro	21	83
Don Giovanni	18	62
Die Fledermaus	13	72
Così Fan Tutte	13	68
Rigoletto	13	62
Die Zauberflöte	12	67
La Traviata	12	54
Carmen	11	53
Faust	8	39
Il Barbiere di Siviglia	8	35
Les Contes d'Hoffmann	8	27

Source: From *Opera News,* November 1992, p. 32.

1.6 MEASURES OF VARIABILITY

Consider the following two data sets:

Set I: 1, 2, 3, 4, 5, 6, 6, 7, 8, 9, 10, 11
Set II: 4, 5, 5, 5, 6, 6, 6, 6, 7, 7, 7, 8

Compute the mean, median, and mode of each of the two data sets. As you see from your results, the two data sets have the same mean, the same median, and the same mode, all equal to 6. The two data sets also happen to have the same number of observations, $n = 12$. But the two data sets are different. What is the main difference between them?

Figure 1.25 shows data sets I and II. The two data sets have the same central tendency (as measured by any of the three measures of centrality), but they have a different **variability.** In particular, we see that data set I is more variable than data set

 FIGURE 1.25  Comparison of Data Sets I and II

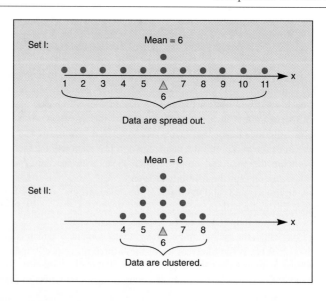

II. The values in set I are more spread out: they lie further away from their mean than do those of set II.

There are several measures of variability or **dispersion.** We have already discussed one such measure—the interquartile range. (Recall that the interquartile range is defined as the difference between the upper quartile and the lower quartile.) The interquartile range for data set I is 5.5, and the interquartile range of data set II is 2 (show this). The interquartile range is one measure of the dispersion or variability of a set of observations. Another such measure is the *range.*

> The **range** of a set of observations is the difference between the largest observation and the smallest observation.

The range of the observations in Example 1.1 is: Largest number – Smallest number = 24 − 6 = 18. The range of the data in set I is 11 − 1 = 10, and the range of the data in set II is 8 − 4 = 4. We see that, conforming with what we expect from looking at the two data sets, the range of set I is greater than the range of set II. Set I is more variable.

The range and the interquartile range are measures of the dispersion of a set of observations; the interquartile range is more resistant to extreme observations. There are also two other, more commonly used, measures of dispersion. These are the *variance* and the square root of the variance—the *standard deviation.*

The variance and the standard deviation are more useful than the range and the interquartile range, because, like the mean, they use the information contained in all the observations in the data set or population. (The range contains information only on the distance between the largest and smallest observations, and the interquartile range contains information only about the difference between upper and lower quartiles.)

> The **variance** of a set of observations is the average squared deviation of the data points from their mean.

When our data constitute a sample, the variance is denoted by s^2, and the averaging is done by dividing the sum of the squared deviations from the mean by $n - 1$.

A sample variance may be found by using the following formula:

$$s^2 = \frac{\text{Sum } (x_i - \bar{x})^2}{n - 1}$$

Recall that $\bar{x}$ is the sample mean, the average of all the observations in the sample. Thus, the numerator in the sample variance formula is equal to the sum of the squared differences of the data points x_i (where $i = 1, 2, \ldots, n$) from their mean, $\bar{x}$. These differences are squared because if we just compute the deviations from the mean and then average them, we get zero (prove this with any of the data sets). Therefore, when seeking a measure of the variation in a set of observations, we square the deviations from the mean; this removes the negative signs, and thus the measure is not equal to zero. We divide this numerator by the denominator $n - 1$ to get a kind of average of the items summed in the numerator, because only $n - 1$ of our n data points are free to vary. The nth is predictable, since we already know the mean. That is, the number of *degrees of freedom* associated with the sample variance is $n - 1$. (Note, however, that the summation in the numerator extends over all n data points, not just $n - 1$ of them.)

When we have an entire population at hand, we denote the total number of observations in the population by N; the variance is denoted by σ^2; and the averaging is done by dividing by N. (σ is the Greek letter sigma; we call the variance *sigma-squared*.)

We define the population variance as follows:

$$\sigma^2 = \frac{\text{Sum } (x_i - \mu)^2}{N}$$

where μ is the population mean.

Unless noted otherwise, we will assume that all our data sets are samples and do not constitute entire populations and will use the appropriate equation for the variance. Let's discuss the *standard deviation*.

The **standard deviation** of a set of observations is the (positive) square root of the variance of the set.

The standard deviation of a sample is the square root of the sample variance, and the standard deviation of a population is the square root of the variance of the population.[2]

A sample standard deviation may be found by using the following equation:

$$s = \sqrt{s^2} = \sqrt{\frac{\text{Sum } (x_i - \bar{x})^2}{n - 1}}$$

The equation for a population standard deviation is:

$$\sigma = \sqrt{\sigma^2} = \sqrt{\frac{\text{Sum } (x_i - \mu)^2}{N}}$$

The population variance, σ^2, and the population standard deviation, σ, are population parameters. The sample variance, s^2, and the sample standard deviation, s, are sample statistics.

Why would we use the standard deviation when we already have its square, the variance? Because the standard deviation is a more meaningful measure. The variance

---

[2]A note about calculators: If your calculator is designed to compute means and standard deviations, find the key for the standard deviation. Typically, there will be two such keys. Consult your owner's handbook to be sure you are using the key that will produce the correct computation for a sample (division by $n - 1$) versus a population (division by N).

Calculations Leading to the Sample Variance in Example 1.4   **TABLE 1.5**

x	$(x - \bar{x})$		$(x - \bar{x})^2$
6	6 – 15.85	=–9.85	97.0225
9	9 – 15.85	=–6.85	46.9225
10	10 – 15.85	=–5.85	34.2225
12	12 – 15.85	=–3.85	14.8225
13	13 – 15.85	=–2.85	8.1225
14	14 – 15.85	=–1.85	3.4225
14	14 – 15.85	=–1.85	3.4225
15	15 – 15.85	=–0.85	0.7225
16	16 – 15.85	= 0.15	0.0225
16	16 – 15.85	= 0.15	0.0225
16	16 – 15.85	= 0.15	0.0225
17	17 – 15.85	= 1.15	1.3225
17	17 – 15.85	= 1.15	1.3225
18	18 – 15.85	= 2.15	4.6225
18	18 – 15.85	= 2.15	4.6225
19	19 – 15.85	= 3.15	9.9225
20	20 – 15.85	= 4.15	17.2225
21	21 – 15.85	= 5.15	26.5225
22	22 – 15.85	= 6.15	37.8225
24	24 – 15.85	= 8.15	66.4225
		0	378.5500

is a *squared* quantity; it is an average of squared numbers. By taking its square root, we "unsquare" the units and get a quantity denoted in the original units of the problem (e.g., dollars instead of dollars squared, which would have little meaning in most applications). If the observations differ from the mean by one unit or more, the variance tends to be large because it is in squared units. The mathematical properties of the variance simplify some computations, but the standard deviation is more easily interpreted.

Let's find the variance and the standard deviation of the data in Example 1.4. It is convenient to carry out hand computations of the variance by use of a table. After doing the computation, we will show a shortcut that will help in the calculation. Table 1.5 shows how the mean, $\bar{x}$, is subtracted from each of the values, and the results are squared and added together. (For convenience, the data were ordered from smallest to largest.) At the bottom of the last column we find the sum of all squared deviations from the mean. Finally, the sum is divided by $n - 1$, giving s^2, the sample variance. Taking the square root gives us s, the sample standard deviation.

By the formula for variance, the variance of the sample is equal to the sum of the third column in the table, 378.55, divided by $n - 1$: $s^2 = 378.55/19 = 19.923684$. The standard deviation is the square root of the variance: $s = \sqrt{19.923684} = 4.4635954$, or, using two-decimal accuracy, $s = 4.46$.[3]

---

[3]In quantitative fields such as statistics, there is always the problem of decimal accuracy. How many digits after the decimal point should we carry? This question has no easy answer; everything depends on the required level of accuracy. As a rule, we will use only two decimals, since this suffices in most applications. In some procedures, such as regression analysis, it is recommended that more digits be used in computations (these computations, however, are usually done by computer).

If you have a calculator with statistical capabilities, you may avoid having to use a table such as Table 1.5. If you need to compute by hand, there is a shortcut formula for computing the variance and the standard deviation.

Shortcut formula for the sample variance:

$$s^2 = \frac{\text{Sum } (x_i^2) - [\text{Sum } (x_i)]^2/n}{n - 1}$$

Again, the standard deviation is just the square root of the quantity in the equation above. We will now demonstrate the use of this computationally simpler formula with the data of Example 1.4. We will then use this simpler formula and compute the variance and the standard deviation of the two data sets we are comparing (set I and set II).

As before, a table will be useful in carrying out the computations. The table for finding the variance will have a column for the data points, x, and a column for the squared data points, x^2. Table 1.6 shows the computations for the variance of the data in Example 1.4.

We have:

$$s^2 = \frac{\text{Sum } (x_i)^2 - [\text{Sum } (x_i)]^2/n}{n - 1} = \frac{5{,}403 - (317)^2/20}{19} = \frac{5{,}403 - 100{,}489/20}{19}$$

$$= 19.923684$$

The standard deviation is obtained as before: $s = \sqrt{19.923684} = 4.46$. Using the same procedure demonstrated in Table 1.6, we find the following quantities leading to the variance and the standard deviation of set I and of set II. Both are assumed to be samples, not populations.

Set I: Sum of $x = 72$, Sum of $x^2 = 542$, $s^2 = 10$, and $s = \sqrt{10} = 3.16$
Set II: Sum of $x = 72$, Sum of $x^2 = 446$, $s^2 = 1.27$, and $s = \sqrt{1.27} = 1.13$

As expected, we see that the variance and the standard deviation of set II are smaller than those of set I. While each has a mean of 6, set I is more variable. That is, the values in set I vary more about their mean than do those of set II, which are clustered more closely together. Let's now look at how we can compute the variance of a *population*. We use the equation that divides by the size of the population, N, rather than by $n - 1$, as in the case of the sample variance. Assuming that the departure data in Example 1.2 constitute an entire population, we find the population variance as

$$\sigma^2 = \frac{\text{Sum } (x_i - \mu)^2}{N} = \frac{1}{29} [(88.1 - 86.748)^2$$
$$+ (89.4 - 86.748)^2 + \ldots + (89.9 - 86.748)^2] = 9.968$$

and the standard deviation is

$$\sigma = \sqrt{9.968} = 3.157$$

Since most data we will see constitute samples rather than entire populations, this computation will not commonly be done.

The following relationship is sometimes useful between the mean and the standard deviation.

TABLE 1.6

Shortcut Computations for the Variance in Example 1.4

x	x^2
6	36
9	81
10	100
12	144
13	169
14	196
14	196
15	225
16	256
16	256
16	256
17	289
17	289
18	324
18	324
19	361
20	400
21	441
22	484
24	576
317	5,403

The Empirical Rule

If the distribution of the data is mound-shaped—that is, if the histogram of the data is more or less symmetric with a single mode or high point—the following rules will apply:

1. Approximately 68 percent of the observations will be within one standard deviation of the mean.
2. Approximately 95 percent of the observations will be within two standard deviations of the mean.
3. A vast majority of the observations (all of them, or almost all of them) will be within three standard deviations of the mean.

For the data set in Example 1.4, even though the distribution of the data set is not perfectly symmetric, the empirical rule holds approximately (especially its last two parts). The mean is 15.85 and the standard deviation is 4.46. The two points that are one standard deviation on either side of the mean are $15.85 - 4.46 = 11.39$ and $15.85 + 4.46 = 20.31$. We see that 14 out of our 24 data points lie within this range. This is 58.3 percent of our observations. The two points that lie two standard deviations on either side of the mean are $15.85 - 2(4.46) = 6.93$ and $15.85 + 2(4.46) = 24.77$. We see that 23 out of our 24 data points lie within this range, or 95.83 percent of our observations. The three standard deviation limits around the mean are 2.47 and 29.23. One hundred percent of our observations lie within this range.

---

65. Explain why we need measures of variability and what information these measures convey.

66. What is the most important measure of variability and why?

67. What is the computational difference between the variance of a sample and the variance of a population?

68. Find the range, the variance, and the standard deviation of the data set in Problem 13 (assumed to be a sample).

69. The standard deviation is often used as a measure of *volatility*. Which of the currencies in Problem 57 has the largest standard deviation? Do you believe that these standard deviations can be used to compare volatilities *across* currencies? Explain. Use the MINITAB command DESCRIBE to find the standard deviations of the currencies.

70. Find the standard deviation of the number of irrigated farms per state in Problem 59. Compare the standard deviation for 1982 with that for 1987. Assume a population in each case, not a sample.

71. What are the variance and the standard deviation of the number of opera productions listed in Problem 64? Can the empirical rule be used here? If so, use it and state your conclusions.

72. Find the mean, variance, and standard deviation of the following data. These data are the wealth, in billions of U.S. dollars, of the richest people on earth, as reported recently in *Fortune* magazine. Is the distribution mound-shaped so the empirical rule may apply? Use MINITAB in the analysis.

25.0, 20.0, 8.7, 7.5, 7.4, 6.0, 5.7, 5.5, 5.0, 5.0, 4.4, 4.0, 4.0, 3.6, 3.4, 3.1, 3.0, 3.0, 2.9, 2.8, 2.8, 2.5, 2.5, 2.5, 2.4, 2.4, 2.4, 2.2, 2.0, 2.0, 2.0, 1.9, 1.8, 1.7, 1.6, 1.5, 1.5, 1.5, 1.5, 1.4, 1.3, 1.3, 1.3, 1.3, 1.2, 1.2, 1.2, 1.2, 1.1, 1.1, 1.1., 1.0, 1.0, 1.0, 1.0, 1.0, 1.0, 1.0, 1.0, 1.0.

PROBLEMS

73. Interpret the following MINITAB results (ignore TRMEAN, SEMEAN).

MTB > DESCRIBE C1

	N	MEAN	MEDIAN	TRMEAN	STDEV	SEMEAN
C1	17	48.6	9.0	48.5	47.8	11.6

	MIN	MAX	Q1	Q3
C1	1.0	99.0	4.5	99.0

74. The following table shows ranges for caffeine content in drinks and medicines. Use the ideas discussed in this section to comment on the meaning of the information in this display.

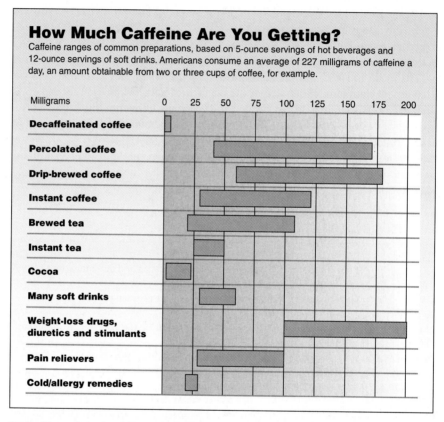

Reprinted by permission from "Headache? You Skipped Your Coffee," *The New York Times,* October 15, 1992, p. A18, with data from the U.S. Food and Drug Administration.

1.7 CAUTION, CARE WITH NUMBERS, AND ETHICS

We must use care when handling numbers. We have already seen that displaying numbers graphically can be misleading and that we should be careful when looking at graphical displays of information. Another potential pitfall is the numbers themselves—and what we do with them. In this chapter, we have made references to outliers: data points that look different from the rest. A number that is much greater than all the rest, or one that is much smaller than the rest, is an outlier. If an outlier is due to gross error—for example, a coder typed the numbers 95, 96, and then 4 instead of 94—then clearly the outlier should be removed (or, better, corrected if possible).

The assumption that some "error" is implicit in all outliers is the underlying philosophy behind many statistical methods designed to be *resistant*. For example, the *trimmed mean* (TRMEAN) reported in the MINITAB output for the DESCRIBE command is a mean obtained after removing the highest 5 percent and the lowest 5 percent of the data points and averaging the remaining data.

Such methods, when used indiscriminately, can be very dangerous. Sometimes an outlier is not due to error—the outlier may have a story to tell us. This was very well demonstrated in 1984, when scientists around the world began reporting troubling results about the depletion of atmospheric ozone levels above Antarctica. The Goddard Space Center had done a survey of atmospheric ozone around the world at that time and, surprisingly, found no unusual information—entirely missing what later became known as the ozone hole over Antarctica. The Goddard scientists regularly fed results into the computer, and the computer was programmed to delete any measurement below 180 Dobson units, because at that time no one thought it at all possible to have an ozone level in such a low range. Any such measurements, if made, were automatically assumed to be in error and thus discarded from the analysis. Hence, all the observations made in the ozone hole in Antarctica were automatically deleted, and no one at Goddard was aware of the hole until other scientists reported it. (For more on this story, see S. Roan, *Ozone Crisis,* New York: Wiley, 1989, pp. 129–33.)

Ethics

Statisticians collect data, analyze data, and report their findings. These activities entail great responsibility. Often, decisions that affect society are made based on the information reported as a result of a statistical analysis, and wrong decisions may cost society dearly. When dealing with data collection, analysis, and reporting, we must be careful, for there is the potential for abuse.

Monday, August 9, 1993

Psychiatrist Convicted of Fraud—'Made Up' Drug Study Data

New York Times

Minneapolis

A nationally known child psychiatrist has been convicted by a federal jury of falsifying data in a $250,000 drug study at the University of Minnesota.

Dr. Barry Garfinkel, who is widely respected for his work on suicide among teenagers, was found guilty Thursday after a two-week trial in which a former assistant testified that he had told her to "make up" data.

The verdict in U.S. District Court makes Garfinkel one of the few American scientists ever convicted of a crime in connection with a study of an experimental drug regulated by the Food and Drug Administration.

Garfinkel, 46, was accused of falsifying reports during a study of Anafranil, an antidepressant that he tested from 1986 to 1989 on patients with obsessive-compulsive disorders. His research was part of a national study to prove that Anafranil was safe and effective enough to be sold in the United States.

Continued

Monday, August 9, 1993 (continued)

The drug was later approved by the FDA, but without most of Garfinkel's data, which were scrapped after the accusations came to light.

Garfinkel maintained that he had never set out to defraud anyone and had merely made honest mistakes while running the study.

But Assistant U.S. Attorney Andrew Luger said Garfinkel treated the study "like a joke," faking reports on patient examinations that had either never taken place or had been conducted by people with no medical training.

Subjects in the study were supposed to come in for regular medical visits to determine whether the drug was helping or harming them.

Garfinkel blamed his study coordinator, Michelle Rennie, for filling out the false reports. Rennie, in turn, testified that Garfinkel had ordered her to make up data on patient examinations that were never conducted.

Lies and Rigged 'Star Wars' Test Fooled the Kremlin, and Congress

By TIM WEINER
Special to The New York Times

WASHINGTON, Aug. 17 — Officials in the "Star Wars" project rigged a crucial 1984 test and faked other data in a program of deception that misled Congress as well as the intended target, the Soviet Union, four former Reagan Administration officials said.

The deception program was designed to feed the Kremlin half-truths and lies about the project, formally known as the Strategic Defense Initiative, the former Administration officials said. It helped persuade the Soviets to spend tens of billions of dollars to counter the American effort to develop a space-based shield against nuclear attack proposed by former President Ronald Reagan in 1983, they said.

But the deceptive information originally intended for consumption in the Kremlin also seeped into closed briefings that helped persuade Congress to spend more money on strategic defense, the former Reagan Administration officials said. All would speak only on the condition that they not be named, and several still hold sensitive military and intelligence posts.

"It Was Used Improperly"

One military officer who described the deception program said it had overstepped its boundaries. "It wasn't designed to deceive Congress," he said. "It was used improperly."

The former Administration officials cited what they said was a clear example of a rigged test that misled Congress and the Kremlin. In June 1984, they said, project officials conducted the fourth attempt to hit a target missile launched from California with an interceptor missile launched from the Pacific.

The first three tests in the series had failed. It was crucial that the fourth succeed, a scientist with the project said. "We would lose hundreds of millions of dollars in Congress if we didn't perform it successfully," he said. "It would be a catastrophe."

To insure that the missile defense program would be seen as a success, the test was faked, the former Reagan Administration officials said.

"We rigged the test," the scientist said. "We put a beacon with a certain frequency on the target vehicle. On the interceptor, we had a receiver." In effect, the scientist said, the target was talking to the missile, saying: "Here I am. Come get me."

"The hit looked beautiful," the scientist said, "so Congress didn't ask questions."

The preceding two articles demonstrate the problem of outright fraud. Note that in the case described in the second article, no statistical inference was carried out and deception entailed rigging measurements.

SUMMARY

In this chapter we introduced the field of statistics and discussed the basic elements of **data** and **measurements.** We saw how data are collected—measurements obtained through designed experiments or by other means. **Scales of measurement** were defined for observations. From weakest to strongest, **nominal, ordinal, interval,** and **ratio** scales may apply to a measurement. We defined two important concepts, a **population** and a **sample,** and we saw the importance of a random sample for statistical inference. We defined **data distributions** and saw a variety of methods for displaying data and their distributions. These included **pie charts, bar graphs,** the **histogram, stemplots, boxplots,** and others. We cautioned that graphs may sometimes be misleading. We saw how a data distribution can be further analyzed to discover its **percentiles,** as well as special percentiles called **quartiles,** and the **median**—the observation that would lie in the middle of the data set. We also defined the **mean** as a measure of centrality and compared it with the median. We noted that the mean contains information from all the observations but that the median was resistant to the influence of **outliers.** We also saw a third measure of centrality, the **mode** or modes of the distribution. Then, measures of variability of a distribution were presented: the **variance** and the **standard deviation,** as well as the **range** and the **interquartile range** (seen earlier). We discussed the **empirical rule,** which, for mound-shaped data distributions, ties in the mean and the standard deviation with percentiles of the distribution. Finally, we advocated caution when handling data and discussed ethics in the context of statistics.

KEY FORMULAS

Sample mean:
$$\frac{\text{Sum of all observations}}{n}$$

Sample standard deviation:
$$\sqrt{\frac{\text{Sum } (x_i - \bar{x})^2}{n - 1}}$$

Population mean:
$$\frac{\text{Sum of all observations}}{N}$$

Population standard deviation:
$$\sqrt{\frac{\text{Sum } (x_i - \bar{x})^2}{N}}$$

ADDITIONAL PROBLEMS

75. Twenty randomly chosen people are shown a television commercial and asked to rank it as to overall appeal on a scale of 0 to 100. The results are given below.

   89, 75, 59, 96, 88, 71, 43, 62, 80, 92, 76, 72, 67, 60, 79, 85, 77, 83, 87, 53

Find the mean, variance, and standard deviation of the sample of ratings.

76. Presented in the following list are data on government expenditures by Western Hemisphere countries, expressed as a percentage of a nation's gross domestic product.

Argentina	12.7	Haiti	87.4
Bahamas	13.2	Honduras	14.6
Belize	14.0	Jamaica	21.5
Bolivia	13.6	Martinique	26.0
Brazil	8.9	Mexico	10.8
Canada	19.4	Netherlands Antilles	21.3
Chile	12.4	Nicaragua	11.6
Colombia	8.0	Panama	15.3
Costa Rica	18.3	Paraguay	7.3
Cuba	2.1	Peru	12.5
Dominican Republic	7.8	Puerto Rico	17.1
Ecuador	15.0	Surinam	16.4
El Salvador	15.1	Trinidad and Tobago	16.8
Guadeloupe	32.3	Uruguay	13.5
Guatemala	7.7	U.S.A.	20.7
Guyana	28.9	Venezuela	13.1

Construct a box plot for the data on the listed countries. Are there any outliers (or suspected outliers)?

77. The following data are annual salary figures, in thousands of dollars (to the nearest thousand), for a random sample of executives in a particular industry.

58, 72, 39, 108, 59, 66, 48, 112, 65, 72, 77, 69, 80, 92, 50, 95, 78, 44, 107, 90, 82, 70, 88, 71, 67.

Draw a histogram of the frequency distribution of the executive salaries. Find the mean salary and the standard deviation of the salaries.

78. The table "Who's Hot and Who's Not" shows data on U.S. market share for Japanese car makers.

Who's Hot and Who's Not

	1990 U.S. Market Share*	Percent Change from Previous Year
Toyota	7.64%	17.5
Honda	6.12	14.3
Nissan	4.49	−1.5
Mazda	2.52	7.2
Mitsubishi	1.38	34.0
Subaru	0.80	−14.9
Isuzu	0.80	−2.4
Suzuki	0.15	−28.6
Daihatsu	0.11	—

*Vehicles sold under Japanese nameplates, including cars and light trucks.

Source: Reprinted from *Business Week,* January 21, 1991, p. 37.

Draw a bar graph of these data and a pie chart showing the total 1990 percentage of Japanese cars in the U.S. market for the period of this study.

79. Find the median, the interquartile range, and the 45th percentile of the following data.

    23, 26, 29, 30, 32, 34, 37, 45, 57, 80, 102, 147, 210, 355, 782, 1,209

    Also draw a box plot using MINITAB (or another package), and identify outliers.

80. Analyze the following display.

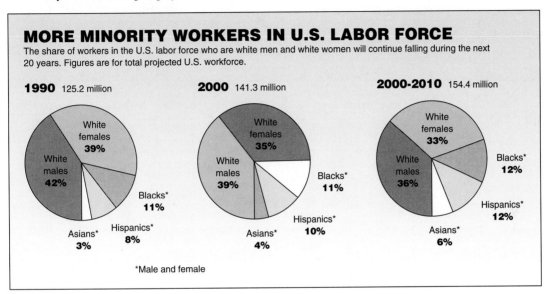

Reprinted from the *San Francisco Chronicle,* August 10, 1993, p. A2.

81. Comment on the following display.

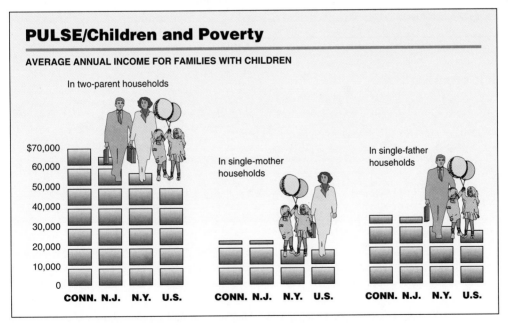

Reprinted from *The New York Times,* November 9, 1992, p. B1.

82.   Comment on and interpret/critique the graphs about truck manufacturing.

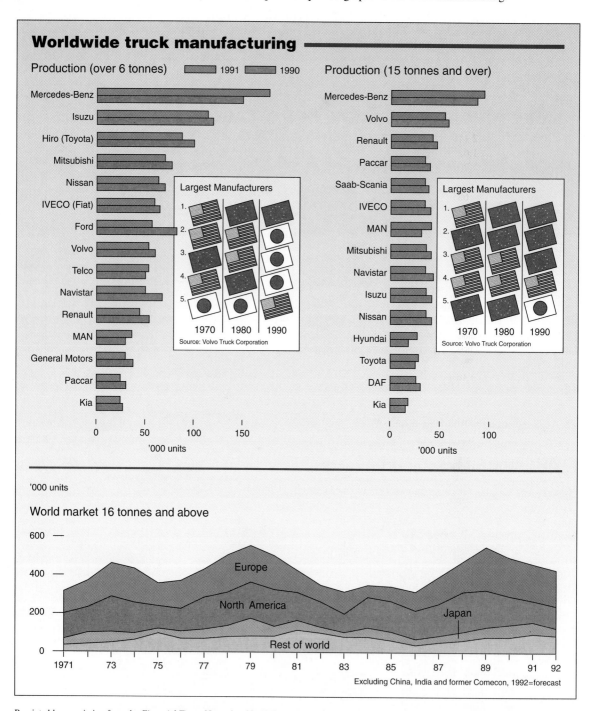

Reprinted by permission from the *Financial Times*, November 23, 1992, p. I, with data from Mercedes-Benz and Volvo Truck Corporation.

83.   Using data from the table showing complaints against airlines, draw pie charts of the complaint categories, including "Other" for the three years 1986, 1989, and 1991.

Consumer Complaints against U.S. Airlines: 1986 to 1991
(Calendar year data. See source for data on individual airlines)

Complaint Category	Complaints						Percent			Rank		
	1986	1987	1988	1989	1990	1991	1986	1989	1991	1986	1989	1991
Total	10,802	40,985	21,493	10,553	7,703	6,126	100.0	100.0	100.0	(X)	(X)	(X)
Flight problems[1]	3,390	18,019	8,831	4,111	3,034	1,883	31.4	39.0	30.7	1	1	1
Baggage	2,149	7,438	3,938	1,702	1,329	888	19.9	16.1	14.5	2	2	2
Refunds	1,627	3,313	1,667	1,023	701	787	15.1	9.7	12.8	3	3	3
Customer service[2]	702	3,888	2,120	1,002	758	714	6.5	9.5	11.7	5	4	4
Ticketing/ boarding[3]	687	2,458	1,445	821	624	661	6.4	7.8	10.8	6	5	5
Fares[4]	468	937	455	341	312	388	4.3	3.2	6.3	7	7	6
Oversales[5]	849	2,122	1,353	607	399	304	7.9	5.8	5.0	4	6	7
Advertising	122	344	141	89	96	96	1.1	0.8	1.6	9	9	8
Smoking	311	888	546	232	74	30	2.9	2.2	0.5	11	10	10
Tours	33	90	37	22	29	23	0.3	0.2	0.4	11	10	10
Credit	40	101	35	19	5	10	0.4	0.2	0.2	10	11	11
Other	424	1,387	925	584	342	342	3.9	5.5	5.6	(X)	(X)	(X)

X not applicable. [1] Cancellations, delays etc. from schedule. [2] Unhelpful employees, inadequate meals or cabin service, treatment of delayed passengers. [3] Errors in reservations and ticketing; problems in making reservations and obtaining tickets. [4] Incorrect or incomplete information about fares, discount fare conditions, and availability, etc. [5] All bumping problems, whether or not airline complied with DOT regulations.

Source: The 1992 *Statistical Abstract of the United States.* U.S. Dept. of Transportation, Office of Consumer Affairs, *Air Travel Consumer Report,* monthly.

84. What conclusions can be drawn about incomes from the following display?

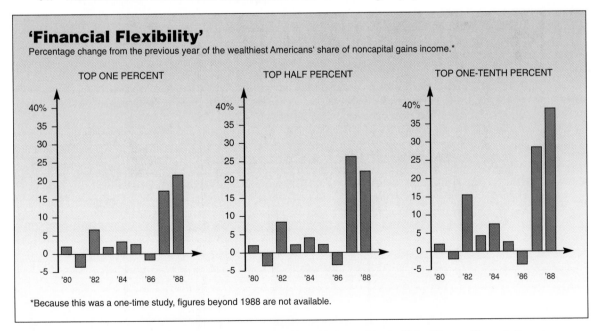

Reprinted from *The New York Times,* September 1, 1993, p. D1, with datza from Feenberg and Poterba, National Bureau of Economic Research.

85. The following display is a time chart of the number of days per year on which the temperature in New York City reached 90 degrees or higher, from the 1870s to the present. The chart also shows a "moving" 10-year average. What is the purpose of the moving average? Based on the displayed information, do you believe that global warming is real? Explain.

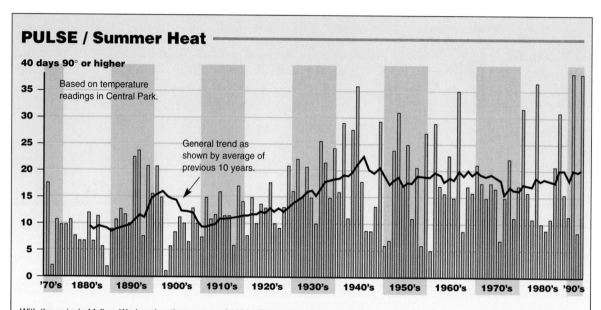

PULSE / Summer Heat

40 days 90° or higher

Based on temperature readings in Central Park.

General trend as shown by average of previous 10 years.

'70's 1880's 1890's 1900's 1910's 1920's 1930's 1940's 1950's 1960's 1970's 1980's '90's

With the arrival of fall on Wednesday, the summer of 1993 will pass into history as one of the hottest on record. One indicator is the number of days 90 degrees or higher, which tied the all-time high of 39 set only two years ago. July alone had 20 such days, a record for the month. And this June, July and August tied the second highest average temperature for the period, 76.9°. Although every summer might seem hotter than the last, the average number of 90 degree-plus days has been relatively steady since the 1940's. The real warming trend of this century occurred between 1910 and the 1940's. A small part of this may be attributed to the increasing amount of heat generated and retained in the growing city, known as a "heat island" effect.

Reprinted from *The New York Times,* September 28, 1993, p. B1, with data from Pennsylvania State University.

86. The display on quality of life provides an unusual graphical comparison of various pieces of information about China and India. Carefully analyze the displayed information using the ideas discussed in this chapter.

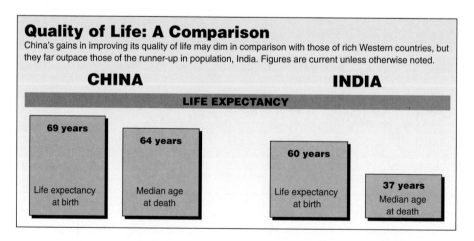

Quality of Life: A Comparison

China's gains in improving its quality of life may dim in comparison with those of rich Western countries, but they far outpace those of the runner-up in population, India. Figures are current unless otherwise noted.

CHINA   **INDIA**

LIFE EXPECTANCY

69 years

64 years

60 years

37 years

Life expectancy at birth

Median age at death

Life expectancy at birth

Median age at death

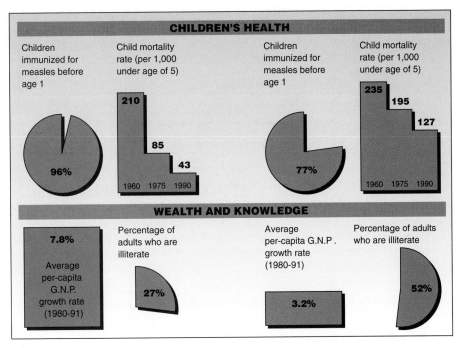

Reprinted from *The New York Times,* September 7, 1993, p. A10, with data from The World Bank.

87. Find the mean and standard deviation of the calories and sodium measurements given in this table. Assume this is a sample.

Supermarket Soups

What you'll find in the best of the cans

Product	Calories	Fat (grams)	Fat (% calories)	Sodium
CANNED SOUPS				
Healthy Choice Chicken Noodle	95 Contains yeast	3	30	435
Campbell's Healthy Request New England Clam Chowder	100 Contains sugar and added color	2	27	490
Progresso Healthy Classics New England Clam Chowder	110 Contains a sulfiting agent and some unfamiliar additives	0	0	480
Bearitos Home Style Minestrone	70 Truly a home recipe with only 5 ingredients; contains yeast	0	0	520
Progresso Healthy Classics Minestrone	120 Label claims 99% fat free; a decent recipe	2	15	490
Campbell's Healthy Request Split Pea w/ Ham	150 The flavor enhancers are acceptable	2	12	400
Healthy Choice Split Pea & Ham	170 Too many preservatives	2	11	500
Health Valley Fat Free Vegetable Barley	65 Healthy ingredients but tastes bland	0	0	245

DRY/INSTANT SOUPS				
Campbell's Low Fat Chicken Oriental Ramen	**145** Many unacceptable ingredients	**1**	**6**	**925**
Fantastic Seranadin' Minestrone	**190** Creative and all natural ingredients	**2**	**9**	**480**
Knorr Hearty Lentil	**220** Contains beef extract and chicken fat; too high in sodium	**0**	**0**	**900**
Sanwa Ramen Pride	**205** More chemicals than food; avoid	**10**	**45**	**805**

Reprinted from *Boston Globe,* February 23, 1994, p. 33, with data from manufacturers and product labels.

88. Using the "Fat (grams)" listing in the previous problem, construct a box plot. Are there any outliers? Do the same for "Fat (% calories)." Are the results the same? Explain.

89. The following data are guesses of the outcome of Super Bowl 1994 by 16 faculty members and staff. Team = Buffalo (B) or Dallas (D); Points = total number of points to be scored in the game.

What kind of variable is "Team"? What is the measurement scale?
What kind of variable is "Points"? What is the measurement scale?

	Team	**Points**		**Team**	**Points**
Alice	B	46	Jay(1)	D	45
Barbara	D	56	Jay(2)	B	49
Carole C.	B	49	Joe	D	68
Dan	D	66	Larry D.	B	60
Dave	D	61	Lynne	B	51
Emmy	D	40	Ned C.	D	45
Erl	B	33	Nick T.	B	45
Jack	B	57	Ralph	D	59

Source: Jack Hegarty, personal communication.

90. Find the mean, median, and mode(s) of "Points" in the previous problem. Also plot the distribution.

91. Use Table 1 in the data appendix (Appendix A). Draw a boxplot for the percentage of the population 65 years and over for all states and DC. Identify any outliers.

INTERVIEW WITH
Melissa S. Anderson

"We have also identified what we call 'questionable research practices' that don't violate any specific research rule, but can clearly lead to serious problems."

Melissa S. Anderson studied mathematics at St. Olaf College in Minnesota and received a masters degree at the University of Iowa. She completed a Ph.D in education and is now on the faculty at the University of Minnesota. She has consulted with the Acadia Institute on research practices and ethics and her research focuses on trends and changes in higher education.

Aczel: Where did you go to school and what type of statistics courses did you take? What did you think of them?

Anderson: I did my undergraduate work at St. Olaf College in mathematics, my masters at the University of Iowa in mathematics, and my Ph.D. at the University of Minnesota in higher education. I took several courses in probability and mathematical statistics from Dick Kleber at St. Olaf. He's legendary at St. Olaf and made everything interesting. Everyone remembers the old mechanical slot machine he brings to class to illustrate statistical principles.

Aczel: What did you do after you graduated from college?

Anderson: I went straight into graduate school after college and then taught in the math department

at St. Olaf for five years. Then I decided to go back to graduate school in a new area, higher education. I'm now a faculty member at the University of Minnesota in higher education. My research here is focused on graduate students, faculty, academic departments, and institutions of higher education. I teach courses on administration in colleges and universities, the history of higher education, the development of academic programs, educational institutions as organizations, policy research, and so on. I continue to study faculty demography in departments (the faculty experience question). I am also involved in the Acadia Institute's Project on Professional Values and Ethical Issues in the Graduate Education of Scientists and Engineers. We have been looking at, among other things, faculty and doctoral students' experiences with academic misconduct in their departments, their values with respect to research and education, and their views of the climates of their departments as places to work. We have been using a new statistical approach, hierarchical linear modeling, as developed by Tony Bryk and Steve Raudenbush, to look at how departments affect the experiences of their students and faculty. It's always exciting and frustrating to explore new statistical methods—it always reminds you how much sympathy students deserve as they struggle with statistical ideas for the first time.

67

Aczel: How did you first get interested in statistics?

Anderson: I always enjoyed stats classes, but I really got interested in statistics when I started work on my dissertation and had to learn a lot about longitudinal analysis on my own. I wanted to see how the range of faculty experience in a department is related to the department's funding *over time*. Building models of change over time is exciting, because so much research deals only with what is going on at a particular point in time.

Aczel: And how did you get involved with the Acadia Institute?

Anderson: When I was in graduate school, one of my professors was working on a project with Judith P. Swazey, the president of Acadia, and I joined the project as a research assistant. Later, I was a postdoctoral fellow on a continuation of the project, funded by the National Science Foundation, and now I am a colleague in the graduate education aspects of the Institute's work.

Aczel: What is the mission of the Acadia Institute?

Anderson: It provides research, consulting, and educational programs. The Acadia Institute is a nonprofit research institute located in Bar Harbor, Maine. It focuses on scientific and medical issues with particular attention to ethics, research, and graduate education.

Aczel: What are the particular issues Acadia is focused on now? Are there any unique studies or concerns as related especially to statistics?

Anderson: Besides the graduate education project in which I've been involved, Acadia is doing work on the federal system for the protection of research subjects and on undergraduate medical education. It is also developing a network for bioethics.

Aczel: We've seen some ethical issues in the medical research field, such as falsification of data, come up in the news lately. What about academic research?

Anderson: We did a survey of 4,000 faculty and doctoral students nationwide in four fields: chemistry, civil engineering, microbiology, and sociology. We asked them if they knew of people in their own departments who had falsified research data. Six percent of the faculty knew of other faculty who had done so, but thirteen percent of the students knew of faculty who had falsified data. Also, about eight percent of the faculty had seen students falsifying data, whereas twice that proportion of students had seen other students doing that. Given the seriousness of this offense, these numbers are really disturbingly high.

Aczel: Are there real ethical issues statistics professionals should be concerned about, or is it just a question of "bad" versus "good" statistics?

Anderson: All researchers must be aware of the ethical standards and practices that apply to their fields. Some of these standards are formal (for example: don't plagiarize, don't fabricate data, don't misuse research funds, don't cause harm to research participants). But we have also identified what we call "questionable research practices" that don't violate any specific research rule, but can clearly lead to serious problems. This category includes, for example, looking the other way when you know someone you work with is using data inappropriately, and skimming over data or statistical results that contradict what you'd like to prove.

Aczel: For students today, what issues in statistics and research will they likely hear about in the future?

Anderson: In my own corner of higher education, there's a lot of excitement about statistical approaches that let us examine changes and trends over time, longitudinally. We're also using new ways of looking at relationships across levels; for instance, we're studying students who are in departments that are in colleges—all these levels complicate the kinds of statistics that are needed. New statistical models that will make this kind of study reasonably easy are just now emerging. It's really fun to explore these new approaches.

The important part about statistics, or any field, is that new approaches usually address deficiencies in what we've understood up to this point. You can look at simple relationships between, say, a student's financial aid and his or her chances of completing college. But the underlying issues, like most of life, are very complicated. If you look at lots and lots of other factors in this relationship (which requires a very complicated model), you see that the simple relationships you thought were clear are actually inaccurate. The point of statistics is to figure out what relationships are really out there in reality, in life—not for any one individual, but for big groups of people. The better we can understand that reality, the better our chances of making good decisions and good policies that affect those broad groups of people.

Aczel: Do you have advice or suggestions to students in their first (and possibly only) statistics course?

Anderson: Ask questions!! Question every step that you don't understand. Question the value and use of every new approach. Ask good questions about how the statistics you are learning are applied. Ask yourself questions to see if you can anticipate what will come next. Ask each other questions to get a different spin on the material. Answers are all well and good, but it's the questions that really move you ahead.

Statistics can be rough. I know of a former college president, now in his 70s, who still wakes up in a panic in the middle of the night, thinking that he's about to face his statistics final exam. I think that asking questions gives students more control over what they are learning and eases the pressure quite a bit.

Aczel: What could educators do better for statistics?

Anderson: Make sure that students know how critical statistics are to virtually any responsible position in our society. Incorrect statistical results or inappropriately interpreted results can have devastating consequences.

What Are the Chances?

2.1 Introduction   72

2.2 Basic Definitions: Events, Sample
Space, and Probabilities   75

2.3 Basic Rules for Probability   81

2.4 Conditional Probability   87

2.5 Independence of Events   92

2.6 Bayes' Theorem   97

A pril 1942 was the worst month in Malta's history. No less than 5,715 Nazi warplanes attacked the Mediterranean island-state the Allies considered crucial to their campaign against Hitler. During that month, 6,728 tons of bombs were dropped on the island, far exceeding the tonnage dropped on Coventry, England. The Maltese suffered greatly throughout the war. In fact, for the collective bravery of the Maltese people during the difficult war years and their perseverance in defending their island against massive German attacks, Malta later received the coveted George Cross from the King of England.

On Wednesday, April 9, 1942, at 4:40 PM, a German warplane, flying high over the great Mosta Church in the center of Malta, dropped a single bomb. The church was unusually full that afternoon, with over 300 people. Coming straight down from the sky, the bomb broke through the very center of the dome. It bounced off the inside walls twice, rolled from one side of the church to the other through masses of people, rolled back to the very center of the church—where it could do the most harm—and came to a stop. It did not explode, and not a single person was injured.

We could conceivably try to calculate the *probability* of such an event, using information on the percentage of German bombs that did not explode (a very small number); the chances that a bomber, from high altitude, would hit the very center of a dome so that the bomb would break through rather than bounce off; and other information. We will do such analyses in this chapter for many other problems. In this case, if we had imagined that such an event might happen and computed its probability when considered alone (rather than as part of a long series of bomb attacks on churches filled with people), we would find that the probability is extremely small. When a dangerous but happily ending event with an extremely small probability occurs, we have a name for it: a miracle.

2.1 INTRODUCTION

Probability is everywhere. It affects all aspects of our lives:

Giants Win Is No Surprise—They Had a 65.635738% Chance

By STEVE RUBENSTEIN

The San Francisco Giants had a 65.6 percent chance of winning yesterday's game, said a statistics professor who worked it out.

"It wasn't that difficult a calculation," said statistics professor Bill Kaigh.

Today, statistics say, the Giants have a 100 percent chance of winning yesterday's game. Numbers don't lie.

"The estimates are derived from principal components and linear and logistic regression analyses of historical data," said Kaigh, a professor at the University of Texas-El Paso, shortly before Will Clark hit a triple to center field to give the Giants the lead.

Five thousand statisticians are in town this week for the meeting of the American Statistical Association. Kaigh, addressing a morning seminar on sports statistics, passed out a 24-page research paper showing exactly why the Giants would probably win yesterday's game, which they did, 10-7.

The formula involves multiplying the Giants' home-game record by the Cincinnati Reds' away-game record subtracted from one and dividing the whole thing by something else with four variables in it.

The answer is actually 65.635738 percent. That explains why Kurt Manwaring hit a two-run homer in the sixth inning.

© San Francisco Chronicle. Reprinted by permission.

A **probability** is a quantitative measure of uncertainty—a number that conveys the strength of our belief in the occurrence of an uncertain event. Since life is full of uncertainty, people have always been interested in evaluating probabilities. The statistician I. J. Good suggested that "the theory of probability is much older than the human species," since the assessment of uncertainty incorporates the idea of learning from experience, which most creatures do.[1]

Among humans the history and lore of probability are often steeped in romantic legend about gambling. In the third book of the great epic of India, the *Mahabarata,* written before AD 400, King Rtuparna demonstrates a knowledge of statistics to Nala, a man possessed by the demigod of dicing. Rtuparna is described as able to estimate the number of leaves on a tree based on the number of leaves on a randomly chosen

---

[1]I. J. Good, "Kinds of Probability," *Science,* no. 129 (February 20, 1959), pp. 443–47.

branch; this form of estimation is the essence of statistical inference. The story seems to prove that long ago Indians knew something about statistics. Rtuparna further says:

I of dice possess the science and in numbers thus am skilled.

Which seems to suggest a familiarity with the laws of probability.[2]

Further investigations will have to be conducted before we can determine to what extent probability and its relationship to statistical inference were known to the Indians of the fourth century. They certainly had some familiarity with the subject.

The rabbis of the early centuries following the destruction of the second Temple in Jerusalem also had some familiarity with the laws of probability. As evidenced in the Talmud (written about the same time as the *Mahabarata*), probability arguments were widely used for determining issues related to dietary laws, paternity, taxes, adultery, and other matters. There are some indications that these early scholars were familiar with the rules of addition and multiplication of probabilities and were able to compare probabilities and make judgments based on their relative magnitudes.[3]

The theory of probability as we know it today, however, was largely developed by European mathematicians such as Galileo Galilei (1564–1642), Blaise Pascal (1623–1662), Pierre de Fermat (1601–1665), Abraham de Moivre (1667–1754), and others.

As in India, the development of probability theory in Europe is often associated with gamblers, who pursued their interests in the famous European casinos, such as the one at Monte Carlo (shown here). Many books on probability and statistics tell the story of the Chevalier de Mére, a French gambler who enlisted the help of Pascal in an effort to obtain the probabilities of winning at certain games of chance, leading to much of the European development of probability.

Courtesy of Amir Aczel.

---

[2]See Ian Hacking, *The Emergence of Probability* (New York: Cambridge University Press, 1975), p. 7.

[3]See Nachum L. Rabinovitch, "Probability in the Talmud," *Biometrika* 56, no. 2 (1969), pp. 437–41.

Today, the theory of probability is an indispensable tool in the analysis of situations involving uncertainty. It forms the basis for inferential statistics as well as for other fields that require quantitative assessments of chance occurrences, such as quality control, management decision analysis, and areas in physics, biology, engineering, and economics.

While most analyses that use the theory of probability today have nothing to do with games of chance, gambling models still provide the clearest examples of probability and its assessment. The reason is that games of chance usually involve dice, cards, or roulette wheels—mechanical devices. If we assume there is no cheating, these mechanical devices tend to produce sets of outcomes that are *equally likely,* and this allows us to compute probabilities of winning at these games.

Suppose that a single die is rolled and that you win a dollar if the number 1 or 2 appears. What are your chances of winning a dollar? Since there are six equally likely numbers (assuming the die is fair) and you win as a result of either of two numbers appearing, the probability that you win is 2/6, or 1/3.

As another example, consider the following situation. An analyst follows the price movements of IBM stock for a period of time and wants to assess the probability that the stock will go up in price in the next week. This is a different type of situation. The analyst does not have the luxury of a known set of equally likely outcomes, where "IBM stock goes up next week" is one of a given number of these equally likely possibilities. Therefore, the analyst's assessment of the probability of the event will be a subjective one. The analyst will base his or her assessment of this probability on knowledge of the situation, guesses, or intuition. Different people may assign different probabilities to this event depending on their experience and knowledge; hence the name *subjective probability.*

Objective probability is probability based on symmetry of games of chance or similar situations. It is also called *classical probability.* This probability is based on the idea that certain occurrences are equally likely (the term *equally likely* is intuitively clear and will be used as a starting point for our definitions). The numbers 1, 2, 3, 4, 5, and 6 on a fair die are each equally likely to occur. Another type of objective probability is long-term *relative-frequency* probability. If, in the long run, 20 out of 1,000 consumers given a taste test for a new soup like the taste, then we say that the probability that a given consumer will like the soup is 20/1,000 = 0.02. If the probability that a head will appear on any one toss of a coin is 1/2, then if the coin is tossed a large number of times, the proportion of heads will approach 1/2. Like the probability in games of chance and other symmetrical situations, relative-frequency probability is objective in the sense that no personal judgment is involved.

Subjective probability, on the other hand, involves personal judgment, information, intuition, and other subjective evaluation criteria. The area of subjective probability—which is relatively new, having been first developed in the 1930s—is somewhat controversial.[4] A physician assessing the probability of a patient's recovery and an expert assessing the probability of success of a merger offer are both making a personal judgment based on what they know and feel about a situation. Subjective probability is also called *personal probability.* One person's subjective probability may very well be different from another person's subjective probability of the same event.

---

[4]The earliest published works on subjective probability are Frank Ramsey's *The Foundation of Mathematics and Other Logical Essays* (London: Kegan Paul, 1931) and the Italian statistician Bruno de Finetti's "La Prévision: Ses Lois Logiques, Ses Sources Subjectives," *Annales de L'Institut Henri Poincaré* 7, no. 1 (1937).

Whatever the kind of probability involved, the same set of mathematical rules holds for manipulating and analyzing probability. We now give the general rules for probability as well as formal definitions. Some of our definitions will involve counting the number of ways in which some event may occur. The counting idea is implementable only in the case of objective probability, although conceptually this idea may apply to subjective probability as well, if we can imagine a kind of lottery with a known probability of occurrence for the event of interest.

2.2 BASIC DEFINITIONS: EVENTS, SAMPLE SPACE, AND PROBABILITIES

In order to understand probability, it is useful to have some familiarity with sets and with operations involving sets.

A **set** is a collection of elements.

The elements of a set may be people, sheep, desks, cars, files in a cabinet, or even numbers. We may define our set as the collection of all sheep in a given pasture, all people in a room, all cars in a given parking lot at a given time, all the numbers between 0 and 1, or all integers. The number of elements in a set may be infinite, as in the last two examples.

A set may also have no elements.

The **empty set** is the set containing *no elements.* It is denoted by $\emptyset$.

We now define the universal set.

The **universal set** is the set containing *everything* in a given context. We denote the universal set by S.

Given a set A, we may define its *complement.*

The **complement** of set A is the set containing all the elements in the universal set S that are *not* members of set A. We denote the complement of A by $\bar{A}$. The set $\bar{A}$ is often called "not A."

A **Venn diagram** is a schematic drawing of sets that demonstrates the relationships between different sets. In a Venn diagram, sets are shown as circles, or other closed figures, within a rectangle corresponding to the universal set, S. Figure 2.1 is a Venn diagram demonstrating the relationship between a set, A, and its complement, $\bar{A}$.

As an example of a set and its complement, consider the following. Let the universal set, S, be the set of all students at a given university. Define A as the set of all students who own a car (at least one car). The complement of A, $\bar{A}$, is thus the set of all students at the university who do *not* own a car.

FIGURE 2.1
A Set, A, and Its Complement, Ā

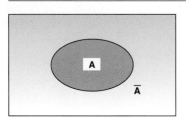

FIGURE 2.2
Sets A and B and Their Intersection

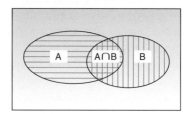

FIGURE 2.3
The Union of A and B

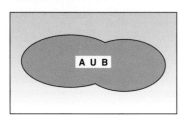

FIGURE 2.4
Two Disjoint Sets

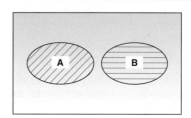

Sets may be related in a number of ways. Consider two sets, A and B, within the context of the same universal set, S. (We say that A and B are *subsets* of the universal set S.) If A and B have some elements in common, we say they intersect.

> The **intersection** of A and B, denoted A ∩ B, is the set containing all elements that are members of *both* A and B.

When we want to consider all of the elements of two sets A and B, we look at their union.

> The **union** of A and B, denoted A ∪ B, is the set containing all elements that are members of *either* A *or* B, *or both*.

As you can see from these definitions, the union of two sets contains the intersection of the two sets. Figure 2.2 is a Venn diagram showing two sets A and B and their intersection, A ∩ B. Figure 2.3 is a Venn diagram showing the union of the same two sets.

As an example of the union and intersection of sets, consider again the set of all students at a university who own a car. This is set A. Now define set B as the set of all students at the university who own a bicycle. The universal set, S, is, as before, the set of all students at the university. A ∩ B is the intersection of A and B—it is the set of all students at the university who own *both* a car and a bicycle. A ∪ B is the union of A and B—it is the set of all students at the university who own a car, a bicycle, or both.

Two sets may have no intersection: they may be **disjoint**. In such a case, we say that the intersection of the two sets is the empty set, ∅. In symbols, when A and B are disjoint, A ∩ B = ∅. As an example of two disjoint sets, consider the set of all students enrolled in a nursing program at a particular university and all the students at the university who are enrolled in an art program. (Assume no student is enrolled in both programs.) A Venn diagram of two disjoint sets is shown in Figure 2.4

In probability theory we make use of the idea of a set and of operations involving sets. We will now provide some basic definitions of terms relevant to the computation of probability. These are an *experiment,* a *sample space,* and an *event.*

An **experiment** is a process that leads to one of several possible **outcomes**. An outcome of an experiment is some observation or measurement.

Drawing a card out of a deck of 52 cards is an experiment. One outcome of the experiment may be that the queen of diamonds is drawn.

A single outcome of an experiment is called a *basic outcome* or an *elementary event.* Any particular card drawn from a deck is a basic outcome. The set of all possible outcomes of an experiment, the set of all elementary events, is the *sample space.*

The **sample space** is the universal set, S, pertinent to a given experiment. The sample space is the set of all possible outcomes of an experiment.

The sample space for the experiment of drawing a card out of a deck is the set of all cards in the deck. The sample space for an experiment of reading the temperature is the set of all numbers in the range of temperatures.

A group of basic outcomes together is defined as an *event.*

An **event** is a subset of a sample space. It is a set of basic outcomes. We say that the event *occurs* if the experiment gives rise to a basic outcome belonging to the event.

For example, the event "an ace is drawn out of a deck of cards" is the set of the four aces within the sample space consisting of all 52 cards. This event occurs whenever one of the four aces (the basic outcomes) is drawn.

The sample space for the experiment of drawing a card out of a deck of 52 cards is shown in Figure 2.5. The figure also shows event A, the event that an ace is drawn.

In this context, for a given experiment we have a sample space with equally likely basic outcomes. When a card is drawn out of a well-shuffled deck, every one of the

FIGURE 2.5   Sample Space for Drawing a Card

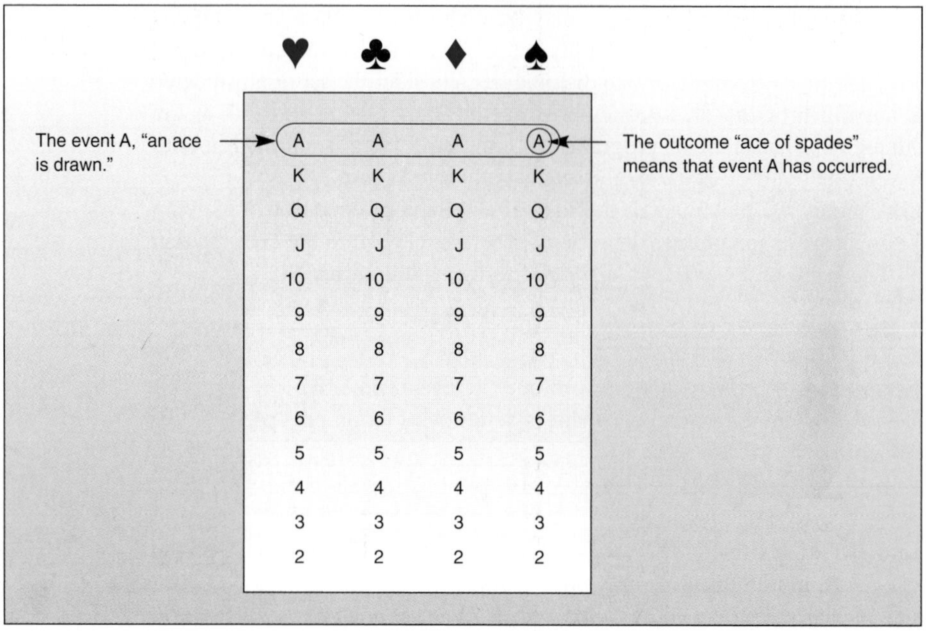

cards (the basic outcomes) is as likely to occur as any other. In such situations, it seems reasonable to define the probability of an event as the *relative size* of the event with respect to the size of the sample space. Since there are four aces and there are 52 cards, the size of A is 4 and the size of the sample space is 52. Therefore, the probability of A is equal to 4/52.

The rule we use in computing probabilities, assuming equal likelihood of all basic outcomes, is as follows:

Probability of A:

$$P(A) = \frac{n(A)}{n(S)}$$

where

$n(A)$ = the number of elements in the set of the event A

$n(S)$ = the number of elements in the sample space, S

The probability of drawing an ace is $P(A) = n(A)/n(S) = 4/52$.

EXAMPLE 2.1   Roulette is a popular casino game. As the game is played in Las Vegas or Atlantic City, the roulette wheel has 36 numbers, numbered 1 through 36, and the number 0 as well as the number 00 (double-zero). What is the probability of winning on a single number that you bet?

Courtesy of The Stock Market/Stephen Frink, 1993.

The sample space, S, in this example consists of 38 numbers (0, 00, 1, 2, 3, . . . , 36), each of which is equally likely to come up. Therefore, using our counting rule, *P* (any one given number) = 1/38.

SOLUTION

Let's now demonstrate the meaning of union and intersection with the example of drawing a card from a deck. Let A be the event that an ace is drawn, and ♥ the event that a heart is drawn. The sample space is shown in Figure 2.6 on page 80. Note that the event A ∩ ♥ is the event that the card drawn is both an ace and a heart (i.e., the ace of hearts). The event A ∪ ♥ is the event that the card drawn is either an ace or a heart or both.

PROBLEMS

1. What are the two main types of probability?
2. What is an event? What is the union of two events? What is the intersection of two events?
3. Define a sample space.
4. Define the probability of an event.
5. Let G be the event that a girl is born. Let F be the event that a baby over 5 pounds is born. Characterize the union and the intersection of the two events.
6. Consider the event that a player scores a point in a game against team A and the event that the same player scores a point in a game against team B. What is the union of the two events? What is the intersection of the two events?
7. Let event A be acceptance to graduate school and event B be getting a job offer. What are the union and intersection of the two events?
8. The Ford Motor Company advertises its cars on radio and on television. The company is interested in assessing the probability that a randomly chosen person is exposed to at least one of these two modes of advertising. If we define event R as the event that a randomly chosen person was exposed to a radio advertisement and event T as the event that the person was exposed to a television commercial, define R ∪ T and R ∩ T in this context.

FIGURE 2.6   The Events A and ♥ and Their Union and Intersection

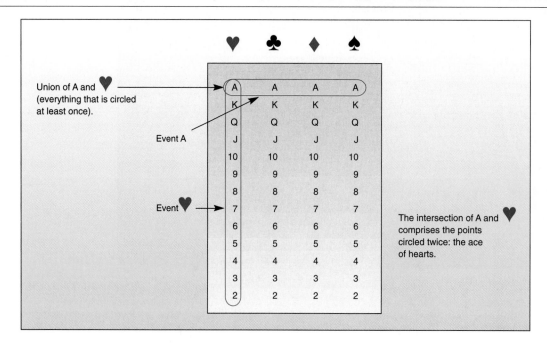

9. A brokerage firm deals in stocks and bonds. An analyst for the firm is interested in assessing the probability that a person who inquires about the firm will eventually purchase stock (event S) or bonds (event B). Define the union and the intersection of these two events.

10. A 1986 article in *Newsweek* by the mathematician John Paulos makes the point that most people have no grasp of the probabilities of events that may affect them: They tend to have great fear of publicized events with small probability, while not worrying at all about events with much higher probability. As an example, Paulos gives the following data: In 1985, 28 million Americans traveled abroad, and 39 of them were killed by terrorists. Based on this information, what is the probability of being killed by terrorists while traveling abroad? Compare this with another statistic reported by Paulos: 1 in 5,300 Americans was killed in an automobile accident in 1985.

11. An airline passenger lost her luggage. When it finally arrived a few days later, she casually asked the ticket agent how often things like this happen. The ticket agent said that, in his experience, a piece of luggage in that particular airport was reported lost about once every two days. He also said that two flights arrive every day, each flight carrying approximately 200 passengers. The average passenger was reported to check in an average of two pieces of luggage. Given this information, what is the probability that you will not find your suitcase (or a particular one of your suitcases, if you check more than one) upon arrival at this airport?

12. The European version of roulette is different from the American version in that the European roulette wheel doesn't have 00. How does this change the probability of winning when betting on a single number? European casinos charge a small admission fee, which is not the case in American casinos. Does this make sense to you, based on your answer to the earlier question?

13. Based on the following article, if you meet a stranger, what is the probability that he or she has murdered someone?

From Trivial Facts Emerge Significant Truths

By LYNN VAN MATRE
TRIBUNE STAFF WRITER

Talk to Les Krantz, and he'll tell you that one of every 1,000 random strangers you encounter has murdered someone.

"Just think about that the next time you're at Wrigley Field," Krantz says with a smile. "And, since only 7 in 10 murders are solved, that means that 3 in 10 aren't and the murderer gets away scot-free."

Reprinted from the *Chicago Tribune,* Sunday, November 14, 1993, p. 1.

14. National Public Radio reported on September 14, 1993, that of 137 Pap smear tests performed at a certain eastern hospital on women who later were found to have cancer, 19 tests erroneously resulted in a negative finding. What is the probability of a false negative at this hospital based on these data?

15. Three of the 16 laboratory mice carried aboard the space shuttle *Columbia* in March 1994 as part of a study of the effects of weightlessness developed an orientation problem. If 1 mouse is randomly selected out of the 16 on board, what is the probability of it exhibiting the problem?

2.3 BASIC RULES FOR PROBABILITY

We have explored probability on a somewhat intuitive level and have seen rules that help us evaluate probabilities in special cases: when we have a known sample space with equally likely basic outcomes (and the extension of this case to infinite sample spaces). We will now look at some general probability rules that hold regardless of the particular situation or kind of probability (objective or subjective). First, let us give a general definition of probability.

> Probability is a measure of uncertainty. The probability of event A is a numerical measure of the likelihood of the event occurring.

The Range of Values

Probability obeys certain rules. The first rule sets the range of values that the probability measure may take.

> For any event, the probability $P(A)$ satisfies:
> $$0 \leq P(A) \leq 1$$

When an event cannot occur, its probability is zero. The probability of the empty set is zero: $P(\varnothing) = 0$. In a deck where half the cards are red and half are black, the probability of drawing a green card is zero because the set corresponding to that event is the empty set: there are no green cards.

Events that are certain to occur have probability 1.00. The probability of the entire sample space S is equal to 1.00: $P(S) = 1.00$. If we draw a card out of a deck, 1 of the 52 cards in the deck will *certainly* be drawn, so the probability of the sample space, the set of all 52 cards, is equal to 1.00.

Within the range of values 0 to 1, the greater the probability, the more confidence we have in the occurrence of the event in question. A probability of 0.95 implies a very high confidence in the occurrence of the event. A probability of 0.80 implies a high confidence. When the probability is 0.5, the event is as likely to occur as it is not to occur. When the probability is 0.2, the event is not very likely to occur. When we assign a probability of 0.05, we believe the event is unlikely to occur, and so on. Figure 2.7 is an informal aid in interpreting probability.

Note that probability is a measure that goes from 0 to 1. In everyday conversation we often describe probability in less formal terms. For example, people sometimes talk about **odds**. If the odds are 1 to 1, the probability is 1/2; if the odds are 1 to 2, the probability is 1/3; and so on. Also, people sometimes say, "The probability is 80 percent." Mathematically, this probability is 0.80.

The Rule of Complements

Our second rule for probability defines the probability of the complement of an event in terms of the probability of the original event. Recall that the complement of set A is denoted by $\bar{A}$.

Probability of the complement:

$$P(\bar{A}) = 1 - P(A)$$

As a simple example, if the probability of rain tomorrow is 0.3, then the probability of no rain tomorrow must be $1 - 0.3 = 0.7$. If the probability of drawing an ace is 4/52, then the probability of the drawn card not being an ace is $1 - 4/52 = 48/52$.

FIGURE 2.7   Interpretation of a Probability

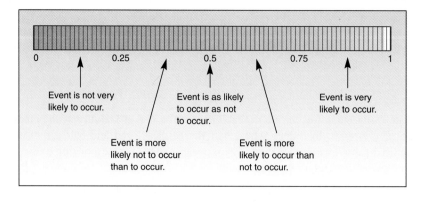

The Rule of Unions

We now state a very important rule, the **rule of unions.** The rule of unions allows us to write the probability of the union of two events in terms of the probabilities of the two events and the probability of their intersection:[5]

The rule of unions:

$$P(A \cup B) = P(A) + P(B) - P(A \cap B)$$

(The probability of the intersection of two events, $P(A \cap B)$, is called their **joint probability.**) The meaning of this rule is very simple and intuitive: When we add the probabilities of A and B, we are measuring, or counting, the probability of their intersection *twice*—once when measuring the relative size of A within the sample space, and once when doing this with B. Since the relative size, or probability, of the intersection of the two sets is counted twice, we subtract it once so that we are left with the true probability of the union of the two events (refer to Figure 2.6). Note that instead of finding the probability of A $\cup$ B by direct counting, as we did earlier in this chapter (see again), we can use the rule of unions: We know that the probability of an ace is 4/52, the probability of a heart is 13/52, and the probability of their intersection—the drawn card being the ace of hearts—is 1/52. Thus, $P(A \cup \heartsuit) = 4/52 + 13/52 - 1/52 = 16/52$, which is exactly what we found from direct counting.

The rule of unions is especially useful when we do not have the sample space for the union of events but do have the separate probabilities. For example, suppose your chance of being offered a certain job is 0.4, your probability of getting another job is 0.5, and your probability of being offered both jobs (i.e., the intersection) is 0.3. Using the rule of unions, your probability of being offered at least one of the two jobs (their union) is 0.4 + 0.5 − 0.3 = 0.6.

Mutually Exclusive Events

When the sets corresponding to two events are disjoint (that is, have no intersection), the two events are called **mutually exclusive** (see Figure 2.4). For mutually exclusive events, the probability of the intersection of the events is zero, because the intersection of the events is the empty set, and we know that the probability of the empty set, $\varnothing$, is zero.

For mutually exclusive events A and B:

$$P(A \cap B) = 0$$

---

[5]The rule can be extended to more than two events. In the case of three events, we have: $P(A \cup B \cup C) = P(A) + P(B) + P(C) - P(A \cap B) - P(A \cap C) - P(B \cap C) + P(A \cap B \cap C)$. With more events, this becomes even more complex.

This fact gives us a special rule for unions of mutually exclusive events. Since the probability of the intersection of the two events is zero, there is no need to subtract $P(A \cap B)$ when computing the probability of the union of the two events. Therefore,

> For mutually exclusive events A and B:
>
> $$P(A \cup B) = P(A) + P(B)$$

This is not really a new rule, since we can always use the rule of unions for the union of two events: If the events happen to be mutually exclusive, we subtract zero as the probability of the intersection.

Continuing our cards example, what is the probability of drawing either a heart or a club? We have: $P(\heartsuit \cup \clubsuit) = P(\heartsuit) + P(\clubsuit) = 13/52 + 13/52 = 26/52 = 1/2$. We need not subtract the probability of an intersection, since no card is a club *and* a heart.

PROBLEMS

16. Assign reasonable numerical probabilities to the three shaded region types in the accompanying chart.

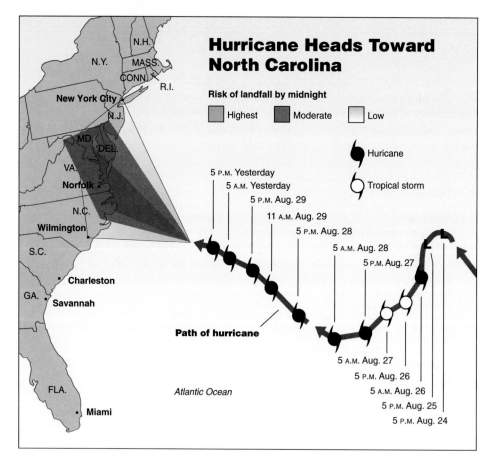

Reprinted from *The New York Times*, August 31, 1993, p. A12, with data from Pennsylvania State University.

17. Assign a reasonable numerical probability to the statement: "Rain is very likely tonight."
18. How likely is an event that has a 0.65 probability? Describe the probability in words.
19. If a team has an 80 percent chance of winning a game, describe its chances in words.
20. ShopperTrak is a hidden electric eye designed to count the number of shoppers entering a store. When two shoppers enter a store together, one walking in front of the other, the following probabilities apply: There is a 0.98 probability that the first shopper will be detected, a 0.94 probability that the second shopper will be detected, and a 0.93 probability that both of them will be detected by the device. What is the probability that the device will detect at least one of two shoppers entering together?
21. A machine produces components for use in cellular phones. At any given time, the machine may be in one, and only one, of three states: operational, out of control, or down. From experience with this machine, a quality-control engineer knows that the probability that the machine is out of control at any moment is 0.02, and the probability that it is down is 0.015.
    a. What is the relationship between the two events "machine is out of control" and "machine is down"?
    b. When the machine is either out of control or down, a repairperson must be called. What is the probability that a repairperson must be called right now?
    c. Unless the machine is down, it can be used to produce a single item. What is the probability that the machine can be used to produce a single component right now? What is the relationship between this event and the event "machine is down"?
22. Following are age and sex data for 20 midlevel managers at a service company: 34 F, 49 M, 27 M, 63 F, 33 F, 29 F, 45 M, 46 M, 30 F, 39 M, 42 M, 30 F, 48 M, 35 F, 32 F, 37 F, 48 F, 50 M, 48 F, 61 F. A manager must be chosen at random to serve on a companywide committee that deals with personnel problems. What is the probability that the chosen manager will be either a woman or over 50 years old, or both? Solve both directly from the data, and by using the law of unions. What is the probability that the chosen manager will be under 30?
23. Suppose that 25 percent of the population in a given area is exposed to a television commercial for Ford automobiles, and 34 percent is exposed to Ford's radio advertisements. Also, it is known that 10 percent of the population is exposed to both means of advertising. If a person is randomly chosen out of the entire population in this area, what is the probability that he or she was exposed to at least one of the two modes of advertising?
24. Suppose it is known that 85 percent of the people who inquire about investment opportunities at a brokerage house end up purchasing stock, and 33 percent end up purchasing bonds. It is also known that 28 percent of the inquirers end up getting a portfolio with both stocks and bonds. If a person is just making an inquiry, what is the probability that he or she will get either stock or bonds, or both (that is, open any portfolio)?
25. See the newspaper article, "Scientists Find Catchy Names Help Ideas Fly," on page 86. A random sample of 100 people were shown the name *chaos* and the name *WIMP*. Of these, 45 people remembered *chaos*, 82 remembered *WIMP*, and 38 remembered both names. What is the probability that a randomly selected person from this group remembered at least one of these two catchy names?
26. A firm has 550 employees; 380 of them have had at least some college education, and 412 of the employees underwent a vocational training program. Furthermore, 357 employees are both college-educated and have had the vocational training. If an employee is chosen at random, what is the probability that he or she is college-educated, has had the training, or both?
27. See the accompanying table of the performance of financial indexes. An investor chooses a financial index for her personal use in evaluating investments. Let us assume that the choice is made at random, with equal probability for each of the indexes in the table below, and without the investor knowing the information in this table.
    a. What is the probability that the index the investor has chosen lost money in the year ending January 1, 1991?

Scientists Find Catchy Names Help Ideas Fly

Instead of Something Dull, Call It "Chaos" or "Nemesis," and the World Will Take Note

BY STEPHEN S. HALL

Most scientific names are dull and pedantic, and usually as impenetrable as the Latin in which Linnaeus named all species known to him. But recently a strange whimsy has started to creep in among the sesquipedalian prose of scientific journals. A few scientists are attaching catchy terms like "chaos" or "chaperone molecules" to the things they discover.

The trend may represent a slap at tradition, or perhaps a shrewd sense that a popular name can win a theory more attention and financing

Even a cursory spin through contemporary journals reveals an abundance of suggestive, metaphoric names. Chemistry has its "buckyballs," spherical frameworks of carbon atoms named after Buckminster Fuller and also called "buckminsterfullerenes." Physics has its "spin glass" and "WIMP's," for weakly interactive massive particles. Besides "chaperone" molecules, which oversee the proper manufacture of proteins in cells, cell biology boasts its "leucine zippers," regions of gene-regulating proteins that link up before interacting with the DNA molecule.

Reprinted from *The New York Times,* October 20, 1992, p. C6.

    b. What is the probability that the chosen index gained at least 200 percent in the 10 years preceding this date?

    c. What is the probability that the index had a yield of at least 5 percent in the last year, gained at least 35 percent in the preceding three years, or both?

    d. Are the event "index lost money in the preceding year" and the event "index had a yield of at least 4 percent in the preceding year" mutually exclusive? Explain.

Percent Gain (or loss) to Jan. 1, 1991

	1 Year	3 Years	5 Years	10 Years	Percent Yield
S&P 500 stock index	(3.1)	48.7	85.7	268.6	3.8
Dow Jones industrial average	(5.6)	38.3	65.8	216.5	3.9
Russell 2000 small-company stock index	(19.5)	16.9	12.7	125.3	2.3
Average equity fund	(5.3)	35.7	58.9	222.2	3.3
Salomon Bros. investment-grade bond index	9.1	34.8	59.6	240.8	8.5
Shearson Lehman Bros. long-term Treasury index	6.3	38.4	66.7	265.0	8.3
Shearson Lehman Bros. municipal bond index*	7.7	32.3	59.3	180.4	6.8

*All figures to Dec. 1, 1990.

Reproduced by permission from *Money* magazine (February 1991).

    28. As part of a student project for the 1994 Science Fair in Orange, Massachusetts, 28 horses were made to listen to Mozart and heavy metal music. The results were as follows: 11 of the 28 horses exhibited some head movements when Mozart was played; 8 exhibited some head movements when the heavy metal was played; and 5 moved their heads when both were played. If a horse is chosen at random, what is the probability the horse exhibited head movements either to Mozart, or to heavy metal, or to both?

2.4 CONDITIONAL PROBABILITY

As a measure of uncertainty, probability depends on information. Thus, the probability you would give the event "IBM stock price will go up tomorrow" depends on what you know about the company and its performance; the probability is *conditional* upon your information set. If you know much about the company, you may assign a different probability to the event than if you know little about the company. We may define the probability of event A *conditional* upon the occurrence of event B. In this example, event A may be the event that the stock will go up tomorrow, and event B may be a favorable quarterly report.

> The **conditional probability** of event A given the occurrence of event B is
>
> $$P(A|B) = \frac{P(A \cap B)}{P(B)}$$
>
> assuming $P(B) \neq 0$.

The vertical line in $P(A|B)$ is read *given,* or *conditional upon.* The probability of event A given the occurrence of event B is defined as the probability of the intersection of A and B, divided by the probability of event B.

A state senate has 100 senators; 60 are Democrats and 40 are Republicans. Of the Democrats, 40 are liberal and 20 are conservative. Of the Republicans, 30 are conservative and 10 are liberal. Given that a senator is a Democrat, what is the probability that he or she is a liberal?

EXAMPLE 2.2

The State Senate

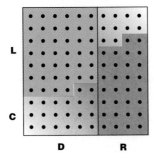

$$P(L|D) = \frac{P(L \cap D)}{P(D)} = \frac{40/100}{60/100} = 0.6666$$

SOLUTION

But we see this directly from the fact that there are 60 Democrats and 40 of them are liberal. This confirms the definition of conditional probability in an intuitive sense.

When two events and their complements are of interest, it may be convenient to arrange the information in a **contingency table.** In Example 2.2 the table would be set up as follows:

	Democrat	Republican	Total
Liberal	40	10	50
Conservative	20	30	50
Total:	60	40	100

Contingency tables help us visualize information and solve problems. There are two other useful forms of the definition of conditional probability.

Variations of the conditional probability formula:

$$P(A \cap B) = P(A|B)P(B)$$

and

$$P(A \cap B) = P(B|A)P(A)$$

These are illustrated in Example 2.3.

EXAMPLE 2.3

A consulting firm is bidding for two jobs, one with each of two large multinational corporations. The company executives estimate that the probability of obtaining the consulting job with firm A, event A, is 0.45. The executives also feel that if the company should get the job with firm A, then there is a 0.90 probability that firm B will also give the company the consulting job. What are the company's chances of getting *both* jobs?

SOLUTION

We are given $P(A) = 0.45$. We also know that $P(B|A) = 0.90$, and we are looking for $P(A \cap B)$, which is the probability that both A and B will occur. From the equation we have $P(A \cap B) = P(B|A)P(A) = 0.90 \times 0.45 = 0.405$.

EXAMPLE 2.4

Twenty-one percent of the executives in a large advertising firm are at the top salary level. It is further known that 40 percent of all the executives at the firm are women. Also, 6.4 percent of all executives are women *and* are at the top salary level. Recently, a question arose among executives at the firm as to whether there is any evidence of salary inequity. Assuming that some statistical considerations (explained in later chapters) are met, do the percentages reported above provide any evidence of salary inequity?

SOLUTION

To solve this problem, we pose it in terms of probabilities and ask whether the probability that a randomly chosen executive will be at the top salary level is approximately equal to the probability that the executive will be at the top salary level *given* the executive is a woman. To answer, we need to compute the probability that the executive will be at the top level given the executive is a woman. Defining T as the event of a top salary and W as the event that an executive is a woman, we get

$$P(T|W) = \frac{P(T \cap W)}{P(W)} = \frac{0.064}{0.40} = 0.16$$

Since 0.16 is smaller than 0.21, we may conclude (subject to statistical considerations) that salary inequity does exist at the firm, because an executive is less likely to make a top salary if she is a woman.

Example 2.4 may incline us to think about the relations among different events. Are different events related, or are they *independent* of each other? In this example, we concluded that the two events, being a woman and being at the top salary level, are related in the sense that the event W made the event T less likely. Section 2.5 quantifies the relations among events and defines the concept of independence.

---

29. A financial analyst believes that if interest rates decrease in a given period, then the probability that the stock market will go up is 0.80. The analyst further believes that interest rates have a 0.40 chance of decreasing during the period in question. Given the above information, what is the probability that the market will go up and interest rates will go down during the period in question?

30. A bank loan officer knows that 12 percent of the bank's mortgage holders lose their jobs and default on the loan in the course of five years. She also knows that 20 percent of the bank's mortgage holders lose their jobs during this period. Given that one of her mortgage holders just lost his job, what is the probability that he will now default on the loan?

31. An express delivery service promises overnight delivery of all packages checked in before 5 PM. The delivery service is not perfect, however, and sometimes delays do occur. Management knows that if delays occur in the evening flight to a major city from which distribution is made, then a package will not arrive on time with probability 0.25. It is also known that 10 percent of the evening flights to the major city are delayed. What percentage of the packages arrive late? (Assume that all packages are sent out on the evening flight to the major city and that all packages arrive on time if the evening flight is *not* delayed.)

32. The following table gives numbers of claims at a large insurance company by kind and by geographical region.

	East	South	Midwest	West
Hospitalization	75	128	29	52
Physician's visit	233	514	104	251
Outpatient treatment	100	326	65	99

Compute column totals and row totals. What do they mean?

 a. If a bill is chosen at random, what is the probability that it is from the Midwest?
 b. What is the probability that a randomly chosen bill is from the East?
 c. What is the probability that a randomly chosen bill is either from the Midwest or from the South? What is the relation between these two events?
 d. What is the probability that a randomly chosen bill is for hospitalization?
 e. Given that a bill is for hospitalization, what is the probability that it is from the South?
 f. Given that a bill is from the East, what is the probability that it is for physician's visit?

PROBLEMS

g. Given that a bill is for outpatient treatment, what is the probability that it is from the West?

h. What is the probability that a randomly chosen bill is either from the East or for outpatient treatment (or both)?

i. What is the probability that a randomly selected bill is either for hospitalization or from the South (or both)?

33. One of the greatest problems in marketing research and other survey fields is the problem of nonresponse to surveys. In home interviews the problem arises when the respondent is not home at the time of the visit or, sometimes, simply refuses to answer questions. A market researcher believes that a respondent will answer all questions with probability 0.94 if found at home. He further believes that the probability that a given person will be found at home is 0.65. Given this information, what percentage of the interviews will be successfully completed?

34. An investment analyst collects data on stocks and notes whether or not dividends were paid and whether or not the stocks increased in price over a given period. Data are presented in the following table.

	Price Increase	No Price Increase	Total
Dividends paid	34	78	112
No dividends paid	85	49	134
Total	119	127	246

a. If a stock is selected at random out of the analyst's list of 246 stocks, what is the probability that it increased in price?

b. If a stock is selected at random, what is the probability that it paid dividends?

c. If a stock is selected at random, what is the probability that it both increased in price and paid dividends?

d. What is the probability that a randomly selected stock neither paid dividends nor increased in price?

e. Given that a stock increased in price, what is the probability that it also paid dividends?

f. If a stock is known not to have paid dividends, what is the probability that it increased in price?

g. What is the probability that a randomly selected stock was worth holding during the period in question; that is, what is the probability that it either increased in price, or paid dividends, or both?

35. A recent cover article in *Business Week* dealt with the salaries of top executives at large corporations. The following table is compiled from data given in four tables in the article and lists the number of firms in the study where the top executive officer made over $1 million a year. The table also lists firms according to whether or not shareholder return was positive during the period in question.

	Top Executive Made More than $1 Million	Top Executive Made Less than $1 Million	Total
Shareholders made money	1	6	7
Shareholders lost money	2	1	3
Total	3	7	10

a. If a firm is randomly chosen from the list of 10 firms studied, what is the probability that its top executive made over $1 million a year?

b. If a firm is randomly chosen from the list, what is the probability that its shareholders lost money during the period studied?

c. Given that one of the firms in this group had negative shareholder return, what is the probability that its top executive made over $1 million?

d. Given that a firm's top executive made over $1 million, what is the probability that the firm's shareholder return was positive?

36. As part of pharmaceutical testing for drowsiness as a side effect of a drug, 100 patients were randomly assigned to one of two groups of 50: a group taking the drug, and a group taking a placebo. Then the number of people in each group who fell asleep in the next hour was recorded. The results were as follows:

	Drug	Placebo
Sleep	18	12
No sleep	32	38

a. If a patient is randomly chosen from the 100 patients in the study, what is the probability that he or she slept in the following hour?

b. Given that a randomly chosen patient was given the placebo, what is the probability that he or she slept in the following hour?

c. Given that a randomly chosen patient slept in the following hour, what is the probability that he or she was given the drug?

37. Discuss the article on longevity with respect to the idea of conditional probability.

Study Challenges Longevity Theory

Some Lifespans May Not Be Preordained by Biological Clock, Report Shows

By GINA KOLATA

In a finding that has surprised experts who study aging, researchers have discovered that life expectancy may not always decrease as organisms grow older. A very few individuals are so long-lived that, past a certain age, their life expectancy actually increases.

The subjects of the experiments, which are preliminary, were fruit flies, but some demographers and experts on aging say the implications could be far reaching. The studies, they say, may help demolish the long-held belief that every species, including humans, has a midnight hour, an age at which the organism essentially falls apart.

The present understanding is that the longer individuals live, the more likely they are to die. That is the basis, for example, of the argument that if cancer were cured tomorrow, people would simply die a little later of something else. In other words, the belief was that there was a sharply defined maximum lifespan that medical advances could help people to approach or attain but not surpass.

Exceeding Expectancies

The new findings, however, indicate that some Methuselahs, among fruit flies, at least, can live at least twice as long as is normal for their species. If the same is true for humans, experts said, it may mean that with better health care and medical treatment, some individuals can live far beyond their biblical fourscore years and 10.

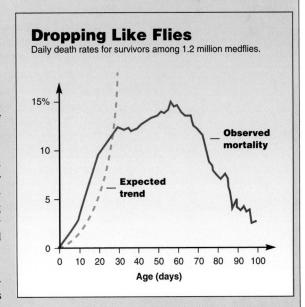

Dropping Like Flies
Daily death rates for survivors among 1.2 million medflies.

Reprinted from *The New York Times*, October 16, 1992, p. A22.

2.5 INDEPENDENCE OF EVENTS

In Example 2.4 we concluded that the probability that an executive made a top salary was lower when the executive was a woman, and we concluded that the two events T and W were *not* independent. We now give a formal definition of statistical independence of events.

Two events, A and B, are said to be *independent* of each other if and only if the following three conditions hold.

Conditions for the **independence of two events** A and B:

$$P(A|B) = P(A)$$

$$P(B|A) = P(B)$$

and, most useful:

$$P(A \cap B) = P(A)P(B)$$

The first two equations have a clear, intuitive appeal. The top equation says that when A and B are independent of each other then the probability of A stays the same even when we know that B has occurred—it is a simple way of saying that knowledge of B tells us nothing about A when the two events are independent. Similarly, when A and B are independent, then knowledge that A has occurred gives us absolutely no information about B and its likelihood of occurring.

The third equation, however, is the most useful in applications. It tells us that when A and B are independent (and only when they are independent), we can obtain the probability of the *joint* occurrence of A and B (i.e., the probability of their intersection) simply by *multiplying* the two separate probabilities. This rule is thus called the **product rule** for independent events. (The rule is easily derived from the first rule, using the definition of conditional probability.)

As an example of independent events, consider the following: Suppose I roll a single die. What is the probability of the number 6 turning up? The answer is 1/6. Now suppose that I told you that I just tossed a coin and it turned up heads. What is *now* the probability that the die will show the number 6? The answer is unchanged, 1/6, because events of the die and the coin are independent of each other. We see that $P(6|H) = P(6)$, which is the first rule above.

In Example 2.2, we found that the probability that a senator is liberal given that he or she is a Democrat is 0.6666. The probability that the senator is liberal is found by direct counting in the sample space: $P(L) = 50/100 = 0.5$. Since $P(L|D) \neq P(L)$, these two events are not independent.

When two events are not independent, neither are their complements. Therefore, Republican and conservative are not independent events (and neither are Republican and liberal, Democrat and conservative in this example). Check these assertions directly. Also show that $P(R \cap L) \neq P(R)P(L)$, which also follows from the lack of independence of the events in this example.

EXAMPLE 2.5

The probability that a consumer will be exposed to an advertisement for a certain product by seeing a commercial on television is 0.04. The probability that the consumer will be exposed to the product by seeing an advertisement on a billboard is 0.06. The two events, being exposed to the commercial and being exposed to the billboard ad, are assumed to be independent. (*a*) What is the probability that the consumer will be exposed to both advertisements? (*b*) What is the probability that he or she will be exposed to at least one of the ads?

(*a*) Since the two events are independent, the probability of the intersection of the two (i.e., being exposed to *both* ads) is: $P(A \cap B) = P(A)P(B) = 0.04 \times 0.06 = 0.0024$. (*b*) We note that being exposed to at least one of the advertisements is, by definition, the union of the two events, and so the rule for union applies. The probability of the intersection was computed above, and we have $P(A \cup B) = P(A) + P(B) - P(A \cap B)$ $= 0.04 + 0.06 - 0.024 = 0.0976$. The computation of such probabilities is important in advertising research. Probabilities are meaningful also as proportions of the population exposed to different modes of advertising, and are thus important in the evaluation of advertising efforts.

SOLUTION

Product Rules for Independent Events

The rules for the union and the intersection of two independent events extend nicely to sequences of more than two events. These rules are very useful in **random sampling**.

Much of statistics involves random sampling from some population. When we sample randomly from a large population, or when we sample randomly with replacement from a population of any size, the elements are independent of one another. For example, suppose that we have an urn containing 10 balls, 3 of them red and the rest blue. We randomly sample one ball, note that it is red, and return it to the urn (this is sampling *with* replacement). What is the probability that a second ball we choose at random will be red? The answer is still 3/10 because the second drawing does not "remember" that the first ball was red. Sampling with replacement in this way ensures independence of the elements. The same holds for random sampling without replacement (i.e., without returning each element to the population before the next draw) *if* the population is relatively large in comparison with the size of the sample. Unless otherwise specified, we will assume random sampling from a large population.

Random sampling from a large population implies independence.

Intersection Rule. The probability of the intersection of several independent events is just the product of the separate probabilities. The rate of defects in corks of wine bottles is very high, 75 percent. Assuming independence, if four bottles are opened, what is the probability that all four corks are defective? Using this rule: P(all 4 are defective) $= P$(first cork is defective) $\times P$(second cork is defective) $\times P$(third cork is defective) $\times P$(fourth cork is defective) $= 0.75 \times 0.75 \times 0.75 \times 0.75 = 0.316$.

If these four bottles were randomly selected, then we would not have to specify independence—a random sample *always* implies independence.

Union Rule. The probability of the union of several independent events—A_1, $A_2, \ldots, A_n$—is given by the following equation:

$$P(A_1 \cup A_2 \cup \ldots \cup A_n) = 1 - P(\bar{A}_1) P(\bar{A}_2) \ldots P(\bar{A}_n)$$

The union of several events is the event that at least one of the events happens. In the example of the wine corks, suppose we want to find the probability that at least one of the four corks is defective. We compute this probability as follows: P(at least one is defective) $= 1 - P$(none are defective) $= 1 - 0.25 \times 0.25 \times 0.25 \times 0.25 = 0.99609$.

EXAMPLE 2.6

Read the accompanying article. Three women (assumed a random sample) in a developing country are pregnant. What is the probability that at least one will die?

Poor Nations' Mothers At Serious Health Risk

In the industrialized world, a woman's odds of dying from problems related to pregnancy are 1 in 1,687. But in the developing world the figure is 1 in 51. The World Bank also says that each year 7 million newborns die within a week of birth because of maternal health problems. The bank and the United Nations are in the midst of an initiative to cut maternal illnesses and deaths in half by the year 2000.

Reprinted from the *San Francisco Chronicle*, August 10, 1993, p. A10.

SOLUTION

P(at least one will die) $= 1 - P$(all 3 will survive) $= 1 - (50/51)^3 = 0.0577$.

EXAMPLE 2.7

A marketing research firm is interested in interviewing a consumer who fits certain qualifications, for example, use of a certain product. It is known that 10 percent of the public in a certain area use the product and would thus qualify to be interviewed. The company selects a random sample of 10 people from the population as a whole. What is the probability that at least 1 of these 10 people qualifies to be interviewed?

SOLUTION

First, we note that if a sample is drawn at random, then the event that any one of the items in the sample fits the qualifications is independent of the other items in the sample. This is an important property in statistics. Let Q_i, where $i = 1, 2, \ldots, 10$, be the event that person i qualifies. Then the probability that at least 1 of the 10 people will qualify is the probability of the union of the 10 events $Q_i(i = 1, \ldots, 10)$. We are thus looking for $P(Q_1 \cup Q_2 \cup \ldots \cup Q_{10})$.

Now, since 10 percent of the people qualify, the probability that person i does not qualify, $P(\bar{Q}_i)$, is equal to 0.90 for each $i = 1, \ldots, 10$. Therefore, the required probability is equal to $1 - (0.9)(0.9) \cdots (0.9)$ (10 times), or $1 - (0.9)^{10}$. This is equal to 0.6513.

Be sure that you understand the difference between *independent* events and *mutually exclusive* events. Although these two concepts are very different, they often cause some confusion when introduced. When two events are mutually exclusive, they are *not* independent. In fact, they are dependent events in the sense that if one happens, the other one cannot happen. The probability of the intersection of two mutually exclusive events is equal to zero. The probability of the intersection of two independent events is *not* zero; it is equal to the product of the probabilities of the separate events.

---

38. The Holly Sugar Company makes sugar cubes. Quality checks revealed that about 1 in 100 cubes is broken. If you reach for the sugar and randomly choose two cubes, what is the probability that at least one of them will be broken? (Assume independence—this assumption is inherent in random sampling.)

39. The chancellor of a state university is applying for a new position. At a certain point in his application process, he is being considered by seven universities. At three of the seven he is a finalist, which means that (at each of the three universities) he is in the final group of three applicants, one of which will be chosen for the position. At two of the seven universities he is a semifinalist, that is, one of six candidates (in each of the two universities). In two universities he is at an early stage of his application and believes there is a pool of about 20 candidates for each position. Assuming that there is no exchange of information, or influence, across universities as to their hiring decisions, and that the chancellor is as likely to be chosen as any other applicant, what is the chancellor's probability of getting at least one job offer?

40. A package of documents needs to be sent to a given destination, and it is important that it arrive within one day. To maximize the chances of on-time delivery, three copies of the documents are sent via three different delivery services. Service A is known to have a 90 percent on-time delivery record, service B has an 88 percent on-time delivery record, and service C has a 91 percent on-time delivery record. What is the probability that at least one copy of the documents will arrive at its destination on time?

41. Using the data from the article "Have a Safe Flight," what is my danger of dying if I need to take 20 (independent) flights in the next few years? 50 flights? 100 flights?

Have a Safe Flight

There is something reassuring about a silver-haired airline captain, with lashings of gold braid on his epaulettes. For most of the time, though, it will not be the captain or even his co-pilot who will actually fly your aircraft. It will be a computer. The crash of a Lauda Air Boeing 767 in Thailand on May 26th with the loss of 233 lives has again raised fears about the increasing automation of airliners. There will be more crashes to come, but that is no reason to take computers out of the cockpit. Pilots, maybe.

New technology is not making flying more dangerous—quite the opposite. If you travelled in an early 1960s Comet IV you

had a one-in-100,000 chance of being killed. Board any jet built in the past ten years and, despite crowded skies and radar screens, your chances of being killed are just one in 500,000. Crashes are so rare that when they happen they are likely to be caused by unusual, even unique, combinations of events.

Reprinted from *The Economist*, June 29, 1991, p. 14.

42. Using the information in the article "Worst News for Politicians," if three preschoolers are randomly chosen, what is the probability that at least one of them thinks that the president lives at home with his parents? What is the probability that at least one thinks that the next president should be Mr. Rogers?

Worst News for Politicians

From *Newsweek*

President Bush seems to have a few problems with the thumb-sucking lobby. A Playskool poll of preschoolers in five American cities has concluded that, while 55 percent of them know George Bush is President, 15 percent think our commander-in-chief is George Washington. There is also evidence that Bush may have an image problem with some individuals in this interest group: 11 percent think he lives at home with his parents.

And when asked what famous person should next be President, 45 percent of the kids said not Dan Quayle, not Colin Powell but . . . Mr. Rogers. Runners-up included Janet Jackson, Arnold Schwarzenegger and Vanna White.

Just wait till they get the vote.

Reprinted from *Parade* magazine, December 29, 1991, p. 6.

43. Are the events "taking the drug" and "sleeping" independent of each other for the patients and the results given in Problem 36?
44. In Problem 32, are the events "hospitalization" and "the claim being from the Midwest" independent of each other?
45. In Problem 34, are "dividends paid" and "price increase" independent events?
46. In Problem 35, are the events "top executive made more than $1 million" and "shareholders lost money" independent of each other? If this is true for all firms, how would you interpret your finding?
47. The accompanying table shows the incidence of malaria and two other similar illnesses. If a person lives in an area affected by all three diseases, what is the probability that he or she will develop at least one of the three illnesses? (Assume that contracting one disease is an event independent from contracting any other disease.)

The Damage Done: Occurrence of Vector-Borne Disease

	Cases (m = million)	Number at Risk, m
Malaria	110m per year	2,100
Schistosomiasis	200m	600
Sleeping sickness	25,000 per year	50

Reprinted by permission from "The Deadly Hitch-Hikers," Science and Technology, *The Economist,* October 31, 1992, pp. 87–88, with data from UNDP, the World Bank, and the WHO.

48. An electronic device has four independent components with a reliability (probability of being functional) of 0.85 each. The device works only if all four components are functional. What is the probability that the device will work when needed?
49. A device similar to the one in problem 48 has three components, but the device works as long as at least one of the components is functional. The reliabilities of the components are 0.96, 0.91, and 0.80. What is the probability that the device will work when needed?
50. A market researcher needs to interview people who drive to work. In the area where the study is undertaken, 75 percent of the people drive to work. If three people agree to be

interviewed, what is the probability that at least one satisfies the requirement? What is the necessary assumption, and how does it apply?

51. When randomly sampling four items from a population, what is the probability that all four elements will come from the top quartile of the population distribution? What is the probability that at least one of the four elements will come from the bottom quartile of the distribution?

52. The display showing the Mazda RX-7 compares options offered in old and new models. For the *old* models, ¼ of the cars made were turbocharged, ¾ standard; all four hood options were equally popular; all nine steering wheel types equally popular; and for the wheels, 1/10 had spoke wheels, the other five options had variable popularity. Ten percent of cars were convertible, the rest coupe; all color and seat options were equally popular. If you drive an old-model Mazda RX-7 silver coupe that is turbocharged and has spoke wheels, what is the probability that the next Mazda of this model you meet on the road will be exactly like yours?

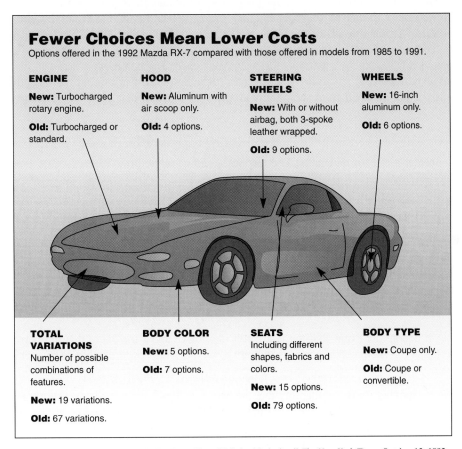

Fewer Choices Mean Lower Costs

Options offered in the 1992 Mazda RX-7 compared with those offered in models from 1985 to 1991.

ENGINE

New: Turbocharged rotary engine.

Old: Turbocharged or standard.

HOOD

New: Aluminum with air scoop only.

Old: 4 options.

STEERING WHEELS

New: With or without airbag, both 3-spoke leather wrapped.

Old: 9 options.

WHEELS

New: 16-inch aluminum only.

Old: 6 options.

TOTAL VARIATIONS

Number of possible combinations of features.

New: 19 variations.

Old: 67 variations.

BODY COLOR

New: 5 options.

Old: 7 options.

SEATS

Including different shapes, fabrics and colors.

New: 15 options.

Old: 79 options.

BODY TYPE

New: Coupe only.

Old: Coupe or convertible.

Reprinted by permission from A. Pollack, "Japan Eases Varieties Marketing," *The New York Times*, October 15, 1992, p. D1, with data from Mazda.

2.6 BAYES' THEOREM

▲ We now present the well-known Bayes' theorem. The theorem allows us to *reverse* the conditionality of events: We can obtain the probability of *B given A* from the probability of *A given B* (and other information).

optional

Bayes' theorem:

$$P(B|A) = \frac{P(A|B)P(B)}{P(A|B)P(B) + P(A|\bar{B})P(\bar{B})}$$

As we see from the theorem, the probability of B given A is obtained from the probabilities of B and $\bar{B}$ and from the conditional probabilities of A given B and A given $\bar{B}$.

The probabilities $P(B)$ and $P(\bar{B})$ are called **prior probabilities** of the events B and $\bar{B}$; the probability $P(B|A)$ is called the **posterior probability** of B. It is possible to write Bayes' theorem in terms of $\bar{B}$ and A, thus giving the posterior probability of $\bar{B}$, $P(\bar{B}|A)$. Incidentally, the terms in the denominator constitute the total probability of event A; the numerator is the joint probability of A and B. Bayes' theorem may be viewed as a means of transforming our *prior* probability of an event, B, into a *posterior* probability of the event B—posterior to the known occurrence of event A.

The use of prior probabilities in conjunction with other information—often obtained from experimentation—has been questioned. The controversy arises in more involved statistical situations where Bayes' theorem is used in mixing the objective information obtained from sampling with prior information that *could* be subjective.

EXAMPLE 2.8

Consider a test for an illness. The test has a known reliability:

1. When administered to an ill person, the test will indicate so with a probability of 0.92.
2. When administered to a person who is not ill, the test will erroneously give a positive result with a probability of 0.04.

Suppose the illness is rare and is known to affect only 0.1 percent of the entire population. If a person is randomly selected for testing from the entire population and the result is positive, what is the posterior probability (posterior to the test result) that the person is ill?

SOLUTION

Let Z denote the event that the test result is positive and I the event that the person tested is ill. The preceding information gives us the following probabilities of events:

$$P(I) = 0.001, P(\bar{I}) = 0.999, P(Z|I) = 0.92, P(Z|\bar{I}) = 0.04$$

We are looking for the probability that the person is ill *given* a positive test result; that is, we need $P(I|Z)$. Since we have the probability with the *reversed* conditionality, $P(Z|I)$, we know that Bayes' theorem is the rule to be used here. Applying the rule to the events Z, I, and $\bar{I}$, we get:

$$P(I|Z) = \frac{P(Z|I)P(I)}{P(Z|I)P(I) + P(Z|\bar{I})P(\bar{I})} = \frac{(0.92)(0.001)}{(0.92)(0.001) + (0.04)(0.999)}$$

$$= 0.0225$$

This result may surprise you. A test with a relatively high reliability (92 percent correct diagnosis when a person is ill and 96 percent correct identification of people who are not ill) is administered to a person, the result is *positive,* and yet the probability that the person is actually ill is only 0.0225!

The reason for the low probability is that we have used *two* sources of information here: the reliability of the test *and* the very small probability (0.001) that a randomly selected person is ill. The two pieces of information were *mixed* by Bayes' theorem, and the posterior probability reflects the mixing of the high reliability of the test with the fact that the illness is rare. The result is perfectly correct as long as the information we have used is accurate. Indeed, subject to the accuracy of our information, if the test should be administered to a large number of people selected randomly from the *entire* population, it would be found that about 2.25 percent of the people in the sample who test positive are indeed ill.

Problems with Bayes' theorem arise when we are not careful with the use of prior information. In this example, suppose the test is administered to people in a hospital. Since people in a hospital are more likely to be ill than people in the population as a whole, the overall-population probability that a person is ill, 0.001, no longer applies. If we should apply this low probability in the hospital, our results would not be correct. This caution extends to all situations where prior probabilities are used: We must always examine the appropriateness of the prior probabilities.

Bayes' theorem may be extended to more than two conditioning events. The resulting extended form of Bayes' theorem is as follows:

Extended Bayes' theorem:

$$P(B_1|A) = \frac{P(A|B_1)P(B_1)}{\text{Sum}[P(A|B_i)P(B_i)]}$$

The theorem gives the probability of one of the sets in the partition, B_1, given the occurrence of event A. A similar expression holds for any of the events B_i.

We demonstrate the use of the equation with the following example. In the solution, we use a table format to facilitate computations.

EXAMPLE 2.9

An economist believes that during periods of high economic growth, the U.S. dollar appreciates with probability 0.70; in periods of moderate economic growth, the dollar appreciates with probability 0.40; and during periods of low economic growth, the dollar appreciates with probability 0.20. During any period of time, the probability of high economic growth is 0.30, the probability of moderate growth is 0.50, and the probability of low economic growth is 0.20. Suppose the dollar has been appreciating during the present period. What is the probability we are experiencing a period of high economic growth?

SOLUTION

Our partition consists of three events: high economic growth (event H), moderate economic growth (event M), and low economic growth (event L). The prior probabilities of the three states are: $P(H) = 0.30$, $P(M) = 0.50$, and $P(L) = 0.20$. Let A denote the event that the dollar appreciates. We have the following conditional probabilities: $P(A|H) = 0.70$, $P(A|M) = 0.40$, and $P(A|L) = 0.20$. Applying the equation using three sets ($i = 1, 2, 3$), we get the following:

TABLE 2.1   Bayesian Revision of Probabilities, Example 2.9

Event	Prior Probability	Conditional Probability	Joint Probability	Posterior Probability	
H	$P(H) = 0.30$	$P(A|H) = 0.70$	$P(A \cap H) = 0.21$	$P(H|A) = \dfrac{0.21}{0.45}$	$= 0.467$
M	$P(M) = 0.50$	$P(A|M) = 0.04$	$P(A \cap M) = 0.20$	$P(M|A) = \dfrac{0.20}{0.45}$	$= 0.444$
L	$P(L) = 0.20$	$P(A|L) = 0.20$	$P(A \cap L) = 0.04$	$P(L|A) = \dfrac{0.04}{0.45}$	$= 0.089$
	Sum $= 1.00$		$P(A) \quad = 0.45$	Sum	$= 1.000$

$$P(H|A) = \frac{P(A|H)P(H)}{P(A|H)P(H) + P(A|M)P(M) + P(A|L)P(L)}$$

$$= \frac{(0.70)(0.30)}{(0.70)(0.30) + (0.40)(0.50) + (0.20)(0.20)} = 0.467$$

We can obtain this answer, along with the posterior probabilities of the other two states, M and L, by using a table. In the first column of the table we write the prior probabilities of the three states H, M, and L. In the second column we write the three conditional probabilities $P(A|H)$, $P(A|M)$, and $P(A|L)$. In the third column we write the joint probabilities $P(A \cap H)$, $P(A \cap M)$, and $P(A \cap L)$. The joint probabilities are obtained by multiplying across in each of the three rows. The sum of the entries in the third column is the total probability of event A. Finally, the posterior probabilities $P(H|A)$, $P(M|A)$, and $P(L|A)$ are obtained by dividing the appropriate joint probability by the total probability of A at the bottom of the third column. For example, $P(H|A)$ is obtained by dividing $P(H \cap A)$ by the probability $P(A)$. The operations and the results are given in Table 2.1

Note that both the prior probabilities and the posterior probabilities of the three states add to 1.00, as required for probabilities of all the possibilities in a given situation. We conclude that, given that the dollar has been appreciating, the probability that our period is one of high economic growth is 0.467, the probability that it is one of moderate growth is 0.444, and the probability that our period is one of low economic growth is 0.089. The advantage of using a table is that we can obtain all posterior probabilities at once. If we use the formula directly, we need to apply it once for the posterior probability of each state.

PROBLEMS

53. A chemical plant has an emergency alarm system. When an emergency situation exists, the alarm sounds with probability 0.95. When an emergency situation does not exist, the alarm system sounds with probability of 0.02. A real emergency situation is a rare event, with probability 0.004. Given that the alarm has just sounded, what is the probability that a real emergency situation exists?

54. When the economic situation is "high," a certain economic indicator rises with probability 0.6. When the economic situation is "medium," the economic indicator rises with probability 0.3. When the economic situation is "low," the indicator rises with probability 0.1. The economy is high 15 percent of the time, it is medium 70 percent of the time, and

it is low 15 percent of the time. Given that the indicator has just gone up, what is the probability that the economic situation is high?

55. An oil explorer orders seismic tests to determine whether oil is likely to be found in a certain drilling area. The seismic tests have a known reliability: When oil does exist in the testing area, the test will indicate so 85 percent of the time; when oil does not exist in the test area, 10 percent of the time the test will erroneously indicate that it does exist. The explorer believes that the probability of existence of an oil deposit in the test area is 0.4. If a test is conducted and indicates the presence of oil, what is the probability that an oil deposit really exists?

56. Before marketing new products nationally, companies often test them on samples of potential customers. Such tests have a known reliability. For a particular product type, it is known that a test will indicate success of the product 75 percent of the time if the product is indeed successful and 15 percent of the time when the product is not successful. From past experience with similar products, a company knows that a new product has a 0.60 chance of success on the national market. If the test indicates that the product will be successful, what is the probability that it really will be successful?

57. A market research field worker needs to interview married couples about use of a certain product. The researcher arrives at a residential building with three apartments. From the names on the mailboxes downstairs, the interviewer infers that a married couple lives in one apartment, two men live in another, and two women live in the third apartment. The researcher goes upstairs and finds that there are no names or numbers on the three doors, so that it is impossible to tell in which of the three apartments the married couple lives. The researcher chooses a door at random and knocks. A woman answers the door. Having seen a woman at the door, what *now* is the probability of having reached the married couple? Make the (possibly unrealistic) assumptions that if the two men's apartment was reached, a woman cannot answer the door; if the two women's apartment was reached, then only a woman can answer; and that if the married couple was reached, then the probability of a woman at the door is 1/2. Also assume a 1/3 prior probability of reaching the married couple. Are you surprised by the numerical answer you obtained?

SUMMARY

In this chapter we introduced the theory of **probability,** which is fundamental to statistical inference. We defined the basic elements of probability: **experiments, outcomes, events,** and the **sample space.** We saw how the probability of an event may be computed in various situations. A distinction was made between two kinds of probability: **objective** and **subjective.** We saw, however, that the same rules apply to handling both kinds of probability. Among the rules for probability we emphasized were the **rule of complements, intersections,** and the important **rule of unions.** We defined **mutually exclusive** events and saw how their probability is computed. We talked about **conditional probability** and how it helps us define the very important concept of **independence** of events. We learned how probabilities of independent events are computed and used. We ended the chapter with the optional topic of **Bayes' theorem,** which allows us to compute the probability of events of reversed conditionality.

KEY FORMULAS

Union of events:
$$P(A \cup B) = P(A) + P(B) - P(A \cap B)$$

Conditional probability:
$$P(A \mid B) = P(A \cap B)/P(B)$$

Product rule for independent events:
$$P(A \cap B) = P(A)P(B)$$

Bayes rule:
$$P(B \mid A) = \frac{P(A \mid B)P(B)}{P(A \mid B)P(B) + P(A \mid \bar{B})P(\bar{B})}$$

ADDITIONAL
PROBLEMS

58. WLDN (AM 1120) is Walden Pond Radio, an offbeat radio station located near Walden Pond in eastern Massachusetts. On January 29, 1991, the following conversation was heard on this radio station.

Weatherman: *And the temperature now is 37 degrees.*
DJ: *But our thermometer outside the studio is not very accurate, is it?*
Weatherman: *You are right, Joe. As soon as the sun hits it, the mercury just bounces all over the place. I would say that there is only a 3 percent chance that the temperature we report at any given time is correct.*

Assume that "correct" means correct to within a degree, and that temperature readings are independent of each other. Use the information above to answer the following two questions.

a. If I listen to the temperature report on this radio station at three different times, what is the probability that I will hear at least one correct temperature?
b. In three times, what is the probability that all three reported temperatures are correct?

59. AT&T was running commercials in 1990 aimed at luring back customers who may have switched to one of the other long-distance phone service providers. One such commercial shows a businessman trying to reach Phoenix and mistakenly getting Fiji, where a Fijian native on a beach responds incomprehensibly in Polynesian. When asked about this advertisement, AT&T admitted that the portrayed incident did not actually take place but added that this was an enactment of something that "could happen."[6] Suppose that 1 in 200 long-distance telephone calls is misdirected. What is the probability that at least one in five attempted telephone calls reaches the wrong number? (Assume independence of attempts.)

60. Refer to the information in the previous problem. Given that your long-distance telephone call is misdirected, there is a 2 percent chance that you will reach a foreign country (such as Fiji). Suppose that I am now going to dial a single long-distance number, what is the probability that I will erroneously reach a foreign country?

61. A fashion designer has been working with the colors green, black, and red in preparing for the coming season's fashions. The designer estimates that there is a 0.3 chance that the color green will be "in" during the coming season, a 0.2 chance that black will be among the season's colors, and a 0.15 chance that red will be popular. Assuming that colors are chosen independently of each other for inclusion in new fashions, what is the probability that the designer will be successful with at least one of her colors?

62. A company president always invites one of her three vice presidents to attend business meetings and claims that her choice of the accompanying VP is random. One of the three has not been invited even once in five meetings. What is the probability of such an occurrence if the choice is indeed random? What conclusion would you reach based on your answer?

63. According to a news report heard on PBS on October 29, 1992, then-President George Bush claimed, "Ninety-two percent of all media reporting is negative." Leaving out the obvious question as to how "negative" should be construed, suppose that media reports are independent of each other. If three reports are heard on the radio, what is the probability that at least one of them is negative (using George Bush's definition of negative)?

64. Consider the figure titled "What Kids Like to Eat." Could the events implicit in the figure be mutually exclusive? Explain. If three kids are randomly chosen, what is the probability that none of the three like hot dogs?

---

[6]While this may seem virtually impossible due to the different dialing procedure for foreign countries, AT&T argues that erroneously dialing the prefix 679 instead of 617, for example, would get you Fiji instead of Massachusetts.

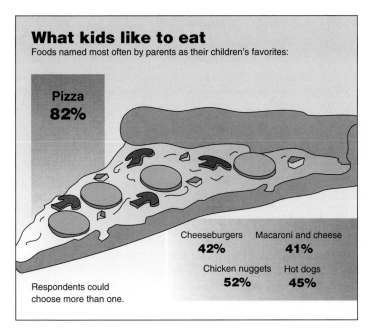

Reprinted from "USA Snapshots," *USA Today*, January 14, 1992, p. 1, with data from a Gallup poll of 1,034 parents of 3- to 11-year-olds.

65. I drive home from the office along one of the routes shown in the accompanying map. Traffic lights in this city are, unfortunately, independent of each other. At each intersection, the numbers shown are the probability of hitting a red light when reaching the intersection driving on a horizontal (i.e., East–West) street. Assume that the probability of hitting a red light when driving on a North–South street is just one minus the given probability. What is my best way home? (Draw it on the map.)

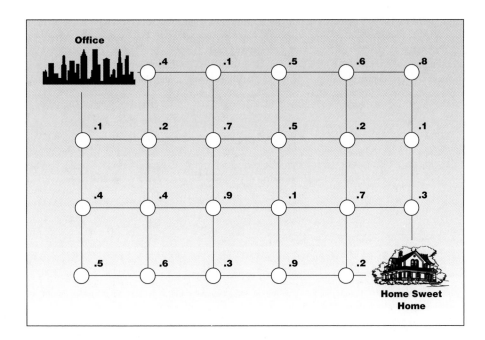

The Wagging Tale

Ten percent of America's dog owners say they are as attached to their dogs as they are to their spouses.

According to a National Family Opinion survey, sponsored by Frosty Paws frozen treats for dogs, people really value their relationships with the family dog—as much as their best friend, said 31.6% of the dog owners polled; their children, 15.1%; their spouses, 10.4%; followed by their neighbors, parents and co-workers. Less than one-third of the respondents said they were not as attached to their dog as to other humans.

According to the survey, 68% of the responding households have only one dog; 21% have two dogs; and 11% keep three or more dogs.

The Wagging Tale 1, no. 1 (1991) (Advertising piece by Brookline Grooming and Pet Supplies, Brookline, MA).

66. Based on *The Wagging Tale:*
    a. If a (dog-owning) household is randomly selected, what is the probability that it has more than one dog?
    b. If four dog owners are randomly selected, what is the probability that at least one of them is attached to the dog as much as to his or her spouse?
67. According to a study reported in the news media in September 1993, 2.1 million Americans suffer from rheumatoid arthritis. Use census data from your library to estimate the population of the United States, and based on these data compute the probability that a randomly selected American newborn will develop the illness in his or her lifetime. What is a major limitation of your estimate of this probability?
68. Describe the trend over the years in the probability of dying on a flight, given the data in the table "Worldwide Airline Fatalities: 1970 to 1990." What is the probability of dying in an airplane crash for a passenger flying 10,000 miles in 1990? Compare it to the same probability in 1980.

Worldwide Airline Fatalities: 1970 to 1990 (Passenger deaths for scheduled air transport operations)

Year	Aircraft fatal accidents	Passenger deaths	Death rate[1]	Year	Aircraft fatal accidents	Passenger deaths	Death rate[1]
1970	29	700	0.29	1981	21	362	0.06
1971	32	884	0.35	1982	26	764	0.13
1972	41	1,209	0.42	1983	20	809	0.13
1973	29	862	0.27	1984	16	223	0.03
1974	21	1,299	0.38	1985	22	1,066	0.15
1975	20	467	0.13	1986[2]	22	546	0.06
1976	24	734	0.19	1987[2]	26	901	0.09
1977	25	516	0.12	1988[2]	28	729	0.07
1978	31	754	0.15	1989[2]	27	817	0.05
1979	31	877	0.16	1990[2 3]	25	495	0.03
1980	22	814	0.14				

[1]Rate per 100 million passenger miles flown. [2]Includes (former) USSR, which began reporting in 1986. [3]Preliminary.

Reprinted from the *Statistical Abstract of the United States, 1992,* Washington DC: U.S. Bureau of the Census, p. 135, with data from the International Civil Aviation Organization, Montreal, Canada, *Civil Aviation Statistics of the World,* annual.

69. Recall from Chapter 1 that the median is the number such that ½ the observations lie above it and ½ the observations lie below it. If a random sample of two items is to be drawn from some population, what is the probability that the population median will lie between these two data points?

70. (*The Von Neumann device.*) Suppose that one of two people is to be randomly chosen, with equal probability, to attend an important meeting. One of them claims that using a coin to make the choice is not fair because the probability that it will land on a "head" or a "tail" is not exactly 0.50. How can the coin still be used for making the choice? (Hint: Toss the coin *twice,* basing your decision on two possible outcomes.) Explain your answer.

71. Is Marilyn right? (Hint: Create a sample space of 16 equally likely points. Each point is a quadruple of *ordered* symbols, e.g., BGGB; then count the number of points leading to each event: 1 girl, 3 boys; 2 girls, 2 boys; etc.)

Ask Marilyn™

By MARILYN VOS SAVANT

If you have four children, they may all be of one sex, there may be three of one sex and one of the other sex, or there may be two of each. Which is most likely?—Adrian R. Beck, Joplin, Mo.

I know it sounds strange, but more families with four children have three of one sex and one of the other sex than any other combination. The chance of having all girls or all boys is 1 in 8; the chances of having two of each are 3 in 8; and the chances of having three girls and one boy (or three boys and one girl) are 4 in 8.

Reprinted from *Parade* magazine, October 2, 1993. (I am indebted to Art O'Leary for bringing it to my attention.)

72. Use Table A.1 in the data appendix (Appendix A). If a state (including DC) is randomly chosen, what is the probability that its population is at least 80 percent white? What is the probability that the population of a randomly chosen state is at least 15 percent black? What is the probability that the population of a randomly chosen state is at least 10 percent Hispanic?

INTERVIEW WITH
Alan Dershowitz

"The question on the other side is, can we ever do better than 90 percent? By the way, I don't think 1 out of every 10 people in jail is innocent. I don't believe that's the case."

Alan Dershowitz is well known for his legal work on the defense teams in such famous cases as Claus Von Bulow, Mike Tyson, Patty Hearst, Michael Milken, and O. J. Simpson. He is currently Felix Frankfurter Professor of Law at Harvard Law School and received his JD from Yale where he graduated first in his class and was editor-in-chief of the Yale Law Journal. His newest book, Contrary to Public Opinion, *was published by Pharos Books in 1992.*

Aczel: As you know . . . I would like to ask you about probability and the law.

Dershowitz: Well, I'm very interested. I wish I had thought to invite you to my class, because in the last week I've been talking about probability. I'll tell you the three contexts in which I was dealing with it. First, I was asking the class whether or not proof beyond a reasonable doubt ought to be given a number, whether jurors ought to be told proof beyond a reasonable doubt means 90 percent or 75 percent or 80 percent. Should jurors have different assessments of what proof beyond a reasonable doubt is and should judges explain to jurors "what does it mean that this person is 90 percent likely? either he did or he didn't do it." And I explain that we have a fingerprint test or a DNA test or a ballistics test—in

these cases all we are ever doing is saying that this fingerprint falls into a category, that if we were to convict all people who had matches like this, 90 percent would be guilty and 10 percent would be innocent. The guy is either guilty or innocent, but I explain to them very simply what probability means I introduce them to Kahnemann and Tversky's arguments about the difficulties of probabilistic thinking. So that was one session in class, whether or not we should allow probabilities to get to the jury. Then, another section dealt with just the problem you're talking about now: unanimous juries, size of juries, peremptory challenges to juries. What's the difference if you have the requirement of 8–4 for conviction, or if you have 12-0. What is the difference if you have 6–2 and you have only 8 people on the jury? Does the defendant or the prosecution benefit from smaller size, from different ratios? Finally I'm going to be spending a lot of time on the issue of statistics in rape and whether it is underreporting or overreporting or both in the context of rape. So my class spends a lot of time on trying to understand simple statistics, simple probability thinking. I think the law is allergic to numbers. The law is terrified of numbers, and you'd rather have jurors being told that proof beyond a reasonable doubt means proof to a moral certainty along with the other instruction that I

107

hate, the one that says you should make this decision in a criminal case as if it was a very important decision in the other important aspects of your life. Well, most of us operate on 51 percent. If you can get 51 percent, why take 49 percent? But in the criminal justice system, if you have 51 percent versus 49 percent, you're supposed to go with the 49 percent. Even if you have 60 percent versus 40 percent, you're supposed to go with the 40 percent. It's very counterintuitive.

You know in the end what I argue—I give them the classic example of the blue buses and the white buses and there is an accident in which a blue bus ran somebody down at a very high speed. We know that whoever was driving the bus is guilty, and whatever company is driving the bus is guilty, and we know nothing more about the bus except that 90 percent of the blue buses in town are owned by company A and 10 percent are owned by company B. So it is clear that it is 90 percent likely—that is, if we were to convict under all these circumstances, 90 percent of the time we'd be right to convict and 10 percent we'd be wrong. I ask the students how many of you would convict under those circumstances and a lot would convict but a lot wouldn't. I ask the students who wouldn't convict because 90 percent is not enough: What would you do with the following case? Let's assume somebody says, "That's the person. I saw him. I am certain of the identification. He's the killer." And then we introduce Elizabeth Laftes and she comes into class and testifies "That's a pretty good identification. It's an 80 percent identification. In 80 percent of the cases that level of identification with that level of knowledge is going to be accurate and in 20 percent of the cases it's going to be wrong." Would you permit conviction there? They all say yes because that's a clinical ID, that's clinical even though you get a statistician coming in and saying, "Sure, we know it's a clinical ID, but the clinician is right only 80 percent of the time." We don't want to hear that.

Aczel: Like "Mr. 100%," Dr. Bradley [*Reversal of Fortune*]. . . . Actually that's an argument the other way—the problem of overconfidence.

Dershowitz: Right. There's a lot of good writing on that, on the overconfidence phenomenon, and going back even to some of the early work by the guy who wrote a book called *Statistical versus Clinical Prediction* in which he assessed the comparative accuracy of clinicians and statisticians in making valid predictions. In the end he came to the conclusion that statisticians are likely to be more accurate only because they are more sensitive to issues of accuracy.

Aczel: Because they're used to thinking of numbers . . .

Dershowitz: Clinicians are saying "sounds good." They never understand that you can make the same mistake 20 times.

The whole notion of probabilistic thinking is key, it seems to me, to the legal system. Whether in the end you want to give juries explicitly probabilistic roles, that's a different question. But surely the law has to be attuned to the fact that it is in a probabilistic business. We never deal with certainty. We know that rules of evidence are designed to create a certain probability and that it's a complex probability because we are very sensitive to type 1 and type 2 errors. I have a section in my class where we deal with predicting the future. I ask basically how many false positives are we willing to tolerate in order to avoid how many false negatives. Or bail decisions. We're going to let somebody out on bail. Is he going to kill, or isn't he going to kill? We're going to reduce somebody's sentence or we're going to increase somebody's sentence. These are all probabilistic determinations. And yet the law is allergic to putting a number on it. We have no jurisprudence of prediction. (I've written fairly extensively about that.)

Aczel: You're unusual in that thinking, aren't you?

Dershowitz: I think so. I want to at least expose the law to the fact that they are doing this. I'm not sure in the end that everything ought to be numbers—I don't want to substitute computers for juries—but I do want the law to know what it is not doing and what it is doing and I want it to face up to the fact, why they're so afraid of numbers.

One reason obviously is that we want a democratic jury and because a democratic jury will have lots of different levels of aptitude and ability and if you gave jurors explicitly probabilistic tasks, the elite members of the jury would play a greater role in the deliberation. You don't want that to happen. You can't have everything in a jury system. A jury system requires lots of compromise with truth.

Aczel: It's my understanding that probability reasoning was used until 1968 with the reversal of *People* v. *Collins* in the California Supreme Court.

Dershowitz: Well, it's the case of *People* v. *Collins* that I teach in my criminal law class. And, no, I don't think that's an accurate assessment. I think that there was no prohibition on using probabilities, but I don't think it's been widely used.

Aczel: I see. It was a bad case.

Dershowitz: I think it's a bad case. I think it's a bad case for two reasons. [Professor Larry Tribe was the law clerk on that case.] It confuses what was wrong in that case. Obviously, the math instructor didn't know what he was doing in terms of the independence of the variables and so on. It's confused in the inherent problem. Now what I do is throw the students a real curve ball because I reverse the facts of *Collins.* I say to the students: Now, what would you do if the following were the situation? The prosecutor in *Collins* says: "Ladies and gentlemen of the jury, use your common sense. Just use your common sense. Black guy, white woman, pony tail, yellow Cadillac. Come on. What's the likelihood that there would be another couple like that? Just use your common sense. This can't be coincidence." The defense attorney then introduces a statistician who says, "Hey, wait a minute. That's not such an uncommon combination of characteristics. Let me tell you how many black men and white women drive yellow cars, and have pony tails. If you think of that statistically, there's probably a one in six chance that there could be another couple like this." Would you allow the defense to use it, to undercut a common sense misconception? Larry [Tribe] won't answer that question. I've tried.

Aczel: Why?

Dershowitz: Because he made a clever argument for the prosecution not being able to use it. When it comes to the defense being denied an opportunity to use math, when I want to show a reasonable doubt, I would like to use statistics because I think jurors are more likely to accept coincidences as absolute certainty than they are really warranted in accepting. There are a lot of coincidences in life. When your two fingerprints with four rings or five rings look the same, I would like to introduce a statistician to say "Wait a minute. First of all, we're not so good at replicating the rings. Maybe it's not exactly the same ring. Second, if each one has a 1 in 10 chance and there's only 10 times 10 times 10 times 10, that's not so improbable when you have an extremely large

population. After all, how was he spotted. He *wasn't* ID'd." . . . I think it's sometimes useful to introduce mathematics—simply, though I wouldn't call it high-level mathematics.

Aczel: So why don't people do it?

Dershowitz: I think, first of all, lawyers aren't trained; they get very nervous about numbers.

Aczel: So it's a matter of education, doing it in the classroom?

Dershowitz: It's a matter of education and doing it in the classroom. I give my students an article (I did a few years ago) on a probabilistic approach. It has some charts, but it's accessible. It's very easy. I wrote it about a subject that the students cared deeply about. The American Bar Association was trying to predict which law students would get in trouble later on so maybe they shouldn't be admitted to law school. And what I did was write a simple article about how complex this notion of analysis, discriminating in this way is, and how much overprediction you need in order to get any significant number. What I tried to argue was that law shouldn't borrow mathematics. Law should use mathematics. We should say: Here's our problem. Now let's think about how math can help us solve that problem.

Aczel: The lawyers have just shied away from it?

Dershowitz: They've traditionally shied away from it. I think after *People* v. *Collins,* Tribe wrote an article called "Trial by Mathematics" in which there was an exchange of a lot of people arguing . . .

Aczel: Mosteller, here at Harvard, wrote about it . . .

Dershowitz: Yes. I, in fact, took Mosteller's course when I first came to Harvard because I wanted to become more sensitive to mathematical models and how I could use math in my own research. For example, when they developed the concept of the xyy chromosome, everybody was saying, "Oh gee, the xyy chromosome is going to be able to predict violence." I did a little article saying, "Wait a minute, let's look at the statistics on this for a minute. Let's see what it does and what it doesn't do."

Aczel: I see.

Dershowitz: The heuristic value of math is powerful.

Aczel: What then is reasonable doubt to you, if you had to put a number on it?

Dershowitz: Well, if I had to put a number on it, I would say proof by a preponderance of evidence really is 51 percent and proof by clear and convincing evidence, I would put at around 75 percent, and I would put proof beyond a reasonable doubt at about 90 percent. But is that enough, though? Does that mean we're willing to have 1 out of every 10 people in jail being innocent? The question on the other side is, can we ever do better than 90 percent? By the way, I don't think 1 out of every 10 people in jail is innocent. I don't believe that's the case.

Aczel: You think it's much less?

Dershowitz: I think it's less.

Aczel: What do you think is that number?

Dershowitz: I would put it at two or three percent. Well, it depends on how you define innocence or guilt. If you are asking me, is it the wrong person who did it, two or three percent. Are you asking me whether, for example, in a date rape case—it's a close case that falls on one side or the other . . .

Aczel: It's a fuzzy-set type problem.

Dershowitz: It's much more there. Are we talking about degrees of guilt? Is it first degree murder or second degree murder? We're getting up there. But on the pure issue: "Is this the wrong person?" two or three percent.

Aczel: So, we're doing quite well.

Dershowitz: We're doing quite well. But it still means for every 10 thousand people in jail, there are 300 who are innocent. That's a lot of people.

Aczel: But it's still better than 10 percent—that's scary.

Dershowitz: If you figure the numbers, there are almost one million people confined in the United States today. That means 30,000 people would be innocent in American prisons. Now, of these people who are innocent, a lot of them did it before and will do it again.

Aczel: So that's where it's a matter of degree?

Dershowitz: No, it's not a matter of degree. They are really innocent. But they are more likely to be picked up. They are more likely to be convicted. They are more likely to be sentenced to jail because they are bad guys. Now, to me, the paradigm case is a case that I'm currently involved in which is Jeffrey McDonald. Jeffrey McDonald is the man who was convicted of killing his whole family—the Marine, the Green Beret. He is either the worst, most horrible criminal imaginable, or the most unbelievably vilified innocent person. There is no gray area. In most cases there is gray. Not in this one.

Aczel: What's the most important thing you want your students to take out of the classroom?

Dershowitz: Passion is a very important thing for me. They're very smart and I want them to approach their roles with a real feeling. In terms of intelligence, there's not much I can teach them. The idea of coming to their task with a real passion is very important. The idea of thinking creatively, of not being bound by existing structures, being able to really think in a way that breaks boundaries. I tell them, "Don't think like a lawyer. Think like a brilliant scientist. Think like a discoverer, like a creator. Think like an explorer, not like a lawyer."

Chance Quantities

Courtesy of The Stock Market/Richard Gross, 1993.

CHAPTER OUTLINE

3.1 Random Variables   115

3.2 Expected Values of Discrete Random
     Variables   124

3.3 The Binomial Distribution   130

3.4 Continuous Random Variables   140

II. An Argument for Divine Providence, taken from the constant Regularity observ'd in the Births of both Sexes. By Dr. John Arbuthnott, Physician in Ordinary to Her Majesty, and Fellow of the College of Physicians and the Royal Society.[1]

A mong innumerable Footsteps of Divine Providence to be found in the Works of Nature, there is a very remarkable one to be observed in the exact Ballance that is maintained, between the Numbers of Men and Women; for by this means it is provided, that the Species may never fail, nor perish, since every Male may have its Female, and of a proportionable Age. This Equality of Males and Females is not the Effect of Chance but Divine Providence, working for a good End, which I thus demonstrate:

Let there be a Die of Two sides, M and F, (which denote Cross and Pile), now to find all the Chances of any determinate Number of such Dice, let the Binome M + F be raised to the Power, whose Exponent is the Number of Dice given; the Coefficients of the Terms will shew all the Chances sought. For Example, in Two Dice of Two sides M + F the Chances are $M^2 + 2 MF + F^2$, that is, One Chance for M double, One for F double, and Two for M single and F single; in Four such Dice there are Chances $M^4 + 4 M^3 F + 6 M^2 F^2 + 4 MF^3 + F^4$, that is, One Chance for M quadruple, One for F quadruple, Four for triple M and single F, Four for single M and triple F, and Six for M double and F double; and universally, if the Number of Dice be n, all their Chances will be expressed in this Series

---

[1] From J. Arbuthnott, "An argument for divine providence, taken from the constant regularity observed in the births of both sexes," *Philosophical Transactions* 27 (1710), pp. 186–90.

$$M^n + M^0 + \frac{n}{1} \times M^{n-1}F + \frac{n}{1} \times \frac{n-1}{2} \times M^{n-2}F^2 + \frac{n}{1} \times \frac{n-1}{2} \times \frac{n-2}{3} \times M^{n-3}F^3 +, \&c.$$

It appears plainly, that when the Number of Dice is even there are as many M's as F's in the middle Term of this Series, and in all the other Terms there are most M's or most F's.

If therefore a Man undertake with an even Number of Dice to throw as many M's as F's, he has all the Terms but the middle Term against him; and his Lot is to the Sum of all the Chances, as the coefficient of the middle Term is to the power of 2 raised to an exponent equal to the Number of Dice: so in Two Dice his Lot is $\frac{2}{4}$ or $\frac{1}{2}$, in Three Dice $\frac{6}{16}$ or $\frac{3}{8}$, in Six Dice $\frac{20}{64}$ or $\frac{5}{16}$, in Eight $\frac{70}{256}$ or $\frac{35}{128}$, &c.

To find this middle Term in any given Power or Number of Dice, continue the Series $\frac{n}{1} \times \frac{n-1}{2} \times \frac{n-2}{3}$, & c. till the number of terms are equal to $\frac{1}{2}$ n. For Example, the coefficient of the middle Term of the tenth Power is $\frac{10}{1} \times \frac{9}{2} \times \frac{8}{3} \times \frac{7}{4} \times \frac{6}{5} = 252$, the tenth Power of 2 is 1024, if therefore A undertakes to throw with Ten Dice in one throw an equal Number of M's and F's, he has 252 Chances out of 1024 for him, that is his Lot is $\frac{252}{1024}$ or $\frac{63}{256}$, which is less than $\frac{1}{4}$.

	Christened.			Christened.			Christened.	
Anno.	Males.	Females.	Anno.	Males.	Females.	Anno.	Males.	Females
1629	5218	4683	1657	3396	3289	1684	7575	7127
30	4858	4457	58	3157	3013	85	7484	7246
31	4422	4102	59	3209	2781	86	7575	7119
32	4994	4590	60	3724	3247	87	7737	7214
33	5158	4839	61	4748	4107	88	7487	7101
34	5035	4820	62	5216	4803	89	7604	7167
35	5106	4928	63	5411	4881	90	7909	7302
36	4917	4605	64	6041	5681	91	7662	7392
37	4703	4457	65	5114	4858	92	7602	7316
38	5359	4952	66	4678	4319	93	7676	7483
39	5366	4784	67	5616	5322	94	6895	6647
40	5518	5332	68	6073	5560	95	7263	6713
41	5470	5200	69	6506	5829	96	7632	7229
42	5460	4910	70	6278	5719	97	8062	7767
43	4793	4617	71	6449	6061	98	8426	7626
44	4107	3997	72	6443	6120	99	7911	7452
45	4047	3919	73	6073	5822	1700	7578	7061
46	3768	3395	74	6113	5738	1701	8102	7514
47	3796	3536	75	6058	5717	1702	8031	7656
48	3363	3181	76	6552	5847	1703	7765	7683
49	3079	2746	77	6423	6203	1704	6113	5738
50	2890	2722	78	6568	6033	1705	8366	7779
51	3231	2840	79	6247	6041	1706	7952	7417
52	3220	2908	80	6548	6299	1707	8379	7687
53	3196	2959	81	6822	6533	1708	8239	7623
54	3441	3179	82	6909	6744	1709	7840	7380
55	3655	3349	83	7577	7158	1710	7640	7288
56	3668	3382						

3.1 RANDOM VARIABLES

Like Marilyn Vos Savant (see Problem 71 in Chapter 2), Dr. John Arbuthnott—over 280 years earlier—was interested in the matter of the probabilities of the number of babies of each sex in a given number of births. In solving Problem 72 in Chapter 2, you constructed a sample space made up of the 16 equally likely points:

BBBB   BBBG   BGGB   GBGG

GBBB   GGBB   BGBG   GGBG

BGBB   GBGB   BBGG   GGGB

BBGB   GBBG   BGGG   GGGG

All of these 16 points are equally likely because when four children are born, the sex of each child is assumed to be independent of those of the other three. Hence the probability of each quadruple (e.g., GBBG) is equal to the product of the probabilities of the four separate, single outcomes—G, B, B, and G—and is thus equal to $(\frac{1}{2})(\frac{1}{2})(\frac{1}{2})(\frac{1}{2}) = \frac{1}{16}$.

Now, let's look at the variable "the number of girls out of four births." This number *varies* among points in the sample space, and it is *random*—given to chance. That'swhy we call such a number a **random variable.**

   A random variable is an uncertain quantity whose value depends on chance.

A random variable has a probability law: A rule that assigns probabilities to the different values of the random variable. The probability law, the probability assignment, is called the **probability distribution** of the random variable. We usually denote the random variable by a capital letter, often X. The probability distribution will then be denoted by $P(X)$.

Look again at the sample space for the sexes of four babies, and remember that our variable is the number of girls out of four births. The first point in the sample space is BBBB; because the number of girls is zero here, $X = 0$. The next four points in the sample space all have one girl (and three boys). Hence, each one leads to the value $X = 1$. Similarly, the next six points in the sample space all lead to $X = 2$; the next four points to $X = 3$; and, finally, the last point in our sample space gives $X = 4$. The correspondence of points in the sample space with values of the random variable is given below:

Sample Space	Random Variable
BBBB	$X = 0$
GBBB	
BGBB	
BBGB	$X = 1$
BBBG	
GGBB	
GBGB	
GBBG	
BGGB	$X = 2$
BGBG	
BBGG	
BGGG	
GBGG	$X = 3$
GGBG	
GGGB	
GGGG	$X = 4$

This correspondence, when a sample space clearly exists, allows us to define a random variable as follows.

A random variable is a function of the sample space.

What is this function? The correspondence between points in the sample space and values of the random variable allows us to determine the distribution function of X as follows: Notice that 1 of the 16 equally likely points of the sample space leads to $X = 0$. Hence, the probability that $X = 0$ is $\frac{1}{16}$. Because 4 of the 16 equally likely points lead to a value $X = 1$, the probability that $X = 1$ is $\frac{4}{16}$, and so forth. Thus, looking at the sample space and counting the number of points leading to each value of X, we find the following probabilities:

$$P(X = 0) = 1/16 = 0.0625$$
$$P(X = 1) = 4/16 = 0.2500$$
$$P(X = 2) = 6/16 = 0.3750$$
$$P(X = 3) = 4/16 = 0.2500$$
$$P(X = 4) = 1/16 = 0.0625$$

The probability statements above constitute the probability distribution of random variable $X =$ the number of girls in four births. Notice how this probability law was obtained simply by associating values of X with sets in the sample space. (For example, the *set* GBBB, BGBB, BBGB, BBBG leads to $X = 1$.) It is useful to write down the probability distribution of X in a table format, but first let's make a small, simplifying notational distinction so that we do not have to write complete probability statements such as $P(X = 1)$ and so on.

As stated earlier, we use a capital letter, such as X, to denote the random variable. But we use a lowercase letter to denote a *particular value* that the random variable can take. For example, $x = 3$ means that some particular set of four births resulted in three girls. Think of X as random and x as known. Before a coin is tossed, the number of heads (in one toss) is an unknown, X. Once the coin lands, we have $x = 0$ or $x = 1$.

Now let's return to the number of girls in four births. We can write the probability distribution of this random variable in a table format, as shown in Table 3.1.

Note an important fact: The sum of the probabilities of all the values of the random variable X must be 1.00. A picture of the probability distribution of the random variable X is given in Figure 3.1. Such a picture is a **probability histogram** for the random variable.

Dr. Arbuthnott was interested in the number of girls (or boys) in any fixed number of births, not necessarily four. Thus his discussion extends beyond this case. In fact,

TABLE 3.1   The Probability Distribution of the Number of Girls in Four Births

Number of Girls, x	Probability, $P(x)$
0	1/16
1	4/16
2	6/16
3	4/16
4	1/16
	16/16 = 1.00

The Probability Distribution of the Number of Girls in Four Births   **FIGURE 3.1**

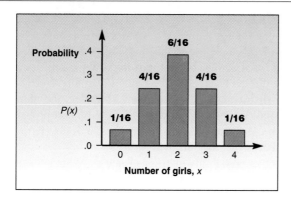

the random variable he describes, which in general counts the number of "successes" (here, a girl is a success) in a fixed number, *n*, of trials, is called a *binomial random variable*. We will study this particular, important random variable in Section 3.3 of this chapter. Dr. Arbuthnott is believed to have been the first to scientifically study the distribution of a binomial random variable and to use it in testing a hypothesis about nature. Hypothesis testing is the topic of Chapter 6. Later on, we will use Dr. Arbuthnott's actual data set reported in his article on births in London in the 1600s and early 1700s. Let's now look at another example.

Figure 3.2 shows the sample space for the experiment of rolling two dice. As can be seen from the sample space, the probability of every pair of outcomes is 1/36. This can be seen from the fact that, by the independence of the two dice, for example: P (6 on Red Die $\cap$ 5 on Green Die) = P (6 on Red Die) $\times$ P (5 on Green Die) = (1/6) $\times$ (1/6) = 1/36, and that this holds for all 36 pairs of outcomes.

EXAMPLE 3.1

Sample Space for Two Dice   **FIGURE 3.2**

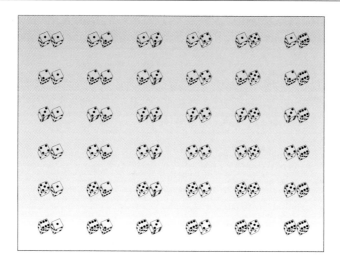

Let X = the sum of the dots on the two dice. Figure 3.3 shows the correspondence between sets in our sample space and the values of X. The probability distribution of X is given in Table 3.2.

The probability distribution allows us to answer various questions about the random variable of interest. Draw a picture of this probability distribution. Such a graph need not be a histogram, used earlier, but can also be a bar graph or column chart of the probabilities of the different values of the random variable. Note from the graph you produced that the distribution of the random variable "the sum of two dice" is symmetric. The central value is $x = 7$, which has the highest probability, $P(7) = 6/36 = 1/6$. This is the *mode,* the most likely value. Thus, if you were to bet on one sum of two dice, the best bet is that the sum will be 7.

We can answer other probability questions, such as: What is the probability that the sum will be at most 5? This is $P(X \leq 5)$. Notice that to answer this question, we require the sum of all the probabilities of the values that are less than or equal to 5:

FIGURE 3.3   Correspondence between Sets and Values of X

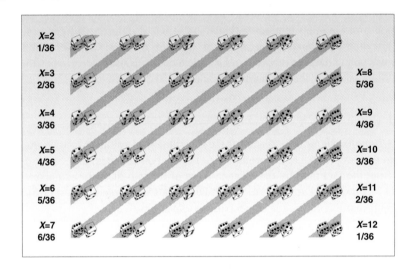

TABLE 3.2   The Probability Distribution of the Sum of Two Dice

x	$P(x)$
2	1/36
3	2/36
4	3/36
5	4/36
6	5/36
7	6/36
8	5/36
9	4/36
10	3/36
11	2/36
12	1/36
	36/36 = 1.00

$$P(2) + P(3) + P(4) + P(5) = 1/36 + 2/36 + 3/36 + 4/36 = 10/36$$

Similarly, we may want to know what the probability is that the sum is *greater than* 9. This is calculated as follows:

$$P(X > 9) = P(10) + P(11) + P(12) = 3/36 + 2/36 + 1/36 = 6/36 = 1/6$$

Most often, unless we are dealing with games of chance, there is no evident sample space. In such situations the probability distribution is often obtained from lists or other data that give us the relative frequency in the recorded past of the various values of the random variable. This is demonstrated in Example 3.2.

EXAMPLE **3.2**

The State of New Hampshire has been concerned over the large number of accidents that have occurred on Saturday nights along a stretch of the Daniel Webster Highway in the southern part of the state. Using data available from the state police, a probability distribution was constructed for the number of accidents that occur on a Saturday from 3:00 PM to midnight. Table 3.3 shows the probability distribution of the random variable X = number of accidents during a Saturday night.

Road Fatalities Most Common Saturday Night

By EARLE ELDRIDGE
GANNETT NEWS SERVICE

WASHINGTON—Almost half of all deaths on the road happen on Saturday between 3 p.m. and midnight, a state-by-state analysis of 1992 vehicle fatalities shows.

Reprinted from *USA Today,* August 9, 1993, p. A1.

The Probability Distribution of the Number of Accidents   TABLE **3.3**

x	P(x)
0	0.1
1	0.2
2	0.3
3	0.2
4	0.1
5	0.1
	1.00

FIGURE 3.4   The Probability Distribution of the Number of Accidents

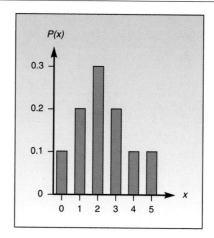

A plot of the probability distribution of this random variable is given in Figure 3.4. When more than two accidents occur on a given night, auxiliary police need to be called in to handle the problems. What is the probability that on a given Saturday night the auxiliary force would have to be called?

$$P(X > 2) = P(3) + P(4) + P(5) = 0.2 + 0.1 + 0.1 = 0.4$$

which is a 40 percent chance.

What is the probability that at least one accident will occur on a Saturday night? $1 - P(0) = 0.9$, a high probability, unfortunately.

Discrete and Continuous Random Variables

Refer back to Example 3.2. Notice that when an accident occurs, the number of accidents, X, jumps up by one. It is impossible to have half an accident or 0.13278 of an accident. The same is true for the number of dots on two dice (you cannot see 2.3 dots, or 5.87 dots, etc.), and of course the number of girls in four births.

> A **discrete random variable** can assume at most a countable number of values.

The values of a discrete random variable do not have to be positive whole numbers, they just have to "jump" from one possible value to the next without being able to have any value in between. The price of a stock, for example, is a discrete random variable because it is reported to the nearest eighth of a dollar: 17 1/8, 17 2/8, 17 3/8, etc. As another example, the amount of money you make on an investment may be $500 or it may be a loss: −$200. At any rate, it can be measured at best to the nearest *cent,* so this variable is discrete.

Discrete and Continuous Random Variables   **FIGURE 3.5**

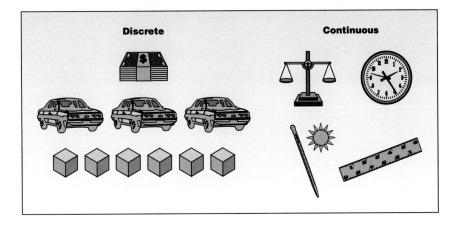

What are continuous random variables, then?

> A **continuous random variable** may take on any value in an interval of numbers (i.e., its possible values are uncountably infinite).

The values of continuous random variables can be measured (at least in theory) to any degree of accuracy. They move *continuously* from one possible value to the next, without having to jump. For example, temperature is a continuous random variable, since it can be measured as $72.00340981136\ldots°$. Weight, height, and time are other examples of continuous random variables.

The difference between discrete and continuous random variables is illustrated in Figure 3.5. Is wind speed a discrete or a continuous random variable?

PROBLEMS

1. The number of telephone calls arriving at an exchange during any given minute between noon and 1:00 PM on a weekday is a random variable with the following probability distribution:

x	$P(x)$
0	0.3
1	0.2
2	0.2
3	0.1
4	0.1
5	0.1

Verify that $P(x)$ is a probability distribution. (What must be the sum of all the probabilities $P(x)$?)

Use the probability distribution to find the probability that between 12:34 and 12:35 PM more than two calls will arrive at the exchange.

2. Typing errors per page for a certain typing pool are known to follow the probability distribution shown here:

x	P(x)
0	0.01
1	0.09
2	0.30
3	0.20
4	0.20
5	0.10
6	0.10

   a. Verify that $P(x)$ is a probability distribution.
   b. Find the probability that at most four errors will be made on a page.
   c. Find the probability that at least two errors will be made on a page.
3. Is the temperature on Mars a discrete or a continuous random variable?
4. Is the amount of gas it takes to fill up your tank after a trip a discrete or a continuous random variable?
5. Is the volume of the juice in a bottle a discrete or a continuous random variable?
6. Is the time you wait in line at the bank a discrete or a continuous random variable?
7. The probability distribution of the number of people who congregate at Times Square in New York for New Year's Eve (to the nearest 100,000) is as follows:[1]

x (in 1,000s)	P(x)
300	0.5
400	0.3
500	0.2

   a. Show that $P(x)$ is a probability distribution.
   b. Find the probability that at least 400,000 people will show up at Times Square next New Year's Eve.
8. The number of intercity shipment orders arriving daily at a transportation company is a random variable X with the following probability distribution:

x	P(x)
0	0.1
1	0.2
2	0.4
3	0.1
4	0.1
5	0.1

   a. Verify that $P(x)$ is a probability distribution.
   b. Use the probability distribution to find the probability that anywhere from one to four shipment orders will arrive on a given day.
   c. When more than three orders arrive on a given day, the company incurs additional costs due to the need to hire extra drivers and loaders. What is the probability that extra costs will be incurred on a given day?

---

[1]Based on data in "New Year's Eve Party," *The New York Times,* January 1, 1993, p. B1.

d. Assuming that the number of orders arriving on different days are independent of each other, what is the probability that no orders will be received over a period of five working days?

9. Returns on investments overseas, especially in Europe and the Pacific Rim, are expected to be higher than those of American markets in the 1990s, and analysts are now recommending investments in international portfolios.[2] One investment consultant believes that the probability distribution of returns (in percent per year) on one such portfolio is as given below:

x (%)	P(x)
9	0.05
10	0.15
11	0.30
12	0.20
13	0.15
14	0.10
15	0.05

a. Verify that $P(x)$ is a proper probability distribution.
b. What is the probability that returns will be at least 12 percent?

10. The number of defects in a machine-made product is a random variable, X, with the following probability distribution:

x	P(x)
0	0.1
1	0.2
2	0.3
3	0.3
4	0.1

a. Show that $P(x)$ is a probability distribution.
b. Find the probability $P(1 < X \leq 3)$.
c. Find the probability $P(1 \leq X \leq 4)$.

11. One six-sided die and one eight-sided die are thrown. Find the probability distribution of the sum of the dots appearing on the two dice.

12. What is the probability distribution of the sum of three six-sided dice?

13. The number of colonies of an organism appearing on a Petri dish in a biological experiment is random. Five percent of the time no colonies will appear; 15 percent of the time one will appear; 22 percent of the time two colonies will appear; 38 percent of the time three colonies will appear; otherwise the number of colonies will be four. Write the probability distribution for the random variable "number of colonies." Use the distribution to determine the probability that anywhere between one and three colonies will appear in a given experiment.

14. Internet is an organization of interlinked computer networks spanning the world. It includes educational institutions, private companies, government agencies, and more. One of Internet's functions is to facilitate the transmission of computer messages (e-mail) throughout the world. The number of messages transmitted on link BBN in western New England between midnight and 5:00 AM in the middle of the week is a random variable with the following probability distribution:

[2]"Your Money," *Working Woman,* March 1991, p. 58.

x	P(x)
0	0.01
1	0.05
2	0.10
3	0.14
4	0.18
5	0.25
6	0.15
7	0.07
8	0.04
9	0.01

a. What is the single most likely number of transmitted messages during this time?

b. Draw a graph of the probability distribution. Is it symmetric?

c. A computer buffer area is activated if more than five messages are to be transmitted overnight. What is the probability that the buffer will be activated on a given night?

15. In a certain card game similar to Blackjack, each face card is worth 10 points and an ace is worth 11 points. The rest of the cards have their stated value as usual. What is the probability distribution of the value of a single card drawn from a well-shuffled deck of the usual 52 cards?

3.2 EXPECTED VALUES OF DISCRETE RANDOM VARIABLES

In Chapter 1, we discussed summary measures of data sets. The most important summary measures discussed were the *mean* and the *variance* (and also the square root of the variance, the *standard deviation*). We saw that the mean is a measure of *centrality,* or location, of the data or population, and that the variance and the standard deviation measure the *variability,* or spread, of our observations.

Like data sets, random variables also have their associated measures of centrality and variability. The mean of the probability distribution of a random variable is called the *expected value* of the random variable (sometimes called the *expectation* of the random variable). The reason for this name is that the mean is the average value of the random variable, and therefore it is the value we "expect" to occur. We denote the mean by two notations: μ for *mean* (as in Chapter 1 for a population) and $E(X)$ for *expected value of X.* In situations where no ambiguity is possible, we will often use μ. In cases where we want to stress the fact that we are talking about the expected value of a particular random variable (here, X), we will use the notation $E(X)$.

The **expected value** of a discrete random variable X is equal to the sum of all values of the random variable, each value multiplied by its probability:

$$\mu = E(X) = \mathrm{Sum}[xP(x)]$$

Suppose a coin is tossed. If it's heads you win a dollar, but if it's tails you lose a dollar. What is the expected value of this game? Intuitively, you know you have an even chance of winning or losing the same amount, so the average or expected value

---

Americans' Life Expectancy Rises to 75.5 Years

WASHINGTON, Aug. 31 (AP)—Life expectancy at birth for Americans in 1991 rose by one-tenth of a year to a record 75.5 years, the Government reported today.

"Women currently are expected to outlive men by an average of 6.9 years, and white persons are expected to outlive black persons by an average of 7 years," the National Center for Health Statistics said in its monthly vital statistics report.

Life expectancy at birth was 78.9 years for women and 72 years for men. It was 79.6 for white females; 73.8 for black females; 72.9 for white males, and 64.6 for black males.

The infant mortality rate of 8.9 deaths per 1,000 live births in 1991 was a record low, as previously reported. The mortality rate for black infants was 17.6 deaths, or more than double the rate of 7.3 deaths among white infants.

A person who turned 50 years old in 1991 could expect to live on average to 79.2 years; a person who turned 65 that year could expect to live to 82.4 years.

The number of accidental deaths, including auto fatalities, has been declining since 1968. The Health and Human Services Secretary, Donna E. Shalala, said the trend showed that prevention works. "The use of safety belts and other safety devices has been on the rise, and we're now seeing how effective injury control programs and highway design can be," she said.

Reprinted by permission of *Associated Press*, September 1, 1993, p. A13.

is zero. Your payoff from this game is a random variable, and we find its expected value from the equation $E(X) = (1)(1/2) + (-1)(1/2) = 0$. The definition of an expected value, or mean, of a random variable thus conforms with our intuition. Incidentally, games of chance with an expected value of zero are called *fair games*.

The idea of the mean, or average, of a random quantity as an expected value is demonstrated in the article on life expectancy.

Return now to the topic with which we opened this chapter. What is the expected number of girls in four births? Using the definition of expected value, we get: $E(X) = 0(1/16) + 1(4/16) + 2(6/16) + 3(4/16) + 4(1/16) = 32/16 = 2$.

Are you surprised by the result? You shouldn't be. We saw in Figure 3.1 that the distribution is symmetric, with the center at 2. The mean, the expected value, is the fulcrum on which the distribution histogram or bar graph balances. The answer also makes intuitive sense. The chance of giving birth to a girl is 50 percent; in four trials—on the *average* (the *mean* is the average)—we expect two girls.

Now let's compute the expected sum of dots on two dice in Example 3.1. Notice again that the distribution is symmetric, centered on the number 7. Thus the mean is 7. We can compute it using a table format, writing the distribution and adding a column for the products $xP(x)$. At the bottom of that column (its sum) we will find the expected value, the mean (see Table 3.4).

Let's also return to Example 3.2 and find the expected value of the random variable involved—the expected number of accidents on a Saturday night. As seen from the computation in Table 3.5, the mean number of accidents on this highway on a Saturday night is $E(X) = 2.3$. This demonstrates an interesting property of means of random variables: The mean does not have to be one of the actual values the random variable can take. There are never 2.3 accidents. This value, however, is the *average*. We expect an average of 2.3 accidents per Saturday night.

TABLE 3.4   Computing the Expected Sum of Dots on Two Dice

x	$P(x)$	$xP(x)$
2	1/36	2/36
3	2/36	6/36
4	3/36	12/36
5	4/36	20/36
6	5/36	30/36
7	6/36	42/36
8	5/36	40/36
9	4/36	36/36
10	3/36	30/36
11	2/36	22/36
12	1/36	12/36
		252/36 = 7

TABLE 3.5   Computing the Expected Number of Accidents on a Saturday Night

x	$P(x)$	$xP(x)$
0	0.1	0
1	0.2	0.2
2	0.3	0.6
3	0.2	0.6
4	0.1	0.4
5	0.1	0.5
	1.00	2.3 ← Mean, $E(X)$

FIGURE 3.6   The Mean of a Discrete Random Variable as a Center of Mass

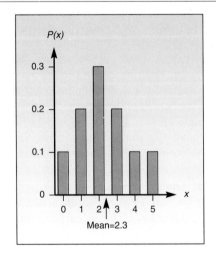

Figure 3.6 demonstrates the mean as the center of mass (the fulcrum) for the distribution of the number of accidents on a Saturday night. (Note that here the distribution is not symmetric.)

Having quantified the center of the distribution of a random variable, we'll now consider the spread of the distribution. The spread is often measured using either the variance or the standard deviation.

The Variance and the Standard Deviation of a Random Variable

The variance of a random variable is the expected squared deviation of the random variable from its mean. The idea is similar to that of the variance of a data set or a population, defined in Chapter 1. Probabilities of the values of the random variable are used as weights in the computation of the expected squared deviation from the mean of a discrete random variable. As with a population, we denote the variance of a random variable by σ^2. Another notation for the variance of X is $V(X)$.

The **variance of a discrete random variable** X is given by the following formula:

$$\sigma^2 = V(X) = E[(X - \mu)^2] = \text{Sum } (x - \mu)^2 P(x)$$

According to this formula, to compute the variance of a random variable, we subtract the mean μ from each value x of the random variable, square the result, multiply it by the probability $P(x)$, and finally add the results for all x.

Let's apply the formula to Example 3.2, to find the variance of the number of accidents on a Saturday night:

$$\sigma^2 = \text{Sum } (x - \mu)^2 P(x)$$

$$= (0 - 2.3)^2(0.1) + (1 - 2.3)^2(0.2) + (2 - 2.3)^2(0.3)$$
$$+ (3 - 2.3)^2(0.2) + (4 - 2.3)^2(0.1) + (5 - 2.3)^2(0.1)$$

$$= 2.01$$

There is, however, a frequently easier way to compute the variance of a discrete random variable. It can be shown mathematically that the formula is equivalent to the following computational form of the variance.

Computational formula for the variance of a random variable:

$$\sigma^2 = V(X) = E(X^2) - [E(X)]^2$$

This formula is a shortcut, and will be the one we use in practice. It states that the variance of a random variable X is equal to the mean of the random variable X^2, minus the squared mean of X.

Now we're ready to compute the variance of the random variable in Example 3.2. This is done in Table 3.6. The first column in the table gives the values of X, the

TABLE 3.6 Computations Leading to the Variance of the Number of Accidents on a Saturday Night Using the Shortcut Formula

x	$P(x)$	$xP(x)$	$x^2P(x)$
0	0.1	0	0
1	0.2	0.2	0.2
2	0.3	0.6	1.2
3	0.2	0.6	1.8
4	0.1	0.4	1.6
5	0.1	0.5	2.5
	1.00	2.3 ← Mean of X	7.3 ← Mean of X^2

second column gives the probabilities of these values, the third column gives the products of the values and their probabilities, and the fourth column is the product of the third column and the first (because we get $x^2P(x)$ by just multiplying each entry $xP(x)$ by x from column one). At the bottom of the third column we find the mean of X, and at the bottom of the fourth column we find the mean of X^2. Finally, we perform the subtraction $E(X^2) - [E(X)]^2$ to get the variance of X:

$$V(X) = E(X^2) - [E(X)]^2 = 7.3 - (2.3)^2 = 2.01$$

This is the same value we found using the other formula for the variance. Note that the shortcut formula holds true for *all* random variables, discrete or otherwise. Once we obtain the expected value of X^2 and the expected value of X, we can compute the variance of the random variable using this equation.

For random variables, as for data sets or populations, the standard deviation is equal to the (positive) square root of the variance. We denote the standard deviation of a random variable X by σ or by $SD(X)$

Formula for the **standard deviation of a random variable:**
$$\sigma = SD(X) = \sqrt{V(X)}$$

In Example 3.2, the standard deviation is: $\sigma = \sqrt{2.01} = 1.418$.

What are the variance and the standard deviation, and how do we interpret their meaning? By definition, the variance is the average squared deviation of the values of the random variable from their mean. Thus, it is a measure of the *dispersion* of the possible values of the random variable about the mean. The variance gives us an idea of the variation or uncertainty associated with the random variable: the larger the variance, the further away from the mean are possible values of the random variable. Since the variance is a squared quantity, it is often more useful to consider its square root—the standard deviation of the random variable. The standard deviation is measured in the original units of the problem and is thus more easily interpreted. When comparing two random variables, the one with the larger variance (standard deviation) is the more variable one. The risk associated with an investment is often measured by the standard deviation of investment returns. When comparing two investments with the same average (*expected*) return, the investment with the higher standard deviation is considered riskier (although a higher standard deviation implies that returns are expected to be more variable—both below *and above* the mean).

Some Properties of Means and Variances

optional

▲ An investor has income from two sources. Investment X gives her a random amount per month, with a certain probability distribution and mean $E(X) = \$350$. Investment B also gives her a random amount, Y, which has a different probability distribution, with mean $E(Y) = \$200$. What is the expected monthly income from both sources? The answer is very simple: $\$350 + \$200 = \$550$. The reason for this is a rule we have for means that says $E(X + Y) = E(X) + E(Y)$, which in this case is equal to $\$350 + \$200 = \$550$.

Example 3.1 demonstrates this property well. The expected value of the sum of the dots on two dice was 7. But let's look at each die *separately*. What is the expected number on a single die? Each of the six numbers 1 to 6 has probability 1/6. The mean, therefore, is found as follows:

$$(1)(1/6) + (2)(1/6) + (3)(1/6) + (4)(1/6) + (5)(1/6) + (6)(1/6) = 3.5$$

This should be intuitively clear since 3.5 is the *center* of the 6 equally likely points. Now, what is the mean of the sum of the two dice? By the rule above, it is just the sum of the two means for the two dice: $3.5 + 3.5 = 7$.

The variance of the sum (or the difference) of two random variables is equal to the sum of the two variances *if* the two random variables are *independent*. If the variance of investment X is 84 and the variance of investment Y is 60, *and the two investments are independent of each other,* then the variance of the total income from both sources is $V(X) + V(Y) = 84 + 60 = 144$. This means that the standard deviation of the total income is the square root of this variance, which is $12.

The mean of any **linear function** of a random variable is just the linear function of the mean of the random variable. What is a linear function? For example, $5X$ is a linear function of the random variable X. So is $2 + X$, and so is $2 + 5X$. All three are lines in the plane where X is the horizontal variable. ($2 + 5X$ is then a line with slope 5 and intercept 2.)

Cost is a linear function of the number of units made. The intercept would then be the fixed cost of the operation, while the slope would be the cost per item produced. Suppose that the fixed cost for producing tires is $10,000 and the variable cost is $20 per tire. Thus, if X tires are made per month, the total cost is $\$10,000 + 20(X)$. The number of tires made per month is a random variable with mean 1,200 and variance 500. What is the mean monthly cost? The rule for the mean of a linear function $aX + b$, where a is the slope and b the intercept, is: $E(aX + b) = aE(X) + b$. Here, therefore, $E(\text{cost}) = 10,000 + 20E(X) = 10,000 + (20)(1,200) = \$34,000$. This property of expected values saves us a lot of computation. We do not have to compute $aX + b$ for each value and then multiply by the probabilities and add. If we know $E(X)$, we simply plug it into the equation.

The variance of a linear function $aX + b$ is equal to $a^2V(X)$. The number multiplying the random variable (the slope of the line) gets squared, then multiplied by the variance of X. In our tire example, since the variance of the number of tires made per month is 500, the variance of the cost per month is $(20)^2(500) = 200,000$; so the standard deviation of the cost is $447.21.

16. What is the expected number of telephone calls in any given minute between noon and 1:00 PM using the probability distribution in Problem 1? What are the variance and the standard deviation of this distribution?

17. What are the mean and the variance of the distribution of typing errors given in Problem 2?

18. What is the average number of people attending the New Year's Eve party at Times Square in New York? (Use the information in Problem 7.)

19. What are the mean and the standard deviation of the number of intercity shipment orders per day based on the probability distribution given in Problem 8?

20. What is the standard deviation of the number of girls in four births?

21. What is the standard deviation of the sum of the dots on two six-sided dice?

22. What is the expected sum of three six-sided dice?

23. What are the mean and the standard deviation of rates of return on overseas investments using the distribution in Problem 9?

24. Based on the information in Problem 10, what is the expected number of defects in a machine-made product?

25. What is the expected sum of the dots on a six-sided die and an eight-sided die?

26. Based on the information in Problem 13, what is the average number of colonies?

27. Using the information in Problem 14, what is the average number of transmitted messages per night?

28. What is the expected value of a card drawn from a regular 52-card deck where each face card is worth 10 points, an ace 11 points, and all other cards are worth their stated number?

▲ 29. The width of a band of material to be cut by machine is 2 inches. The length of a piece of this band is a random variable with mean 5 inches and standard deviation 1 inch. What is the expected area of the band cut by this machine? What is the standard deviation of the area of the band cut by this machine?

▲ 30. At a certain casino, the entrance fee is $15. Once inside, you play a game where you will get $5 times the sum of the dots that will appear on a single roll of two dice. (You can only play once.) What is the expected value of this visit to the casino?

3.3 THE BINOMIAL DISTRIBUTION

Recall yet again the situation of how many girls will be born out of four births. This concerns a sequence of four trials (each birth here is, statistically, a trial), where the probability of success (a girl) is equal in all four trials (0.50 in every trial) and the trials are independent of each other. At the beginning of this chapter, we derived the distribution of this random variable using the sample space.

This experiment, seeing how many successes we can obtain in a sequence of a fixed number of trials, can be generalized. That is, we would not have to compute the probability distribution every time. This experiment is so common, in fact, that its distribution has a special name: the **binomial distribution.** Any random variable that has the following properties is a binomial random variable:

1. Each trial has two possible outcomes, called *success* and *failure.* The two outcomes are mutually exclusive and exhaustive (this means that one outcome, and only one of the two, must occur).

2. The probability of success, denoted by p, remains constant from trial to trial. The probability of failure is denoted by q, where $q = 1 - p$.

3. The n trials are independent. That is, the outcome of any trial does not affect the outcomes of other trials.

As statistical terms, *success* and *failure* do not necessarily have their everyday meaning attached. In fact, when dealing with defective items in a production process, we may define the outcome "item is defective" as a success. Success refers to the occurrence of a particular event (e.g., "item is defective"), and failure refers to the nonoccurrence of the event.

The following are some examples of other binomial random variables:

1. A coin is tossed eight times. Let *H* denote the number of times that heads appear.
2. It is known that 30 percent of the people in a given city prefer to use public transportation. A random sample of 20 people is selected. Let *T* be the number of people in the sample found to prefer public transportation.
3. It is known that 15 percent of the items produced by a machine are defective. A random sample of 12 items is chosen. Let *D* be the number of defective items found in the sample.

In the last two cases, we assume that the population from which drawings are made is very large so that after each one is selected the proportion of people who use public transportation, or items that are defective, remains approximately constant.

What characterizes a binomial distribution is this constancy of the probability of success from trial to trial, and the independence of the trials from one another. Clearly, the distribution of a binomial random variable depends on two parameters: the number of independent trials, *n*; and the constant probability of success in a single trial, *p*. Therefore, we will denote a binomial random variable as: $B(n, p)$. Our opening example, where the random variable *X* is the number of girls in four births, is thus $B(4, 0.5)$. In the first example above, where a coin is tossed eight times and the random variable *H* counts the number of heads, this random variable is $B(8, 0.5)$. In the second example above, *T* is $B(20, 0.3)$; and in the third, the random variable *D* is $B(12, 0.15)$. Why?

In his article published in 1710, reprinted at the beginning of this chapter, Dr. John Arbuthnott appears to have derived the binomial probability distribution. Arbuthnott extended the idea of the number of successes—the number of girls (or boys) in a given number of births for *any* number of births: 2, 3, 4, . . . , 100, and so on. The *probability model* for such an experiment is the same, whatever the number of trials.

The general formula for computing the binomial probabilities, the mathematical rule that gives us the probability distribution, is as follows:

Formula for the **binomial probability distribution**:

$$P(x) = \frac{n!}{x!(n-x)!} p^x q^{n-x}$$

where

p = the probability of success in a single trial
$q = 1 - p$
n = the number of trials
x = the number of successes
$n!$ = *n*-factorial, which is $n(n-1)(n-2) \ldots 1$

An example of $n!$ is $4! = 4 \times 3 \times 2 \times 1 = 24$. (Note also that we define $0! = 1$.)

You may experiment with the formula and verify the probabilities for any number of girls in four births. However, we will not use the formula since the binomial distribution has been tabulated and it is much easier to use the table of its values than to compute using the formula. Furthermore, computer packages will give us values of binomial probabilities as well. Both of these methods will be illustrated in our examples.

The Binomial Table

Appendix B, Table B.1 gives us binomial probabilities for various values of n, the number of trials, and p, the probability of success in a single trial. Part of that table is reproduced in Table 3.7.

TABLE 3.7   Binomial Probabilities for a Few Values of n and p

						p				
n	x	.10	.15	.20	.25	.30	.35	.40	.45	.50
2	0	.8100	.7225	.6400	.5625	.4900	.4225	.3600	.3025	.2500
	1	.1800	.2550	.3200	.3750	.4200	.4550	.4800	.4950	.5000
	2	.0100	.0225	.0400	.0625	.0900	.1225	.1600	.2025	.2500
3	0	.7290	.6141	.5120	.4219	.3430	.2746	.2160	.1664	.1250
	1	.2430	.3251	.3840	.4219	.4410	.4436	.4320	.4084	.3750
	2	.0270	.0574	.0960	.1406	.1890	.2389	.2880	.3341	.3750
	3	.0010	.0034	.0080	.0156	.0270	.0429	.0640	.0911	.1250
4	0	.6561	.5220	.4096	.3164	.2401	.1785	.1296	.0915	.0625
	1	.2916	.3685	.4096	.4219	.4116	.3845	.3456	.2995	.2500
	2	.0486	.0975	.1536	.2109	.2646	.3105	.3456	.3675	.3750
	3	.0036	.0115	.0256	.0469	.0756	.1115	.1536	.2005	.2500
	4	.0001	.0005	.0016	.0039	.0081	.0150	.0256	.0410	.0625
5	0	.5905	.4437	.3277	.2373	.1681	.1160	.0778	.0503	.0312
	1	.3280	.3915	.4096	.3955	.3602	.3124	.2592	.2059	.1562
	2	.0729	.1382	.2048	.2637	.3087	.3364	.3456	.3369	.3125
	3	.0081	.0244	.0512	.0879	.1323	.1811	.2304	.2757	.3125
	4	.0004	.0022	.0064	.0146	.0284	.0488	.0768	.1128	.1562
	5	.0000	.0001	.0003	.0010	.0024	.0053	.0102	.0185	.0312

TABLE 3.8   The Binomial Probability Distribution of H, the Number of Heads Appearing in Five Tosses of a Fair Coin

x	$P(x)$
0	$(5!/0!5!)(0.5)^0(0.5)^5 = 0.031$
1	$(5!/1!4!)(0.5)^1(0.5)^4 = 0.156$
2	$(5!/2!3!)(0.5)^2(0.5)^3 = 0.313$
3	$(5!/3!2!)(0.5)^3(0.5)^2 = 0.313$
4	$(5!/4!1!)(0.5)^4(0.5)^1 = 0.156$
5	$(5!/5!0!)(0.5)^5(0.5)^0 = 0.031$

$$1.000$$

Notice in the column $p = 0.50$, and the set of rows corresponding to $n = 4$, that the probabilities for the different number of girls in four births ($x = 0, 1, 2, 3, 4$) correspond to the probabilities we computed in the first part of this chapter. Table 3.8 shows how the probabilities for the number of heads appearing in five tosses of a fair coin are obtained from the binomial formula. These probabilities are in the binomial table in column $p = 0.50$ and set of rows for $n = 5$.

EXAMPLE 3.3

NEWSWEEK POLL

**Do you have a
favorable opinion
of Hillary Clinton?**

**60% Favorable
20% Unfavorable**

Reprinted from *Newsweek,* February 8, 1993, p. 24.

Suppose that five people are randomly selected from the entire population involved in the accompanying *Newsweek* poll. What is the probability that three or more of them have an unfavorable opinion of Hillary Clinton?

SOLUTION

In Table 3.7, look at the column $p = 0.20$ and the rows corresponding to $n = 5$. We see that for $x = 3$, the probability is 0.0512. This is the probability $P(X = 3)$ for a random variable that is $B(5, 0.20)$. Similarly, under $x = 4$ we find $P(X = 4) = 0.0064$, and under $x = 5$ we find $P(X = 5) = 0.0003$. What we want is $P(X \geq 3) = P(X = 3) + P(X = 4) + P(X = 5) = 0.0512 + 0.0064 + 0.0003 = 0.0579$.

EXAMPLE 3.4

According to new research, chicken collagen, a protein found in the joints of chickens, may help cure rheumatoid arthritis. *The New York Times* reported that 28 people were given this particular protein.[3] Assume that the probability that a single patient will be helped is 0.35, what is the probability that at most 10 patients out of the 28 given the treatment would be helped?

SOLUTION

Assuming patients are independent of each other, the number helped by the collagen is a binomial random variable $B(28, 0.35)$. Instead of a table, we will use MINITAB.

The MINITAB command CDF gives us the *cumulative distribution function of a random variable.* This is the sum of the probabilities of the values of the random variable *from the smallest possible value up to the given value.* The CDF is defined as:

---

[3]Gina Kolata, "New Rheumatoid Arthritis Therapy Shows Promise," *The New York Times,*
September 24, 1993, p. A2.

$$\text{CDF}(x) = P(X \le x) = \text{Sum of } P(i) \text{ for all } i \le x$$

That is, CDF(10) would give us the *sum* of the probabilities $P(X = 0) + P(X = 1) + P(X = 2) + \ldots + P(X = 9) + P(X = 10)$. This is exactly what we want in this example.

The CDF command requires the subcommand specifying the actual distribution, here BINOMIAL. The exact required set of commands, and the output, are shown here:

```
MTB > CDF 10;
SUBC > BINOMIAL 28 0.35.
   K P (X LESS OR = K)
   10.00      0.6160
```

Thus the required probability is 0.616. Now, suppose we wanted to know the probability that 22 or more of the 28 people in the study would be helped by this new cure if the probability that a single patient is helped is 0.35. We know that the sum of all the probabilities of $B(28, 0.35)$ is 1.00, and the computer can give us the cumulative probabilities for any value. So let's get CDF(21) and subtract it from 1.00. This will give us the sum of the probabilities for 22 up to 28, which is what we need.

```
MTB > CDF 21;
SUBC > BINOMIAL 28 0.35.
   K P (X LESS OR = K)
   21.00      1.0000
```

In this case, within the accuracy of the reporting by the computer, the probability is extremely small ($1 - 1.0000 = 0$); that is, within four-decimal accuracy, the answer is zero. It is not likely that 22 or more arthritis patients will be helped if the chance that 1 patient is helped is only 0.35.

Courtesy of Grant Heilman
Photography, Inc./Larry Leferer.

EXAMPLE 3.5

Suppose that all 18 types of smiles are equally likely to occur, so that when you see someone smiling, the probability is $1/18 = 0.0555$ that the smile is the true uplifting kind. If 20 strangers (assumed independent) are smiling in an audience, what is the probability that no more than 3 of them are "really" smiling?

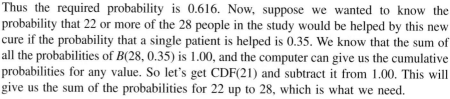

One Smile (Only One) Can Lift a Mood

By DANIEL GOLEMAN

As smiles are beginning to come under precise study, researchers report that not just any happy face will actually make the smiler feel happy. A polite smile, a smile of feigned enjoyment or a grin-and-bear-it grimace will not give the physiological lift that accompanies a genuine grin.

Only one of the 18 or so different kinds of smiles can activate the centers in the brain that regulate pleasant feelings, according to a report in the current issue of Psychological Science. And now researchers have found that even when artificially induced, the smile can produce the same brain changes that occur during spontaneous moments of joy and delight.

The work on smiles is part of a recent effort by scientists studying how emotions are controlled by the brain. "We don't yet know what specific parts of the brain are involved in each emotion," said Dr. Ekman. "We're gathering fundamental knowledge and showing there is a brain pathway that allows you to generate your own emotions."

Reprinted from *The New York Times*, October 26, 1993, p. C11.

Our distribution is $B(20, 0.0555)$, and we need $P(X = 0) + P(X = 1) + P(X = 2) + P(X = 3)$. This is exactly CDF(3). We can get it directly from MINITAB as follows:

```
MTB  > CDF 3;
SUBC > BINOMIAL 20 0.0555.
   K P (X LESS OR = K)
   3.00   0.9775
```

The approximate answer could be obtained using the binomial table with $n = 20$ and $p = 0.06$, which is the closest value in the table to the actual $p = 0.0555$. The answer from the table is 0.9710, which is very close to what we get using the computer.

Mean, Variance, and Shape of the Binomial Distribution

The mean of a binomial distribution is equal to the number of trials, n, times the probability of success in a single trial, p. The variance is equal to the number of trials times p, times q. These formulas and the standard deviation are as follows.

Mean of a binomial distribution:

$$\mu = E(X) = np$$

Variance of a binomial distribution:

$$\sigma^2 = V(X) = npq$$

Standard deviation of a binomial distribution:

$$\sigma = SD(X) = \sqrt{npq}$$

The expected value, the mean, is the long-term average of the random variable's value. It therefore seems very reasonable that the expected value of a binomial random variable should be equal to n times p. With four children, you *expect* an average of two girls. This is because $np = (4)(0.5) = 2$. Following Dr. Arbuthnott's observations on the numbers of boys and girls in any given number of births, the average number of girls in 100 births is $np = 100(0.5) = 50$, which accords with our intuition.

In Example 3.3, our random variable is $B(5, 0.2)$, and so the expected number of people with an unfavorable opinion of Hillary Clinton is $5(0.2) = 1$. The standard deviation is $\sqrt{5(0.2)(0.8)} = 0.894$.

In Example 3.4, the expected number of patients who will be helped by the treatment is $28(0.35) = 9.8$, and the standard deviation is $\sqrt{28(0.35)(0.65)} = 2.523$.

How many "real" smiles do you expect in Example 3.5? There are $np = 20(0.0555) = 1.11$—that is, 1 real smile out of 20 smiling faces. Find the standard deviation.

Refer yet again to the opening of this chapter, this time to Dr. Arbuthnott's data on the actual number of boys and girls born in the years 1629 to 1710. Since the probabilities of a boy and a girl are assumed equal, we expect in the long run roughly equal numbers of boys and girls born every year. We also expect that, since the number will very probably not be *exactly* equal, in about half of the years more boys will be born and in the other half more girls will be born. Thus, in 82 trial years, we would expect $np = 82(0.5) = 41$ years with more girls born than boys, and 41 years with more boys born than girls. Check how close reality is to this expected value. The result is very surprising.

Now, *assume* indeed that we have probability $p = 0.5$ that the number of boys will exceed the number of girls. Our random variable here is $B(82, 0.5)$. Given this assumption, what is the probability of the observed number of years (zero, see table) with girls outnumbering boys? Let's use MINITAB:

```
MTB  > CDF 0;
SUBC > BINOMIAL 82 0.5.
   K P (X LESS    OR = K)
   0.00        0.0000
```

When an observed result has a very small probability given an assumption that we have made, we must go back and seriously question (and even reject) that assumption. This is the key idea behind tests of hypotheses, the topic of Chapter 6. Here we observed more boys every one of the 82 years—this has an extremely small probability ("0.0000" in the MINITAB output) when indeed the probability that boys will outnumber girls is only 0.5. Hence we will reject our original assumption.

FIGURE 3.7   The Binomial Probability Distribution for Various Values of *n* and *p*

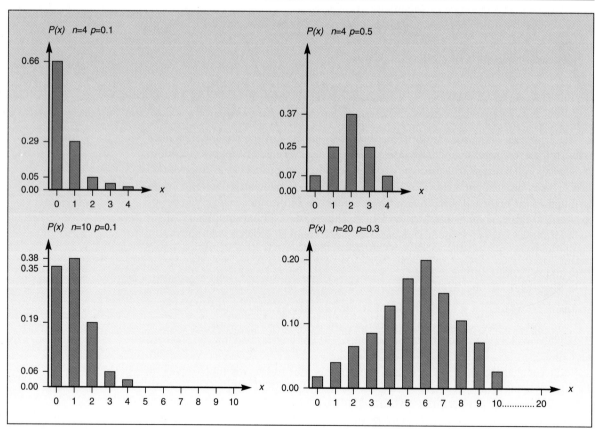

Reprinted from *Time,* February 1, 1993, p. 22.

Actually, it is well known today that slightly more boys than girls are born. Here, a probability distribution led us to this conclusion.

Now let's say we want to represent the binomial distribution graphically. The shape of the probability distribution of a binomial random variable is symmetrical when $p = 0.50$. For small samples, the distribution is skewed to the right when $p < 0.5$ and skewed to the left when $p > 0.5$. As n increases, the distribution becomes more and more symmetric. Figure 3.7 shows the binomial distribution for various values of n and p.

Sampling with or without Replacement

We noted earlier in this chapter that the binomial distribution is appropriate when we are sampling from a large population. As a rule of thumb, N/n should be at least 10. This is true for sampling without replacement: the sample is drawn and not returned into the population. If, on the other hand, we sample an item, note whether it is a success or a failure, and *return* it to the population before the next item is selected for the sample, we are sampling with replacement. When sampling with replacement, the population need not be larger than the sample for the binomial assumption to hold. The reason for this can be seen if we consider a simple case where the population consists of 10 items, 4 of which are successes and 6 failures. For the item first selected, $p = 0.4$. Once the item is returned to the population and has an equal chance of being selected again, the probability of success for the second item is also $p = 0.4$, and the two drawings are independent. Sampling with replacement is not frequently used and will not be discussed further in this book.

PROBLEMS

31. Under what conditions is the number of points scored by a player in a game a binomial random variable?

32. In a hospital ward, 3 out of 10 patients have high blood pressure. If 4 out of 10 are sampled with replacement, is the number of patients with high blood pressure a binomial random variable?

33. Use a computer to find the probability that a binomial random variable with $n = 12$ and $p = 0.21$ will have value of, at most, 7.

34. Use a computer as an aid in finding the probability that a binomial random variable with $n = 20$ and $p = 0.85$ will be equal to 15 or more.

35. With the aid of a table or a computer, find the probability that a binomial random variable with $n = 8$ and $p = 0.6$ will have a value between 3 and 7.

36. Use a table, the formula, or a computer to find the probability that a binomial random variable with $n = 15$ and $p = 0.4$ will have value less than or equal to 5.

37. Every week the Barbados Tourist Board interviews six randomly chosen vacationers on the island about their experience. In general, each vacationer's comments can be classified as mainly positive or mainly negative. The responses are then published in the *Visitor* newspaper. Suppose that only 5 percent of all visitors to Barbados are dissatisfied with their visit. What is the probability that at least two visitors out of the six interviewed will express mainly negative comments?

38. Assume that the data in the figure "Inactive Retirement" on page 138 reflect the entire population in question (rather than a random sample). If 10 retired people were selected at random in 1990, what is the probability that no more than 4 of them were sedentary? If six retired New Yorkers were selected at random in 1986, what is the probability that more than two of them were living without physical activity?

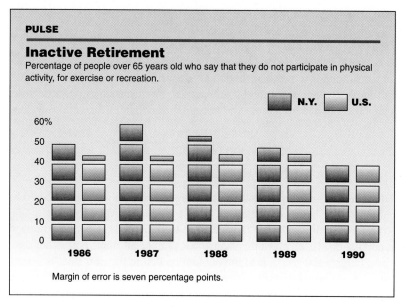

PULSE

Inactive Retirement

Percentage of people over 65 years old who say that they do not participate in physical activity, for exercise or recreation.

Margin of error is seven percentage points.

Reprinted from *The New York Times*, January 14, 1993, p. B1, with data from the *Centers for Disease Control and Prevention, Behavioral Risk Factor Survey.*

39. A new treatment for baldness is known to be effective in 70 percent of the cases treated. Four bald members of the same family are treated; let X be the number of successfully treated members of the family. Is X a binomial random variable? Explain.

40. In a given area, the probability that a person is exposed to an advertisement of a paper company is 0.20. If 10 people are randomly chosen from among the population in the area, what is the probability that at least 5 of them were exposed to the advertisement? What is the probability that at most 2 were exposed to the advertisement?

41. Records of a health insurance company show that 30 percent of its policyholders over 50 years of age submit a claim during the year. Fifteen policyholders over 50 years of age are selected at random. What is the probability that at least 10 will submit a claim during the coming year? What are the mean and standard deviation of the distribution?

42. In *acceptance sampling* in quality control, a lot is accepted only if the number of items that do not conform to production specifications out of a given total is no greater than some preset value. Suppose that out of a large collection of produced items, 20 percent are nonconforming. A random sample of 20 items is selected in an acceptance sampling scheme that requires that no more than 2 items out of 20 be nonconforming. What is the probability that the lot will be rejected?

43. At the post office closest to Graceland, 1 in 10 letters that arrive are addressed to Elvis. In a random sample of eight letters arriving at this post office, what is the probability that at least three are addressed to Elvis?

44. A recent front-page article in the *Boston Globe* implied that cloning living things has already progressed to the stage dramatized in the movie *Jurassic Park*.[4] Cloning is a difficult operation, however, and its success rate is limited. For laboratory animals under

---

[4]R. Saltus, "Embryo Duplication," *Boston Globe,* October 26, 1993, p. 1.

FOR A SONG

Uh-oh. Elvis Presley fans are putting his stamp on willfully misaddressed letters, which, of course, come back stamped *Return to Sender.* Reason: address unknown. No such number, no such zone.

Reprinted from *Time,* February 1, 1993, p. 22.

study it is known that the probability of a successful duplication is 9 percent. In a given experiment, 10 cloning attempts are made. Assuming independence of trials, what is the probability that at least a required minimum of four clones will be successfully attained? What is the mean number of successful clones in this experiment?

Survey on Adultery: "I Do" Means "I Don't"

WASHINGTON, Oct. 18 (AP)—A vast majority of married Americans do not cheat on their spouses, a sex researcher says, disputing claims that adultery is rampant.

About 15 percent of married or previously married Americans have cheated on a spouse, estimates the researcher, Tom W. Smith of the National Opinion Research Center at the University of Chicago. About 21 percent of men and nearly 13 percent of women in 1993 admitted to ever having cheated on a spouse.

Perhaps 3 to 4 percent of husbands and wives have a sexual partner outside their marriage in a given year, said Mr. Smith, director of the center's General Social Survey. About 1,400 people are interviewed in person each year for the national survey. Questions about sexual behavior are answered on a private, written questionnaire.

"There are probably more scientifically worthless 'facts' on extramarital relations than on any other facet of human behavior," Mr. Smith said today in a report at an American Enterprise Institute seminar here.

Mr. Smith said television talk shows and popular magazines commonly report much higher numbers of marital cheaters.

Reprinted by permission of *Associated Press,* October 19, 1993, p. A20.

45. On a popular TV show, when the host asked a "random sample" of eight women in her audience whether they had ever cheated on their husbands, three of them said yes. Using the information in the article about the survey, what is the probability of this event (three

yes answers) or any greater number of yes answers in a sample of eight? Carefully discuss your conclusions as well as the assumptions you made. (Hint: Is the audience of a TV show a random sample of the population at large?)

46.   Using the data in the article in Problem 45, and taking the higher value, 4 percent, for the percentage of men and women who have an affair in a given year, what is the expected number of people who fall into this category in a random sample of 1,000 from the general population?

47.   The probability of contracting malaria when living for one year in an infested area is 3 percent. If 15 people are sent by the Peace Corps to work in such an area of the world, what is the probability of at least two of them contracting the disease during their first year?

3.4 CONTINUOUS RANDOM VARIABLES

Instead of depicting probability distributions by simple graphs, where the height of the point above each value represents the probability of that value of the random variable, we'll use a histogram. We will associate the *area* of each rectangle of the histogram with the probability of the particular value represented. Let's look at a simple example. Let X be the time, measured in minutes, it takes to complete a given task. A histogram of the probability distribution of X is shown in Figure 3.8.

The probability of each value, which equals the area of the rectangle over the value, is written on top of the rectangle. Since the rectangles all have the same base, the height of each rectangle is proportional to the probability. Note that the probabilities add to 1.00, as required. Now suppose that X can be measured more accurately. The distribution of X, with time now measured to the nearest half minute, is shown in Figure 3.9.

Let's continue the process. Time is a *continuous random variable*; it can take on any value measured on an interval of numbers. We may therefore refine our measurement to the nearest quarter minute, the nearest five seconds, or the nearest second, or we can use even more finely divided units. As we refine the measurement scale, the number of rectangles in the histogram increases and the width of each rectangle decreases. The probability of each value is still measured by the area of the rectangle above it, and the total area of all rectangles remains 1.00, as required of all probability distributions. As we keep refining our measurement scale, the discrete distribution of X tends to a continuous probability distribution. The steplike surface formed by the tops of the rectangles in the histogram tends to a smooth function. This

115

FIGURE 3.8   A Histogram of the Probability Distribution of the Time to Complete a Task, with Time Measured to the Nearest Minute

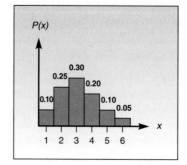

A Histogram of the Probability Distribution of Time to Complete a Task,
with Time Measured to the Nearest Half Minute   **FIGURE 3.9**

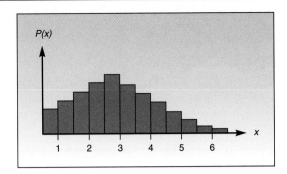

function is denoted by $f(x)$ and is called the *probability density function* of the
continuous random variable X.

The probabilities associated with a continuous random variable X are
determined by the **probability density function** of the random variable.
The function, denoted $f(x)$, has the following properties:

1. $f(x) \geq 0$ for all x.
2.   The probability that X will be between two numbers a and b is
   equal to the area under $f(x)$ between a and b.
3.   The total area under the entire curve of $f(x)$ is equal to 1.00.

Probabilities are still measured as areas under the function. The probability that the
task will be completed in two to three minutes is the area under $f(x)$ between the points
$x = 2$ and $x = 3$. Histograms of the probability distribution of X with our measurement
scale refined further and further are shown in Figure 3.10. Also shown is the density
function, $f(x)$, of the limiting continuous random variable X. The density function is
the limit of the histograms as the number of rectangles approaches infinity and the
width of each rectangle approaches zero.

For a continuous random variable, the probability of occurrence of any *given* value
is zero. We see this from property 2 of the above definition of the probability density
function, noting that the *area* under a curve between a point and itself is the area of a
line, which is zero. *For a continuous random variable, nonzero probabilities are
associated only with intervals of numbers.*

We now give an example of a continuous random variable. The simplest continuous
distribution is the *uniform* distribution over an interval. A continuous random variable
has the uniform distribution over an interval I if it is equally likely to be in one
subinterval of I as in any other subinterval of the same length. The density of a
uniform random variable over the interval $I = [0,5]$ is as follows:

$$f(x) = \begin{cases} 1/5 & \text{for } 0 \leq x \leq 5 \\ 0 & \text{otherwise} \end{cases}$$

FIGURE 3.10 Histograms of the Distribution of Time to Complete a Task as Measurement Is Refined to Smaller and Smaller Intervals of Time, and the Limiting Density Function, $f(x)$

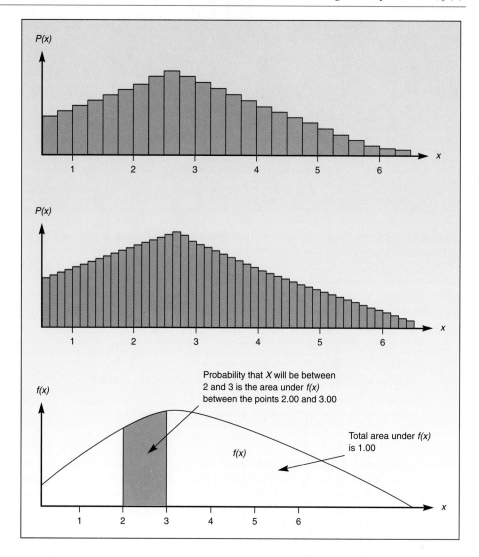

This density is shown in Figure 3.11. Note that the entire area under the curve is equal to the area of the rectangle with base 5 and height 1/5 and is therefore equal to 1.00, as required. Also, $f(x) \geq 0$ for all x. To find the probability that X will be between 1 and 3, $P(1 \leq X \leq 3)$, we need to compute the area under $f(x)$ between 1 and 3. This is the area of the rectangle with base $3 - 1 = 2$ and height 1/5. The probability is thus 2/5. This is also shown in the figure.

The expected value of a continuous random variable is the average value that would result from a repeated drawing of values of the random variable. It is the center of the "mass" associated with the area under the graph of the density function if we assume this area is covered by some material, such as cardboard, of uniform thickness. This is demonstrated in Figure 3.12. Using this intuitive principle, we find the mean of the uniform [0,5] distribution to be $E(X) = 2.5$.

The Uniform [0,5] Density **FIGURE 3.11**

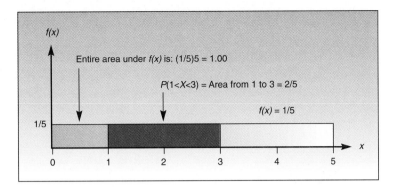

The Mean of a Continuous Distribution as the Center of the Mass under the Density Function **FIGURE 3.12**

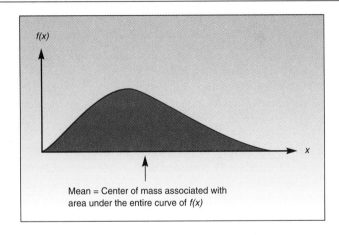

EXAMPLE 3.6

Many calculators have a function that generates random numbers. Typically, when the Rand # key is pushed, the calculator gives a number drawn from the uniform probability distribution [0, 1]. What is the probability that a random number that is generated this way will fall between 0.2 and 0.4?

SOLUTION

We will use MINITAB. The command CDF can be used to find probabilities for uniform continuous random variables. For *continuous random variables, the CDF is the area under the curve, from a given point all the way to the left.* By specifying CDF 0.2 for a uniform random variable between 0 and 1, we are asking for the area under the density from 0.2 all the way to 0. We then do the same for 0.4. Both are shown here:

```
MTB > CDF 0.2;
SUBC > UNIFORM 0 1.
   0.2000   0.2000
MTB > CDF 0.4;
SUBC > UNIFORM 0 1.
   0.4000   0.4000
```

FIGURE 3.13 _____ The Uniform [0,1] Density

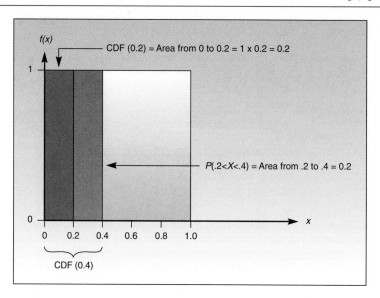

To find the probability that the uniform random variable will be between 0.2 and 0.4, we subtract the smaller area from the larger one: $0.4 - 0.2 = 0.2$, which is the answer. Figure 3.13 shows this situation.

Is it likely that a random number generator will give you eight consecutive numbers all between 0.2 and 0.4? Assuming independence, the answer is $(0.2)^8 = 0.00000256$, which is a very small number. If you should get eight such consecutive numbers, therefore, you may question whether the generator actually gives random numbers.

The MINITAB command CDF may of course be used to compute more difficult uniform distribution probabilities. Here is an example that computes the probability $P(X < 3)$ for a uniform [0,8] random variable:

```
MTB  > CDF 3;
SUBC > UNIFORM 0 8.
  3.0000    0.3750
```

Most continuous random variables with which we will work have density functions more complicated than the uniform density. Areas under the densities of these random variables, needed for probability calculations, are obtainable from tables such as the ones in Appendix C. Chapter 4 is devoted to the most important continuous distribution we will use: the normal distribution.

SUMMARY

In this chapter we introduced **random variables**—quantities that are determined by chance. We classified a random variable as **discrete** or **continuous,** depending on the type of values they may take. We defined the **probability distribution** of a random variable: the law that assigns probabilities to values of the random variable. We saw how the probabilities of events relevant to the random variable may be computed using the probability distribution. We defined the **mean,** also called the **expected value,** of a random variable, as well as its **variance** and **standard deviation.** We defined the expected value of a random variable as the mean of its probability distribution. We defined an important discrete probability distribution: the **binomial distribution,** and we saw how probabilities for a binomial random variable may be found in tables or computed directly. Finally, we introduced continuous random variables and saw that the probability distribution of a continuous random variable is its **density function.**

KEY FORMULAS

Mean of a discrete random variable:
$$\mu = \text{Sum}[xP(x)]$$

Standard deviation of a discrete random variable:
$$\sigma = \sqrt{E(X^2) - [E(X)]^2}$$

PROBLEMS

48. What is the main difference between a continuous random variable and a discrete one?
49. How do we obtain probabilities of intervals of numbers for a continuous random variable?
50. Let X be a continuous random variable. Explain why $P(X = a) = 0$ for any given number a.
51. Let X have the probability density function:

$$f(x) = \begin{cases} (1/2)x & \text{for } 0 \le x \le 2 \\ 0 & \text{otherwise} \end{cases}$$

   a. Sketch the probability density function.
   b. Show that $f(x)$ is a probability density function.
   c. Find the probability that X will be between 0 and 1.
52. Write the density of a uniform [0,7] random variable. (Hint: the total area under $f(x)$ must equal 1.00.)
53. Use the density of Problem 52 to find the probability that X will be between 0 and 2.
54. What is the expected value of the uniform [0,7] random variable of Problem 52?
55. For the random variable of Problem 52 find the probability that X will be greater than 1.5.

ADDITIONAL PROBLEMS

56. The incidence of Down's syndrome in babies has been found to have a strong relation to the age of the mother. Based on the information in a recent article, the probability distribution for the age of the mother of a baby who has Down's Syndrome has been derived.

Age	P(age)	Age	P(age)
20	0.003	41	0.032
25	0.004	42	0.050
30	0.005	43	0.067
35	0.009	44	0.085
36	0.011	45	0.110
37	0.013	46	0.140
38	0.016	47	0.180
39	0.020	48	0.230
40	0.025		

Data from M. D'Alton and A. DeCherney, "Prenatal Diagnosis," *New England Journal of Medicine,* January 14, 1993, pp. 114–18.

What is the average age of mothers of babies with Down's syndrome? What is the standard deviation of the age of these mothers?

57. A recent article on unconventional medicine in America implies the following probability distribution for the number of visits per year to an herbal medicine provider:

Number of Visits	Probability
0	0.80
1	0.04
2	0.04
3	0.03
4	0.03
5	0.02
6	0.01
7	0.01
8	0.01
9	0.01

Data from D. Eisenberg et al., "Unconventional Medicine in the United States," *New England Journal of Medicine,* January 28, 1993, pp. 246–52.

   a. Is this a proper probability distribution?
   b. What is the probability that a randomly chosen person goes to an herbal medicine provider at least six times a year?
   c. What is the average number of visits per year to an herbal medicine provider?

58. An article implies that 40 percent of all Mediterranean dishes are based on the "hot-cold" contrast.[5] If five Mediterranean dishes are selected at random, what is the probability that at least two are hot-cold?

59. A triangular continuous distribution is one where the density function looks like a triangle. If the length of the base of the triangle is 2, what must be its height?

60. The following is the probability distribution of the weight, in tons, of asteroids hitting earth:

Weight	Probability
5	0.4
6	0.3
7	0.2
8	0.1

Find the mean and the standard deviation of the weight of asteroids hitting earth.

61. Twenty percent of all plants in the Amazon are known to have medicinal value. If 10 plants are randomly selected, what is the probability that at least 2 have medicinal value?

---

[5]L. Grivetti, "Nutrition Past—Nutrition Today," *Nutrition Today,* May/June 1992, p. 15.

62.

Some Experts Discount Reports of Danger From Cellular Phones

Even Optimists Concede Long-Term Data Are Incomplete

Fears over cellular phones were first fanned last month when David Reynard of St. Petersburg, Fla., appeared on the Larry King program to discuss his lawsuit attributing his wife's fatal brain tumor to the cellular phone he had bought for her.

Since then, three others have come forward, contending that they also believe that their brain tumors were fostered by their cellular phones. As evidence, they said their tumors arose in the part of the brain nearest to where their telephone antennas pressed against their skulls, and they said they had used their phones hours every day.

Dr. David Perlmutter, a neurologist in Naples, Fla., who has cared for two of the patients, expressed alarm at the coincidence of the tumor sites.

Effects of Microwave Radiation

"I'm not blaming their brain tumors on cellular phones," he said. "I'm saying there's a possible danger here, and we need to open our eyes and demand proof of safety."

Cellular telephones pick up and transmit signals through their own antennas, which may be attached to the handset itself or are mounted on a car, sometimes within inches of a passenger's head. They operate at a radio frequency of 840 to 880 megahertz, or millions of cycles per second, which is at the lowest end of the microwave portion of the electromagnetic spectrum.

Biologists have long known that microwave radiation at very high power can cause severe heat damage to the body and cataracts. But cellular phones are far too weak to heat up tissue; the power coursing through them is equal to about one-tenth the wattage of a dim light bulb.

But when the antenna is near the head of a caller or a car passenger, it does not take much wattage to throw the radio frequencies into the brain, the body's most delicate tissue. What exactly happens to the molecules of brain tissue as they are bobbled and wiggled around 880 million times a second is unclear. And to be fair, even in the most avid user, phone radiation makes up only a small fraction of the brain's exposure to radio waves.

Noting the dearth of relevant data, Dr. Elizabeth Jacobson, acting director for the Center for Devices and Radiological Health at the Federal Food and Drug Administration, said that while the agency saw nothing to fear in cellular phones and no reason for people to stop using them, nonetheless, "We can't give a blanket assertion the phones are safe." She said the findings were incomplete and often contradicted one another.

Reprinted from *The New York Times*, February 2, 1993, p. C3, with data from Dr. A. W. Guy/University of Washington.

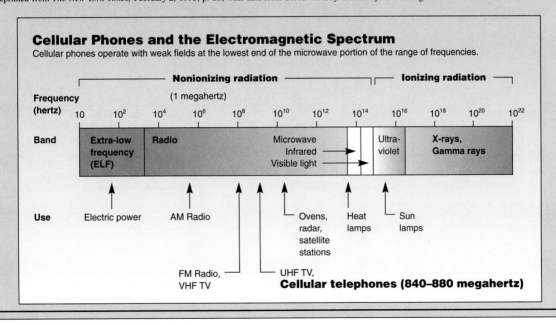

Cellular Phones and the Electromagnetic Spectrum
Cellular phones operate with weak fields at the lowest end of the microwave portion of the range of frequencies.

Determining the causes of cancer is a complex and difficult undertaking. We will not attempt it here. However, let us assume that the human brain can be divided into eight roughly equal parts. What is the probability that four brain-cancer patients (all heavy users of cellular phones) will all have their tumors occur in the same eighth of the brain? What are the assumptions we make here? Can any conclusion be drawn based on this information?

63.

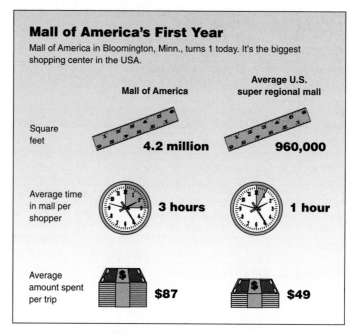

Mall of America's First Year

Mall of America in Bloomington, Minn., turns 1 today. It's the biggest shopping center in the USA.

	Mall of America	Average U.S. super regional mall
Square feet	4.2 million	960,000
Average time in mall per shopper	3 hours	1 hour
Average amount spent per trip	$87	$49

Reprinted from *USA Today*, August 11, 1993, p. 1B, with data from the Mall of America, Urban Land Institute, Stillerman Jones & Co.

    a.   Describe the reported averages in a way that relates to particular random variables.

▲ *b.*   Assume that the amount a consumer spends at a shopping mall, *Y*, is a linear function of the random variable *X*, where *X* is the number of hours spent at the mall. What is this linear function?

▲ *c.*   Does the same linear function in *b* apply to Mall of America and to the "average" U.S. super-regional mall? Explain.

64.   Weekly Nielsen ratings for the television show "Dateline NBC" is a random variable with the following probability distribution:

x	P(x)
5	0.1
6	0.1
7	0.2
8	0.3
9	0.2
10	0.1

From information in "What Audiences Are Watching," *The New York Times*, October 25, 1993, p. D6.

    a.   What is the probability that on a given week the rating for this show will be at least 7?

    b.   What is the mean rating for this show?

65. In the article about bias in Britain, assume the figure for bias is exactly 1 in 3. If a random sample of eight Britons is taken, what is the probability that at least two will be found to profess a bias against Arab or Pakistani neighbors? What is the expected number of people in the sample with such a bias?

British Poll Says Gypsies Face the Most Bias

By WILLIAM E. SCHMIDT

LONDON, Oct. 24—Nearly one in three Britons would prefer not to have Arabs or Pakistanis as neighbors, and two of three say they do not want to live near Gypsies, according to a nationwide survey on British attitudes toward minorities.

The survey was conducted by Gallup of Great Britain on behalf of the American Jewish Committee, which has sponsored similar surveys in the United States and elsewhere in Europe.

David Singer, a Jewish Committee official, said some of the findings, particularly the deep-seated antipathy toward Gypsies, reflected the results of previous surveys in other parts of Europe.

The British survey consisted of face-to-face interviews with 959 people from Sept. 2 to Sept. 7. Those interviewed were a racially representative sample of men and women over the age of 16. The margin of error in the survey is plus or minus three percentage points. Nonwhites, including Indians and Pakistanis, make up about 5 percent of the 57 million people in Britain.

The findings of the survey coincide with growing tension over race relations in London. Last month, a candidate with the rightist British National Party, which campaigns on the platform of "rights for whites," won a local election in a racially mixed working-class neighborhood in East London that has been the scene of growing racial tension.

Reprinted from *The New York Times*, October 25, 1993, p. A7.

66. Refer to the same article above on bias in Britain. If a random sample of 15 Britons is taken, what is the probability that at least 10 of them do not want to live near Gypsies? What are the mean and the standard deviation of people in this sample with such a bias?

67. In the last few years, the proportion of people who get melanoma (a skin cancer) during the summer in sunny Australian cities is a random variable with a continuous uniform distribution from 1 to 3 percent. What is the mean percentage of people who get melanoma in one summer?

68. In Problem 67, what is the probability that in a particular summer the percentage of people who get melanoma will be 1 1/2 percent or less?

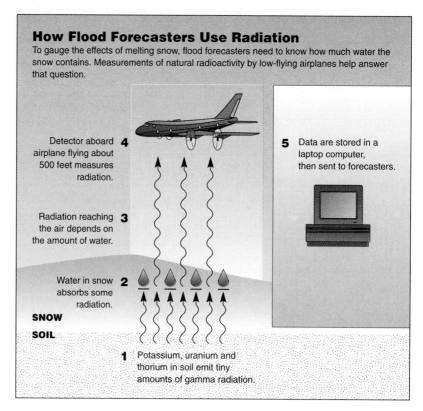

How Flood Forecasters Use Radiation

To gauge the effects of melting snow, flood forecasters need to know how much water the snow contains. Measurements of natural radioactivity by low-flying airplanes help answer that question.

4 Detector aboard airplane flying about 500 feet measures radiation.

3 Radiation reaching the air depends on the amount of water.

2 Water in snow absorbs some radiation.

SNOW

SOIL

1 Potassium, uranium and thorium in soil emit tiny amounts of gamma radiation.

5 Data are stored in a laptop computer, then sent to forecasters.

Reprinted from *USA Today,* March 17, 1994, p. 14A, with data from the National Weather Service Hydrology Office, Minneapolis.

69. The amount of radiation measured by the National Weather Service airplane in a given area is a random variable. Is this random variable discrete or continuous?

70. Suppose that the National Weather Service airplane's probability of successfully forecasting an impending flood in a given area during one flight is 0.85. If 10 independent flights take place, what is the probability that at least 7 of them will correctly forecast the flood?

INTERVIEW WITH
Arnold Zellner

"In statistics, there isn't much guidance with respect to methods for producing unusual and ugly facts."

Arnold Zellner received a PhD from the University of California at Berkeley in economics after a bachelors in physics from Harvard. He is a past president and Fellow of the American Statistical Association and was founding editor of the Journal of Business and Economic Statistics. *He teaches at the University of Chicago and does research in Bayesian inference and economic models and forecasting.*

Note: Professor Zellner is also an active and expert representative of Bayesian approaches to statistical problems. Bayes theorem (covered in optional section 2.6) allows for the use of prior probabilities to improve the accuracy of some statistical analyses. How often and how applicable the Bayesian approach is, as compared to the traditional or frequentist approach, is another debated topic in statistics.

Aczel: In your view is there a schism between the two approaches [Bayesian and frequentist], the two philosophies, or is it disintegrating in a way?

Zellner: Frankly, I don't believe that many of us are expert in the philosophy of science and thus sometimes our disagreements on concepts of probability, causality, axiom systems, and so on are not well founded and more emotional than rational. What saves the day, I think, are pragmatic considerations.

For example, we take a problem and work it from the Bayesian and non-Bayesian points of view and compare solutions. If one's better than the other, I'll use the better solution.

Aczel: You've been very active in the Bayesian Society; tell us what is currently happening with the group.

Zellner: We just had the first world meeting of the International Society for Bayesian Analysis (ISBA) with over 200 attending sessions on Friday and Saturday (August 93). Stu Hunter, President of ASA, addressed the banquet and wished the group well. As I mentioned earlier, we concentrated on Bayesian analysis in the natural, biological and social sciences as well as on technical statistical aspects of Bayesian analysis. We had sessions on Bayesian analysis in astronomy, physics, psychology, econometrics, the earth sciences and many other areas. In the earth sciences, a paper on earthquakes was featured, quite appropriate for a meeting held in San Francisco. There was also a session on Bayesian analysis on Wall Street. Jose Quintana, a vice president at Chase-Manhattan Bank in New York, reported on his work using Bayesian time series analysis techniques to track stock prices and form portfolios. Also, workers at the investment company Goldman-Sachs are employing Bayesian portfolio analysis to help clients with their investment decisions. In this and the other areas that I mentioned, workers are employing Bayesian techniques to learn from data, make decisions, and predict. In many of the papers presented, the performance of Bayesian and non-Bayesian techniques was compared. Also, we had a number of sessions devoted to development of new Bayesian statistical techniques. We're looking forward to having a good meeting in Alicante at a resort hotel. We'll work hard and also have a good time. As you know, the old saying is that Bayesians have more fun.

Aczel: But some people would say that they're still stuck with their priors, right?

Zellner: Oh, that's been mentioned before. Usually, that's what people say, "stuck," but what they don't realize is that in many problems in business, engineering, and other areas, people use judgment. Of course, if you don't have data and have to make decisions and forecasts, judgment is all that you have. By use of priors, we are just formalizing, and hopefully improving, what is already being done. If you're in business and you don't have good judgment, you're out of business very quickly. Thus engineers, businessmen, scientists, and policy makers have background information that they want to formalize and use to solve all kinds of problems. The prior helps them to achieve this goal.

Aczel: Would the Bayesian approach give you some edge in forecasting that would give you an advantage over somebody else in the sense that your forecast of tomorrow's price may be better than anybody else's?

Zellner: Well, as I said before, forecasting stock prices well is very hard. But there are many who do forecast. For example, Jose Quintana, a vice president at the Chase-Manhattan Bank in New York, reported (at the International Society for Bayesian Analysis) on his use of Bayesian state-space models for forecasting returns in world markets and the use of the forecasts in forming portfolios. He has compared his forecasts with those provided by the traditional random walk models, and so on.

Aczel: How does he do in comparison with random walk forecasts?

Zellner: He thinks that he does better. You can ask him for his papers and judge for yourself. In our work

on forecasting growth rates of real GNP or GDP for eighteen countries, we definitely can beat random walk model forecasts. In this work, we have used Bayesian shrinkage forecast techniques. Use of the (prior) information that the countries' coefficient vectors are not too far different in value has improved our forecasts. Many others have also found this to be the case.

Aczel: What is this shrinkage? Do you reduce the standard error somehow?

Zellner: You are essentially pulling in outlying forecasts toward a meanlike forecast. In terms of overall performance, this gives you improvement in the overall countries' mean square error of forecast, but not necessarily for each and every country. And in empirical forecasting year by year, you can see the effect when you do the calculations. In an article that I wrote with Franz Palm, we tried to explain why a simple average of forecasts would work better than individuals' forecasts. It may be that individual biases are being averaged out by the simple averaging.

Aczel: How about if they are all biased?

Zellner: That's what probably happened in 1982. The forecasters didn't expect a change in policy that probably produced the −2.5 percent growth rate in the U.S. economy. Many forecasts were in the vicinity of (positive) 2 percent, perhaps all biased upward because of a lack of information about a policy change. However, my colleague Walter Fackler somehow came up with a December 1981 forecast for 1982 of −1.1 percent or −1.2 percent GNP growth. On checking his annual forecasts against those of the Blue Chip Company sample, his RMSE was somewhat smaller. However, as Steven McNees of the Federal Reserve Bank of Boston has shown, no single forecaster in the Blue Chip Company sample has beaten the simple averages on all the variables being forecasted. That simple averages of forecasts can perform that well in actual forecasting is remarkable.

Aczel: Going back to the philosophy of science for a moment and to statistics and economics, what is your view? Do you favor applications first and theory second?

Zellner: Sometimes the theory comes first and the applications later, or as in the Stein shrinkage problem, the applications were there in the 1920s and the theory came later. I do not know whether he knew about these applications. In current research, theorists are finding it difficult to produce good, dependable macroeconomic theory. Thus, we have developed relatively simple models that forecast reasonably well and then determine which macroeconomic theories can explain why. Here we are going from the application to the theory. In physics, many times experiments produce unusual results that challenge theorists. For example, no ether drift in the Michelson-Morley experiment that Einstein explained later after attempts by Lorenz and others. Thus sometimes you get the empirical result and then the theoretical result or new statistical method. Other times, the theory comes first and the empirical validation comes later. Another thing that is important in this area is unusual facts and what Huxley called a tragedy of science, namely the destruction of a beautiful hypothesis or theory by an ugly fact.

Aczel: Hit by the real world . . .

Zellner: That's right. I think that we need to produce more unusual and ugly facts since this prompts many people to think about new theories to explain them. For example, the fact that there was no ether drift was unusual. In statistics, there isn't much guidance with respect to methods for producing unusual and ugly facts. It's an area that needs more attention in my opinion. Many of the Nobel prizewinners in economics, including Friedman, Kuznets, Modigliani, Tobin, and others did much work to produce and explain unusual facts. For example, Kuznets discovered that the percentage of income saved did not change very much over decades in the face of large increases in income, an ugly fact that raised considerable doubts about the Keynesian theory that predicted the percentage of income saved would increase rather than stay relatively constant at about 8 to 10 percent. After this finding, these wonderful theorists, Friedman, Modigliani, Tobin, and others produced new theories of consumption and saving to explain Kuznets' unusual and ugly fact. It's very important to produce unusual facts. In an essay in my book, *Basic Issues in Econometrics,* I put forward eight procedures for producing unusual facts. I don't see this topic emphasized in textbooks.

Aczel: You think it is important?

Zellner: Yes, very important.

Aczel: Tell us about yourself, your early days and what attracted you to different areas, statistics and others.

Zellner: I received a PhD at Berkeley years ago. My brother, Norman, did a PhD at Berkeley too, in agricultural economics, while I did mine in economics. I did my undergraduate work at Harvard and got a degree in physics. I took an introductory course in economic statistics and several courses in economics but found courses in physics and math more appealing. After my undergraduate work, I spent a summer working as a physicist at the Naval Ordnance Testing Station at Inyokern on the Mojave Desert in California before beginning graduate work in physics at Berkeley. My brother and his friends in agricultural economics were working on interesting problems using econometric and statistical methods that were of great interest to me. So, I decided to switch to economics with special emphasis on econometrics and statistics. Ever since, I've taught econometrics and economic statistics and pursued research mainly in econometrics and statistics, first at the University of Washington in Seattle, then at the University of Wisconsin in Madison, and since 1966 at the University of Chicago. It turned out that working in econometrics and statistics has been quite enjoyable and rewarding, my cup of tea, you might say.

CHAPTER

4

The Bell-Shaped Curve

Courtesy of Debra Gross Aczel.

4.1 The Normal Probability
    Distribution   156

4.2 The Standard Normal
    Distribution   159

4.3 The Transformation of Normal
    Random Variables   167

4.4 The Relationship between X and Z
    and the Use of the Inverse
    Transformation   173

4.5 The Normal Distribution as an
    Approximation to Other Probability
    Distributions   176

4.6 Normal Data   178

A t the southern end of the Aegean, just as it becomes the Sea of Crete, lies the volcanic island of Santorini, shrouded in mystery and the base of many myths. Around the dawn of civilization, the volcano of Santorini (also known as Thera) erupted in a tremendous explosion—possibly the most cataclysmic in the human experience. Since that devastating eruption, believed to have destroyed the well-developed Bronze Age culture of the Minoans in the eastern Mediterranean, the island has been repopulated many times and today is one of the most popular tourist attractions of Greece. Whitewashed barrel-vaulted Cycladic houses, beautifully decorated in blue, have been built right on the rim of the ancient volcano, now dormant, and tourists can take a donkey ride from the town all the way down to the water-filled caldera. Cruise ships sail right into the caldera, offering stunning views of this unique island, whose beautiful villages encircle the visitor.

The natives of Santorini had always felt that their island was somehow inhabited by ghosts of earlier civilizations, but nobody knew for sure. In 1967, the great Greek archaeologist Spyridon Marinatos began excavating on Santorini. Marinatos believed that the island was actually Plato's mythical lost continent of Atlantis. Modern day scholars tend to agree—it is believed that Santorini is the only place on Earth where entire towns sank to the bottom of the sea, as in Plato's description of the loss of Atlantis. What Marinatos needed was a "body"—the actual remains of a city or town destroyed by the Bronze Age eruption—a kind of Pompeii. Over the years, various excavations had taken place without result. It was due to Marinatos's genius that in 1967 he actually found what so many had looked for. In the southwestern corner of the island, close to the waterline, he uncovered an entire city that predated Pompeii by over 1,500 years. The ancient city, which the excavators named Akrotiri, had two-storied houses, with beautiful frescoes depicting everyday life and religious ceremonies and large ornamented containers of wine and food staples. But no remains of people or any valuables such as gold were ever found. What happened to the inhabitants of Akrotiri will forever remain a mystery. Some say they are still stranded on some desert island to which they fled to escape the imminent explosion on Santorini.

The discovery of Akrotiri raised tremendous interest in the scientific community. One of the key questions scientists have tried to answer is: When exactly did this Bronze Age eruption take place? Accurate dating of the explosion on Santorini would contribute greatly to our understanding of the chronology of events in the Mediterranean cradle of civilization. For example, it is well known that the Minoan palace-civilization on Crete declined around 1500 BC, but that the great palace of King Minos at Knossos survived for about another 80 years. Based on artistic motifs and anthropological reasons, scientists had generally believed that the fateful eruption also occurred around 1500 BC, but nobody knew for sure. As in many cases, *statistics* offered a solution.

4.1 THE NORMAL PROBABILITY DISTRIBUTION

We'll begin this chapter by showing how statistics can be used to solve a problem such as the one concerning the date of the volcanic eruption on Santorini, from the chapter opener.

Archaeologists use a scientific method called carbon dating. Along with the common, nonradioactive form of carbon occurring in nature, we also find small amounts of radioactive forms. The radioactive element carbon 14 is useful in dating any ancient artifacts that contain carbon (objects made of wood or other organic matter). Since carbon 14 decays over time, more of it existed in the past than today.

Jars found at Akrotiri.

Courtesy of Amir Aczel.

Simply stated, carbon dating consists of comparing the amount of radioactive carbon in an artifact with the amount of radioactive carbon in our present atmosphere, leading us to an extrapolated estimate of the age of the artifact. This estimate, however, is not a single number, but rather a realization of a *random variable.* The random variable has a *probability distribution* that looks like a bell. This bell-shaped curve is called the **normal probability distribution.**[1] In the middle of the curve, where we find the highest point of the function, is the *mean* of the distribution. The normal distribution is shown in Figure 4.1.

In 1988, H. N. Michael and P. P. Betancourt reported their results of carbon dating they performed on the remains of barley and other food material in jars found in the rubble at Akrotiri.[2] Based on their estimates, we get a normal distribution for the date of the eruption with a mean of 1619 BC and a standard deviation of 25 years. This distribution is shown in Figure 4.2.

The Normal Probability Distribution   **FIGURE 4.1**

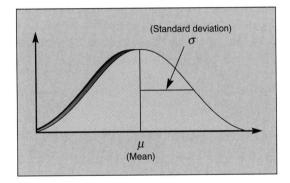

The Normal Distribution for the Date of the Eruption on Santorini   **FIGURE 4.2**

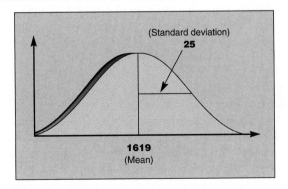

---

[1]The normal distribution is also called the *Gaussian distribution,* named after the German mathematician Gauss, although it is now known to have been discovered before his time.

[2]H. N. Michael and P. P. Betancourt, "The Thera Eruption: Further Arguments for an Early Date," *Archaeometry* 30 (1988), pp. 169–74.

For any normal distribution, the probability that the random variable with the given distribution will fall within *one standard deviation* of the mean is 0.6826 (68.26 percent). Similarly, the probability that the random variable will have a value within *two standard deviations* of its mean is about 0.9544 (95.44 percent).[3] Thus, we can be 95.44 percent confident, based on the carbon dating results, that the true date of the eruption on Santorini is anywhere from $1619 - 2(25) = 1569$ to $1619 + 2(25) = 1669$ years BC. Therefore, based on the carbon dating, the originally hypothesized date of 1500 BC is not very likely to be correct.

The Shape of the Normal Distribution

The normal probability distribution is very commonly used in statistics because of a rule called the **central limit theorem,** which states that, under certain conditions, as the number of random factors affecting some phenomenon increases, the distribution of the sum (or average) of these factors approaches the normal distribution. The formula for the normal density, $f(x)$, with mean μ and standard deviation σ, is given below. It is provided simply for general interest—we will not use it directly in any way. If you like to do your own programming on a computer, using a language such as Basic or Fortran, you can program this function, choosing values for the mean and the standard deviation, and observe the shape of the computer-produced graph.

The probability density function of a normal random variable with mean μ and standard deviation σ:

$$f(x) = (1/\sqrt{2\pi}\sigma)e^{-(x-\mu)^2/2\sigma^2} \qquad \text{for} \quad -\infty < x < \infty$$

where
$e = 2.718...$
$\pi = 3.141...$

The normal curve is symmetric. It slopes downward on both sides of the mean, getting steeper for a while, then slowing down—until it is almost level with the horizontal axis. (The curve, however, never quite reaches the horizontal axis; it just continues to get closer and closer to it as we move to infinity on the right side and to negative infinity on the left side.)

A nice analogy for understanding the interesting shape of the normal curve is to think of a perfectly mound-shaped ski mountain. As a skier starts off at the top in either direction, left or right, he or she begins to accelerate downward. After moving down quite fast, the skier starts to slow down his or her descent at the point where the slope starts to decrease. That particular start-to-slow-down point is a distance of exactly one standard deviation (σ) on the horizontal scale from the top of the mountain. On the horizontal scale, the top of the mountain is the mean (μ). All this is demonstrated in Figure 4.3.

---

[3] These probability statements are the equivalents of the empirical rule of Chapter 1 for a normal random variable. They will be formalized later in this chapter.

Bell Mountain

FIGURE 4.3

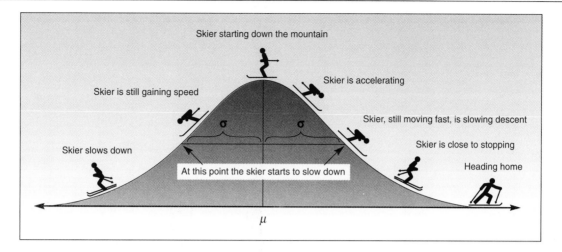

Figure 4.4 shows several normal distributions with different means and standard deviations. *The mean of the distribution is a measure of its location:* It tells us where on the number line the distribution is centered. (Since the normal density is symmetrical, with one peak in the center, the *mean, μ,* is also the *mode* and the *median* of the probability distribution; thus, μ is also the point where the density function is highest and the point that splits the area under the curve in half. The area to each side of μ is equal to 0.5.) The *standard deviation* of the distribution is a measure of the spread, or *variability,* of the distribution. When the standard deviation is large, the density function is wide and, hence, also low (because the total area under the curve must equal 1.00). When the standard deviation is small, the curve is narrow and high.

Notation

We will denote the distribution, mean, and variance of a normal random variable by a simple notational statement:

$$X \text{ is } N(\mu, \sigma)$$

This will mean that the random variable X has a normal distribution (N) with mean μ and standard deviation σ.

For example, X is N(5.7, 2) means that the random variable X has a normal distribution with mean 5.7 and standard deviation 2.

4.2 THE STANDARD NORMAL DISTRIBUTION

Since there are infinitely many possible normal distributions (choose any number for the mean, μ, and any number for the standard deviation, σ, and you have another normal distribution), one of them was selected as our *standard.* Probabilities associated with intervals defined by different values of the standard normal random variable have been computed and are given in tables. A transformation then allows us to apply the tabulated probabilities to *any* normal random variable. The standard normal random variable has a special name, Z (rather than the general name X we use for other random variables).

FIGURE 4.4   Normal Distributions with Different Means and Standard Deviations

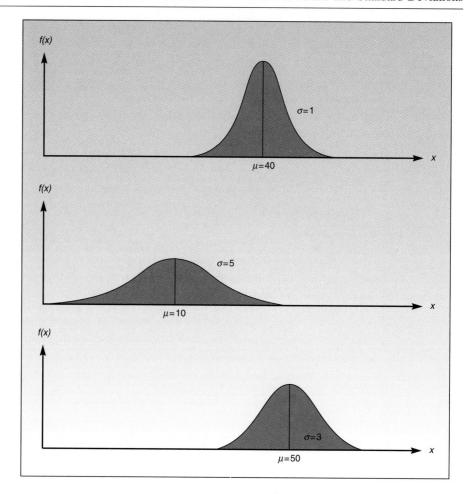

> We define the **standard normal random variable,** Z, as the normal
> random variable with mean $\mu = 0$ and standard deviation $\sigma = 1$.

In the notation just established, we say:

$$Z \text{ is } N(0, 1)$$

A graph of the standard normal density function is given in Figure 4.5.

Finding Probabilities of the Standard Normal Distribution

Probabilities of intervals are *areas* under the density $f(z)$ over the intervals in question.
From the range of values $-\infty < x < \infty$, we see that any normal random variable is
defined over the entire real line. Thus, intervals of interest are sometimes *semi-infinite*
intervals, such as a to ∞ or b to $-\infty$ (where a and b are numbers). While such intervals

The Standard Normal Density Function   **FIGURE 4.5**

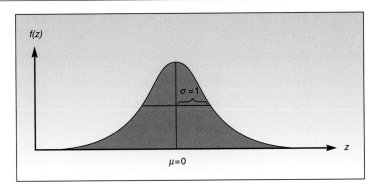

The Table Area, TA, for a Point z of the Standard Normal Distribution   **FIGURE 4.6**

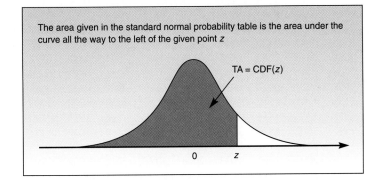

have infinite length, the probabilities associated with them are finite; they are, in fact, no greater than 1.00, as required of all probabilities. The reason for this is that the area in either of the "tails" of the distribution (the two narrow ends of the distribution, extending toward $-\infty$ and $+\infty$) becomes very small very quickly as we move away from the center of the distribution.

Tabulated areas under the standard normal density are probabilities of intervals extending from $-\infty$ to points z to its right. Table B.2 in Appendix B gives areas under the standard normal curve between $-\infty$ and point z. The total area under the normal curve is equal to 1.00. The **table area** associated with a point z is thus equal to the value of the cumulative distribution function, CDF(z).

The table area, TA, is shown in Figure 4.6. Appendix Table B.2 is reproduced here as Table 4.1. Let's see how the table is used in obtaining probabilities for the standard normal random variable. In the following examples, refer to Figure 4.6 and Table 4.1.

1. Let's find the probability that the value of the standard normal random variable will be less than 1.56. That is, we want $P(Z < 1.56)$. In Figure 4.6, substitute 1.56 for the point z on the graph. In Table 4.1, we're looking for the area in the row labeled 1.5 and the column labeled .06; we thus find the probability 0.9406.

TABLE 4.1

Standard Normal Probabilities

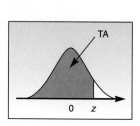

z	.00	.01	.02	.03	.04	.05	.06	.07	.08	.09
−3.4	.0003	.0003	.0003	.0003	.0003	.0003	.0003	.0003	.0003	.0002
−3.3	.0005	.0005	.0005	.0004	.0004	.0004	.0004	.0004	.0004	.0003
−3.2	.0007	.0007	.0006	.0006	.0006	.0006	.0006	.0005	.0005	.0005
−3.1	.0010	.0009	.0009	.0009	.0008	.0008	.0008	.0008	.0007	.0007
−3.0	.0013	.0013	.0013	.0012	.0012	.0011	.0011	.0011	.0010	.0010
−2.9	.0019	.0018	.0018	.0017	.0016	.0016	.0015	.0015	.0014	.0014
−2.8	.0026	.0025	.0024	.0023	.0023	.0022	.0021	.0021	.0020	.0019
−2.7	.0035	.0034	.0033	.0032	.0031	.0030	.0029	.0028	.0027	.0026
−2.6	.0047	.0045	.0044	.0043	.0041	.0040	.0039	.0038	.0037	.0036
−2.5	.0062	.0060	.0059	.0057	.0055	.0054	.0052	.0051	.0049	.0048
−2.4	.0082	.0080	.0078	.0075	.0073	.0071	.0069	.0068	.0066	.0064
−2.3	.0107	.0104	.0102	.0099	.0096	.0094	.0091	.0089	.0087	.0084
−2.2	.0139	.0136	.0132	.0129	.0125	.0122	.0119	.0116	.0113	.0110
−2.1	.0179	.0174	.0170	.0166	.0162	.0158	.0154	.0150	.0146	.0143
−2.0	.0228	.0222	.0217	.0212	.0207	.0202	.0197	.0192	.0188	.0183
−1.9	.0287	.0281	.0274	.0268	.0262	.0256	.0250	.0244	.0239	.0233
−1.8	.0359	.0351	.0344	.0336	.0329	.0322	.0314	.0307	.0301	.0294
−1.7	.0446	.0436	.0427	.0418	.0409	.0401	.0392	.0384	.0375	.0367
−1.6	.0548	.0537	.0526	.0516	.0505	.0495	.0485	.0475	.0465	.0455
−1.5	.0668	.0655	.0643	.0630	.0618	.0606	.0594	.0582	.0571	.0559
−1.4	.0808	.0793	.0778	.0764	.0749	.0735	.0721	.0708	.0694	.0681
−1.3	.0968	.0951	.0934	.0918	.0901	.0885	.0869	.0853	.0838	.0823
−1.2	.1151	.1131	.1112	.1093	.1075	.1056	.1038	.1020	.1003	.0985
−1.1	.1357	.1335	.1314	.1292	.1271	.1251	.1230	.1210	.1190	.1170
−1.0	.1587	.1562	.1539	.1515	.1492	.1469	.1446	.1423	.1401	.1379
−0.9	.1841	.1814	.1788	.1762	.1736	.1711	.1685	.1660	.1635	.1611
−0.8	.2119	.2090	.2061	.2033	.2005	.1977	.1949	.1922	.1894	.1867
−0.7	.2420	.2389	.2358	.2327	.2296	.2266	.2236	.2206	.2177	.2148
−0.6	.2743	.2709	.2676	.2643	.2611	.2578	.2546	.2514	.2483	.2451
−0.5	.3085	.3050	.3015	.2981	.2946	.2912	.2877	.2843	.2810	.2776
−0.4	.3446	.3409	.3372	.3336	.3300	.3264	.3228	.3192	.3156	.3121
−0.3	.3821	.3783	.3745	.3707	.3669	.3632	.3594	.3557	.3520	.3483
−0.2	.4207	.4168	.4129	.4090	.4052	.4013	.3974	.3936	.3897	.3859
−0.1	.4602	.4562	.4522	.4483	.4443	.4404	.4364	.4325	.4286	.4247
−0.0	.5000	.4960	.4920	.4880	.4840	.4801	.4761	.4721	.4681	.4641

2. Let's find the probability that Z will be less than −2.47. Figure 4.7 shows the required area for the probability $P(Z < -2.47)$. From the table, we find:

$$P(Z < -2.47) = 0.0068$$

3. Find $P(1 < Z < 2)$. The required probability is the area under the curve between the two points 1 and 2. This area is shown in Figure 4.8. The table gives us the area under the curve to the left of 1, and the area under the curve to the left of 2. Areas are additive; therefore, $P(1 < Z < 2) = $ TA(for 2.00) − TA(for 1.00) = 0.9772 − 0.8413 = 0.1359.

4. What is the probability that the standard normal random variable, Z, will *exceed* 1.56? From the table (as shown in item 1 above), we know that the

z	.00	.01	.02	.03	.04	.05	.06	.07	.08	.09
0.0	.5000	.5040	.5080	.5120	.5160	.5199	.5239	.5279	.5319	.5359
0.1	.5398	.5438	.5478	.5517	.5557	.5596	.5636	.5675	.5714	.5753
0.2	.5793	.5832	.5871	.5910	.5948	.5987	.6026	.6064	.6103	.6141
0.3	.6179	.6217	.6255	.6293	.6331	.6368	.6406	.6443	.6480	.6517
0.4	.6554	.6591	.6628	.6664	.6700	.6736	.6772	.6808	.6844	.6879
0.5	.6915	.6950	.6985	.7019	.7054	.7088	.7123	.7157	.7190	.7224
0.6	.7257	.7291	.7324	.7357	.7389	.7422	.7454	.7486	.7517	.7549
0.7	.7580	.7611	.7642	.7673	.7704	.7734	.7764	.7794	.7823	.7852
0.8	.7881	.7910	.7939	.7967	.7995	.8023	.8051	.8078	.8106	.8133
0.9	.8159	.8186	.8212	.8238	.8264	.8289	.8315	.8340	.8365	.8389
1.0	.8413	.8438	.8461	.8485	.8508	.8531	.8554	.8577	.8599	.8621
1.1	.8643	.8665	.8686	.8708	.8729	.8749	.8770	.8790	.8810	.8830
1.2	.8849	.8869	.8888	.8907	.8925	.8944	8962	.8980	.8997	.9015
1.3	.9032	.9049	.9066	.9082	.9099	.9115	.9131	.9147	.9162	.9177
1.4	.9192	.9207	.9222	.9236	.9251	.9265	.9279	.9292	.9306	.9319
1.5	.9332	.9345	.9357	.9370	.9382	.9394	.9406	.9418	.9429	.9441
1.6	.9452	.9463	.9474	.9484	.9495	.9505	.9515	.9525	.9535	.9545
1.7	.9554	.9564	.9573	.9582	.9591	.9599	.9608	.9616	.9625	.9633
1.8	.9641	.9649	.9656	.9664	.9671	.9678	.9686	.9693	.9699	.9706
1.9	.9713	.9719	.9726	.9732	.9738	.9744	.9750	.9756	.9761	.9767
2.0	.9772	.9778	.9783	.9788	.9793	.9798	.9803	.9808	.9812	.9817
2.1	.9821	.9826	.9830	.9834	.9838	.9842	.9846	.9850	.9854	.9857
2.2	.9861	.9864	.9868	.9871	.9875	.9878	.9881	.9884	.9887	.9890
2.3	.9893	.9896	.9898	.9901	.9904	.9906	.9909	.9911	.9913	.9916
2.4	.9918	.9920	.9922	.9925	.9927	.9929	.9931	.9932	.9934	.9936
2.5	.9938	.9940	.9941	.9943	.9945	.9946	.9948	.9949	.9951	.9952
2.6	.9953	.9955	.9956	.9957	.9959	.9960	.9961	.9962	.9963	.9964
2.7	.9965	.9966	.9967	.9968	.9969	.9970	.9971	.9972	.9973	.9974
2.8	.9974	.9975	.9976	.9977	.9977	.9978	.9979	.9979	.9980	.9981
2.9	.9981	.9982	.9982	.9983	.9984	.9984	.9985	.9985	.9986	.9986
3.0	.9987	.9987	.9987	.9988	.9988	.9989	.9989	9989	.9990	.9990
3.1	.9990	.9991	.9991	.9991	.9992	.9992	.9992	.9992	.9993	.9993
3.2	.9993	.9993	.9994	.9994	.9994	.9994	.9994	.9995	.9995	.9995
3.3	.9995	.9995	.9995	.9996	.9996	.9996	.9996	.9996	.9996	.9997
3.4	.9997	.9997	.9997	.9997	.9997	.9997	.9997	.9997	.9997	.9998

probability that Z will be less than 1.56 is 0.9406. Now, the *total* area under the normal curve is 1.00, so the area to the *right* of 1.56 must be equal to $1 - 0.9406 = 0.0594$. (Recall that the probability that any continuous random variable, such as the normal, will *equal* any given number is 0. This is why we don't need to worry about the equal sign in any of our inequalities and may use strict inequalities.)

In cases where we need probabilities based on values with greater than second-decimal accuracy, we may use a linear interpolation between two probabilities obtained from the table. For example, $P(Z < 1.645)$ is found as the midpoint between the two probabilities $P(Z < 1.64)$ and $P(Z < 1.65)$. This is found, using the table, as the midpoint of 0.9495 and 0.9505, which is 0.95.

FIGURE 4.7   Finding the Probability that Z Is Less than -2.47

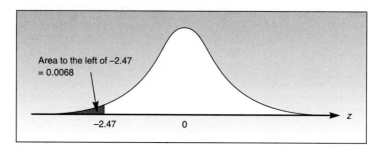

FIGURE 4.8   Finding the Probability that Z Is between 1 and 2

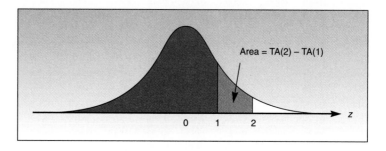

The computer package MINITAB can be used to find probabilities for normal random variables. We demonstrate the use of MINITAB in finding the area to the left of $Z = 1.96$ using the command **CDF.** This command gives the area to the left of any specified point.

```
MTB > CDF 1.96;
SUBC> NORMAL 0 1.
   1.9600  0.9750
```

Finding Values of Z Given a Probability

In many situations, instead of finding the probability that a standard normal random variable will be within a given interval, we may be interested in the reverse: finding an interval with a given probability. Consider the following examples:

1. Find a value z of the standard normal random variable such that the probability that the random variable will have a value less than z is 0.90. We look *inside* the table for the value closest to 0.90; we do this by searching through the values inside the table, noting that they increase from 0 to numbers close to 1 as we go down the columns and across rows. The closest value we find to 0.90 is the table area 0.8997. This value corresponds to 1.28 (row 1.2 and column .08). This is illustrated in Figure 4.9, and Figure 4.10.

2. Find the value of the standard normal random variable that cuts off an area of 0.80 to its *right*. The table area, to the left of the point, is therefore $1 - 0.80 = 0.20$. We now scan the inside of Table 4.1 for an area as close as we can get to 0.20. The closest we get is 0.2005, and this corresponds to $z = -0.84$.

Using the Normal Table to Find a Value, Given a Probability 0.9   **FIGURE 4.9**

z	.00	.01	.02	.03	.04	.05	.06	.07	↓ .08	.09
0.0	.5000	.5040	.5080	.5120	.5160	.5199	.5239	.5279	.5319	.5359
0.1	.5398	.5438	.5478	.5517	.5557	.5596	.5636	.5675	.5714	.5753
0.2	.5793	.5832	.5871	.5910	.5948	.5987	.6026	.6064	.6103	.6141
0.3	.6179	.6217	.6255	.6293	.6331	.6368	.6406	.6443	.6480	.6517
0.4	.6554	.6591	.6628	.6664	.6700	.6736	.6772	.6808	.6844	.6879
0.5	.6915	.6950	.6985	.7019	.7054	.7088	.7123	.7157	.7190	.7224
0.6	.7257	.7291	.7324	.7357	.7389	.7422	.7454	.7486	.7517	.7549
0.7	.7580	.7611	.7642	.7673	.7704	.7734	.7764	.7794	.7823	.7852
0.8	.7881	.7910	.7939	.7967	.7995	.8023	.8051	.8078	.8106	.8133
0.9	.8159	.8186	.8212	.8238	.8264	.8289	.8315	.8340	.8365	.8389
1.0	.8413	.8438	.8461	.8485	.8508	.8531	.8554	.8577	.8599	.8621
1.1	.8643	.8665	.8686	.8708	.8729	.8749	.8770	.8790	.8810	.8830
→1.2	.8849	.8869	.8888	.8907	.8925	.8944	.8962	.8980	.8997	.9015
1.3	.9032	.9049	.9066	.9082	.9099	.9115	.9131	.9147	.9162	.9177
1.4	.9192	.9207	.9222	.9236	.9251	.9265	.9279	.9292	.9306	.9319
1.5	.9332	.9345	.9357	.9370	.9382	.9394	.9406	.9418	.9429	.9441
1.6	.9452	.9463	.9474	.9484	.9495	.9505	.9515	.9525	.9535	.9545

Finding z Such that $P(Z < z) = 0.9$ **FIGURE 4.10**

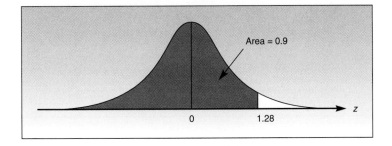

3. Find a 0.99 probability interval, symmetric about 0, for the standard normal random variable. The required area between the two z values that are equidistant from 0 on either side is 0.99. Therefore, the area under the curve between 0 and the positive z value is 0.99/2 = 0.495. We now look in our normal probability table for the area closest to 0.995 (we add 0.5 to 0.495 to account for the area to the left of zero). The area 0.995 lies exactly between the two areas 0.9949 and 0.9951, corresponding to $z = 2.57$ and $z = 2.58$. Therefore, a simple linear interpolation between the two values gives us $z = 2.575$. This is correct to within the accuracy of the linear interpolation. The answer, therefore, is $z \pm 2.575$. This is shown in Figure 4.11.

The area to the right of a positive z is called a **tail area.** This area, and the origin of its name, are demonstrated in Figure 4.12.

The computer package MINITAB can be used very easily for finding values that satisfy given probabilities. The command for doing so is **INVCDF,** the "inverse" of the CDF. Let's use MINITAB to find the value of z such that the area to its left is 0.75. This is shown below.

```
MTB > INVCDF 0.75;
SUBC> NORMAL 0 1.
   0.7500  0.6745
```

FIGURE 4.11   A Symmetric 0.99 Probability Interval about 0 for a Standard Normal Random Variable

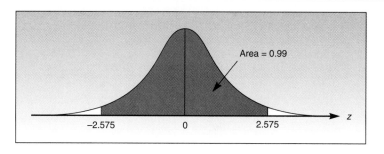

FIGURE 4.12   The Tail and Its Area

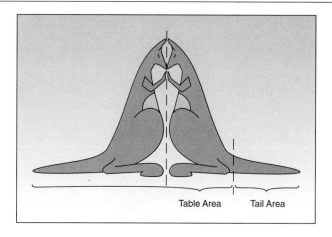

PROBLEMS

1. Find the following probabilities: $P(-1 < Z < 1)$, $P(-1.96 < Z < 1.96)$, $P(-2.33 < Z < 2.33)$, $P(Z < 2.58)$, $P(-3 < Z < 3)$.

2. What is the probability that a standard normal random variable will be between the values -2 and 1?

3. Find the probability that a standard normal random variable will have a value between -0.89 and -2.66.

4. Find the probability that a standard normal random variable will have a value greater than 3.02.

5. Find the probability that a standard normal random variable will be between 2 and 3.

6. Find the probability that a standard normal random variable will have a value less than or equal to -2.5.

7. Find the probability that a standard normal random variable will be greater in value than -2.33.

8. Find the probability that a standard normal random variable will have a value between -2 and 15.

9. Find the probability that a standard normal random variable will have a value less than -45.

10. Find the probability that a standard normal random variable will be between -0.01 and 0.05.

11. A sensitive measuring device is calibrated so that errors in the measurements it provides are normally distributed with mean 0 and variance 1.00. Find the probability that a given error will be between -2 and 2.

12. The deviation (in degrees) of a magnetic needle away from the magnetic pole in a certain area in northern Canada is a normally distributed random variable with mean 0 and standard deviation 1.00. What is the probability that the absolute value of the deviation from the north pole at a given moment will be more than 2.4 degrees?

13. Is it likely that a standard normal random variable will have value less than -4? Explain.

14. Find a value such that the probability that the standard normal random variable will be above it is 0.55.

15. Find a value of the standard normal random variable cutting off an area of 0.575 to its left.

16. Find a value of the standard normal random variable cutting off an area of 0.50 to its right. (Do you need the table for this probability? Explain.)

17. Find z such that $P(Z > z) = 0.48$.

18. Find two values, equidistant from 0 on either side, such that the probability that a standard normal random variable will be between them is 0.40.

19. Find two values of the standard normal random variable, z and $-z$, such that $P(-z < Z < z) = 0.95$.

20. Find two values of the standard normal random variable, z and $-z$, such that the two corresponding "tail areas" of the distribution (the area to the right of z and the area to the left of $-z$) add to 0.01.

21. Find two values defining tails of the normal distribution with an area of 0.05 each.

22. Use a computer to find the probability that the standard normal random variable will have a value less than 1.3.

23. Use a computer to find the value of the standard normal distribution that has an area of 0.25 to its left.

4.3 THE TRANSFORMATION OF NORMAL RANDOM VARIABLES

The importance of the standard normal distribution derives from the fact that any normal random variable may be transformed to the standard normal random variable. Let's say we want to transform X, where X is $N(\mu, \sigma)$, into the standard normal random variable Z, which is $N(0,1)$. Look at Figure 4.13. Here we have a normal random variable, X, with mean $\mu = 50$ and standard deviation $\sigma = 10$. We want to transform this random variable to a normal random variable with $\mu = 0$ and $\sigma = 1$. Imagine the distribution of X as a balloon with the shape of a bell. The air in the balloon is the area under the curve, the probability. The total amount of air in the balloon—the total area under the curve—is equal to 1.00. The balloon is centered over the value 50, the mean of the distribution. The width of the balloon, its standard deviation, is equal to 10. We want to transform this distribution to one with mean 0 and standard deviation 1. How can we do this?

We move the balloon from its center of 50 to a center of 0. This is done by *subtracting* 50 from all the values of X. Thus, we shift the balloon 50 units back so that its new center is 0. Next we need to make the width of the distribution, its standard deviation, equal to 1. This is done by squeezing the balloon so that its width decreases from 10 to 1. Because the total amount of air in the balloon (the total probability under the curve) must remain 1.00, the balloon must grow upward as it is squeezed to a width of 1 to maintain the same amount of air. This is shown in Figure 4.13. Mathematically, squeezing the curve to make the width 1 is equivalent to dividing the random variable by its standard deviation. Under the assumption of flexibility, the volume of the balloon (area under the curve) adjusts so that the total remains the same. *All probabilities* (areas under the curve) *adjust accordingly*. The mathematical transformation from X to Z is thus achieved by first subtracting μ from X and then dividing the result by σ.

FIGURE 4.13

Transforming a Normal Random Variable with Mean 50 and Standard Deviation 10 into the Standard Normal Random Variable

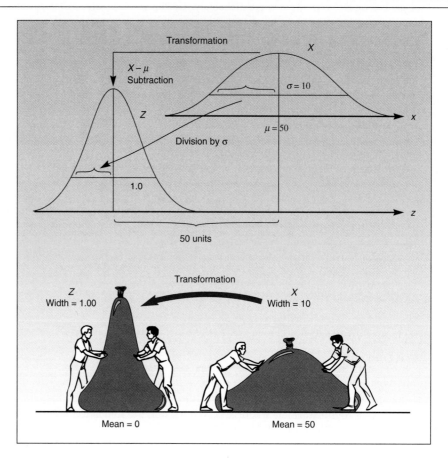

The transformation of X to Z:

$$Z = \frac{X - \mu}{\sigma}$$

The transformation of this equation takes us from a normal random variable X with mean μ and standard deviation σ to the standard normal random variable. We also have an opposite, or *inverse,* transformation, which takes us from the standard normal random variable Z to the normal random variable X with mean μ and standard deviation σ.

The inverse transformation of Z to X:

$$X = \mu + Z\sigma$$

You can verify mathematically that the first equation does the opposite of the second one. Note that multiplying the random variable Z by the number σ increases the width of the balloon from 1 to σ, thus making σ the new standard deviation. Adding μ makes μ the new mean of the random variable. The actions of multiplying and then adding are the opposite of subtracting and then dividing. We note that the two transformations, one an inverse of the other, turn a *normal* random variable into another *normal* random variable. If a transformation is carried out on a random variable that is not normal, the result will not be a normal random variable.

Using the Normal Transformation

Let's consider our random variable X with mean 50 and standard deviation 10, X is N(50,10). Suppose we want to find the probability that X is greater than 60. That is, we want to find $P(X > 60)$. We cannot evaluate this probability directly, but if we can transform X to Z, we will be able to find the probability in the Z table, Table B.2 in Appendix B. The required transformation is $Z = (X-\mu)/\sigma$. To carry out the transformation, we will substitute Z for X in the probability statement, $P(X > 60)$. If, however, we carry out the transformation on one side of the probability inequality, we must also do it on the other side. In other words, transforming X into Z requires us also to transform the value 60 into the appropriate value of the standard normal distribution. We transform the value 60 into the value $(60-\mu)/\sigma$. The new probability statement is:

$$P(X > 60) = P\left(\frac{X - \mu}{\sigma} > \frac{60 - \mu}{\sigma}\right) = P\left(Z > \frac{60 - \mu}{\sigma}\right)$$

$$= P\left(Z > \frac{60 - 50}{10}\right) = P(Z > 1)$$

Why does the inequality still hold? We subtracted a number from each side of an inequality; this does not change the inequality. In the next step we divided both sides of the inequality by the standard deviation, σ. The inequality does not change because we can divide both sides of an inequality by a positive number, and a standard deviation is always a positive number. (Recall that dividing by 0 is not permissible; and dividing, or multiplying, by a negative value would reverse the direction of the inequality.) From the transformation, we find that the probability that a normal random variable with mean 50 and standard deviation 10 will have a value greater than 60 is exactly the probability that the standard normal random variable Z will be greater than 1. The latter probability can be found using Table 2. We find $P(X > 60) = P(Z > 1)$ $= 1 - 0.8413 = 0.1587$. Examples 4.1 and 4.2 will help clarify this process.

EXAMPLE 4.1

An Italian automaker believes that the number of kilometers one of its engine models will go until it fails is normally distributed with mean 160,000 kilometers and standard deviation 30,000 kilometers. What is the probability that a given engine of this type will last anywhere from 100,000 to 180,000 kilometers before it needs to be replaced?

SOLUTION

Figure 4.14 shows the normal distribution for X, that is, N(160,000, 30,000) and the required area on the scale of the original problem and on the transformed z scale. We have the following (where the probability statement inequality has three sides and we carry out the transformation on all three sides).

FIGURE 4.14   Probability Computation for Example 4.1

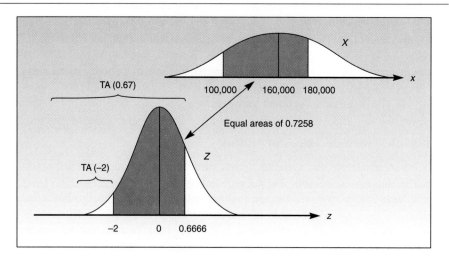

$$P(100{,}000 < X < 180{,}000) = P\left(\frac{100{,}000 - \mu}{\sigma} < \frac{X - \mu}{\sigma} < \frac{180{,}000 - \mu}{\sigma}\right)$$

$$= P\left(\frac{100{,}000 - 160{,}000}{30{,}000} < Z < \frac{180{,}000 - 160{,}000}{30{,}000}\right)$$

$$= P(-2 < Z < 0.67) = 0.7486 - 0.0228 = 0.7258$$

Thus, there is a 0.7258 chance that a given engine of this kind will last anywhere from 100,000 to 180,000 kilometers before it needs to be replaced.

EXAMPLE 4.2   The concentration of impurities in a semiconductor used in the production of microprocessors for computers is a normally distributed random variable with mean 127 parts per million and standard deviation 22. A semiconductor is acceptable only if its concentration of impurities is below 150 parts per million. What proportion of the semiconductors are acceptable for use?

SOLUTION X is N(127,22), and we need $P(X < 150)$. Using the equation, we get the following:

$$P(X < 150) = P\left(\frac{X - \mu}{\sigma} < \frac{150 - \mu}{\sigma}\right) = P\left(Z < \frac{150 - 127}{22}\right)$$

$$= P(Z < 1.05) = 0.8531$$

Thus, 85.3 percent of the semiconductors are acceptable for use. This also means that the probability that a randomly chosen semiconductor will be acceptable for use is 0.8531. This solution is illustrated in Figure 4.15.

Probability Computation for Example 4.2

FIGURE 4.15

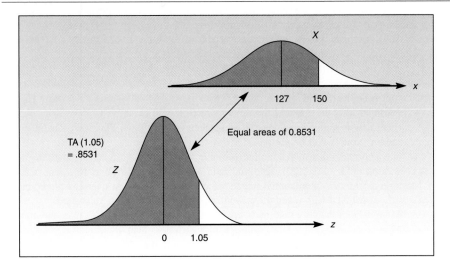

Now we summarize the transformation procedure used in computing probabilities of events associated with a normal random variable X that is $N(\mu, \sigma)$.

Transformation formulas of X to Z, where a and b are numbers:

$$P(X < a) = P\left(Z < \frac{a - \mu}{\sigma}\right)$$

$$P(X > b) = P\left(Z > \frac{b - \mu}{\sigma}\right)$$

$$P(a < X < b) = P\left(\frac{a - \mu}{\sigma} < Z < \frac{b - \mu}{\sigma}\right)$$

To use MINITAB to obtain probabilities for any normal random variable, we can specify the mean and the standard deviation in the subcommand. The probabilities so obtained are more accurate than the ones we get from the table, where we have been rounding the numbers to two decimals in order not to have to interpolate in the table. The computer has a more accurate method of computation. The following shows the computation of probabilities for Example 4.1 and Example 4.2:

```
MTB > CDF 180000;
SUBC> NORMAL 160000 30000.
  1.80E+05   0.7475
MTB > CDF 100000;
SUBC> NORMAL 160000 30000.
  1.00E+05   0.0228
MTB > CDF 150;
SUBC> NORMAL 127 22.
  150.0000   0.8521
```

PROBLEMS

24. For a normal random variable with mean 674 and standard deviation 55, find the probability that its value will be below 600.

25. Let X be a normally distributed random variable with mean 410 and standard deviation 2. Find the probability that X will be between 407 and 415.

26. If X is normally distributed with mean 500 and standard deviation 20, find the probability that X will be above 555.

27. For a normally distributed random variable with mean -44 and standard deviation 16, find the probability that the value of the random variable will be above 0.

28. A normal random variable has mean 0 and standard deviation 4. Find the probability that the random variable will be above 2.5.

29. Let X be a normally distributed random variable with mean $\mu = 16$ and standard deviation $\sigma = 3$. Find $P(11 < X < 20)$. Also find $P(17 < X < 19)$ and $P(X > 15)$.

30. The time it takes an international telephone operator to place an overseas phone call is normally distributed with mean 45 seconds and standard deviation 10 seconds.
   a. What is the probability that my overseas call will go through in less than one minute?
   b. What is the probability that I will get through in less than 40 seconds?
   c. What is the probability that I will have to wait more than 70 seconds for my call to go through?

31. The number of votes cast in favor of a controversial proposition is believed to be approximately normally distributed with mean 8,000 and standard deviation 1,000. The proposition needs at least 9,322 votes in order to pass. What is the probability that the proposition will pass?

32. Under the system of floating exchange rates, the rate of foreign money to the U.S. dollar is affected by many random factors, and this leads to the assumption of a normal distribution of small daily fluctuations. In a certain period, the rate of German marks per U.S. dollar is believed to have a mean of 2.06 and a standard deviation of 0.08. Find the following.
   a. The probability that tomorrow's rate will be above 2.07.
   b. The probability that tomorrow's rate will be below 2.065.
   c. The probability that tomorrow's exchange rate will be between 2.00 and 2.20.

33. The production level at a plant is believed to be approximately normally distributed with mean 134,786 items per week and standard deviation 13,000. Find the probability that weekly production will exceed 150,000 items. Also find the probability that production will drop below 100,000 units in a given week. Suppose that labor disputes arise and production during the week drops below 80,000. Management blames the union for deliberately slowing down production, while the union claims that production is within acceptable levels of variation. Given the assumptions of normality and the stated mean and standard deviation, do you believe the union? Explain.

34. Certain diet drinks are stated to have, on the average, only 5 calories per serving. If the caloric content of servings is normally distributed with the stated 5 calories as the mean and with standard deviation of 0.5 calorie, what is the probability that a given serving will contain more than 7 calories?

35. Use a computer to find the probability that a normal random variable with mean 17 and standard deviation 5 will have a value below 10.

36. Use a computer as an aid in finding the probability that a normal random variable with mean 125 and standard deviation 14 will exceed 130.

37. Two normal random variables are *independent* of each other. One has mean 150 and standard deviation 10; the other has mean 140 and standard deviation 15.
   a. What is the probability that *both* random variables will exceed 160?
   b. What is the probability that *at least one* of the two random variables will exceed 160?

38. The rate of return on one investment is normally distributed with mean 11 percent and standard deviation 4 percent. Another investment has a rate of return that is normally distributed with mean 9 percent and standard deviation 3 percent. The two investments are independent of each other. What is the probability that both investments will yield at least 7 percent return each? What is the probability that at least one of the two investments will yield at least 6 percent return?

4.4 THE RELATIONSHIP BETWEEN *X* AND *Z* AND THE USE OF THE INVERSE TRANSFORMATION

In this section, we'll look more closely at the relationship between *X*, a normal random variable with mean μ and standard deviation σ, and the standard normal random variable. The fact that the standard normal random variable has mean 0 and standard deviation 1 has some important implications. When we say that *Z* is greater than 2, we are also saying that *Z* is more than 2 *standard deviations above its mean.* This is so because the mean of *Z* is 0 and the standard deviation is 1; hence, $Z > 2$ is the same event as $Z > [0 + 2(1)]$.

Now consider a normal random variable *X* with mean 50 and standard deviation 10. Saying that *X* is greater than 70 is exactly the same as saying that *X* is 2 standard deviations above its mean. This is so because 70 is 20 units above the mean, 50, and 20 units = 2(10) units, or two standard deviations of *X*. Thus, the event $X > 70$ is the same as the event $X >$ (two standard deviations above the mean). This event is identical to the event $Z > 2$. Indeed, this is what results when we carry out the transformation:

$$P(X > 70) = P\left(\frac{X - \mu}{\sigma} > \frac{70 - \mu}{\sigma}\right) = P\left(Z > \frac{70 - 50}{10}\right) = P(Z > 2)$$

Normal random variables are related to each other by the fact that the probability that a normal random variable will be above (or below) its mean a certain number of standard deviations is exactly equal to the probability that any other normal random variable will be above (or below) its mean the same number of (its) standard deviations. In particular, this property holds for the standard normal random variable. The probability that a normal random variable will be greater than (or less than) *z* units above its mean is the same as the probability that the standard normal random variable will be greater than (less than) *z*. The change from a *z value* of the random variable *Z* to *z standard deviations* above the mean for a given normal random variable *X* should suggest to us the inverse transformation:

$$x = \mu + z\sigma$$

That is, the value of the random variable *X* may be written in terms of the number (*z*) of standard deviations (σ) it is above or below the mean (μ). Three examples are useful here. We know from the standard normal probability table that the probability that *Z* is greater than -1 and less than 1 is 0.6826 (show this). Similarly, we know that the probability that *Z* is greater than -2 and less than 2 is 0.9544. Also, the probability that *Z* is greater than -3 and less than 3 is 0.9974. These probabilities may be applied to *any* normal random variable as follows.

> 1. The probability that a normal random variable will be within a distance of *one standard deviation* from its mean (on either side) is 0.6826, or *approximately* 0.68.
> 2. The probability that a normal random variable will be within *two standard deviations* of its mean is 0.9544, or *approximately* 0.95.
> 3. The probability that a normal random variable will be within *three standard deviations* of its mean is 0.9974.

These probability statements are the equivalents of the *empirical rule* of Chapter 1.

We use the inverse transformation when we want to get from a given probability to the value or values of a normal random variable X. Examples 4.3 and 4.4 illustrate the inverse transformation.

EXAMPLE 4.3

The amount of fuel a jetliner consumes on a flight between two cities is a normally distributed random variable, X, with mean $\mu = 5.7$ tons and standard deviation $\sigma = 0.5$. Carrying too much fuel is inefficient because it slows the plane. If, however, too little fuel is loaded on the plane, an emergency landing may be necessary. The airline would like to determine the amount of fuel to load so that there will be a 0.99 probability that the plane will arrive at its destination.

FIGURE 4.16   The Solution of Example 4.3

Courtesy of American Airlines.

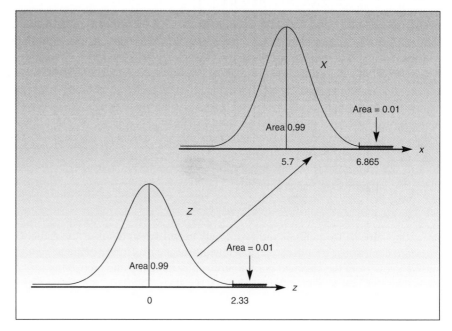

SOLUTION

We have X is N(5.7,0.5). First, we must find the value z such that $P(Z < z) = 0.99$. Following our methodology, we find that the required table area is TA $= 0.99$ and the corresponding z value is 2.33. Transforming the z value to an x value, we get: $x = \mu + z\sigma = 5.7 + (2.33)(0.5) = 6.865$. Thus, the plane should be loaded with 6.865 tons of fuel to give a 0.99 probability that the fuel will last throughout the flight. The transformation is shown in Figure 4.16.

Scores on a psychological test are normally distributed with mean 2,450 and standard deviation 400. Find two values such that there will be a 0.95 probability that a given score will fall between them.

EXAMPLE 4.4

Here X is N(2,450, 400). From the section on the standard normal random variable, we know how to find two values of Z such that the area under the curve between them is 0.95 (or any other area). We find that $z = 1.96$ and $z = -1.96$ are the required values. We now need to use the inverse transformation equation. Since there are *two* values, one the negative of the other, we may combine them in a single transformation:

$$x = \mu \pm z\sigma$$

Applying this special formula we get: $x = 2,450 \pm (1.96)(400) = 1,666$ and $3,234$.

Summary of the procedure of obtaining values of a normal random variable, given a probability:

1. Draw a picture of the normal distribution in question and the standard normal distribution.
2. In the picture, shade in the area corresponding to the probability.
3. Use the table to find the z value (or values) that give the required probability.
4. Use the transformation from Z to X to get the appropriate value (or values) of the original normal random variable.

The MINITAB command INVCDF will help us find the two required values in Example 4.4. We know that if the area between the two required values is 0.95, then the area to the left of the right-hand point is 0.975, and the area to the left of the left-hand point is 0.025. We thus need the INVCDF of these two table areas for a normal random variable with mean 2,450 and standard deviation 400:

```
MTB > INVCDF 0.975;
SUBC> NORMAL 2450 400.
   0.9750  3.23E+03  ← (This is the exponential notation for 3,230.)
MTB > INVCDF 0.025;
SUBC> NORMAL 2450 400.
   0.0250  1.67E+03  ← (This is exponential notation for 1,670.)
```

---

39. If X is a normally distributed random variable with mean 120 and standard deviation 44, find a value x such that the probability that X will be less than x is 0.56.

PROBLEMS

40. For a normal random variable with mean 16.5 and standard deviation 0.8, find a point of the distribution such that 85 percent of the values of the random variable will be above it.
41. In an effort to try to detect signals from extraterrestrial civilizations, NASA launched a large radio telescope in 1992.[4] It was determined that the average frequency to be scanned

---

[4]"ET, Phone Us," *Newsweek,* October 12, 1992, pp. 66–72.

should be 1,500 megaherz, and that frequencies should be randomly selected according to a normal probability distribution with this mean and a standard deviation of 500 megaherz. What are the 90 percent probability bounds for the frequencies to be scanned?

42. Find two values of the normal random variable with mean 88 and standard deviation 5 lying symmetrically on either side of the mean and covering an area of 0.98 between them.

43. For X that is N(32,7), find two values, x_1 and x_2, symmetrically lying on each side of the mean, with $P(x_1 < X < x_2) = 0.99$.

44. If X is a normally distributed random variable with mean -61 and standard deviation 22, find the value such that the probability that the random variable will be above it is 0.25.

45. Let X be a normally distributed random variable with mean 600 and variance 10,000. Find two values x_1 and x_2 such that $P(X > x_1) = 0.01$ and $P(X < x_2) = 0.05$.

46. Pierre operates a currency exchange office at Orly Airport in Paris. His office is open at night when the airport bank is closed, and most of his customers are returning American tourists who need to change their remaining French francs back to U.S. dollars. From experience, Pierre knows that the demand for dollars on any given night during high season is approximately normally distributed with mean $25,000 and standard deviation $5,000. If Pierre carries too much cash in dollars overnight, he pays a penalty: interest on the cash. On the other hand, if he runs short of cash during the night, he must send a person downtown to an all-night financial agency to get the required cash. This, too, is costly to him. Therefore, Pierre would like to carry an amount x overnight such that the probability that demand will be above the amount is equal to 0.15. Can you help Pierre find the required amount of dollars to carry?

47. The demand for high-grade gasoline at a service station is normally distributed with mean 27,009 gallons per day and standard deviation 4,530. Find two values that will give a symmetric 0.95 probability interval for the amount of high-grade gasoline demanded daily.

48. The percentage of protein in a certain brand of dog food is a normally distributed random variable with mean 11.2 percent and standard deviation 0.6 percent. The manufacturer would like to state on the package that the product has a protein content of at least x_1 percent and no more than x_2 percent. The company wants its statement to be true for 99 percent of the packages sold. Determine the values x_1 and x_2.

49. An article on music education reports that students' responses to videotaped practica were found to be normally distributed with a mean of 31 seconds and a standard deviation of 8 seconds.[5] Give symmetric 95 percent probability bounds for all response times.

50. Use a computer to find the point of the normal distribution with mean 19 and standard deviation 6 that cuts off an area of 0.40 to its left.

51. Use a computer to find the point of the normal distribution with mean 253 and standard deviation 20 that has an area of 0.85 to its left.

52. Let X be a normally distributed random variable with mean 97 and standard deviation 10. Find x such that $P(102 < X < x) = 0.05$.

4.5 THE NORMAL DISTRIBUTION AS AN APPROXIMATION TO OTHER PROBABILITY DISTRIBUTIONS

optional

▲ The normal distribution is, by its very nature, the limiting probability distribution in many situations. In fact, the majority of the probability distributions discussed in this book tend to approach the normal distribution as some quantity relevant to the distribution, such as sample size, increases indefinitely. For example, the *t* distribution and the chi-square distribution—two important probability distributions used in statistics and introduced in the next chapters—tend toward the normal distribution as the sample size from which they arise increases.

---

[5]E. Price, "Sequential Patterns of Music Instruction and Learning to Use Them," *Journal of Research in Music Education,* Spring 1992, pp. 14–29.

Approximating a Binomial Distribution with $n = 7$ and $p = 0.5$
by a Normal Distribution (Example 4.5)

FIGURE 4.17

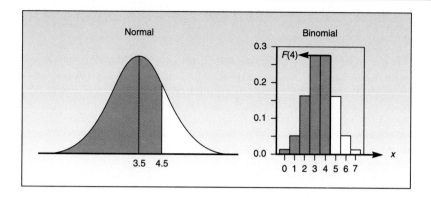

As n, the number of trials in the binomial experiment, increases, the discrete binomial probability distribution tends toward the continuous normal probability distribution. This means that, for large n, we may approximate a binomial probability with the probability obtained for a normal random variable with the same mean and the same standard deviation as the binomial. Example 4.5 illustrates.

EXAMPLE 4.5

Suppose that the proportion of people who use a certain product is $p = 0.50$. From a random sample of seven people, we want to evaluate the probability that at most four of them will be product users. We want to do this both directly and by way of a normal approximation. The situation is demonstrated in Figure 4.17, where the sum of the weights to the left of and including 4 is shown. Also shown is the normal approximation to the required probability: the area under the normal curve to the left of the point 4.5. (We use 4.5 because we want to include 4 but not 5, and the best way to do this—using a continuous approximation to a discrete distribution[6]—is to draw the line centered between the two points 4 and 5.)

The normal distribution we use has a mean and a standard deviation identical to those of the given binomial. We have $\mu = np = (0.5)(7) = 3.5$ and $\sigma = \sqrt{np(1 - p)} = \sqrt{(7)(0.5)(0.5)} = 1.323$. The normal approximation to the binomial probability is:

$$P(X < 4.5) = P\left(Z < \frac{4.5 - \mu}{\sigma}\right)$$

$$= P\left(Z < \frac{4.5 - 3.5}{1.323}\right) = P(Z < 0.76) = 0.7764$$

The true probability, obtained from the binominal table, is 0.7734. Thus, the approximation, even with a small number of trials ($n = 7$), is quite good.

[6]This is sometimes called a *continuity correction*.

In the following problems, use a normal distribution to compute the required probabilities. Whenever you can, compare your answers with the true binomial probabilities obtained from the binomial table, Table B.1 in Appendix B. In each problem, also state the assumptions necessary for a binomial distribution, and indicate whether the assumptions are reasonable.

53. The manager of a restaurant knows from experience that 70 percent of the people who make reservations for the evening show up for dinner. The manager decides one evening to overbook and accept 20 reservations when only 15 tables are available. What is the probability that more than 15 parties will show up?

54. An advertising research study indicates that 40 percent of the viewers exposed to an advertisement try the product during the following four months. If 100 people are exposed to the ad, what is the probability that at least 20 of them will try the product in the following four months?

55. A computer system contains 45 identical microchips. The probability that any microchip will be in working order at a given time is 0.80. A certain operation requires that at least 30 of the chips be in working order. What is the probability that the operation will be carried out successfully?

56. Sixty percent of the managers who enroll in a special training program will successfully complete the program. If a large company sends 28 of its managers to enroll in the program, what is the probability that at least half of them will pass?

57. A large state university sends recruiters throughout the state in order to recruit graduating high school seniors to enroll in the university. From the university's records, it is known that 25 percent of the students who are interviewed by the recruiters actually enroll. If last spring the university recruiters interviewed 1,889 graduating seniors, what is the probability that at least 500 of them will enroll this fall?

4.6 NORMAL DATA

In statistics, we deal with data. The idea of a random variable in statistics is therefore often linked with the analysis of data. We have already seen a good example of how this works in the context of the normal distribution in the *empirical rule.*

A random variable is a kind of "black box." It randomly produces observations based on some predetermined probability law—its probability distribution. A population of observations or measurements, or a sample of such measurements, is often assumed to have been produced by a random variable. This can be demonstrated as follows:

$$\text{RANDOM VARIABLE} \rightarrow \text{Data: } x_1, x_2, x_3, x_4, x_5, x_6, x_7, x_8 \ldots$$

If a data set has indeed been produced by a normal random variable, then if we should look at a histogram or other graph of the relative frequencies of occurrence of the values in our data set, this graph should somehow mimic the graph of the density of a normally distributed random variable.

The empirical rule is just a restatement of this fact using particular percentiles of the normal distribution. More specifically, since there is a 0.6826 probability that a normal random variable will fall within one standard deviation of the mean (check this using the Z table), roughly 68 percent of observations that were produced by such a random variable will fall within one standard deviation of their mean. Similarly, roughly 95 percent of the observations will fall within two standard deviations of the mean and 99.7 percent will be within three standard deviations if data were derived from a normal distribution. There is nothing magical about these percentages; they simply come from the standard normal distribution table. One could equally find other percentiles of the normal distribution and apply them to data sets that one believes are roughly mound-shaped—that is, possibly having been derived as realizations of a normal random variable.

For example, if I have a data set of 1,000 observations believed to be normally distributed, and their mean is 155 and their standard deviation is 10, what percentage of them do I expect to fall below 175? The point 175 is two standard deviations above the mean. Thus its z value is 2, because $(175 - 155)/10 = 2$. From the normal distribution table I find that the area to the left of 2.00 is 0.9772, so roughly 977 of my 1,000 observations would be below 175.

But how can we tell whether observations follow a normal distribution? There are statistical tests that will be able to make such a determination. At this point, however, we will look at a simpler, visual way of assessing the normality of our observations. We will use the MINITAB command NSCORES. We take the data set in column C1 and compute their "nscores," against which we then plot the data. If the graph looks like a straight line, then the data set is at least roughly normally distributed.

EXAMPLE 4.6

The weight of a diamond is measured in carats. In this context (as opposed to gold), a carat is ¹⁄₂₀ of a gram. The following data are carat measurements of diamonds arriving at an auction house in Antwerp, Belgium (the measurements have been ordered from smallest to largest):

0.6, 0.8, 1, 1.2, 1.3, 1.5, 1.7, 1.7, 1.8, 1.9, 1.9, 2, 2, 2, 2, 2.1, 2.2, 2.2, 2.2, 2.3, 2.3, 2.5, 2.8

In trying to evaluate diamond prices, some interest centers on the carat distribution of the stones. Is the distribution close to normal?

SOLUTION

We will use MINITAB.

```
MTB > SET C1
DATA> .6 .8 1 1.2 1.3 1.5 1.7 1.7 1.8 1.9 1.9 2 2 2 2 2.1 2.2 2.2 2.2 2.3 2.3 2.5 2.8
DATA> END
MTB > NSCORES C1 put in C2
MTB > PLOT C1 C2
```

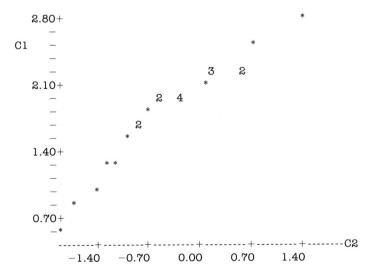

Courtesy Tino Hammid.

Since the plot does not look like a straight line, the distribution of these data is probably not close to normal.

PROBLEMS

The following data sets, from United Nations publications, are mean age at first marriage for men and women in different countries.

	Males		Females		Difference between the Sexes	
Country	Mean Age at First Marriage	Average Annual Change, 1950–85	Mean Age at First Marriage	Average Annual Change, 1950–85	In Age at First Marriage	Average Annual Change
Asia and the Middle East						
Brunei	26.1	.01	25.0	.20	1.1	−.19
Hong Kong	29.2	.02	26.6	.19	2.6	−.17
Indonesia	24.8	.07	21.1	.13	3.7	−.06
Japan	29.5	.08	25.8	.04	3.7	.04
Korea	27.8	.11	24.7	.14	3.1	−.03
Malaysia	26.6	.07	23.5	.15	3.1	−.08
Nepal	21.5	.07	17.9	.07	3.6	.0
Philippines	25.3	.01	22.4	.01	2.9	.0
Singapore	28.4	.10	26.2	.26	2.2	−.16
Thailand	24.7	.01	22.7	.05	2.0	−.04
Bangladesh	23.9	−.01	16.7	.04	7.2	−.05
India	23.4	.09	18.7	.11	4.7	−.02
Pakistan	24.9	.09	19.8	.10	5.1	−.01
Sri Lanka	27.9	.03	24.4	.12	3.5	−.09
Algeria	25.3	−.02	21.0	.03	4.3	−.05
Cyprus	26.3	.11	24.2	.09	2.1	.02
Egypt	26.9	.05	21.4	.08	5.5	−.03
Iraq	25.2	−.06	20.8	.01	4.4	−.07
Iran	24.2	−.07	19.7	.12	4.5	−.19
Israel	26.1	.02	23.5	.10	2.6	−.08
Jordan	26.8	.10	22.8	.12	4.0	−.02
Kuwait	25.2	.01	22.4	.18	2.8	−.17
Morocco	27.2	.09	22.3	.17	4.9	−.08
Syria	25.7	.02	21.5	.09	4.2	−.07
Tunisia	27.8	.02	24.3	.18	3.5	−.16
Turkey	23.6	.06	20.7	.07	2.9	−.01
North America, Oceania, and Europe						
United States	25.2	.05	23.3	.08	1.9	−.03
Canada	25.2	.0	23.1	.02	2.1	−.02
Australia	25.7	.01	23.5	.09	2.2	−.08
New Zealand	24.9	−.03	22.8	.02	2.1	−.05
Denmark	28.4	.06	25.6	.13	2.8	−.07
Finland	27.1	.04	24.6	.06	2.5	−.02
Norway	26.3	−.05	24.0	.03	2.3	−.08
Sweden	30.0	.10	27.6	.19	2.4	−.09
Ireland	24.4	−.23	23.4	−.11	1.0	−.12
England	25.4	−.02	23.1	.03	2.3	−.05
Austria	27.0	−.02	23.5	−.03	3.5	.01
Belgium	24.8	−.05	22.4	−.03	2.4	−.02
France	26.4	.0	24.5	.05	1.9	−.05
West Germany	27.9	.01	23.6	−.03	4.3	.04
Luxembourg	26.2	−.08	23.1	−.05	3.1	−.03

Country	Males		Females		Difference between the Sexes	
	Mean Age at First Marriage	Average Annual Change, 1950–85	Mean Age at First Marriage	Average Annual Change, 1950–85	In Age at First Marriage	Average Annual Change
North America, Oceania, and Europe						
Netherlands, The	26.2	−.04	23.2	−.05	3.0	.01
Switzerland	27.9	−.01	25.0	.01	2.9	−.02
Greece	27.6	−.07	22.5	−.11	5.1	.04
Italy	27.1	−.05	23.2	−.05	3.9	.0
Portugal	24.7	−.08	22.1	−.08	2.6	.0
Spain	26.0	−.10	23.1	−.11	2.9	−.01
Bulgaria	24.5	.03	20.8	−.01	3.7	.04
Czechoslovakia	24.7	−.08	21.7	−.04	3.0	−.04
East Germany	25.4	−.01	21.7	−.07	3.7	.06
Hungary	24.8	−.06	21.0	−.05	3.8	−.01
Poland	25.9	.02	22.8	.04	3.1	−.02
Romania	24.9	.04	21.1	.08	3.8	−.04
USSR	24.2	.0	21.8	.07	2.4	−.07
Yugoslavia	26.1	.07	22.2	.0	3.9	.07
Caribbean, Central America, and South America						
Cuba	23.5	−.09	19.9	−.08	3.6	−.01
Dominican Republic	26.1	.02	19.7	.05	6.4	−.03
Haiti	27.3	−.04	23.8	.06	3.5	−.10
Trinidad	27.9	.05	22.3	.12	5.6	−.07
Costa Rica	25.1	−.03	22.2	.01	2.9	−.04
El Salvador	24.7	−.03	19.4	−.01	5.3	−.02
Guatemala	23.5	−.02	20.5	.06	3.0	−.08
Honduras	24.4	−.05	20.0	.16	4.4	−.21
Mexico	24.1	−.02	20.6	−.03	3.5	.01
Nicaragua	24.6	−.08	20.2	.01	4.4	−.09
Panama	25.0	.01	21.3	.10	3.7	−.09
Argentina	25.3	−.07	22.9	−.01	2.4	−.06
Bolivia	24.5	.0	22.1	−.02	2.4	.02
Brazil	25.3	−.09	22.6	−.04	2.7	−.05
Chile	25.7	−.04	23.6	.0	2.1	−.04
Colombia	25.9	−.04	22.6	.03	3.3	−.07
Ecuador	24.3	−.04	21.1	.0	3.2	−.04
Paraguay	26.0	−.02	21.8	.03	4.2	−.05
Peru	25.7	.01	22.7	.05	3.0	−.04
Uruguay	25.4	−.13	22.4	−.03	3.0	−.10
Venezuela	24.8	−.05	21.2	.10	2.6	−.15
Africa						
Benin	24.9	.0	18.3	.07	6.6	−.07
Central African Republic	23.3	.04	18.4	.07	4.9	−.03
Congo	27.0	.13	21.9	.18	5.1	−.05
Ghana	26.9	.06	19.4	.15	7.5	−.09
Kenya	25.5	.08	20.3	.11	5.2	−.03
Mali	27.3	.05	16.4	.01	10.9	.04
Liberia	26.6	.03	19.4	.12	7.2	−.09
Mauritius	27.5	.06	23.8	.15	4.7	−.09
Mozambique	22.7	−.04	17.6	−.06	5.1	.02
Réunion	28.1	.03	25.8	.10	2.3	−.07
Senegal	28.3	.02	18.3	.05	10.0	−.03
South Africa	27.8	.02	25.7	.10	2.1	−.08
Togo	26.5	.07	17.6	−.07	8.9	.0
Tanzania	24.9	.07	19.1	.11	5.8	−.04
Zambia	25.1	.06	19.4	.11	5.7	−.05

Reprinted from T. Bergstrom and M. Bagnoli, "Courtship as a Waiting Game," *Journal of Political Economy* 101, no. 1 (1993), pp. 185–201.

problems 58–63

58. Is the mean age at first marriage for men in Asia and the Middle East normally distributed?
59. Is the same variable in Problem 58 normally distributed for the data of North America, Oceania, and Europe?
60. Answer questions 58 and 59 for women in these regions.
61. Combine all the data in the table for men only and check for normality.
62. Combine all the data in the table for women only and check for normality.
63. Look at some of the data sets in Chapter 1 and check for normality.

SUMMARY

In this chapter, we introduced a very important continuous random variable, the **normal random variable** (also called **Gaussian random variable**). We saw how probabilities for a normal random variable may be obtained from a table of the areas under the standard normal density. We saw how such probabilities may apply to any normal random variable with any mean and any standard deviation by applying the **normal transformation.** Similarly, we learned how to find values for any normal random variable that will satisfy specific probability statements. We saw how the normal distribution can be used as an approximation for the distribution of other random variables. We discussed normally distributed data sets, and saw how they related to the **empirical rule** of Chapter 1.

KEY FORMULAS

The normal transformation:

$$Z = \frac{X - \mu}{\sigma}$$

The inverse normal transformation:

$$x = \mu + z\sigma$$

ADDITIONAL PROBLEMS

64. Scores on a management aptitude examination are believed to be normally distributed with mean 650 (out of a total of 800 possible points) and standard deviation 50. What is the probability that a randomly chosen manager will achieve a score above 700? What is the probability that the score will be below 750?
65. An article on children's learning skills reports that the average score on solving a complicated problem was found to be 2.7.[7] If scores are normally distributed with this mean and a standard deviation of 0.5, what is the probability that a given child will score above 3.0?
66. Viliuisk encephalomyelitis is a progressive neurological disorder that is found only among the Iakut people of Siberia. The symptoms of this illness—fever, headache, and somnolence—tend to fade an average of four and a half weeks after they appear, and there

---

[7]L. English, "Children's Use of Domain-Specific Knowledge and Domain-General Strategies in Novel Problem Solving," *British Journal of Educational Psychology* 62 (1992), pp. 203–16.

is a standard deviation of one week.[8] Assuming a normal distribution, give 90 percent probability bounds for the time the symptoms fade.

67. A serious cause of danger to marine environments is the increasing demand in the United States and abroad for tropical marine aquariums. Unlike in freshwater aquariums, the fish and other elements in marine aquariums, such as small reefs and plants, are all taken from the wild. Outside their natural environment, a majority of the beautiful marine creatures die before they reach their destinations.[9] The average time a particular fish can survive in transit outside its natural environment is 58 hours, and the standard deviation is 18 hours. Give an upper 99 percent probability bound for the number of hours a fish can survive. Assume a normal distribution.

68. In an experiment to find ways to encourage people to stop smoking, the method used was to try to force subjects to inhale cigarette smoke as quickly as possible. The average number of puffs a subject could take in this experiment before having to quit was 51, and the standard deviation was 5.[10] Assuming a normal distribution, and that these results extend to all people, what is the probability that a randomly selected person could take more than 60 rapid puffs?

69. An article on the relation of cholesterol levels in human blood to aging reports that the average cholesterol level for women aged 70–74 was found to be 230 mg/dl.[11] If the standard deviation was 20 mg/dl and the distribution normal, what is the probability that a given woman in this age group would have a cholesterol level over 200 mg/dl?

70. Models of stock options pricing make the assumption of a normal distribution. An analyst believes that the price of an IBM stock option with expiration date December 12, 1995, is a normally distributed random variable with mean $8.95 and variance 4. The analyst would like to determine a 0.90 probability lower limit for the price of the option. Find the required limit.

71. Weekly rates of return (on an annualized basis) for certain securities over a given period of time are believed to be normally distributed with mean 8.00 percent and variance 0.25. Give two values, x_1 and x_2, such that you are 95 percent sure that annualized weekly returns will be between the two values.

72. The impact of a television commercial, measured in terms of excess sales volume over a given time period, is believed to be approximately normally distributed with mean 50,000 and variance 9,000,000. Find 0.99 probability bounds on the volume of excess sales that would result from a given airing of the commercial.

73. A travel agency believes that the number of people who sign up for tours to Hawaii during the Christmas–New Year holiday season is an approximately normally distributed random variable with mean 2,348 and standard deviation 762. For reservation purposes, the agency's management wants to find the number of people such that the probability is 0.85 that at least that many people will sign up. They also need 0.80 probability bounds on the number of people who will sign up for the trip.

74. A loan manager at a large bank believes that the percentage of her customers who default on their loans during each quarter is an approximately normally distributed random variable with mean 12.1 percent and standard deviation 2.5 percent. Give a lower bound, x, with 0.75 probability that the percentage of people defaulting on their loans is at least x. Also give an upper bound, x', with 0.75 probability that the percentage of loan defaulters is below x'.

---

[8]L. Goldfarb and D. Gajdusek, "Viliuisk Encephalomyelitis in the Iakut People of Siberia," *Brain* 115 (1992), pp. 961–78.

[9]M. Derr, "Raiders of the Reef," *Audubon,* March/April 1992, pp. 48–56.

[10]M. Jarvik et al., "Measuring the Stop Point of Rapid Smoking," *Behavior Research Methods, Instruments & Computers* 24, no. 3 (1992), pp. 420–22.

[11]C. Newschaffer, T. Bush, and W. Hale, "Aging and Total Cholesterol Levels: Cohort, Period, and Survivorship Effects," *American Journal of Epidemiology* 136, no. 1 (1992), pp. 23–31.

75. The Tourist Delivery Program was developed by several European automakers such as Mercedes and Volvo. In this program, a tourist from outside Europe—most are from the United States—may purchase an automobile in Europe and drive it in Europe for as long as six months, after which the manufacturer will ship the car to the tourist's home destination at no additional cost. In addition to the time limitations imposed, some countries impose mileage restrictions so that tourists will not misuse the privileges of the program. In setting the limitation, some countries use a normal distribution assumption. It is believed that the average number of miles driven by a tourist in the program is normally distributed with mean 4,500 and standard deviation 1,800. If a country wants to set the mileage limit at a point such that 80 percent of the tourists in the program will want to drive fewer miles, what should the limit be?

76. The number of newspapers demanded daily in a large metropolitan area is believed to be an approximately normally distributed random variable. If more newspapers are demanded than are printed, the paper suffers an opportunity loss, in that it could have sold more papers, and also a loss of public goodwill. On the other hand, if more papers are printed than will be demanded, the unsold papers are returned to the newspaper office at a loss. Suppose that management believes it is most important to guard against the first type of error, unmet demand, and would like to set the number of papers printed at a level such that 75 percent of the time, demand for newspapers will be below that point. How many papers should be printed daily if the average demand is 34,750 papers and the standard deviation of demand is 3,560?

77. Arrival times of airplanes at their destination airports are normally distributed. Air traffic controllers must deal with these random arrivals in arranging for a free runway for a landing. The arrival time depends on many factors, such as weather conditions, flight duration, and flight altitude. Suppose that for a certain type of flight, arrival time is a normally distributed random variable with mean 42.0 minutes after the hour and standard deviation 2.4 minutes. If an air traffic controller wants to hold a runway for the plane until a time such that she can be 98 percent sure that the plane will not arrive any later, until what time should the controller wait?

▲ 78. Pollution has caused the death of large numbers of trees in German forests. The proportion of dying trees in a certain region is as high as 0.30. Sometimes it happens that so many trees die, there are not enough trees to prevent erosion of the soil, and this causes the destruction of the entire forest. In a particular location, there are 238 trees. Scientists believe that if pollution should kill as many as 50 trees, all 238 trees in the location will eventually die. Assuming that the probability that any single tree will die from pollution is indeed 0.30, and that trees are affected independently of each other, what is the probability that out of 238 trees at least 50 will die, thus endangering the forest?

79. The amount of electricity demanded in a midsized town is normally distributed. If the utility should decide to produce exactly the mean electric demand for the month, what is the probability that demand will be less than production?

80. In setting warranties on appliances, manufacturers want to set the time limit in such a way that few appliances will have to be repaired at the manufacturer's expense, yet they want the buyer to have some degree of protection against malfunctions over a period of time after the purchase. A manufacturer of a certain appliance would like to set the expiration time of the warranty at such a level that 90 percent of the appliances made will remain in working order throughout the period. What should the time period be if the life of the appliance is normally distributed with mean 38 months and standard deviation 11 months?

INTERVIEW WITH
Bud Goode

"When the Green Bay Packers pay $14,000,000 to Reggie White, the ex-Eagle defensive end, they are getting a player who earns 16 sacks in a 16-game season."

Bud Goode lives in Studio City, California, and is active as a statistical consultant to many of the National Football League coaches. His statistical services have been purchased by 21 of 28 teams. Goode has written for The New York Times, Sports Illustrated, The Washington Post, Inside Sports, "PRO" *magazine, and the Super Bowl program. In August 1993, he was recognized by the American Statistical Association for his pioneering work in statistical analysis of sports and he is currently active in teaching statistics concepts with sports at Belmont High School in Los Angeles, California.*

Aczel: How did you first become interested in statistics and what about the field caught your attention?

Goode: I took my first course in introductory statistics at Occidental College in 1941 as a freshman in the Department of Physical Education. The department head, Dr. Carl Trieb, turned out 9 of the top 10 scores in the annual test for phys-ed teachers (in the Los Angeles City Unified School District). Simple reason: His brother, Dr. Martin Trieb, ran the Boys' Phys Ed Department.

The course was called "Tests and Measurements." Our text was the little red book by Professor R. A. Fisher. I still have it. The concepts of central ten-

dency, variation, and correlation were so appealing I thought of giving up religion for statistics. The war came along and when I returned, having been blessed with a full disability pardon from the service in 1946, I would no longer fill the physical requirements of phys ed. I used that as an excuse to move to another major (psychology—and more statistics).

Aczel: Did you ever expect that the ASA would honor your work as they did in August of 1993?

Goode: No. Although I had started my research circa 1961 (the head writer on the Groucho Marx "You Bet Your Life" TV show was responsible for this new direction), I had, indeed, been doing basic statistical research, using an old 1403 computer and later the IBM 7094 at UCLA, but I did not know how much my work had penetrated the ASA hallways. It was a surprise when Stu called me.

Aczel: When did you find out?

Goode: ASA President Elect, Stu Hunter, called before the Boston convention in 92 and advised that he was considering some kind of luncheon with me speaking for 30 minutes . . . but I couldn't make it to Boston because of modest health problems. He hoped I could do both the Boston and San Francisco conventions. I obviously made the trip to 'Frisco.

Aczel: You are interested in statistics education at the elementary level. How did that get started?

Goode: My interest in early schooling/teaching statistics was part of my life: I had earned a teaching credential [Calif. State General Secondary Credential, Science] circa 1950. I have always been a frustrated teacher. When we had the 50th anniversary of Belmont High School graduating class, my close childhood friend, three-star General Thomas Howard Tackaberry, III, flew out from his home in North Carolina. We were both class officers in 41. The principal, Dr. Martha Helm [*sic*] spoke to us and when she learned my clients (in what I call my retirement) were NFL coaches, she asked if I would visit the school as a speaker.

I did, and have been going back every Friday since, as a volunteer teacher. I teach introductory statistics, making the Monday Night Game of the Week a homework assignment. It works. We move on to the reasons why students are successful in school, and why some folks are more successful in life than others.

Aczel: Let's turn to sports—football in particular—when did you start analyzing football?

Goode: I started the sports/football analysis in 1960, with a prediction of the Rose Bowl in January 1961. This was a result of the Groucho Marx head writer invitation to bid on a marketing/promotion account for a computer service bureau.... When asked by Groucho's head writer (the late Hy Friedman) to bid on a marketing/promotion account for the computerized service bureau he was taking public, I suggested we do promotional pieces on "... computer looks at Rose Bowl statistics ... and predicts the winner."

That suggestion sold the idea to the board of directors of a small holding company. The computer company went public, and I started to do the marketing and promotion.

What happened? Six of the seven local TV stations brought their camera crews to the computer company and did interviews; 13 radio station interviews and a page in the Sunday *Herald/Examiner* saturated the Los Angeles area.

It was an immediate success.

In all fairness, we did not use the computer at the time. We did scattergrams of the stats that came out of the Football News and used variables where the scattergrams pictured strong relationships with our three measures of success—points scored on offense, points allowed on defense, and the difference between the two (the winning margin). It wasn't until the company (Universal Data Processing) moved from the IBM 1401 level to the 1403 that we had a statistical package to work with.

In the mid-60s I realized that I was able to "plant" sports page stories on every major event where the stats were seen through the electronic eyes of the hardware. If I could give the idea away for free, could I also sell it? So I took the idea to the *L.A. Times* sports editor. He turned it down. "I don't like statistics features," he said.

Okay. There was still a second newspaper in town, the *L.A. Herald/Examiner*, where the then managing editor, Don Goodenow, said, "Great idea!" and bought it immediately. From there it was sold to the *Long Beach Independent*. Managing Editor, the late Miles Sines, also thought it was a novel idea to look at sports stats in an objective manner. He bought the weekly column and told the Des Moines Register and the Tribune Syndicate about it.

They sold it nationally in places like the *Boston Globe, Cleveland Plain Dealer, Minneapolis Tribune* . . . and I sold the *Washington Post, Oakland Tribune, Chicago Today,* and later the *New York Times*. From 1963–1971, "Bud Goode's Computer Corner on Sports" was seen in 36 metro papers on a weekly basis (not just football, but baseball, basketball, golf, Indy 500—wherever there was reliable data).

Parenthetically, my dear wife, Betty, kept another newspaper syndicate check for a given season—$3.77. Syndicates have a funny way of keeping books and find ways to add "additional expenses," which are deducted from the author's receipts. Three bucks for a season's work! It doesn't seem fair.

Aczel: What was the most "catchy" statistic you kept in that early period?

Goode: In looking at football I created a correlation matrix and used several columns as criteria—points scored (column 43), points allowed (column 44), winning margin (column 123), and points by offense only and defense only. In the beginning I had not "transformed" many (or any) of the variables to create efficiency measures. So when I first was able to produce a correlation matrix, I felt a great deal of pleasure (power!). Here was a 70 × 70 matrix of correlations, and among the largest correlates with winning margin I found the interception.

So I walked into Art Linkletter's office (Link was one of my three bread and butter clients in showbiz) and knowing Art to be an avid sports fan, I asked, "Which is the most important passing stat?"

He replied, "The bomb!"

"Wrong," I said. "It's the interception!" And I proceeded to explain the correlation matrix.

Sometime later, when I began using the BMDP statpack, and transformed a number of variables, the "bomb"—which measures yards per pass attempt—emerged as the significant variable on both offense and defense. I apologized to Linkletter at lunch. We see one another on his birthday almost every year, and laugh about the anecdote.

Aczel: Now, you indicated that NFL teams contact you regularly. Which teams and for what kind of data?

Goode: Twenty-one of the 28 teams have been clients at one time or another. Coach Knox has been the longest lived (Rams, Buffalo Bills, Seattle Seahawks, and Rams again this year). Coach Paul Wiggin, who was my contact at the Saints, and now at the Vikes, called one day and said, "Bud, I'm looking at the stats for three teams. Do you have team A, B, and C as clients?"

"Yes, Coach."

"Bud, you're changing the game."

It may be true. The short passing game, which I have long thought filled the role of a running play, does add predictive and explanatory variance. So I encourage my clients to use the short ball control game and consider these dump passes as running plays.

Aczel: What particular NFL statistics did you come up with or originate that a fan would recognize?

Goode: First, we must consider the errors of omission and commission made by radio/TV analysts. For example, fans recognize that turnovers are important. They may not know, however, that the interception is more than twice as important as the lost or recovered fumble. The reason: When a team is leading they tend to run the ball more to move the clock and protect the lead. So there are more interceptions when a team falls behind and goes to the air to score in the shortest possible time.

By comparison, the fumble is linked to the running game, and the running game is linked to the team with the lead. So it is not so much discovering or creating new stats. Determining the importance of

each stat *vis à vis* our three key criterion measures adds understanding.

Creating unit values, for example, is one way to measure the importance of a stat. Each variable contributes some point value to the offense, defense, and winning margin. The unit value is computed by multiplying the correlation coef. by the ratio of the two standard deviations. It shows the impact of a unit change in the independent variable on the criterion measure. In this sense, the interception is worth (rule of thumb) five points in the winning margin, while the lost or recovered fumble is valued at two–three points (depending on the year).

Indeed, teams have gone to the Super Bowl and ranked last in the league on lost or recovered fumbles (Dolphins are one example).

When Vince Lombardi joined the Packers the run/pass ratio almost doubled in his first season moving from 1–to–1 (one running play for each passing play) to almost 2–to–1. And the Packers were winners.

The stat that *Sports Illustrated* emphasized in their 1974 profile, yards per pass attempt, is my favorite Goode stat. Joe Marshall, once pro football editor at *SI,* who wrote the piece, said, "Bud wants his tombstone to read, 'Here lies Goode. He told the pro football world about the importance of Yards per Pass Attempt.' "

When the Saints were a client, they improved from 4.3 yards per toss in 1977 to a league-leading high of 7.1. Adding one yard per attempt increases the winning margin by 3½ points per game (unit value). So the Saints' increase in their winning margin (3.5 ×

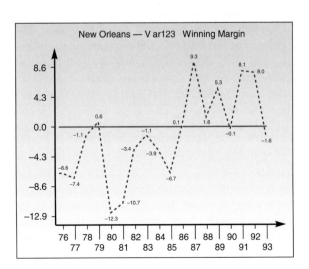

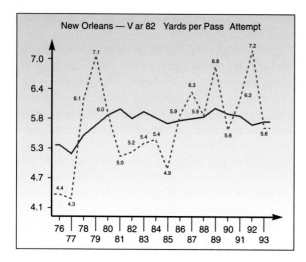

New Orleans — V ar 82   Yards per Pass   Attempt

2.8) should add a theoretical 9.8 points to the margin (they actually moved from −7.4 to +.6, a change of 8 points) . . . and went 8−8 for the first time in their long history. (See the New Orleans Saints charts.)

The qb sack is also a stat which I emphasized and which fans are familiar with. One sack is worth 3 points in the winning margin (plus or minus a fractional change from year to year). This is a heavy impact on the margin and probably makes the sack the most underappreciated by the fans.

When the Green Bay Packers pay $14,000,000 to Reggie White, the ex-Eagle defensive end, they are getting a player who earns 16 sacks in a 16-game season (or one sack per game). This sets the performance limit for the best man at the position. Some of my research shows results that are counterintuitive. As an example, the correlations between fumbles and the criterion measures are smaller than the correlations between interceptions and the criteria. The reason: The lost fumble correlates with the number of running plays. And since running the ball (a clock-eating stat) correlates with being ahead on the scoreboard, the lost fumble, for a strong running team, may be a sign of strength.

Aczel: What types of statistical techniques do you use?

Goode: We do a weekly summary of stats. We use basics like the mean, median, trimmed mean, standard error, standard deviation, coefficient of variation, maximum, minimum, range, correlation and correlation squared (can be interpreted as a percentage relationship between the dependent and indepen-

dent variables), the significance of the difference between two means (*t*-test). Other reports compute a correlation matrix, regression equations using Monte Carlo simulation.

When there are rules changes which impact these descriptive stats, the changes are reflected in the correlations—when the range increases, for example, as a result of clock changes (more plays per game), the range and standard deviation change.

Aczel: Have you tried to correlate winning records with the franchise data levels?

Goode: Franchise data level? I'm not sure I understand.

Aczel: Well, which teams do the best and the worst?

Goode: The St. Louis Cardinals (then Phoenix Cardinals and now Arizona Cardinals) have historically been weak on defensive passing efficiency (defined in a statistical sense by variable 83, opponent yards per pass attempt allowed). Passing efficiency on offense and defense is one of the vectors in a seven-dimensional space. If a team is consistently weak in an important area it is impossible to hide the weakness from the scouting report or from Goode's computer-aided stat analysis. The weakness shows up.

In their wisdom, the Cardinals, puny on pass defense, in one recent season, drafted a punter, not a defensive back. In addition, after three seasons of continued improvement under their recent coach, they fired him (yet Cards were improved every year under his direction).

Aczel: What kind of "errors" do you have to deal with regularly in the data you get? How have you resolved those?

Goode: I have two editing programs that clean the data. It takes time, but most of the changes are caught.

Aczel: Are there better and worse sources of data?

Goode: Data come from Elias Sports. In the beginning, data for college football came from NCAA and other sources.

Aczel: Are there statistical lessons for students to learn from your sports experience?

Goode: Absolutely. It is the main reason I have been a volunteer teacher at Belmont. Belmont is the largest high school in California, 90 percent Latino. I have written the coaching LEGENDS in Spanish so the non–English-speaking students can understand the reports.

My approach is to teach them introductory statistics tied to a hot button subject (sports—making the Monday Night Game of the Week, for example, a homework assignment). Two Hispanic students entered the North American Rockwell Annual Computer Sciences Competition this semester and won one of the 15 awards. This is the first time in the 70-year history of Belmont that Latino kids entered and won anything.

As a result the school and the L.A. Unified School District/Industry-aided group have funded the program. They are buying us a new 486 with wings, and I've asked BMDP to load their package (done) that we will begin using this fall.

They learn that statistics are ideas about numbers: First, the idea of an average; second, the idea of hot and cold variation around the average; third, the idea of the difference between averages; and finally, not a difference, but a likeness—how are two variables associated, linked, correlated?

From these concepts they learn that there is a correlation between education and income; an average performance in school and a strong or weak performance; a change or difference in performance if they do more than 15 minutes of homework; a difference in the interest rate they pay the bank or credit card and the interest the bank pays them.

Let's Take a Sample

AP/World Wide Photos

5.1 Statistical Inference   192

5.2 Sampling Distributions   199

5.3 The Sampling Distribution of the Sample Mean   200

5.4 Confidence Intervals   205

5.5 Confidence Intervals for μ When σ Is Unknown—The t Distribution   214

5.6 The Sampling Distribution of the Sample Proportion, $\hat{P}$ 222

5.7 Sample Size Determination   228

5.8 Epilogue: The Polls   234

N othing demonstrates the raw power of statistics more clearly than the presidential election polls. Consider the behavior of former president George Bush while campaigning unsuccessfully for reelection in 1992. For months the national polls consistently showed the president lagging by 8 or 10 percentage points behind challenger Bill Clinton. Throughout this period, Bush dismissed most polls as meaningless and even called people who conduct them "nutty pollsters." Then on October 28, 1992, a Gallup poll conducted for CNN and *USA Today* showed Clinton leading Bush by only 2 percentage points. At once Bush hailed what he called "good news in the national polls."[1] Unfortunately for Bush, his change in attitude toward statistical polling was not enough to change the outcome of the election.

---

[1]M. Kranish and S. Lehigh, "Bush Hails New Poll," *Boston Globe,* October 29, 1992, p. 1.

5.1 STATISTICAL INFERENCE

A well-conducted poll is an excellent example of **statistical inference.** To make the inference, we draw a *random sample* from some large population (to continue the example in the chapter opener, the population of all voters in the United States); measure a characteristic of the sample (the percentage of those who say they'll vote for a certain candidate); and then apply what we found in the sample to the entire population. This is done by specifying a "margin of error" for the sample finding, which, when added to and subtracted from the sample result, gives us a range of values. We can be highly confident that this range contains the actual percentage of those in the entire population who will vote for the candidate. Such a range, called a *confidence interval,* reflects the fact that the sample percentage will usually not be exactly equal to the population percentage, but will likely be in the neighborhood of values around it. As we will see in this chapter, the sample percentage is a *random variable,* and probability theory is used in setting the boundaries of the confidence interval.

It was the implied construction of such a confidence interval, based on the results of a Gallup poll, that gave George Bush hope of winning the 1992 election. The poll indicated a sample percentage of 38 percent for Bush versus 40 percent for Clinton; it had a margin of error of ±3 percent. To draw an inference about the percentage of voters for Bush in the entire population, we compute:

$$38\% \pm 3\% = [35\%, 41\%]$$

an interval that contains 40 percent—the sample finding for Clinton. This implies that Bush was within the margin of error of Clinton and that, statistically, either of them at that time could have won the election.[2] This is shown in Figure 5.1. A later Gallup poll predicted the outcome of the election more accurately.[3]

Interestingly, it was George H. Gallup, founder of the Gallup organization, who in 1936 launched modern scientific polling and became the statisticians' David to the Goliath of an overconfident magazine named the *Literary Digest.* Gallup had an ingenious idea. He offered newspapers and magazines a subscription to his new polling service with a money-back guarantee. The subscriptions would pay for the expensive polling of the many thousands of voters Gallup planned to interview during the 1936 presidential campaign, although if his forecast of the election results was wrong, the money-back guarantee would bankrupt him.

Gallup aimed to give every voter in the United States an equal chance of being selected for the random sample, which included only several thousand voters. This is exactly how we try to conduct statistical polls today, but in 1936 such statistically based polling was not yet widespread. Gallup's competitors selected much larger samples with less care about trying to make them truly representative. As the November election drew near, Gallup's confidence in his new, scientific method of polling grew shaky. The reason for his anxiety was that his poll was showing a clear lead for incumbent Franklin Delano Roosevelt over the Republican challenger,

---

[2]Since there is statistical uncertainty about Clinton's 40 percent as well, *joint* statistical inference about the two population percentages is possible. The simple analysis above, however, is the one commonly implied when poll results are reported.

[3]Of course, presidential elections are not decided by the popular vote but rather by the electoral college. Thus, even if Bush had tied Clinton in the population percentages, he still could have lost by failing to capture states that would give him the required total number of electors.

115

FIGURE 5.1

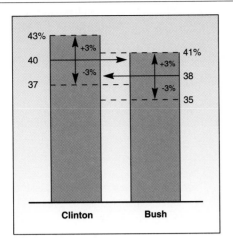

Clinton's sample percentage is inside of the interval for Bush's percentage, and vice versa. (The intervals are obtained by adding and subtracting the "margin of error.")

Governor Alfred Landon of Kansas. This prediction was the opposite of what the well-known *Literary Digest* was predicting. The *Digest* had successfully predicted the results of all presidential elections since the magazine's inception. This gave the *Digest* enormous clout, and its predictions were reported on the front pages of many important newspapers around the country. Gallup's subscribers were therefore getting quite nervous about reporting his predictions. They were putting pressure on the young pollster: Could he be doing it all wrong?

The *Literary Digest*'s sample was huge: 10 million voters! These voters, however, were not a random sample from the entire population of voters in the United States. The magazine drew its sample using automobile registrations, telephone numbers, and the magazine's own readership. Of course, in 1936, relatively few people had cars or phones, and those who did tended to be wealthier and to vote Republican. And the magazine itself was conservative. Within the restricted universe of the *Digest*'s sample, Governor Landon was by far the favorite candidate. What occurred was a sampling *bias*. The sample was biased away from what would have been obtained if all voters in the entire country had been polled, since within the entire country, FDR was the favorite. Thus, while the *Digest*'s sample was enormous in size, it was not representative of the nation as a whole. The biased sampling scheme of the *Literary Digest* and the correct, scientific sampling scheme used by Gallup (whose sample was unbiased and *much* smaller) are shown in Figure 5.2.

Just before the 1936 election day, *The New York Times* prominently displayed the *Digest*'s predictions of a Landon landslide victory on its front page. After the election, the actual election results appeared on the front page (see accompanying articles).

As a result of this great embarrassment, the vaunted *Literary Digest* had to close its doors right after the 1936 election. This is why you likely haven't heard of it, even though it was once as widely read as *Time* and *Newsweek* are today. As for Gallup, a struggling, unknown pollster in those days, he is now a household name.[4] The moral of this story is that it may be worthwhile to pay close attention to the study of statistics.

---

[4]For more on this story, see D. W. Moore, *The Superpollsters* (New York: Four Walls, 1992), pp. 31–72.

FIGURE 5.2

a. Digest Sample. This map shows traditionally Republican states in dark. The *Digest* poll was biased, as if it were drawn from the dark-colored states only.

b. A Correct Random Sample.

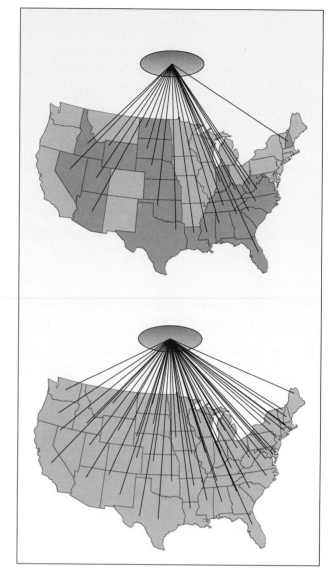

Digest Poll Gives Landon 32 States Landon Leads 4–3 in Last Digest Poll

Final Tabulation Gives Him 370

Electoral Votes to 161 for

President Roosevelt

Governor Landon will win the election by an electoral vote of 370 to 161, will carry thirty-two of the forty-eight States, and will lead President Roosevelt about four to three in their share of the popular vote, if the final figures in *The Literary Digest* poll, made public yesterday, are verified by the count of the ballots next Tuesday.

Roosevelt's Plurality is 11,000,000 History's Largest
46 States Won by President, Maine and Vermont by
Landon, Many Phases to Victory

Democratic Landslide Looked

Upon as Striking Personal

Triumph for Roosevelt

By ARTHUR KROCK

As the the count of ballots cast Tuesday in the 1936 Presidential election moved toward completion yesterday, these facts appeared:

Franklin Delano Roosevelt was re-elected President, and John N. Garner Vice President, by the largest popular and electoral majority since the United States became a continental nation—a margin of approximately 11,000,000 plurality of all votes cast, and 523 votes in the electoral college to 8 won by the Republican Presidential candidate, Governor Alfred M. Landon of Kansas. The latter carried only Maine and Vermont of the forty-eight States of the Union. . . .

The New York Times, Thursday, November 5, 1936. Copyright © 1936 by the New York Times Company. Reprinted by permission.

One last question remains: If the *Digest*'s methodology was so faulty, how was it possible for the magazine to successfully predict the results of all the previous elections? The answer seems to be that before 1936 the electorate was not split along socioeconomic lines. If a sampling bias exists but is not related to the issue we are sampling, the bias may not affect the results. However, we can never be quite sure whether a bias will affect our results; therefore, selecting our sample in an unbiased, random way will act as insurance against potential inferential problems. What worked well by chance in earlier elections failed the *Digest* in 1936.

Today, polls are generally conducted quite well. Figure 5.3 shows the results of various polls conducted during the last few presidential campaigns. This figure is very important. Notice that each poll result, shown as a colored dot, lies above or below the actual election result; that most of the poll results lie relatively close to the actual election result (the population percentage the poll is estimating); and that the various polls predicting an election result seem *centered* around the actual result.

Remember that a **random sample** is one in which each element or person in the population has an equal chance of being selected.

In a random sample, the following are true:

1. The sample proportion, denoted by $\hat{P}$, is a random variable.
2. The mean of the random variable $\hat{P}$ (the center of its probability distribution) is the population proportion, p.
3. The variance (or spread) of the probability distribution of $\hat{P}$ decreases as the sample size, n, increases.
4. As the sample size, n, increases, the probability distribution of $\hat{P}$ approaches a *normal distribution*.

FIGURE 5.3   Poll Results from Recent Presidential Campaigns

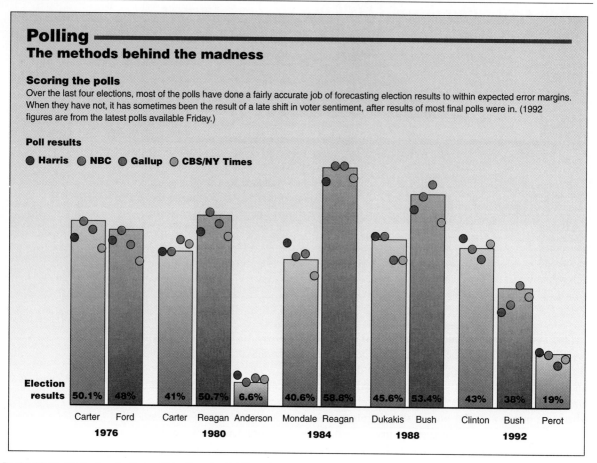

Reprinted by permission from the *Boston Globe,* November 2, 1992, p. 1.

Item 4 on the above list is due to the important and well-known *central limit theorem,* which we use extensively in obtaining probabilities for values of a sample statistic. The probability distribution of a sample statistic such as the sample proportion is called the *sampling distribution* of the statistic. (The sampling distribution is explained in Section 5.2 of this chapter.) The central limit theorem states that in the limit, as the sample size increases, the sampling distribution becomes normal. The sampling distribution of $\hat{P}$, described in the boxed list above, is shown in Figure 5.4. This is a theoretical extension of what we see in Figure 5.3 based on actual, but few, polls. In Figure 5.4, we see what a very large number of (unbiased, representative) poll results, all based on large samples (of, say, 1,000 voters each) would look like.

Sampling Methods and Statistical Inference

We began this chapter with an example of statistical inference, the science of drawing conclusions about a population based on information in a random sample selected from the population. Inferential methods, by which we draw conclusions about a larger

Sampling Distribution of $\hat{P}$

FIGURE 5.4

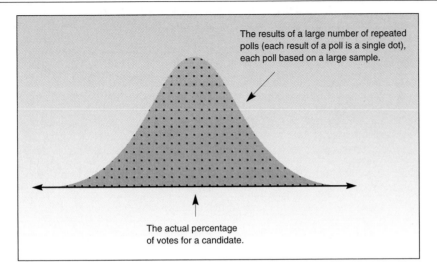

The results of a large number of repeated polls (each result of a poll is a single dot), each poll based on a large sample.

The actual percentage of votes for a candidate.

"parent" population, are thus different from statistical methods that describe actual data.

The science of inference allows us to draw conclusions about a population if that population can be proportionally represented by a sample, ideally a smaller-scale exact replica of the population. The sample should mirror what we have in the population but on a scale that is both manageable and not exceedingly expensive. Typically, from a large population of, say, 10 million, we would draw a sample of 1,000 or so. We would want this sample to contain as much information as possible about the entire population from which it was drawn.

To get good representation of the population, we need to draw the sample at *random,* which means, as we said earlier, that each element in the population would have an equal chance of being selected for the sample. One such method is to select a **simple random sample (SRS),** whereby from a numbered list (called a *frame*) of all the elements of the population we randomly select (using random-number tables or a computer) a sample of size *n.* Throughout this book, we'll assume that a random sample is a simple random sample, although other sampling schemes are available. Some of these are described below.

A **stratified random sample** is a random sample selected from a population that has easily discernible strata. Each stratum is a subpopulation that is roughly homogeneous with respect to what we want to measure. For example, a town has several neighborhoods, some wealthy and some less so. If we are sampling to estimate average income, we know that the averages in different neighborhoods are different from each other but that income levels are similar within a neighborhood. Instead of selecting a simple random sample, we may sometimes find it more efficient to select random samples from the neighborhoods of the town, with each subsample in proportion to the size of the neighborhood from which it is drawn. We then combine the subsamples to yield a more efficient estimate of the overall average income than what we might get from a simple random sample.

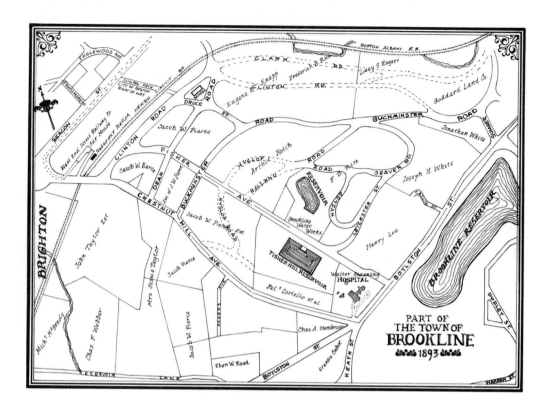

Look at the accompanying map of Brookline. Here a stratified random sampling scheme might have us select separate random samples from the neighborhood below Chestnut Hill Avenue; the neighborhood bounded by Chestnut Hill Avenue, Clinton Road, and Buckminster Road; and the neighborhood bordered by Boylston Street, Chestnut Hill Avenue, and Buckminster Road.

A **cluster sampling scheme** is a sampling method where we randomly select *clusters* of elements rather than the elements themselves. This may make our sampling more efficient. In the example of the Brookline map above, on the northwest corner of the map Beacon Street has many large apartment buildings. It may be useful when sampling from that part of town not to select *people* at random, but rather to select buildings. Once the interviewer is in one of the buildings selected at random, he or she may want to interview several people who live there. The cluster sampling scheme allows us to do so correctly.

A **systematic sampling scheme** is one that allows us to make use of a particular arrangement of elements in the population. Suppose that we want to select a random sample from the population of all files in a filing cabinet. Instead of using simple random sampling—which would require giving every file a number, then randomly selecting from the set of numbers—we can take advantage of the way files are already ordered in the cabinet. We may select the first file at random (following certain rules) and then take every fifth or eighth, or in general every kth, file thereafter.

These and other more advanced sampling schemes are described in advanced books on statistics. The point of this subsection is simply to make you aware that simple random sampling is not the only sampling method in use, although in this book every random sample is assumed to be an SRS.

1. Briefly describe the sampling schemes discussed in this section.
2. Analyze a local or national poll reported in your newspaper. Add and subtract the reported margin of error (sometimes called *sampling error*) to see what values are included in the resulting interval. Are any of these values meaningful in your particular case?
3. Discuss the polling results shown in Figure 5.3.
4. Why is the sample proportion unlikely to be exactly equal to the population proportion?
5. What is the purpose of polling? Why are polls important?
6. Why does a poll have a margin of error?

5.2 SAMPLING DISTRIBUTIONS

Recall from Chapter 1 that in elementary statistics, we deal with two major kinds of situations: *qualitative* and *quantitative.* In a qualitative analysis, we are interested in the proportion of the population (or, in popular use, its percentage) that satisfies a certain characteristic. The percentage of voters for a certain candidate is a qualitative situation; every element (person or item) in the entire population either has the quality of interest (vote for the candidate) or not. We are interested in the proportion (or equivalently, the percentage) of the population that has the quality of interest. We take a random sample and measure the sample proportion (e.g., a proportion of $\hat{P} = 0.35$ translates to a sample percentage of 35 percent). The sample proportion is a random variable because the sample is randomly selected from the entire population. Sometimes, by chance, more items with the characteristic of interest appear in the sample than their proportion in the population, and sometimes fewer. *On the average,* however, when many samples are taken, the proportion of the sample is equal to that of the population. As we said earlier, the sample proportion, as a random variable, will have a normal distribution if the sample size is large enough. This probability distribution is called the **sampling distribution** of the sample proportion.

> The mean of the distribution of the sample proportion, $\hat{P}$, is the population proportion, p, and its standard deviation is $\sqrt{p(1 - p)/n}$.

In contrast, in a quantitative situation our interest is not in whether elements of the population possess a quality of interest. Rather, we associate a *quantity* with each element of the population. In such cases, we are interested in the *average* of this quantity in the entire population; in other words, we're interested in the *population mean.* We estimate the population mean by the **sample mean.** Like the sample proportion, the sample mean is also a random variable and has a sampling distribution, which we'll discuss in Section 5.3.

7. What is a sampling distribution?
8. What is the sampling distribution of the population proportion when the sample size is large?
9. If the population proportion is $p = 0.5$, what is the mean of the sampling distribution of the sample proportion, $\hat{P}$?
10. If the population proportion is $p = 0.5$, what is the standard deviation of the sampling distribution of the sample proportion, $\hat{P}$?

5.3 THE SAMPLING DISTRIBUTION OF THE SAMPLE MEAN

Figure 5.5 shows a population frequency distribution, particular randomly selected values from the population (which constitute a given random sample), the population mean, and the sample mean. As you can see, the sample mean is a random quantity that estimates the population mean and (hopefully) falls relatively close to it.

When we sample from a population with mean μ and standard deviation σ, the sample mean has a distribution with the same *center*, μ, as the population but with a *standard deviation* that is $1/\sqrt{n}$ the size of the standard deviation of the population distribution. The central limit theorem applies to the sample mean as well as to the sample proportion.

When the sample size is large enough, the sampling distribution of $\overline{X}$ is normal.

Figure 5.6 shows the population distribution and the sampling distribution of the sample mean when the sample is large.

52

158

> The *central limit theorem*: When sampling from a population with mean μ and finite standard deviation σ, the sampling distribution of the sample mean, $\overline{X}$, will tend to a normal distribution with mean μ and standard deviation $\sigma/\sqrt{n}$ as the sample size n becomes large. For a large enough n,
>
> $$\overline{X} \text{ is } N(\mu, \sigma/\sqrt{n})$$

The central limit theorem is remarkable because it states that the distribution of the sample mean $\overline{X}$ tends to a normal distribution *regardless* of the distribution of the

FIGURE 5.5 A Population Distribution, a Random Sample from the Population, and Their Respective Means

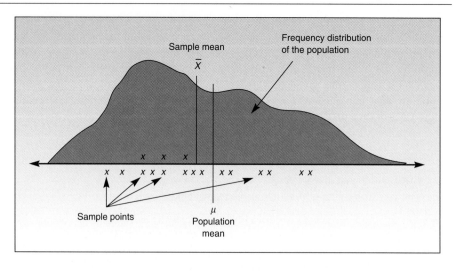

Sample mean
$\overline{x}$

Frequency distribution
of the population

Sample points

μ
Population
mean

A (Nonnormal) Population Distribution and the Normal Sampling Distribution
of the Sample Mean When a Large Sample Is Used

FIGURE 5.6

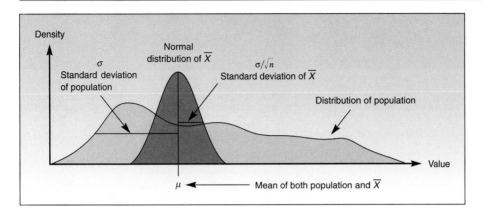

population from which the random sample is drawn. The theorem allows us to make probability statements about the possible range of values the sample mean may take. It allows us to compute probabilities of how far away $\overline{X}$ may be from the population mean it estimates. For example, using our rule of thumb for the normal distribution, we know that the probability that the distance between $\overline{X}$ and μ will be less than $\sigma/\sqrt{n}$ is approximately 0.68. This is so because, as you remember, the probability that the value of a normal random variable will be within one standard deviation of its mean is 0.6826; here our normal random variable has mean μ and standard deviation $\sigma/\sqrt{n}$. Other probability statements can be made as well; we will see their use shortly.

First, however, we'll ask an obvious question: When is a sample size, n, large enough so that we may apply the theorem? The central limit theorem says that, *in the limit,* as n increases toward infinity, the distribution of $\overline{X}$ becomes a normal distribution (regardless of the distribution of the population). The *rate* at which the distribution approaches a normal distribution does depend, however, on the shape of the distribution of the parent population. If the population itself is normally distributed, the distribution of $\overline{X}$ is normal for *any* sample size n. On the other hand, for population distributions that are very different from a normal distribution, a relatively large sample size is required to achieve a good normal approximation for the distribution of $\overline{X}$. Figure 5.7 shows several parent population distributions and the resulting sampling distributions of $\overline{X}$ for different sample sizes.

Since we often do not know the shape of the population distribution, it would be useful to have some general rule of thumb telling us when a sample is large enough so that we may apply the central limit theorem:

*In general, a sample of 30 or more elements is considered large enough for
the central limit theorem to take effect.*

We emphasize that this is a general, and somewhat arbitrary, rule. A larger minimum sample size may be required for a good normal approximation when the population distribution is very different from a normal distribution. By the same token, a smaller minimum sample size may suffice for a good normal approximation when the population distribution is close to a normal distribution.

FIGURE 5.7

The Effects of the Central Limit Theorem: The Distribution of $\overline{X}$
for Different Populations and Different Sample Sizes

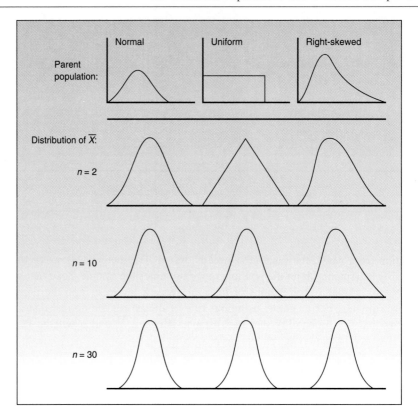

EXAMPLE 5.1

In a very large group of gifted children, the average IQ is 120 and the standard deviation is 15. If 100 children are randomly selected from this group, what is the probability that the average IQ of the children in the sample will be less than 117?

SOLUTION

In solving problems such as this one, we use the techniques of Chapter 4. There we used μ as the mean of the normal random variable and σ as its standard deviation. Here our random variable $\overline{X}$ is normal (at least approximately so, by the central limit theorem, because our sample size is large) and has mean μ. Note, however, that the standard deviation of our random variable $\overline{X}$ is $\sigma/\sqrt{n}$ and not just σ. We proceed as follows:

$$P(\overline{X} < 117) = P\left(Z < \frac{117 - \mu}{\sigma/\sqrt{n}}\right)$$

$$= P\left(Z < \frac{117 - 120}{15/\sqrt{100}}\right) = P(Z < -2) = 0.0228$$

EXAMPLE 5.2

The Stock Market/Chuck Savage, 1993.

At a midwestern college, undergraduate students spend an average of $21.00 per week on entertainment and the population standard deviation is $7.00. If a random sample of 144 undergraduates is selected, what is the probability that the sample average will be between $20.00 and $22.00?

$$P(20 \leq \overline{X} \leq 22) = P\left(\frac{20 - \mu}{\sigma/\sqrt{n}} \leq Z \leq \frac{22 - \mu}{\sigma/\sqrt{n}}\right)$$

SOLUTION

$$= P\left(\frac{20 - 21}{7/\sqrt{144}} \leq Z \leq \frac{22 - 21}{7/\sqrt{144}}\right) = P(-1.71 \leq Z \leq 1.71) = 0.9564 - 0.0436 = 0.9128$$

Note that when sampling from a population that is *itself normal,* no central limit theorem is necessary and the sample mean is immediately normal (regardless of the sample size). Since the standard deviation of $\overline{X}$ is $\sigma/\sqrt{n}$, the normal distribution of $\overline{X}$ gets narrower and narrower. This is demonstrated in Figure 5.8 on page 204. The sampling distribution of $\overline{X}$ *always* gets narrower as the sample size increases (regardless of whether the population distribution is normal or not).

---

11. What is the sampling distribution of the sample mean?
12. State the central limit theorem as it applies to the sample mean.
13. When sampling from a population with standard deviation $\sigma = 50$ using a sample of size $n = 100$, what is the standard deviation of the distribution of the sample mean?

PROBLEMS

FIGURE 5.8

A Normally Distributed Population and the Sampling Distribution
of the Sample Mean for Different Samples Sizes

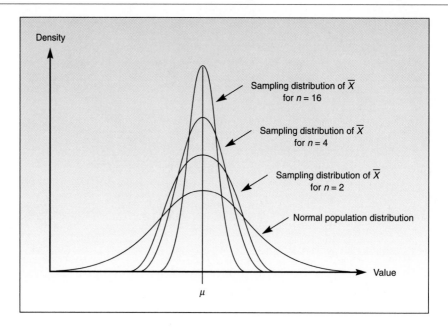

14. If the population mean is 1,247, the population variance is 10,000, and the sample size is 100, what is the probability that $\overline{X}$ will be less than 1,230?

15. Japan's birthrate is believed to be 1.57 births per woman. Assume that the population standard deviation is 0.4. If a random sample of 200 women is selected, what is the probability that the sample mean will fall between 1.52 and 1.62?

16. When sampling from a population with standard deviation $\sigma = 55$, using a sample of size $n = 150$, what is the probability that $\overline{X}$ will be at least eight units away from the population mean μ?

17. The Colosseum, once the most popular monument in Rome, dates from about AD 70. Since then earthquakes have caused considerable damage to the huge structure, and engineers are currently trying to make sure the building will survive future shocks. The Colosseum can be divided into several thousand small sections. Suppose that the average section can withstand a quake measuring 3.4 on the Richter scale with a standard deviation of 1.5. A random sample of 100 sections is selected and tested for the maximum earthquake force they can withstand. What is the probability that the average section in the sample can withstand an earthquake measuring at least 3.6 on the Richter scale?

18. It has been suggested that an investment portfolio selected randomly by throwing darts at the stock market page of *The Wall Street Journal* may be a sound (and certainly well diversified) investment.[5] Suppose that you own such a portfolio of 16 stocks randomly selected from all stocks listed on the New York Stock Exchange (NYSE). On a certain day, you hear on the news that the average stock on the NYSE rose 1.5 points. Assuming that the standard deviation of stock price movements that day was 2 points and assuming stock price movements were normally distributed around their mean of 1.5, what is the probability that the average stock in your portfolio increased in price?

---

[5]See the very readable book by Burton G. Malkiel, *A Random Walk Down Wall Street,* 5th ed. (New York: W. W. Norton, 1991).

5.4 CONFIDENCE INTERVALS

> A **confidence interval** is a range of numbers believed to include an unknown population parameter. Associated with the interval is a measure of the confidence we have that the interval does indeed contain the parameter of interest.

The sampling distribution of a statistic gives a *probability* associated with a range of values the statistic may take. After the sampling has taken place and a particular estimate has been obtained, this probability is transformed into a *level of confidence* for a range of values that may contain the unknown parameter.

In this section, we will see how to construct confidence intervals for the population mean, μ, when the population standard deviation, σ, is known. Then we will alter this unrealistic situation and see how a confidence interval for μ may be constructed without knowledge of σ.

Confidence Interval for the Population Mean When the Population Standard Deviation Is Known

The central limit theorem tells us that when we select a large random sample from any population with mean μ and standard deviation σ, the sample mean, $\overline{X}$, is (at least approximately) normally distributed with mean μ and standard deviation $\sigma/\sqrt{n}$. If the population itself is normal, $\overline{X}$ is normally distributed for any sample size. Recall that the standard normal random variable Z has a 0.95 probability of being within the range of values −1.96 and 1.96. (You may check this using Table B.2 in Appendix B.) Transforming Z to the random variable $\overline{X}$ with mean μ and standard deviation $\sigma/\sqrt{n}$, we find that—before the sampling—there is a 0.95 probability that $\overline{X}$ will fall within the interval:

$$\mu \pm 1.96 \frac{\sigma}{\sqrt{n}}$$

Once we have obtained our random sample, we have a particular value, $\bar{x}$. This particular $\bar{x}$ either lies within the range of values specified by the above equation or does not lie within this range. Since we do not know the (fixed) value of the population parameter μ, we have no way of knowing whether $\bar{x}$ is indeed within the range given in the above equation. Since the random sampling has already taken place and a particular $\bar{x}$ has been computed, we no longer have a random variable and may no longer talk about probabilities. We do know, however, that since the presampling probability that $\overline{X}$ will fall in the interval in the above equation is 0.95, about 95 percent of the values of $\bar{x}$ obtained in a large number of repeated samplings will fall within the interval. Since we have a single value $\bar{x}$ that was obtained by this process, we may say that we are *95 percent confident that $\bar{x}$ lies within the interval.* This idea is demonstrated in Figure 5.9.

Consider a particular $\bar{x}$ and note that the distance between $\bar{x}$ and μ is the same as the distance between μ and $\bar{x}$. Thus, $\bar{x}$ falls inside the interval $\mu \pm 1.96\sigma/\sqrt{n}$ if and only if μ happens to be inside the interval $\bar{x} \pm 1.96\sigma/\sqrt{n}$. In a large number of repeated trials, this would happen about 95 percent of the time. We therefore call the interval

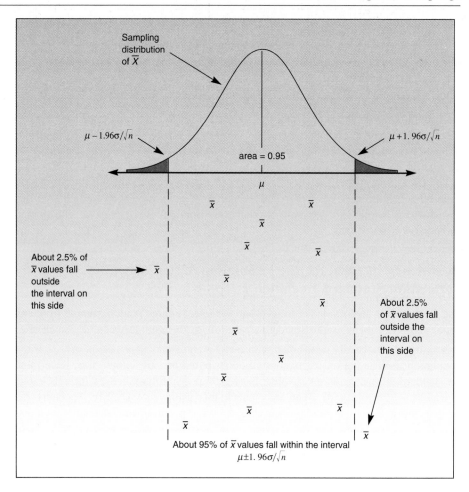

$\overline{x} \pm 1.96\sigma/\sqrt{n}$ a 95 percent confidence interval for the unknown population mean μ. This is demonstrated in Figure 5.10.

Instead of measuring a distance of $1.96\sigma/\sqrt{n}$ on either side of μ (an impossible task since μ is unknown), we measure the same distance of $1.96\sigma/\sqrt{n}$ on either side of our *known* sample mean, $\overline{x}$. Since, *before the sampling,* the random interval $\overline{X} \pm 1.96\sigma/\sqrt{n}$ had a 0.95 probability of capturing μ, *after the sampling* we may be percent confident that our particular interval $\overline{x} \pm 1.96\sigma/\sqrt{n}$ indeed contains the population mean μ. We cannot say that there is a 0.95 probability that μ is inside the interval, because the interval $\overline{x} \pm 1.96\sigma/\sqrt{n}$ is not random, and neither is μ. The population mean μ is unknown to us but is a fixed quantity—not a random variable.[6]

---

[6]We are using what is called the *classical* interpretation of confidence intervals. An alternative view, the Bayesian approach, allows us to treat an unknown population parameter as a random variable. As such, the unknown population mean μ may be stated to have a 0.95 *probability* of being within an interval.

The Construction of a 95 Percent Confidence Interval for the Population Mean μ **FIGURE 5.10**

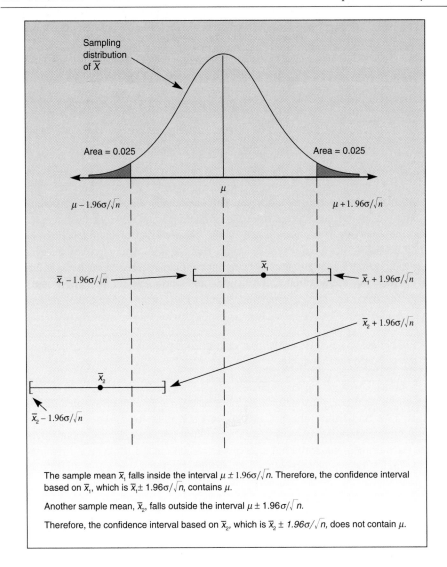

The sample mean $\overline{x}_1$ falls inside the interval $\mu \pm 1.96\sigma/\sqrt{n}$. Therefore, the confidence interval based on $\overline{x}_1$, which is $\overline{x}_1 \pm 1.96\sigma/\sqrt{n}$, contains μ.

Another sample mean, $\overline{x}_2$, falls outside the interval $\mu \pm 1.96\sigma/\sqrt{n}$.

Therefore, the confidence interval based on $\overline{x}_2$, which is $\overline{x}_2 \pm 1.96\sigma/\sqrt{n}$, does not contain μ.

Either μ lies inside the confidence interval (in which case the probability of this event is 1.00) or it does not (in which case the probability of the event is 0). We do know, however, that 95 percent of all possible intervals constructed in this manner will contain μ. Therefore, we may say that we are 95 percent *confident* that μ lies in the particular interval we have obtained.

A 95 percent confidence interval for μ when σ is known and sampling is done from a normal population, or a large sample is used:

$$\overline{x} \pm 1.96 \frac{\sigma}{\sqrt{n}}$$

To compute a 95 percent confidence interval for μ, all we need to do is substitute the values of the required entities in the equation. Suppose, for example, that we are sampling from a normal population, in which case the random variable $\overline{X}$ is normally distributed for any sample size. We use a sample of size $n = 25$, and we get a sample mean $\bar{x} = 122$. Suppose we also know the population standard deviation: $\sigma = 20$. Let's compute a 95 percent confidence interval for the unknown population mean, μ. Using the above equation, we get:

$$\bar{x} \pm 1.96 \frac{\sigma}{\sqrt{n}} = 122 \pm 1.96 \frac{20}{\sqrt{25}} = 122 \pm 7.84 = [114.16, 129.84]$$

Thus, we may be 95 percent confident that the unknown population mean μ lies anywhere between the values 114.16 and 129.84.

In most applications, the 95 percent confidence interval is used. There are, however, many other possible levels of confidence. You may choose any level of confidence you wish, find the appropriate z value from the standard normal table, and use it instead of 1.96 in the equation to get an interval of the chosen level of confidence. Using the standard normal table, we find, for example, that for a 90 percent confidence interval we use the z value 1.645, and for a 99 percent confidence interval we use $z = 2.58$ (or, using an accurate interpolation, 2.576). Let's now formalize the procedure and make some definitions.

We define $z_{\alpha/2}$ as the z value that cuts off a right-tail area of $\alpha/2$ under the standard normal curve.

For example, 1.96 is $z_{\alpha/2}$ for $\alpha/2 = 0.025$ because $z = 1.96$ cuts off an area of 0.025 to its right. (We find from Table B.2 in Appendix B that, for $z = 1.96$, TA = 0.9750; therefore, the right-tail area is $\alpha/2 = 0.025$.) Now consider the two points 1.96 and -1.96. Each of them cuts off a tail area of $\alpha/2 = 0.025$ in the respective direction of its tail. The area between the two values is therefore equal to $1 - \alpha = 1 - 2(0.025) = 0.95$. The area under the curve excluding the tails, $1 - \alpha$, is called the **confidence coefficient.** (And the combined area in both tails, α, is called the **error probability.** This probability will be important to us in Chapter 6.) The confidence coefficient multiplied by 100, expressed as a percentage, is the **confidence level.**

A $(1 - \alpha)100\%$ confidence interval for μ when σ is known and sampling is done from a normal population, or with a large sample:

$$\bar{x} \pm z_{\alpha/2} \frac{\sigma}{\sqrt{n}}$$

Thus, for a 95 percent confidence interval for μ we have:

$$(1 - \alpha)100\% = 95\%$$

$$1 - \alpha = 0.95$$

$$\alpha = 0.05$$

$$\frac{\alpha}{2} = 0.025$$

From the normal table, we find $z_{\alpha/2} = 1.96$. This is the value we substitute for $z_{\alpha/2}$ in the above equation.

For example, suppose we want an 80 percent confidence interval for μ. We have $1 - \alpha = 0.80$ and $\alpha = 0.20$; therefore, $\alpha/2 = 0.10$. We now look in the standard normal table for the value of $z_{0.10}$, that is, the z value that cuts off an area of 0.10 to its right. We have: TA = $1.00 - 0.10 = 0.90$, and from the table we find $z_{0.10} = 1.28$. The confidence interval is therefore $\bar{x} \pm 1.28\sigma/\sqrt{n}$. This is demonstrated in Figure 5.11.

Let's compute an 80 percent confidence interval for μ using the information presented earlier. We have $n = 25$ and $\bar{x} = 122$. We also assume $\sigma = 20$. To compute an 80 percent confidence interval for the unknown population mean, μ, we use the equation and get:

$$\bar{x} \pm z_{\alpha/2}\frac{\sigma}{\sqrt{n}} = 122 \pm 1.28\frac{20}{\sqrt{25}} = 122 \pm 5.12 = [116.88, 127.12]$$

Comparing this interval with the 95 percent confidence interval for μ we computed earlier, we note that the present interval is *narrower*. This is an important property of confidence intervals.

When sampling from the same population, using a fixed sample size, the higher the confidence level, the wider the interval.

Intuitively, a wider interval has more of a presampling chance of "capturing" the unknown population parameter. If we want a 100 percent confidence interval for a parameter, the interval must be $[-\infty, \infty]$. The reason for this is that 100 percent confidence is derived from a presampling probability of 1.00 of capturing the parameter, and the only way to get such a probability using the standard normal distribution is by allowing Z to be anywhere from $-\infty$ to ∞. If we are willing to be more realistic (nothing is certain) and accept, say, a 99 percent confidence interval, our interval would be finite and based on $z = 2.58$. The width of our interval would then be $2(2.58\sigma/\sqrt{n})$. If we further reduce our confidence requirement to 95 percent, the width of our interval would be $2(1.96\sigma/\sqrt{n})$. Since both σ and n are fixed, the 95 percent interval must be narrower. The more confidence you require, the more you need to sacrifice in terms of a wider interval.

The Construction of an 80 Percent Confidence Interval for μ **FIGURE 5.11**

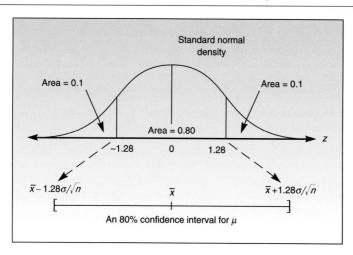

If you want both a narrow interval *and* a high degree of confidence, you need to buy a large amount of information—take a large sample—because, as you recall, the larger the sample size, *n*, the narrower the interval. This makes sense in that if you buy more information, you will have less uncertainty.

When sampling from the same population, using a fixed confidence level, the larger the sample size, n, *the narrower the confidence interval.*

Suppose that the 80 percent confidence interval developed earlier was based on a sample size $n = 2,500$ instead of $n = 25$. Assuming that $\bar{x}$ and σ are the same, the new confidence interval should be 10 times as narrow as the previous one (because $\sqrt{2,500} = 50$, which is 10 times as large as $\sqrt{25}$). Indeed, the new interval is:

$$\bar{x} \pm z_{\alpha/2}\frac{\sigma}{\sqrt{n}} = 122 \pm 1.28\frac{20}{\sqrt{2,500}} = 122 \pm 0.512 = [121.49, 122.51]$$

This interval has width $2(0.512) = 1.024$, while the width of the interval based on a sample of size $n = 25$ is $2(5.12) = 10.24$. This demonstrates the value of information.

When the population standard deviation, σ, is not known, then theoretically the normal distribution should not be used. Instead, we should use the *t* distribution, which we discuss in Section 5.5. However, when the population standard deviation, σ, is unknown and the sample size is large (in most cases this means 30 elements or more), we will still use the normal distribution in setting confidence intervals. We use the sample standard deviation, *s*, instead of the unknown standard deviation of the population.

When we estimate the standard deviation of a statistic using a data set, the standard deviation of the statistic is called its **standard error.** Thus, $s/\sqrt{n}$ is the standard error of $\overline{X}$.

A large-sample $(1 - \alpha)100\%$ confidence interval for μ:

$$\bar{x} \pm z_{\alpha/2}\frac{s}{\sqrt{n}}$$

We demonstrate the use of the above equation in Example 5.3.

EXAMPLE 5.3

An economist wants to estimate the average amount in checking accounts at banks in a given region. A random sample of 100 accounts gives $\bar{x} = \$357.60$ and $s = \$140.00$. Give a 95 percent confidence interval for μ, the average amount in any checking account at a bank in the given region.

SOLUTION

We find the 95 percent confidence interval for μ as follows:

$$\bar{x} \pm z_{\alpha/2}\frac{s}{\sqrt{n}} = 357.60 \pm 1.96\frac{140}{\sqrt{100}} = [330.16, 385.04]$$

Thus, based on the data and the assumption of random sampling, the economist may be 95 percent confident that the average amount in checking accounts in the area is anywhere from $330.16 to $385.04.

Computers may be used to give us confidence intervals for population parameters such as the population mean. In MINITAB, the command ZINTERVAL is used to give confidence intervals of a specified confidence level (the default being 95 percent) for the population mean when σ is known, or when σ is assumed to be known because the sample size is large. When σ is known, state its value after the confidence level in the command. This is demonstrated in Example 5.4.

EXAMPLE 5.4

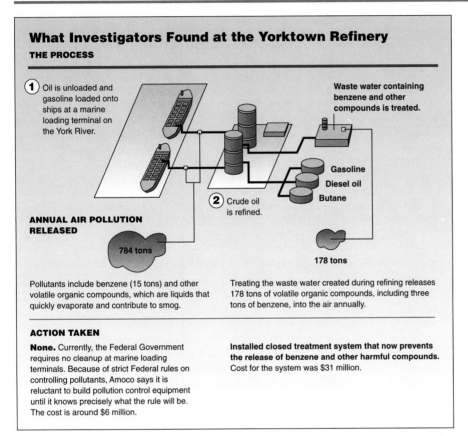

What Investigators Found at the Yorktown Refinery
THE PROCESS

1 Oil is unloaded and gasoline loaded onto ships at a marine loading terminal on the York River.

Waste water containing benzene and other compounds is treated.

2 Crude oil is refined.

Gasoline
Diesel oil
Butane

ANNUAL AIR POLLUTION RELEASED

784 tons

178 tons

Pollutants include benzene (15 tons) and other volatile organic compounds, which are liquids that quickly evaporate and contribute to smog.

Treating the waste water created during refining releases 178 tons of volatile organic compounds, including three tons of benzene, into the air annually.

ACTION TAKEN

None. Currently, the Federal Government requires no cleanup at marine loading terminals. Because of strict Federal rules on controlling pollutants, Amoco says it is reluctant to build pollution control equipment until it knows precisely what the rule will be. The cost is around $6 million.

Installed closed treatment system that now prevents the release of benzene and other harmful compounds. Cost for the system was $31 million.

Reprinted by permission from *The New York Times*, November 29, 1993, p. B7, with data from the Environmental Protection Agency and Amoco.

Suppose that the Environmental Protection Agency wants to estimate the *average* benzene air pollution released per day on the York River loading terminal. The standard deviation of the population is believed to be 0.025 tons. The population is believed to be normal. The data, a random sample of 10 daily measurements, all in tons, are as follows:

$$0.04, \ 0.03, \ 0.04, \ 0.06, \ 0.05, \ 0.09, \ 0.01, \ 0.09, \ 0.07, \ 0.08$$

Give a 95 percent confidence interval for the average daily release of benzene at the terminal, in tons.

SOLUTION

```
MTB > SET C1
DATA > 0.04 0.03 0.04 0.06 0.05 0.09 0.01 0.09 0.07 0.08
DATA > END
MTB > ZINTERVAL 95 0.025 C1
```

con't

THE ASSUMED SIGMA = 0.0250

	N	MEAN	STDEV	SE MEAN	95.0 PERCENT C.I.
C1	10	0.05600	0.02675	0.00791	(0.04048, 0.07152)

PROBLEMS

19. What is a confidence interval, and why is it useful? What is a confidence level?

20. Explain why in classical statistics it makes no sense to describe a confidence interval in terms of probability.

21. Explain how the postsampling confidence level is derived from a presampling probability.

22. Suppose that you computed a 95 percent confidence interval for a population mean. The user of the statistics claims your interval is too wide to have any meaning in the specific use for which it is intended. Discuss and compare two methods of solving this problem.

23. A wine importer needs to report the average percentage of alcohol in bottles of French wine. From previous experience with different kinds of wine, the importer believes the population standard deviation is 1.2 percent. The importer randomly samples 60 bottles of the new wine and obtains a sample mean $\bar{x} = 9.3$ percent. Give a 90 percent confidence interval for the average percentage of alcohol in all bottles of the new wine.

24. A mining company needs to estimate the average amount of copper ore per ton mined. A random sample of 50 tons gives a sample mean of 146.75 pounds. The population standard deviation is assumed to be 35.2 pounds. Give a 95 percent confidence interval for the average amount of copper in the population of tons mined. Also give a 90 percent confidence interval and a 99 percent confidence interval for the average amount of copper per ton.

25. The manufacturer of batteries used in small electrical appliances wants to estimate the average life of a battery. A random sample of 82 batteries yields $\bar{x} = 34.2$ hours and $s = 5.9$ hours. Give a 95 percent confidence interval for the average life of a battery.

26. A telephone company wants to estimate the average length of long-distance calls on weekends. A random sample of 50 calls gives a mean $\bar{x} = 14.5$ minutes and standard deviation $s = 5.6$ minutes. Give a 95 percent confidence interval and a 90 percent confidence interval for the average length of a long-distance phone call on a weekend.

27. According to the Environmental Protection Agency, the average concentration of lead in the blood of children in the United States is 6 micrograms per deciliter.[7] If this estimate is based on a random sample of 1,500 children randomly selected from the entire country and the sample standard deviation was 5 micrograms per deciliter, give a 95 percent confidence interval for the average blood lead level of all children in the United States.

28. It has been asserted that the average American eats 2,500 calories a day.[8] If this is the sample mean based on 514 randomly selected Americans and the standard deviation is 1,230 calories a day, give a 90 percent confidence interval for the daily caloric intake of all Americans.

29. A study of stress and its effects on health involved a random sample of 90 people. The study showed a drop of a person's diastolic blood pressure from an average of 87 millimeters of mercury (mmHg) while the person was with strangers to an average of 84 mmHg while alone, and a further reduction in blood pressure to an average of 83 mmHg while with family members.[9] If the standard deviations were 7, 8, and 7, respectively, give 90 percent confidence intervals for the population blood pressure levels in each of the three conditions.

---

[7]*EPA Journal* 17, no. 2 (March/April 1991), p. 54. I am indebted to Ray Ledoux of the EPA in Boston for this reference.

[8]C. Suplee, "Two Skeptical Views on Dieting," *International Herald Tribune,* June 27, 1991, p. 1.

[9]D. Goleman, "New Light on How Stress Erodes Health," *The New York Times,* December 15, 1992, p. C1.

30. A study by the Center for Science in the Public Interest reported in 1993 that food from Chinese restaurants was found to have higher fat content than previously believed, and certainly more than hamburgers. A random sample of 200 portions of moo shoo pork were analyzed, giving a sample mean of 64 grams of fat and a sample standard deviation of 18 grams. Give a 95 percent confidence interval for the average fat content of all moo shoo pork servings of this size. (The study reported the average Big Mac to have 20 grams of fat.)

Mozart Makes the Brain Hum, a Study Finds

By MALCOLM W. BROWNE

Can it be that the music of Mozart is not only exalting but can also improve intelligence?

An experiment on students at the University of California at Irving suggests that listening to 10 minutes of Mozart's piano music significantly improves performance in intelligence tests taken immediately afterward. The finding is being reported today in the British scientific journal *Nature* by researchers from the university.

The researchers found that after students listened to Mozart's Sonata for Two Pianos in D Major (K. 448), as performed by Murray Perahia and Radu Lupu, their test scores were a mean of eight or nine points higher than the scores the same students achieved after listening to a recorded message suggesting that they imagine themselves relaxing in a peaceful garden or to silence. The effect was only temporary, however.

One researcher, Dr. Frances H. Rauscher, said in an interview that all the students were asked about their tastes in music, and that although some liked Mozart and some did not, their test scores generally improved after the music session, with no measurable differences attributable to varied tastes.

"Listening to such music may stimulate neural pathways important to cognition," Dr. Rauscher said, adding, "Incidentally, Mozart himself often scribbled numbers and mathematical expressions on his manuscript scores."

Thirty-six students, half of them men and half of them women, took part in the experiment. After each listening period they were given standard nonverbal I.Q. tests of spatial reasoning, involving questions about the geometry of paper objects shown as they would look after being folded or cut.

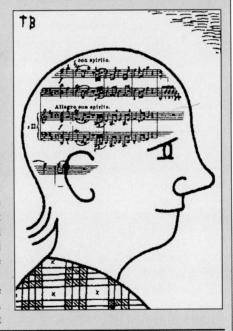

Reprinted by permission from *The New York Times,* October 14, 1993, p. B9.

31. Read the article on the connection between music and intelligence. Suppose that the increases in the IQ test scores (in points) for the 36 students in the sample are as follows:

8, 7, 5, 10, 12, 5, 4, 10, 8, 4, 3, 8, 13, 2, 20, 11, 13, 6, 11, 9, 3, 14, 15, 10, 7, 5, 3, 10, 5, 4, 11, 14, 7, 3, 9, 12.

Use MINITAB (or another computer program) to give a 95 percent confidence interval for the average increase in IQ test scores following a person's listening to a similar musical piece.

32. The following are the height measurements, in inches, of a sample of 42 Siberian cranes observed in their winter grounds in India in 1994. Use the ZINTERVAL command in MINITAB, or another computing package, to construct a 90 percent confidence interval for the height of all Siberian cranes that migrate to India. Assume the following data constitute a random sample from this population:

    29, 33, 28, 31, 30, 30, 27, 34, 28, 35, 30, 29, 26, 27, 35, 37, 30, 21, 26, 32, 31, 36, 34, 29, 25, 29, 31, 34, 31, 31, 27, 26, 37, 35, 35, 30, 28, 27, 34, 31, 30, 29.

33. Suppose you have a confidence interval based on a sample of size n. Using the same level of confidence, how large a sample is required to produce an interval of half the width?

34. The width of a 95 percent confidence interval for μ is 10 units. If everything else stays the same, how wide would a 90 percent confidence interval be for μ?

5.5 CONFIDENCE INTERVALS FOR μ WHEN σ IS UNKNOWN—THE t DISTRIBUTION

In constructing confidence intervals for μ, we assume a normal population distribution or a large sample size (for normality via the central limit theorem). Until now, we have also assumed a known population standard deviation. This assumption was necessary for theoretical reasons so that we could use standard normal probabilities in constructing our intervals.

In real sampling situations, however, the population standard deviation, σ, is rarely known. The reason for this is that both μ and σ are population parameters. When we sample from a population with the aim of estimating its unknown mean, it is quite unlikely that the other parameter of the same population, the standard deviation, will be known.

The t Distribution

As we mentioned earlier, when the population standard deviation is not known, we may use the sample standard deviation, S, in its place. *If the population is normally distributed,* the standardized statistic

$$t = \frac{\overline{X} - \mu}{S/\sqrt{n}}$$

has a **t distribution** with $n - 1$ degrees of freedom. The degrees of freedom of the distribution are the degrees of freedom associated with the sample standard deviation, S (as explained in Chapter 1). The t distribution is also called *student's distribution,* or *student's* t *distribution.*

What is the origin of the name *student*? W. S. Gossett was a scientist at the Guinness brewery in Dublin, Ireland. In 1908, Gossett discovered the distribution of the quantity in the above equation. He called the new distribution the t distribution. The Guinness brewery, however, did not allow its workers to publish findings under their own names. Therefore, Gossett published his findings under the pen name Student. As a result, the distribution became known also as student's distribution.

The t distribution is characterized by its degrees-of-freedom parameter, df. For any integer value, df $= 1, 2, 3, \ldots$, there is a corresponding t distribution. The t distribution resembles the standard normal distribution, Z: It is symmetric and bell shaped. The t distribution, however, is flatter than Z in the middle part and has wider tails.

Courtesy of Guiness America, Inc.

The mean of t is the same as the mean of Z (for df > 1), but the variance of t is larger than the variance of Z. As df increases, the variance of t approaches 1.00, which is the variance of Z. Having wider tails and a larger variance than Z is a reflection of the fact that the t distribution has a greater inherent uncertainty—remember that σ is unknown and is estimated by the *random variable S*. The t distribution thus reflects the uncertainty in two random variables, $\overline{X}$ and S, while Z reflects only an uncertainty due to $\overline{X}$. The greater uncertainty in t (which makes confidence intervals based on t wider than those based on Z) is the price we pay for not knowing σ and having to estimate it from our data. As df increases, the t distribution approaches the Z distribution. Figure 5.12 shows several t distributions, with different numbers of degrees of freedom, along with the standard normal distribution, which is their limit as df goes to infinity.

Values of t distributions for selected tail probabilities are given in Table B.3 in Appendix B (reproduced here as Table 5.1). Since there are infinitely many t distributions—one for every value of the degrees-of-freedom parameter—the table contains probabilities for only some of these distributions. For each distribution, the table gives values that cut off given areas under the curve to the *right*. The t table is thus a table of values corresponding to right-tail probabilities.

Let's consider an example. A random variable with a t distribution with 10 degrees of freedom has a 0.10 probability of exceeding the value 1.372. It has a 0.025 probability of exceeding the value 2.228, and so on for the other values listed in the table. Since the t distributions are symmetric about zero, we also know, for example, that the probability that a random variable with a t distribution with 10 degrees of freedom will be less than -1.372 is 0.10. These facts are demonstrated in Figure 5.13.

As we noted earlier, the t distribution approaches the standard normal distribution as the df parameter approaches infinity. The t distribution with "infinite" degrees of freedom is defined as the standard normal distribution. The last row in the appendix table (Table 5.1) corresponds to df $= \infty$, the standard normal distribution. Note that the value corresponding to a right-tail area of 0.025 in that row is 1.96, which we recognize as the appropriate z value. Similarly, the value corresponding to a right-tail

115

FIGURE 5.12

Several t Distributions Showing the Convergence to the Standard Normal Distribution as the Degrees of Freedom Increase

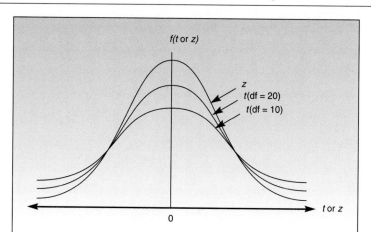

TABLE 5.1

Values and Probabilities of t Distributions

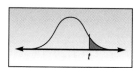

Degrees of Freedom	$t_{.100}$	$t_{.050}$	$t_{.025}$	$t_{.010}$	$t_{.005}$
1	3.078	6.314	12.706	31.821	63.657
2	1.886	2.920	4.303	6.965	9.925
3	1.638	2.353	3.182	4.541	5.841
4	1.533	2.132	2.776	3.747	4.604
5	1.476	2.015	2.571	3.365	4.032
6	1.440	1.943	2.447	3.143	3.707
7	1.415	1.895	2.365	2.998	3.499
8	1.397	1.860	2.306	2.896	3.355
9	1.383	1.833	2.262	2.821	3.250
10	1.372	1.812	2.228	2.764	3.169
11	1.363	1.796	2.201	2.718	3.106
12	1.356	1.782	2.179	2.681	3.055
13	1.350	1.771	2.160	2.650	3.012
14	1.345	1.761	2.145	2.624	2.977
15	1.341	1.753	2.131	2.602	2.947
16	1.337	1.746	2.120	2.583	2.921
17	1.333	1.740	2.110	2.567	2.898
18	1.330	1.734	2.101	2.552	2.878
19	1.328	1.729	2.093	2.539	2.861
20	1.325	1.725	2.086	2.528	2.845
21	1.323	1.721	2.080	2.518	2.831
22	1.321	1.717	2.074	2.508	2.819
23	1.319	1.714	2.069	2.500	2.807
24	1.318	1.711	2.064	2.492	2.797
25	1.316	1.708	2.060	2.485	2.787
26	1.315	1.706	2.056	2.479	2.779
27	1.314	1.703	2.052	2.473	2.771
28	1.313	1.701	2.048	2.467	2.763
29	1.311	1.699	2.045	2.462	2.756
30	1.310	1.697	2.042	2.457	2.750
40	1.303	1.684	2.021	2.423	2.704
60	1.296	1.671	2.000	2.390	2.660
120	1.289	1.658	1.980	2.358	2.617
∞	1.282	1.645	1.960	2.326	2.576

Table Probabilities for a Selected t Distribution (df = 10)   **FIGURE 5.13**

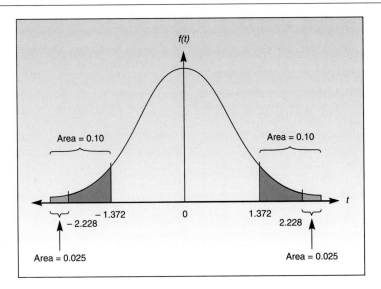

area of 0.005 is 2.576, and the value corresponding to a right-tail area of 0.05 is 1.645. These, too, are values we recognize for the standard normal distribution. Look upward from the last row of the table to find cutoff values of the same right-tail probabilities for t distributions with different degrees of freedom. Suppose, for example, that we want to construct a 95 percent confidence interval for μ using the t distribution with 20 degrees of freedom. We may identify the value 1.96 in the last row (the appropriate z value for 95 percent) and then move up in the same column until we reach the row corresponding to df = 20. Here we find the required value $t_{\alpha/2} = t_{0.025} = 2.086$.

A $(1 - \alpha)100\%$ confidence interval for μ when σ is not known (assuming a normally distributed population):

$$\bar{x} \pm t_{\alpha/2}\frac{s}{\sqrt{n}}$$

where $t_{\alpha/2}$ is the value of the t distribution with $n - 1$ degrees of freedom that cuts off a tail area of $\alpha/2$ to its right.

After reading the article, "Picking Up Mammals' Deep Notes," on page 218, imagine that the scientists wanted to estimate the average frequency of elephant calls. This event is a rare one—only 12 calls were observed, with the following frequencies, in hertz (Hz):

EXAMPLE 5.5

$$14, 16, 17, 17, 24, 20, 32, 18, 29, 31, 15, 35$$

Using these data, and assuming a normal distribution for the population of possible call frequencies, give a 95 percent confidence interval for the average elephant call frequency.

Picking Up Mammals' Deep Notes

By JANE E. BRODY
ITHACA, N.Y.

Watching elephants from a tower in the African savanna seems a long way from the music studies Katy Payne pursued here as an undergraduate student at Cornell University in the 1950s. But she insists that her career as a field biologist, which has produced remarkable discoveries about the language of the world's largest mammals, whales and elephants, has followed "a natural, logical line," and that her knowledge of music has been an essential ingredient in her studies of animal communication.

By recording 12 calls emitted by Amboseli elephants, ... and by coordinating these infrasounds with observations of the animals' behavior, Ms. Payne and Dr. Poole identified a greeting rumble, contact call and answer, a "let's go" rumble, a rumble uttered by males during periods of heightened sexuality, a female chorus that replies to this rumble, and a song sung by females in heat, which may be repeated like a lovesick refrain for up to 40 minutes.

Reprinted by permission from *The New York Times*, November 9, 1993, p. C1.

SOLUTION

Since the sample size is $n = 12$, we need to use the t distribution with $n - 1 = 11$ degrees of freedom. In Table B.3, Appendix B in the row corresponding to 11 degrees of freedom and the column corresponding to a right-tail area of 0.025 (this is $\alpha/2$), we find: $t_{0.025} = 2.201$. (We could also have found this value by moving upward from 1.96 in the last row.) Using this value, we construct the 95 percent confidence interval as follows:

$$\bar{x} = 22.23, \ s = 7.51$$

$$\bar{x} \pm t_{\alpha/2}\frac{s}{\sqrt{n}} = 22.33 \pm 2.201\frac{7.51}{\sqrt{12}} = [17.56, 27.10]$$

The scientists may be 95 percent confident that the average frequency of elephant calls for this particular subspecies is anywhere from 17.56 Hz to 27.10 Hz.

Looking at the t table, we note the convergence of the t distributions to the Z distribution—the values in the rows preceding the last get closer and closer to the corresponding z values in the last row. Although the t distribution is the correct distribution to use whenever σ is not known (assuming the population is normal), when df is *large* we may use the standard normal distribution as an adequate approximation to the t distribution. Thus, instead of using 1.98 in a confidence interval based on a sample of size 121 (df = 120), we may just use the z value 1.96, and not bother consulting the t table.

For small samples (fewer than 30 items), we will use the t distribution as demonstrated above. For large samples, we will use the Z distribution as an adequate approximation. Remember, however, that this division of large versus small samples is arbitrary.

Whenever σ is not known (and the population is assumed normal), the correct distribution to use is the t distribution with n − 1 degrees of freedom.

If you wish, you may always use the more accurate values obtained from the t table (when such values can be found in the table) rather than the standard normal approximation. In this chapter and elsewhere we will assume that the population satisfies, at least approximately, a normal distribution assumption. For large samples, this assumption is less crucial.

For cases where the population standard deviation is unknown, and the population may be assumed normal, the MINITAB command TINTERVAL is used in constructing confidence intervals for the population mean μ. Here again, if a confidence level is not specified in the command, the default of 95 percent is automatically assumed. Incidentally, both the TINTERVAL and the ZINTERVAL commands are "clever" and will be able to correctly interpret the confidence level when it is specified either as a percentage (e.g., 90) or as a decimal (e.g., .90). The TINTERVAL command is demonstrated in Example 5.6.

EXAMPLE 5.6

The Hoxne Hoard is a cache of Late Roman coins accidentally discovered by a farmer who was using a metal detector to search for a lost hammer in Somerset, Britain, in 1992. The hoard contained 14,000 gold coins, with an estimated total value far exceeding the budget for new purchases by any museum. The hoard also contained 12 rare bronze coins minted in Trier in the late fourth century AD. Archaeologists wanted to use these coins to estimate the average zinc content of all bronze coins of this mintage. The zinc percentage data for the bronze coins in the hoard are given below. The coins are assumed to be a random sample from a normal population. We use the MINITAB command TINTERVAL to give a 95 percent confidence for the average population zinc percentage.

SOLUTION

```
MTB > SET C1
DATA > 5.5 4.9 5.2 6.4 3.8 5.7 6.3 5.6 6.0 4.8 4.0 6.2
DATA > END
MTB > TINTERVAL C1

          N   MEAN  STDEV  SE MEAN   95.0 PERCENT C.I.
   C1    12   5.367  0.858    0.248    ( 4.821,    5.912)
```

PROBLEMS

35. What is the purpose of the t distribution?
36. What assumptions must be made when we use the t distribution?
37. What happens to the t distribution as the number of degrees of freedom increases?
38. Refer to the article "Professors at Work" on the next page. Suppose that the standard deviation of the hours worked by professors in the sample is 8 hours. Construct a 95 percent confidence interval for the average number of hours worked by a tenured or tenure-track professor at the University of Oregon.

Professors at Work

From the outside, being a university professor looks like a cushy job: two or three courses a term, flexible office hours, summers off. And make no mistake—being a university professor *is* a great job. But not because of the free time.

The UO [University of Oregon] took a random sample of 20 tenured or tenure-track faculty members to find out just what it is they *do* all day, only to discover they do it evenings and weekends as well—an average 58 hours a week.

Not surprisingly, instruction alone involves about 32 hours of a UO professor's time each week: eight hours in the classroom plus 24 hours preparing lectures, advising students and grading papers. More than 17 hours a week are devoted to research, which itself involves teaching graduate students outside the classroom. Another nine hours fall to professional service, including administrative chores.

For their efforts, UO faculty members earn an average $11,100 *less* than their peers at comparable institutions. And they do it with far fewer support staff: just five administrators and managers per 100 faculty members at the UO, compared to almost 15 at the University of California, Berkeley, nearly 20 at the University of Washington—and more than 31 at the University of Michigan.

But what about those summer vacations? Many faculty members keep up their research over the break, even though they aren't getting paid. They enjoy their subjects, after all—and they need a head start preparing next year's lectures.

Reprinted by permission from *Old Oregon* Magazine, © 1993 University of Oregon, all rights reserved.

39. A new treatment for hepatitis is being tested for effectiveness. There is need for statistical estimation to determine the average number of days from beginning of treatment until the patient's health is improved. A random sample of 11 patients is selected and the number of days till cured is recorded. Data are as follows:

$$4, 4, 3, 8, 5, 6, 7, 12, 5, 3, 8$$

Construct the appropriate 95 percent confidence interval.

40. Assume the data in the table on medieval vessels constitute a random sample and that the population is normally distributed with respect to all the variables listed. Construct a 90 percent confidence interval for the mean length of a vessel.

Estimated Dimensions and Cargo Capacity of Selected Medieval Vessels

Vessel	Type	Date (AD, approx.)	Length (m)	Max. Breadth (m)	Depth of Hold (m)	Approx. Cargo Capacity (tonnes)
1. Skuldelev 1	Nordic	1000	16.3	4.5	2.1	24
2. Hedeby 3	Nordic	1060	c. 25	5.7	2.5	40
3. Lynaes	Nordic	1150	c. 25	6.0	2.5	60
4. Dublin (TG.6/9)	Nordic	1200	c. 25	6.5	2.5	60
5. Bergen Great Ship	Nordic	1240	c. 30	9.5	3.7	155
6. Bremen Cog	Cog	1380	22.7	7.6	4.3	120
7. R. Hamble	Nordic	1418	c. 40	c. 15	c. 6.5	1,400
8. *Santa Maria*	Nao	1492	23.6	7.97	4.7	106

Reprinted from Sean McGrail, "The Future of the Designated Wreck Site in the R. Hamble," *International Journal of Nautical Archaeology* 22, no. 1 (1993), pp. 45–51.

41. Using the data in the medieval vessels table, construct a 99 percent confidence interval for the average vessel depth of hold.

42. Again using the medieval vessels table, give a 95 percent confidence interval for the average maximum breadth of a medieval vessel.

43. Verify the reported means and standard deviations in the table showing data on a treatment for high cholesterol (and correct them if they are wrong). Assuming a normal population of pretreatment total cholesterol level, and that the reported results are a random sample, construct a 95 percent confidence interval for the population mean.

Demographic and Clinical Data before Bezafibrate-Lovastatin Treatment

Patient Number	Sex	Age	Height (cm)	Body Mass Index (kg/m²)	Pretreatment* Total Cholesterol (mg/dl)	Pretreatment* Triglycerides (mg/dl)	Pretreatment* LDL Cholesterol (mg/dl)	Pretreatment* HDL Cholesterol (mg/dl)	Underlying Diseases
1	F	60	156	28	302	283	184	60	CAD, HBP
2	M	61	173	28	330	302	239	31	NIDDM, CAD
3	M	56	163	24	285	333	187	31	NIDDM, CAD
4	M	47	186	24	431	260	365	34	—
5	M	45	169	28	377	297	278	40	CAD
6	M	56	167	23	328	290	259	34	CAD, HBP
7	M	41	170	27	286	250	202	34	HBP
8	M	58	174	28	354	453	227	36	CAD, HBP, NIDDM
9	M	63	172	28	433	272	348	31	CAD, HBP
10	M	64	155	27	339	362	242	25	CAD, HBP
Mean		54.1±	170±	26.5±	346.5±	310.2±	253.10±	35.7±	
±SD		7.81	8.21	1.5	53.44	75.73	62.3	9.7	

LDL = low-density lipoprotein; HDL = high-density lipoprotein; CAD = coronary artery disease; HBP = high blood pressure; NIDDM = noninsulin-dependent diabetes mellitus.

*Lipid levels after treatment with bezafibrate and diet but before the start of combination therapy.

Reprinted from Daniel Yeshurun et al., "Treatment of Severe, Resistant Familial Combined Hyperlipidemia with a Bezafibrate-Lovastatin Combination," *Clinical Therapeutics* 15, no. 2 (1993), p. 389.

44. Again referring to the cholesterol treatment table, and assuming a normal population and a random sample, construct a 90 percent confidence interval for the population mean pretreatment LDL cholesterol level.

45. Making the same assumptions as in Problem 43, construct a 90 percent confidence interval for the population mean pretreatment HDL cholesterol level.

Demographic Characteristics and History of Illness for Study Subjects

Variable	Schizophrenia ($n = 30$) M	Schizophrenia ($n = 30$) SD	Schizoaffective Disorder ($n = 18$) M	Schizoaffective Disorder ($n = 18$) SD
Continuous Variables				
Age	27.7	6.16	27.5	7.29
Education (years)	11.8	1.77	12.1	1.93
Socioeconomic status	4.9	0.42	4.8	0.44
Age of onset	21.3	5.27	17.8	4.49
Duration of illness (years)	7.2	7.97	10.0	8.02
Duration of current hospitalization (days)	32.8	14.06	29.1	14.40

Reprinted from Kim T. Mueser et al., "Expressed Emotion, Social Skill, and Response to Negative Affect in Schizophrenia," *Journal of Abnormal Psychology* 102, no. 3 (1993), pp. 339–51.

46. Refer to the table with data on schizophrenia patients. For schizoaffective disorder, give a 95 percent confidence interval for average age of onset.

47. As in Problem 46, for schizoaffective disorder, give a 95 percent confidence interval for average duration of illness.

48. The following are systolic blood pressure measurements for 25 patients, assumed to be a random sample of people with hypertension problems:

152, 149, 187, 165, 148, 195, 200, 208, 199, 167, 180, 185, 205, 198, 188, 171, 179, 206, 202, 190, 179, 188, 190, 206, 179

Use MINITAB (the TINTERVAL command) or another computing package to construct a 90 percent confidence interval for the mean systolic pressure for all patients with this disorder.

5.6 THE SAMPLING DISTRIBUTION OF THE SAMPLE PROPORTION, $\hat{P}$

The sampling distribution of the sample proportion, $\hat{P}$, is derived from the binomial distribution with parameters n and p, where n is the sample size and p is the population proportion. Recall that the binomial random variable, X, counts the number of successes in n trials. Since $\hat{P} = X/n$ and n is fixed (determined before the sampling), the distribution of the number of successes, X, gives us the distribution of $\hat{P}$.

In the first section of this chapter, we saw how the sample proportion—viewed as a percentage—is used in polling. As in all large-sample applications, we used the central limit theorem, which allows us to use the limiting normal distribution instead of the binomial.

199

As the sample size, n, *increases, the sampling distribution of* $\hat{P}$ *approaches a normal distribution with mean* p *and standard deviation* $\sqrt{p(1-p)/n}$.

In order for us to use the normal approximation for the sampling distribution of $\hat{P}$, the sample size needs to be large. A commonly used rule of thumb says that the normal approximation to the distribution of $\hat{P}$ may be used only if both np and $n(1 - p)$ are greater than 5, although some statisticians recommend both terms be greater than 9 or even 10. We demonstrate the use of the theorem with Example 5.7.

EXAMPLE 5.7

Twenty-five percent of the patients treated by a new drug for AIDS show signs of improvement within six weeks. If a random sample of 100 patients is selected and treated by the new drug, what is the probability that at least 20 percent of them will show signs of improvement within six weeks?

SOLUTION

We need $P(\hat{P} \geq 0.20)$. Since $np = 100(0.25) = 25$ and $n(1 - p) = 100(0.75) = 75$, both numbers large enough, we may use the normal approximation to the distribution of $\hat{P}$. The mean of $\hat{P}$ is $p = 0.25$, and the standard deviation of $\hat{P}$ is $\sqrt{p(1-p)/n} = 0.0433$. We have:

$$P(\hat{P} \geq 0.20) = P\left(Z \geq \frac{0.20 - 0.25}{0.0433}\right) = P(Z \geq -1.15) = 0.8749$$

Large-Sample Confidence Intervals for the Population Proportion, p

The mean of the sampling distribution of $\hat{P}$ is the population proportion, p, and the standard deviation of the distribution of $\hat{P}$ is $\sqrt{pq/n}$, where $q = 1 - p$. Since the standard deviation of the estimator depends on the unknown population parameter, its value is also unknown to us. It turns out, however, that for large samples we may use our actual estimate $\hat{p}$ instead of the unknown parameter p in the formula for the standard deviation. We will therefore use $\sqrt{\hat{p}\hat{q}/n}$ as our estimate of the standard error of $\hat{P}$.

A large-sample $(1 - \alpha)100\%$ confidence interval for the population proportion, p:

$$\hat{p} \pm z_{\alpha/2}\sqrt{\frac{\hat{p}\hat{q}}{n}}$$

where the sample proportion, $\hat{p}$, is equal to the number of successes in the sample, x, divided by the number of trials (the sample size), n, and $\hat{q} = 1 - \hat{p}$.

Example 5.8 demonstrates how to use the above formula.

Scientists studying evolution recently discovered that the blackcap, a European migratory bird, has changed its migration pattern in the very short time (in evolutionary terms) of only 40 years. The bird, whose breeding origin is southern Norway, has extensive summer breeding grounds in Germany and in border areas of Austria, Switzerland, and France (see map). In 1950 and earlier, all blackcaps were known to migrate from these areas to winter grounds in the western Mediterranean areas of northern Spain. As of the early 1990s, a significant proportion of the birds have been

EXAMPLE 5.8

V. Hasselblad/Vireo.

found to migrate to new wintering grounds in a totally different direction: England. Scientists now believe that this change in route is already genetically encoded in the birds, rather than being a chance aberration. Proving this theory required statistical inference. Some birds were bred in captivity, then put in covered cups and allowed to attempt their migration.[10] If out of 100 birds randomly selected for this experiment, 34 were found to move in the direction from Germany to England and 66 were found to move in the direction of Spain, estimate the proportion of all blackcaps that would migrate to the new grounds in England.

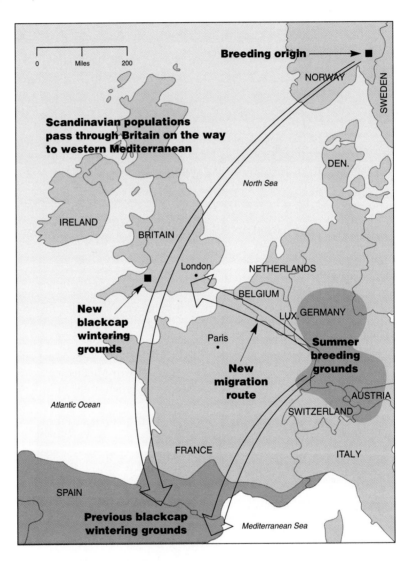

SOLUTION

We have $x = 34$ and $n = 100$, so our sample estimate of the proportion is $\hat{p} = x/n = 34/100 = 0.34$. We now use the formula to obtain the confidence interval for the population proportion, p. A 95 percent confidence interval for p is:

---

[10]C. K. Yoon, "Bird-Watching Biologists See Evolution on the Wing," by *The New York Times,* reprinted by permission December 22, 1992, p. C1.

$$\hat{p} \pm z_{\alpha/2}\sqrt{\frac{\hat{p}\hat{q}}{n}} = 0.34 \pm 1.96\sqrt{\frac{(0.34)(0.66)}{100}}$$

$$= 0.34 \pm 1.96(0.04737) = 0.34 \pm 0.0928$$

$$= [0.2472, 0.4328]$$

The scientists may be 95 percent confident that the percentage of all blackcaps that would go to England is anywhere from 24.72 percent to 43.28 percent. Since the entire interval is above 20 percent, genetic-evolutionary theory leads the scientists to the conclusion that the change in route is not temporary or accidental but rather genetically transmitted.

Suppose the scientists in Example 5.8 are not happy with the width of the confidence interval. What can be done about it? This is a problem of *value of information,* and it applies to all estimation situations. As we stated earlier, for a fixed sample size, the higher the confidence you require, the wider the confidence interval. The sample size is in the denominator of the standard error term, as we have also seen in the case of estimating μ. If we should increase n, the standard error of $\hat{P}$ will decrease, and there will be less uncertainty about the parameter being estimated. If the sample size cannot be increased but you still want a narrower confidence interval, you must reduce your confidence level. Thus, for example, if the scientists agree to reduce the confidence level to 90 percent, z will be reduced from 1.96 to 1.645, and the confidence interval will shrink to:

$$0.34 \pm 1.645(0.04737) = 0.34 \pm 0.07792 = [0.2621, 0.4179]$$

The scientists may be 90 percent confident that anywhere from 26.21 percent to 41.79 percent of the birds would go to the new winter grounds in Britain. This interval is narrower, but the confidence level is also lower. Suppose the scientists want *both* a narrow interval and a high confidence of 95 percent. How could that be achieved? By *raising the sample size.* Suppose that a random sample of 200 birds was selected and that the sample result was the same proportion. That is, $x = 68$, $n = 200$, and $\hat{p} = x/n = 0.34$. What would be a 95 percent confidence interval in this case? Using our formula, we get:

$$\hat{p} \pm z_{\alpha/2}\sqrt{\frac{\hat{p}\hat{q}}{n}} = 0.34 \pm 1.96\sqrt{\frac{(0.34)(0.66)}{200}} = [0.2743, 0.4057]$$

This interval is considerably narrower than our first 95 percent confidence interval, which was based on a sample of 100.

We should mention that when estimating proportions using small samples, the binomial distribution may be used in forming confidence intervals. Since the distribution is discrete, it may not be possible to construct an interval with an exact, prespecified confidence level such as 95 percent or 99 percent. We will not demonstrate the method here.

---

49. What is the sampling distribution of the sample proportion?   *PROBLEMS*
50. How does the central limit theorem apply to the sample proportion?

51. A maker of portable exercise equipment, designed for health-conscious people who travel too frequently to use a regular athletic club, wants to estimate the proportion of traveling businesspeople who may be interested in the product. A random sample of 120 traveling businesspeople indicates that 28 of them may be interested in purchasing the portable fitness equipment. Give a 95 percent confidence interval for the proportion of all traveling businesspeople who may be interested in the product.

52. Based on the displayed information about AIDS, construct a 95 percent confidence interval for the proportion of the population who would be afraid of contracting AIDS from a transfusion. How does your result relate to the reported "sampling error"?

If you had to receive a blood transfusion at a hospital, would you be worried about AIDS: tainted blood?

Yes	73%
No	25%

Reprinted from *Time* magazine, December 13, 1993, p. 30.

53. Use the information in the graph "Favoured Drugs" to construct a 95 percent confidence interval for the proportion of all British 16-year-olds who have used cannabis.

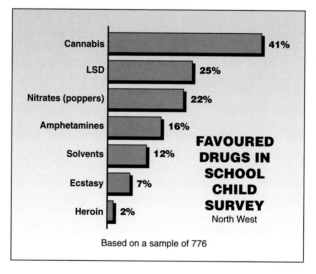

Reprinted from the *London Times,* November 15, 1993, p. 8, with data from the University of Manchester "Alcohol and Drug Use among North West Youth" survey (Base: 776 16-year-olds).

54. Based on the same data on drug use as in Problem 53, give a 95 percent confidence interval for the proportion of all British 16-year-olds who have used LSD.

55. Repeat Problem 54 for nitrates.

56. A study showed that one in three college students admitted to cheating at least once on an exam. This finding is based on a random sample of 3,630 college students.[11] Assuming the reported result is exact (exactly one in three students), give a 99 percent confidence interval for all college students who cheat at least once.

---

[11]R. Morin, "Honesty May No Longer Be the Best Policy," *Washington Post National Weekly Edition,* December 7–13, 1992, p. 36. I am indebted to President J. Cronin of Bentley College for this reference.

57. In 1990, 3,000 volunteers assisted scientists in projects in 50 countries, and 80 percent of all these volunteers were single.[12] Assuming that the 3,000 volunteers constitute a random sample of people who would assist in research abroad, give a 95 percent confidence interval for the proportion of single people who would volunteer.

58. In 1983, the federal government claimed that the state of Alaska had overpaid 20 percent of the Medicare recipients in the state. The director of the Alaska Department of Health and Social Services planned to check this claim by selecting a random sample of 250 recipients of Medicare checks in the state and determining the number of overpaid cases in the sample. Assuming the federal government's claim is correct, what is the probability that less than 15 percent of the people in the sample will be found to have been overpaid?

59. For a fixed sample size, what is the value of the true population proportion, p, that maximizes the variance of the sample proportion, $\hat{P}$? (Hint: Try several values of p on a grid between 0 and 1.)

60. An advertisement for Citicorp Insurance Services, Inc., claims "one person in seven will be hospitalized this year." Suppose you keep track of a random sample of 180 people over an entire year. Assuming Citicorp's advertisement is correct, what is the probability that fewer than 10 percent of the people in your sample will be found to have been hospitalized (at least once) during the year? Explain.

61. A quality-control analyst wants to estimate the proportion of imperfect jeans in a large warehouse. The analyst plans to select a random sample of 500 pairs of jeans and note the proportion of imperfect pairs. If the actual proportion in the entire warehouse is 0.35, what is the probability that the sample proportion will deviate from the population proportion by more than 0.05?

62. Before the crown prince of Japan, Naruhito, finally found a bride after years of search, the Japanese magazine *Bunshun* polled 100 young women on the streets of Tokyo about the attractiveness of the crown prince's hairstyle. Only five women in the sample found the prince's present hairstyle attractive.[13] Give a 95 percent confidence interval for the proportion of all young women in Japan that favored the crown prince's hairstyle. What assumptions must you make in the inference?

63. In a linguistics study, a random sample of 220 English speakers were taught a Dutch dialect. After eight weeks of study, student tests showed that 121 of them revealed an absence of T-voicing.[14] Give a 90 percent confidence interval for the proportion of all English-speaking people who, after completing the eight-week course, would show absence of T-voicing.

64. Supposedly the IRS looks for "nice" numbers in taxpayers' returns for further investigation, which may lead to an audit. Thus, for example, a return reporting an expense item of $3,000 would lead to initial scrutiny while numbers such as $3,137.12 would not.[15] A random sample of 10,000 returns is selected by the IRS out of a population of returns where 10 percent contain round numbers. What is the probability that at least 800 returns in the sample will contain round numbers?

65. A Gallup/CNN public opinion poll reported on January 22, 1993, used a sample of 400 randomly chosen people. The survey was aimed at determining public support for Zoë Baird, President Clinton's nominee for attorney general. Seventy-two people in the sample indicated support. Give a 95 percent confidence interval for the population proportion of all Americans who at that time supported the nominee.

66. In a quality control sample of 1,200 randomly selected color computer monitors, 52 were found to emit low-level radiation. Give a 99 percent confidence interval for the proportion of all screens of this type that emit low-level radiation.

---

[12]"How to Survive with One Paddle in a Two-Paddle Boat," *Condé Nast Traveler,* March 1992, p. 132.

[13]"People," *The International Herald Tribune,* June 15–16, 1991, p. 20.

[14]J. Chambers, "Dialect Acquisition," *Language* 68, no. 4 (1992), p. 696.

[15]*Tax Guide for College Teachers* (College Park, MD: Academic Information Service, 1991).

5.7 SAMPLE SIZE DETERMINATION

One of the most common questions a statistician is asked before any actual sampling takes place is: "How large should my sample be?" From a *statistical* point of view, the best answer to this question is: "Get as large a sample as you can afford. If possible, 'sample' the entire population." If you need to know the mean or proportion of a population, and you can sample the entire population (i.e., carry out a census), you will have all the information and will know the parameter exactly. Clearly, this is better than any estimate but it is unrealistic in most situations due to financial constraints, time constraints, and other limitations. "Get as large a sample as you can afford" is the best answer if money and time are no object, because the larger the sample, the smaller the standard error of our statistic—and the smaller the standard error, the less uncertainty with which we have to contend. This is demonstrated in Figure 5.14.

When the sampling budget is limited, however, the question often is how to find the *minimum* sample size that will satisfy some precision requirements. In such cases, you should explain to the designer of the study that he or she must first give you answers to the following three questions:

1. How close do you want your sample estimate to be to the unknown parameter? The answer to this question is denoted by *D* (for "distance").
2. What do you want the confidence level to be so that the distance between the estimate and the parameter is less than or equal to *D*?
3. What is your estimate of the variance (or standard deviation) of the population in question?

Only after you have answers to all three questions can you give an answer as to the minimum required sample size. Often the statistician is told: "How can I give you an estimate of the variance? I don't know. You are the statistician." In such cases, try to get from your client some idea about the variation in the population. If the population is approximately normal and you can get 95 percent bounds on the values in the

FIGURE 5.14   The Standard Error of a Statistic as a Function of Sample Size

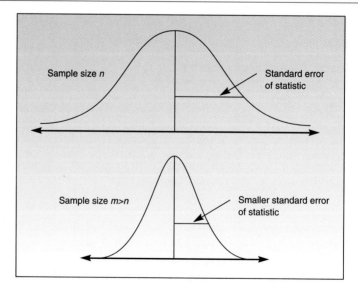

population, divide the difference between the upper and lower bounds by 4; this will give you a rough guess of σ. Or you may take a small, inexpensive pilot survey and estimate σ by the sample standard deviation. Once you have obtained the three required pieces of information, all you need to do is substitute the answers into the appropriate one of the following equations.

Minimum required sample size in estimating the population mean, μ:

$$n = \frac{z^2_{\alpha/2}\sigma^2}{D^2}$$

Minimum required sample size in estimating the population proportion, p:

$$n = \frac{z^2_{\alpha/2}\, pq}{D^2}$$

The two equations above are derived from the formulas for the corresponding confidence intervals for these population parameters based on the normal distribution. In the case of the population mean, D is the half-width of a $(1 - \alpha)100\%$ confidence interval for μ, and therefore

$$D = z_{\alpha/2}\frac{\sigma}{\sqrt{n}}$$

This is used in obtaining the boxed equation above for the minimum sample size for estimating the population mean. The second boxed equation, for the minimum required sample size in estimating the population proportion, is derived in a similar way. Note that the term pq acts as the population variance. In order to use the equation, we need a guess of p, the unknown population proportion. Any prior estimate of the parameter will do. When none is available, we may take a pilot sample, or—in the absence of any information—we use the value $p = 0.5$. This value maximizes pq and thus assures us a minimum required sample size that will work for any value of p.

EXAMPLE 5.9

A marketing research firm wants to conduct a survey to estimate the average amount spent on entertainment by each person visiting a popular resort. The people who plan the survey would like to be able to determine the average amount spent by all people visiting the resort to within $120, with 95 percent confidence. From past operation of the resort, an estimate of the population standard deviation is $\sigma = \$400$. What is the minimum required sample size?

SOLUTION

Using the equation for the minimum required sample size, we find:

$$n = \frac{z^2_{\alpha/2}\sigma^2}{D^2}$$

We know that $D = 120$, and σ^2 is estimated at $400^2 = 160,000$. Since we want a 95 percent confidence, $z_{\alpha/2} = 1.96$. Using the equation, we get:

$$n = \frac{(1.96)^2 160,000}{120^2} = 42.684$$

Therefore, the minimum required sample size is 43 people. (We cannot sample 42.684 people, so we go to the next higher integer.)

EXAMPLE 5.10

The manufacturers of a sports car want to estimate the proportion of people in a given income bracket who are interested in the model. The company wants to know the population proportion, p, to within 0.10 with 99 percent confidence. Current company records indicate that the proportion p may be around 0.25. What is the minimum required sample size for this survey?

SOLUTION

Using the equation for the minimum sample size, we get:

$$n = \frac{z^2_{\alpha/2} pq}{D^2} = \frac{(2.576)^2(0.25)(0.75)}{0.10^2} = 124.42$$

The company should therefore obtain a random sample of at least 125 people.

PROBLEMS

Want a Room With a View? Idea May Be in the Genes

By WILLIAM K. STEVENS

Can humans be truly human and truly fulfilled in a world of glass and concrete set apart from nature, surrounded by cultural artifacts and pursuits, enclosed in electronic cocoons where much of reality comes from the television screen and the computer display?

Not in a million years, according to a new hypothesis. It holds that eons of evolution, during which humans constantly and intimately interacted with nature, have imbued Homo sapiens with a deep, genetically based emotional need to affiliate with the rest of the living world. Meeting this need, according to what is called the biophilia hypothesis, may be as important to human well-being as forming close personal relationships.

Reprinted by permission from *The New York Times,* November 30, 1993, p. C1.

67. Let's say that after reading the accompanying article, scientists want to estimate the proportion of all humans with a desire for a room with a view (being close to nature). They want to determine that proportion to within 0.1 with 95 percent confidence. There is no estimate as to what the actual proportion may be. What is the minimum required sample size?

68. Read the article on the new computerized Graduate Record Examination (GRE). Let's say that researchers at the Educational Testing Service would like to estimate the average GRE

score that would be obtained by students taking the new computerized test to within 30 points with 95 percent confidence. The standard deviation of scores is believed from previous experience to be about 100 points. What is the minimum required sample size?

No. 2 Pencil Fades as Graduate Exam Moves to Computer

By MICHAEL WINERIP

The Educational Testing Service, creators of the examinations that give Americans the jitters—the S.A.T., G.R.E., P.S.A.T.—today takes a major step toward eliminating the standardized paper and pencil test with the introduction of a new computerized version of the Graduate Record Examination.

Though paper and pencil will remain an option for now, by the 1996–97 school year, all 400,000 students who take the G.R.E. each year for admission to graduate school will do it on a computer.

Instead of sitting in a room with hundreds of other people on one of five annual test dates, students will be able to go to a computer center and take the G.R.E. on any of several days during the week, for a total of more than 150 days a year.

Reprinted by permission from *The New York Times,* November 15, 1993, p. A1.

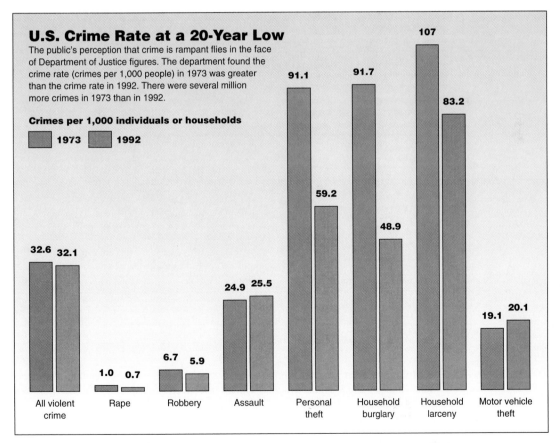

U.S. Crime Rate at a 20-Year Low

The public's perception that crime is rampant flies in the face of Department of Justice figures. The department found the crime rate (crimes per 1,000 people) in 1973 was greater than the crime rate in 1992. There were several million more crimes in 1973 than in 1992.

Crimes per 1,000 individuals or households

1973   1992

Reprinted by permission from the *Chicago Tribune,* November 14, 1993, p. 5, with data from the Department of Justice.

69. An industry that has been growing steadily in recent years is that of firms that create names for new businesses. These firms constantly test the appeal of names being considered for use by their clients. One such firm wants to gather a random sample to test the appeal of a potential name for a high-tech company. The name maker wants to estimate the proportion of people who will respond favorably to the name to within 0.10 of the true population proportion with 99 percent confidence. The company has a rough idea that the population proportion may be around 0.75. What is the minimum required sample size?

70. The department of justice wants to take a random sample of counties around the country to estimate the rates (crimes/1,000 population) for the various crimes reported in the display on page 231. Using an overall standard deviation guess of 25 crimes per 1,000 population, and a desire to estimate the average crime rates to within 10 per 1,000 population with 90 percent confidence, how many counties should be sampled?

71. A public relations consultant wants to estimate the proportion of rock 'n' rollers with unusual parents (see the display titled "Career Kids"). This is to be estimated to within 5 percent with 90 percent certainty. A guess of the population percentage is 70 percent. What is the minimum required sample size?

Career Kids PARENTS

Rebellion being one of the cornerstones of the rock 'n' roll tradition, it is not surprising to discover that some of the biggest names in the industry are the prodigy of the strictest upbringings. Preacher parents have given us Alice Cooper, Marvin Gaye, and Tori Amos, and military families have blessed us with armed-forces brats such as Jim Morrison, Ann and Nancy Wilson, and Michael Stipe.

But what of other schools of parenting?

In the creative corner is 'Bangle' Susannah Hoffs, whose mother was a film director, and David Crosby, whose father was a cinematographer; also Paul McCartney, Neneh Cherry, Elvis Costello, Kim Wilde, Prince, and Bob Seger, whose fathers were all musicians; 'Beastie Boy' Adam Horovitz whose father, Israel, was a playwright; and Michael Nesmith, whose father invented Liquid Paper.

Blue-collar parents of rock 'n' rollers include: the milkmen fathers of Sting and Kate Bush; the forklift operator father of Chris Isaak; and the truck driver dads of Elvis Presley, Big Daddy Kane, and Captain Beefheart.

For unusual professions, the father of Stuart Copeland (of The Police) wins the prize for being the most exciting: He was a spy.

Actually, most of my family are in oil

Reprinted by permission from *Hot Air,* October–December 1993, p. 9.

72. Find the minimum required sample size for estimating the average number of designer shirts sold per day to within 10 units with 90 percent confidence if the standard deviation of the number of shirts sold per day is about 50.

73. Find the minimum required sample size of accounts if the proportion of accounts in error is to be estimated to within 0.02 with 95 percent confidence. A rough guess of the proportion of accounts in error is 0.10.

FIGURE 5.15

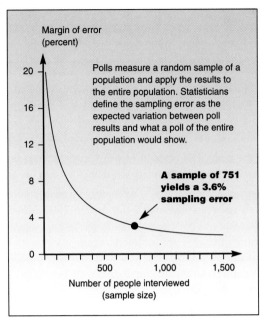

Reprinted courtesy of *The Boston Globe*, "Polling" graphic by Neil Pinchin which was published November 2, 1993.

FIGURE 5.16

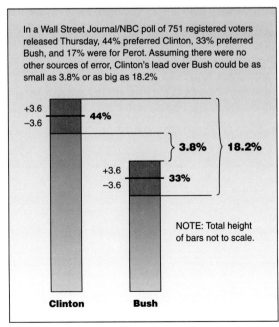

Reprinted courtesy of *The Boston Globe*, "Applying sampling error" graphic by David Butler which was published November 2, 1992.

5.8 EPILOGUE: THE POLLS

As we have seen, the width of a confidence interval for the population proportion decreases as the sample size increases. The standard deviation of the sample proportion is $\sqrt{pq/n}$ and thus depends on the unknown population proportion, p. As you may have found in your solution to one of the problems, this standard deviation is maximized if $p = q = 0.5$. Thus, using this value gives an overall upper bound for the standard error of the sample proportion.

Let's return once again to the subject of polling. The margin of error reported by pollsters is the half-width of a 95 percent confidence interval for the population proportion (expressed as a percentage). It is that plus-or-minus something we use for obtaining a 95 percent confidence interval for the population percentage of voters for a candidate. Hence the margin of error is $1.96 \sqrt{pq/n}$. Since p (and q) are unknown, pollsters use the upper bound obtained in using $p = q = 0.5$. This gives us, approximately, margin of error $= 1/\sqrt{n}$. The margin of error and its relation to sample size are demonstrated for polls conducted during the 1992 presidential campaign in Figure 5.15.

Figure 5.16, reprinted from the same source, again shows a media interpretation of the poll results (this time a more realistic poll than the Gallup poll we discussed in the first section of this chapter). Compare this figure with Figure 5.1.

SUMMARY

In this chapter we introduced the ideas of **statistical sampling** and **sampling distributions.** We learned how to obtain a **simple random sample** from a given population, and we mentioned other statistical sampling schemes. We learned that sample statistics can be used to estimate population parameters, and we saw how the possible values of a statistic lead to its sampling distribution. We talked about the sampling distribution of the sample mean and the sampling distribution of the sample proportion. We saw how the **central limit theorem** assures us that these distributions approach a normal distribution as the sample size gets large. We learned how a pre-sampling probability statement about a statistic based on its sampling distribution can be used in constructing a **confidence interval** for the unknown population parameter the statistic is estimating. This was done in the case of the population mean and the population proportion. We introduced the *t* **distribution,** also known as **student's distribution.** We saw how this distribution is used in constructing confidence intervals for the population mean when the population standard deviation is unknown.

KEY FORMULAS

Confidence interval for the population mean:

$$\bar{x} \pm z_{\alpha/2} \frac{\sigma}{\sqrt{n}}$$

when σ is not known: $\bar{x} \pm t_{\alpha/2} \dfrac{s}{\sqrt{n}}$

Large-Sample Confidence interval for the population proportion:

$$\hat{p} \pm z_{\alpha/2} \sqrt{\hat{p}\hat{q}/n}$$

74. A new study shows that fat in food may cause heart attacks within hours of consuming the meal. The study looked at a random sample of 170 men aged 40 to 49. In this sample, blood hypercoagulation (a condition that may cause arterial clogging) occurred within an average of 6.5 hours after a fatty meal was consumed.[16] Assuming the sample standard deviation was 3.8 hours, give a 95 percent confidence interval for the number of hours it takes for hypercoagulation to develop following a fatty meal in the entire population of men aged 40 to 49.

75. In a random sample of 3,349 residents of the South Pacific island of Moorea, 1,900 expressed objection to the intention of a Japanese company to build a large golf course on their island.[17] Give a 90 percent confidence interval for the percentage of all Mooreans who opposed the golf course.

76. In a study of hypnotic effects, a random sample of 86 people was studied. Out of these people, 56 were identified as highly hypnotizable.[18] Give a 99 percent confidence interval for the proportion of highly hypnotizable people in the entire population from which this sample was selected.

Defining Appropriate Dress in the Workplace

By JULIE HATFIELD

She is a constant distraction to the male employees, with her spiked heels, fishnet stockings and plunging neckline. Her makeup is more appropriate for a nightclub, and her skirt is so short and tight that her colleagues wonder how she ever sits down.

If this profile fits anyone in your office, then someone should tell that person to go home and change. Consultants who counsel management on how to curb sexual harassment in the workplace all have the same thing to say about the responsibility of employees—mostly women employees—to dress appropriately.

"If it occurs to you that it might suggest sexuality, don't wear it," said Arthur Bauer, president of American Media Inc., an Iowa company that produces educational films on sexual harassment.

Not everyone agrees with that position, but a recent survey of 1,769 psychiatrists suggests that many of them do believe in a link between "provocative" clothing and sexual harassment and sex crimes.

The survey, of members of the American Psychiatric Association and the American Association of Medical Specialists, was conducted by Donna Vali, an independent researcher, and Dr. Nicholas Rizzo, a forensic psychiatrist who worked for 20 years in the Massachusetts court system.

The survey included these statements and responses:

• Female attire that appears to the male to invite direct sex attention tends to increase the risk of sex crimes. Yes: 63 percent; no: 21 percent; the rest undecided.

• A male may interpret sexually teasing attire as uncaring and unfair. This may result in thoughts of revenge against the female who brought on the distress, sometimes expanded to hostility against females in general. Yes: 85 percent; no: 11 percent.

• Parents of young females who wish to minimize the risk of sex crime should consider what the girl's attire may be signaling as interpreted by males. Yes: 88 percent; no: 9 percent.

Reprinted by permission from the *Boston Globe,* January 16, 1992, p. 31.

---

[16]"Fatty Meals Pose Risk within Hours," *The New York Times,* January 21, 1993, p. A18.

[17]"Bunkered in Moorea," *The Economist,* June 22, 1991, p. 69.

[18]R. Atkinson and H. Crawford, "Hypnotic Susceptibility and Visuospatial Skills," *American Journal of Psychology* 105, no. 4 (Winter 1992), pp. 527–39.

77. For each of the results reported in the article on appropriate dress in the workplace, provide a 95 percent confidence interval for the population proportion of all American psychiatrists who hold these opinions.

78. In a study of side effects of drugs to combat hypertension, 10 out of a random sample of 132 patients treated with the drugs suffered from dizziness.[19] Give a 95 percent confidence interval for the population proportion of all patients who would suffer from dizziness.

79. In a random sample of 240 couples suffering from infertility, it was found that 48 were infertile due to an interaction between the two members rather than due to one of them alone.[20] Give a 99 percent confidence interval for the proportion of all couples suffering from infertility where the cause of the infertility is due to interaction between the two members of the couple.

80. It is estimated that the average household charitable contribution to cultural institutions is $193. Assume that this sample mean is based on a sample of size 500 and that the sample standard deviation is $78. Give a 95 percent confidence interval for the population mean contribution for all households.

81. For advertising purposes, the Beef Industry Council needs to estimate the average caloric content of 3-ounce top loin steak cuts. A random sample of 14 pieces gives a sample mean of 212 calories and a sample standard deviation of 38 calories. Give a 95 percent confidence interval for the average caloric content of a 3-ounce cut of top loin steak. Also give a 98 percent confidence interval for the average caloric content of a cut.

82. The major Italian magazine *L'Espresso* conducted an extensive study of the behavior of Europeans using a random sample of 10,000 respondents from the European Community. Among the findings of the study was that the average European spends, per day:[21]

> 8 hours, 3 minutes sleeping (s.d. = 1 1/3 hours)
>
> 3 hours, 27 minutes watching TV (s.d. = 2 hours)
>
> 2 hours, 10 minutes with friends (s.d. = 25 minutes)

Give 95 percent confidence intervals for the population averages.

83. Out of a random sample of 70 patients treated with beta carotene, 34 showed the side effect of conjunctivitis.[22] Give an 80 percent confidence interval for the population proportion of all patients who would develop this side effect in response to treatment with beta carotene.

84. Using the table on discrimination on page 237, for the recruitment category, give a 95 percent confidence interval for the proportion of women who felt they were treated the same. Do this, too, for the hiring category. Do the same for the legal assignments and settlements categories. Do the same for the courtroom, pretrial, and research categories.

85. What is the minimum required sample size for determining the proportion of defective items in a production process if the proportion is to be known to within 0.05 with 90 percent confidence? No guess as to the value of the population proportion is available.

86. How many test runs of an automobile are required for determining its average miles per gallon (mpg) rating on the highway to within 2 mpg with 95 percent confidence, if a guess is that the variance of the population of mpg is about 100?

87. A company that conducts surveys of current jobs for executives wants to estimate the average salary of an executive at a given level to within $2,000 with 95 percent

---

[19]C. Espinel et al., "Enalapril and Verapamil in the Treatment of Isolated Systolic Hypertension in the Elderly," *Clinical Therapeutics* 14, no. 6 (November–December 1992), pp. 835–45.

[20]S. McDaniel, J. Hepworth, and W. Doherty, "Medical Family Therapy with Couples Facing Infertility," *American Journal of Family Therapy,* December 1992, pp. 101–22.

[21]G. Invernizzi, "Il Tempo Ritrovato," *L'Espresso,* January 10, 1993, p. 139.

[22]S. Lippman et al., "Beta Carotene in Treatment to Prevent Oral Carcinogenesis," *New England Journal of Medicine* 328, no. 1 (January 7, 1993), p. 15.

Discrimination—Comparing Own Treatment to the Treatment of Men (Percentages)

Treatment	Treated the Same	Treated Differently	
		Beneficial	Discriminatory
Recruitment	52.4	27.1	20.5*
Hiring	45.0	35.0	20.0
On-the-job			
Rewards			
Salary	57.7	1.5	40.8
Promotion	61.5	5.7	32.8
Activities			
Legal assignments	61.2	3.8	34.9
Settlements	58.0	10.2	32.3
Courtroom	59.4	9.4	31.1
Pretrial	65.6	6.2	28.1
Research	70.0	3.8	25.5

NOTE: $N = 200$. The number of responses to these items varies slightly. Not all items, particularly those listed under activities, are relevant to jobs in all work settings.

*Rows may not add to 100 percent because of rounding.

Reprinted from Janet Rosenberg, Harry Perlstadt, and William R. F. Phillips, "Now That We Are Here: Discrimination, Disparagement, and Harrassment at Work and the Experience of Women Lawyers," *Gender & Society* 7, no. 3 (September 1993), pp. 415–33.

confidence. From previous surveys it is known that the variance of executive salaries is about 40 million. What is the minimum required sample size?

88. A ski resort operator wants to estimate the average length of stay of skiers at the resort. A random sample of 17 skiers gives a sample mean stay of 2.5 days and a sample standard deviation of 1.1 days. Give a 90 percent confidence interval for the average length of stay of all skiers at this resort.

89. Interpret the following computer output:

```
MTB > TINTERVAL 90 C1

            N     MEAN   STDEV   SE MEAN   90.0 PERCENT C.I.
C1         48   53.604   3.120     0.450   ( 52.848,   54.360)
```

Compare the results with the ones you would get if you used the normal distribution instead of the t distribution.

90. In a random sample of 2,000 switches used in video display apparatus, 5 switches are found to be defective. Give a 99 percent confidence interval for the proportion of defective switches in the entire (large) population. How do you interpret this particular confidence interval?

91. A random sample of 128 condominiums in Minneapolis reveals a sample mean price of $136,080 and a sample standard deviation of $29,100. Give a 98 percent confidence interval for the average price of a condominium in Minneapolis.

92. An express delivery service needs to estimate the average weight of a package sent overnight. A random sample of 20 packages is selected. The weights, in pounds, are as follows:

0.4, 1.2, 3.7, 1.1, 2.3, 5.0, 0.2, 4.1, 0.8, 6.8, 9.8, 7.5, 3.5, 4.7, 3.3, 2.8, 1.0, 5.9, 4.6, 0.1

Give a 95 percent confidence interval for the average weight of all packages sent overnight.

93. Tests of methane concentrations over the western Pacific and Antarctic oceans revealed a sample mean of 1.75 parts per million (ppm) and a standard deviation of 0.5 ppm. This is

based on 38 measurements.[23] Construct a 95 percent confidence interval for the concentration of methane in the entire region. Reprinted here is a map showing the location of the observations in the sample. Comment on the method used to obtain these data.

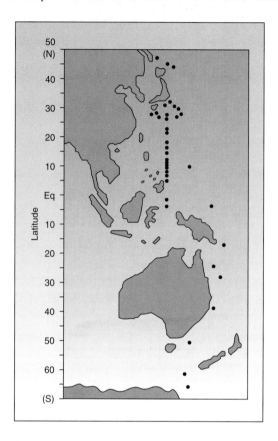

94. Recall that the median of a population is the point such that half the population values lie below it and half the population values lie above it.[24] Suppose that a random sample of two elements is selected from a population. We want to construct a confidence interval for the population *median*. (If we assume the population distribution is symmetric, however, this is also a confidence interval for the population mean.) We define the confidence interval as [the smaller of the two sample values, the larger of the two sample values]. What is the confidence level of such a confidence interval?

95. Generalize the procedure of Problem 94 as follows: We select a random sample of size *n* and define the confidence interval for the population median as [lowest sample value, highest sample value]. What is the confidence level of such an interval? What is one big limitation of this procedure?

---

[23]H. Matsueda et al., "Atmospheric Methane over the Western Pacific and the Antarctic Ocean," *Geochemical Journal* 26 (1992), pp. 21–28.

[24]This problem and the next one are harder than the rest but instructive.

INTERVIEW WITH
Bradley Efron

"Statistics is the science of information gathering, especially when the information arrives in little pieces instead of big ones."

Bradley Efron is Professor and Chairman of the Department of Statistics at Stanford University and has been awarded the Ford Prize, MacArthur Prize, and the Wilks Medal for his research work in computer applications in statistics, particularly with techniques known as the bootstrap *and the* jackknife. *He has been invited lecturer or keynote speaker over 50 times, and his latest book,* An Introduction to the Bootstrap, *was published by Chapman & Hall in 1993.*

[Editor's note: the *bootstrap* and the *jackknife* refer to computer-intensive statistical methods that help harness the power of modern computers in estimating sampling distributions of statistics and are used in statistical inference.]

Aczel: Professor Efron, you, probably more than anyone else, have brought the power of modern computers into statistics. How do you see the future of this marriage?

Efron: First of all, I think [John] Tukey deserves a lot of credit for bringing computers in. At least to my attention. And I believe we are heading toward a world in which people won't even notice that that is the way it is being done. That is, all statistics will be done with a lot of computing, just because you won't have to make a lot of assumptions or theoretical calculations. People already do it, the Box-Jenkins method for example. I think it will just take over after a period of time, that more and more of the theory will be packaged in a way that the consumer doesn't have to think through the theory each time.

Aczel: Do you think it (modern computers and software) will replace our profession (statistics), eventually?

Efron: No. No, I believe it will expand our profession. We will be the people who package the theory. It's sort of like saying: Do you think that computers will put the computer programmers out of business? No, there will be more of us: People will use statistics much more, because they will feel better about it. They will feel like it's not such an alien subject, and there will be much more demand for it.

Aczel: I want to ask you a question about computationally intensive methods in statistics. . . .

Efron: I believe I—Persi [Diaconis] and I—invented the term *computationally intensive statistics.* I have never found a previous reference before our *Scientific American* article . . . so I claim that, ha ha. . .

Aczel: Good! You deserve it! Did you devise the bootstrap [an important computationally intensive method] because you had an application in mind, or were you pursuing theory?

239

Efron: What actually happened was that Rupert Miller in our department was working on the jackknife. He had written a paper called the "Trustworthy Jackknife" in which he tried to figure out when the jackknife method gave dependable variance estimates. The Jackknife was considered very mysterious. It worked, but nobody could figure out why it worked. And sometimes it didn't work. What I thought was that the jackknife must be a differential, local kind of approximation for something else. And so when I started looking for the something else I came up with the bootstrap.

Aczel: What areas of statistics do you see as becoming more important to society in the future?

Efron: OK, now that's a good question. I'm not sure it's areas. Do you mean subject areas like survival analysis? Survival analysis has been a wonderful statistical success story. As I go around the country I have been struck with how prosperous and successful the biostat departments are as opposed to the pure stat departments. And I think that biostat is really moving into the forefront of the national consciousness, or international consciousness, and it'll be as much a part of discourse as things about health cures are now. That is, biostatistics will naturally be a part of newspaper stories about controls and treatments and how the significance level is going, and things like that. So I think that will be tremendously important. I work in biostat and biological kinds of things so I don't get to see a large part of the other kinds of applications. The history since 1900 has been that statistics just takes over field after field in terms of being the methodology of choice, and I think that'll continue. I've done papers with people in astronomy and physics lately, and they're starting to use statistics a lot more for the simple reason that they have to be efficient now. It's hard to say, but I don't see any area where it's being resisted much.

Aczel: What are some of the more interesting problems and applications you've worked on?

Efron: The one I enjoyed most recently was one on Hubble's law. There's a debate in the astronomical community. Hubble's law is that the further a galaxy is away the faster it recedes in a linear way, that space is expanding linearly. Some very good astronomers doubt that. I worked on a set of galaxy redshifts with Vahi Petrosian at Stanford to see whether it was true or not. Much to my surprise, after a long time

Hubble's law worked out to be pretty good. I didn't believe it. It just didn't look right to me. But it worked out really well.

I get a lot of good problems in biostatistical consulting. I work for drug companies. I am designing a trial for a company that makes male potency drugs, and it's a rough trial to design. The spurious effects are very easy to have happen, and so it was a lot of fun designing the trial and I hope it will be fun to analyze it too. One of the most fun data sets I ever worked on was that Shakespeare data set—the "how many words did Shakespeare know" problem with Ron Thisted which was a philological kind of data set. So they come up everywhere.

Aczel: Is Hubble's law, by the way, related to relativity? What does relativity predict?

Efron: When Hubble announced his law, Einstein had proved earlier that one of the stable forms for the universe was a constant expansion. And then since everybody told him that was impossible, he recanted. Then Hubble showed that that was what seemed to be happening. That's sort of the accepted majority view. However, there are some other solutions that other people concur are the right ones. I do not know the physics well enough to say.

Aczel: And the statistics prove Hubble's law?

Efron: The statistics at least were very consistent with Hubble's law and not with the other ones. And I was quite surprised because there's been a lot of good criticism of it. The data set was 476 galaxies collected from a bunch of different surveys—it's hard to do redshifts. It sure didn't look like it was going to work, but it did.

Aczel: I want to ask you about the foundations of statistics. Do you think there are holes in the foundations of statistics the way there are in set theory?

Efron: I don't think the holes are the way they are in set theory in that set theory is presumably perfectable, whereas statistics seems much further from that goal yet. I find myself returning again and again as a basis of statistics to things that are just examples like normal translation problems. If you can reduce a problem to a normal translation problem you sort of know the answer. So if you can reduce any problem to that you have the answer. Or confidence intervals. If you can get the kind of confidence property—that seems like the right answer, so if you get something else that approximately does that, that must be pretty good. The only complete theory of statistics is the Bayesian theory and even though it's unassailable it somehow misses part of the story, which is that you can't use it as an actual driving theory for complicated problems. You always are then forced to do something too complicated, and make up your mind on things you have no opinions on. So somehow Bayesian theory is wonderful but it doesn't tell the whole story. Frequentist theory is shot full of contradictions but it seems to work so well. Trying to reconcile those two things. . . one of my hopes is that computer intensive statistics will make it easier to reconcile philosophical issues because there won't be so many technical problems.

Aczel: You're talking about Bayesian versus Frequentist methods?

Efron: Yes.

Aczel: So you think computationally intensive methods will. . .

Efron: Yes, that's just a hope. I use them that way at certain times. The two philosophies don't seem that separate when you can . . . I'm working on empirical Bayes kinds of things again now. And I've been using the bootstrap on empirical Bayes, starts bringing the two things together. They don't seem as separate as they [did] to me a long time ago. Maybe I'm just softening up.

Aczel: You started out as a math major. What made you change—or was it a change—into statistics?

Efron: My own history was that my Dad was a truck driver and was an amateur mathematician, and also was the sports statistician for the bowling league and the baseball league at home (St. Paul). And so I grew up with a lot of numbers around and a fairly sophisticated Dad who knew a lot about how to calculate things. I thought I was going to be a mathematician. I went to CalTech, and I think I would have stayed a mathematician if mathematics was like it was a hundred years ago where you computed things, but I have no talent at all for modern abstract mathematics. And so I wanted to go into something that was more computational. After CalTech I came to Stanford. And statistics was definitely better.

Aczel: You made the transition OK?

Efron: Actually, they didn't have statistics courses at CalTech, but some faculty member let me read Cramer's book. It was a reading course. And I really loved that book. I read it from front to back. I don't see anybody else do it the way that Cramer did. I still have that book, completely marked up.

Aczel: Can you say something about statisticians who influenced you in your early years?

Efron: Yes, easily. At Stanford, there was Rupert Miller and Lincoln Moses, [who] ran the biostat program. I'd gone in as a mathematician. It was in the biostat program that I learned statistics as a living day-to-day kind of thing that helps people. It was a big eye opener seeing how they did statistics.

Aczel: You are the recipient of the prestigious MacArthur Award, among many other distinctions. Can you tell us something about the project that you were doing with the MacArthur?

Efron: Well, the MacArthur Award was not an award for doing projects. They specifically say that it isn't. They specifically say they won't say anything to make you do anything. They want the award to be a prize, not an impediment to your future work. And they're very nice people to deal with in that regard. So, I never did anything directly. However, it had a big effect on me in the following sense that I never expected. People both inside the university and outside the university suddenly paid a lot more attention to what I had to say.

Aczel: Is there no Nobel Prize in statistics? They consider it part of math [which has no Nobel Prize]?

Efron: Statistics does not have big prizes. And it's probably a good thing. It hasn't had a good effect, say, on physics or math to have the Field medals or Nobel Prizes. They tend to produce a lot of big egos and squabbling. Statistics has been, in a sense, a sort of intellectual working man's field, where you work for the love of the field and not for fame, because there sure isn't any.

Aczel: What is statistics to you—is it mathematics, computer science, philosophy, all of the above, or something else?

Efron: My definition, that is in our book, is that statistics is the science of information gathering, especially when the information arrives in little pieces rather than in one or two big pieces.

Aczel: Like a puzzle?

Efron: It's really quite amazing that there can be a theory of statistics. You might think that you couldn't have theories about things . . . I mean, like astronomy is about stars and geology is about rocks, well statistics is about information and information gathering. It's not clear that that's a "thing" in the same way that rocks are rocks.

Aczel: You mean like a concept of a number or something like that?

Efron: Yes, it's pretty philosophical. The philosophers usually say that statistics is impossible. They say that you can't learn from experience. They can always think of counterexamples. But we live in a world where the examples outnumber the counterexamples by quite a bit.

Aczel: Where do you stand on estimation versus hypothesis testing and criticism of p-values? Like when a student asks "why is it alpha .05?" or "what happens when my p-value is .06, why is it *not* significant but when it's .04 it is?" Or in general, what can you say about hypothesis testing, confidence intervals, the future. . . ?

Efron: Estimation is putting things together and testing is pulling things apart. Analysis versus synthesis. I've always liked confidence intervals because they let you straddle the middle ground. There are times for all those things. Hypothesis testing is immensely efficient if you're really just trying to know whether an effect is positive or not positive. It's the only thing you can use when you've got $n = 10$, a lot of times. But it isn't nearly so interesting when you've got a bigger n and you really want to know more. It's usually fairly clear that the hypothesis is true or not true. The question is, how true. And then estimation and confidence intervals come more . . . I use confidence intervals a lot more. I think most good applied statisticians use confidence intervals more than they show up in the books. I wish they were in the books. The reason they are not in the books so much is they are much harder to explain than either estimation or hypothesis testing. But that's one of those things that I hope the future will bring us to—a much easier acceptance of things like intervals be-

cause you'll be able to get them out of the automatic machine and then you can spend your time understanding what they are instead of this horrible . . . The math is quite involved.

Aczel: They will be more accurate, right, if you use the bootstrap? I'm really fascinated with the idea that the bootstrap gives you any required level of accuracy and correctness, especially replicated bootstrap, and as the sample size grows, and that you're really doing better than normal theory, t-distribution.

Efron: Yes, you're doing one step better, and that one step is a big step.

Aczel: From 1 over root n to 1 over n . . .

Efron: Yes, I think that's very important. And the thing that I'm working on—and a lot of people are working on—is trying to make that really dependable so you really get it—and it really works—every time. Prepackaging this theory isn't so easy. It's hard work.

Aczel: One more question. You mentioned earlier a new book you wrote about the bootstrap.

Efron: Yes. Now that our little book is out—[Robert] Tibshirani and I have a little book called *An Introduction to the Bootstrap,* published by Chapman and Hall, 1993—it will make it easier to understand the bootstrap. What's happening is that the bootstrap has had its run, its first run, at the theory world, and now I think it will go into applications. The applied people have always seemed happier with it than the theory people.

Aczel: Can you say something about your outside interests? You're a human being and a very interesting one, not just a statistician.

Efron: I love statistics. I live right at the Stanford campus. I go in all the time. I don't work hard every day, though. I work every day. In the evenings I like to go to movies with my girlfriend, and I go to almost all movies. I'm interested in astronomy and science. I claim that I'm an amateur scientist as well as a professional scientist. And I love science. I think it's the greatest thing people ever thought of. So I guess my hobby is science, too. I like to sit around and talk . . . Movies, stat, and science.

6

Trial by Probability

6.1 Introduction   247

6.2 Statistical Hypothesis Testing   249

6.3 Tests about the Population Mean, μ, When the Population Standard Deviation, σ, Is Known   257

6.4 Tests about the Population Mean, μ, When the Population Standard Deviation, σ, Is Unknown   274

6.5 Large-Sample Tests for the Population Proportion   284

6.6 How Hypothesis Testing Works   292

6.7 The Probability of a Type II Error and the Power of the Test   297

N ot long after the Norman Conquest of England, the Royal Mint was established in London. The mint has been in constant operation from its founding to this very day, producing for the crown gold and silver coins (and in later periods, coins from cheaper metals). Sometime during the reign of Henry II (1154–1189), a mysterious ceremony called the "trial of the pyx" was initiated.

The word *pyx* is Old English for "box," and the ceremony was an actual trial by jury of the contents of a box. The ancient trial had religious overtones, and the jurors were all members of the Worshipful Company of Goldsmiths. The box was thrice locked and held under guard in a special room, the Pyx Chamber, in Westminster Abbey. It was ceremoniously opened at the trial, which was held once every three or four years.

What did the pyx contain, and what was the trial? Every day, a single coin of gold (or silver, depending on what was being minted) was randomly selected by the minters and sent to Westminster Abbey to be put in the pyx. In three or four years, the pyx contained a large number of coins. For a given type of coin, say a gold sovereign, the box also contained a royal standard, which was the exact desired weight of a sovereign. At the trial, the contents of the box were carefully inspected, counted, and later some coins were assayed. The total weight of all gold sovereigns was recorded. Then the weight of the royal standard was multiplied by the number of sovereigns in the box and compared with the actual total weight of the sovereigns. A given tolerance was allowed in the total weight, and the trial was declared a success if the total weight was within the tolerance levels established above and below the computed standard.

The trial was designed so that the king or queen could maintain control of the use of the gold and silver ingots furnished to the mint for coinage. If, for example, coins were too heavy, then the monarch's gold was being wasted. A shrewd merchant could then melt down such coins and sell them back to the mint at a profit. This actually happened often enough that such coins were given the name "come again guineas" as they would return to the mint in melted-down form, much to the minters' embarrassment. On the other hand, if coins contained too little gold, then the currency was being debased and would lose its value. In addition, somebody at the mint would then be illegally profiting from the leftover gold.

When the trial was successful, a large banquet would be held in celebration. We may surmise that when the trial was not successful, well, the Tower of London was not too far away. The trial of the pyx is practiced (with modifications) to this day. Interestingly, the famous scientist and mathematician Isaac Newton was at one time (1699 to 1727) master of the mint. In fact, one of the trials during Newton's tenure was not successful, but he survived.[1]

Enlargement of Henry III Silver Penny of 1248 (Long Cross Type).

Gold and Silver Trial Plate of 1649. In the first year of the Commonwealth the Goldsmiths' Company were charged with preparing trial plates of crown gold (22 carats) and silver. This is the gold plate with the reverse impression of a half-broad (10s).

[1] Adapted from S. Stigler, "Eight Centuries of Sampling Inspection: The Trial of the Pyx," *Journal of the American Statistical Association,* September 1977, pp. 493–500.

6.1 INTRODUCTION

Figure 6.1 demonstrates how the trial of the pyx was carried out throughout history—and still is today. As in any trial carried out within the legal systems of Britain, the United States, and many other countries, the defendant (the mint) is at first assumed innocent. This assumption is called the *null hypothesis*. As the trial begins, in our example of gold sovereigns, the crown believes the null hypothesis that the average weight of all coins minted during the period in question is exactly 1 troy ounce. The pyx contains a *random sample* of coins selected from all coins of this type minted in this period. If the null hypothesis is true and the average weight of all coins is indeed 1 troy ounce, then the average weight of the coins in the large random sample of coins in the pyx should not deviate too much. How much is too much? This is where the tolerance, or remedy, comes in. Modern statistical theory tells us that for large samples the sample mean has a roughly normal distribution with the mean being the population mean and the standard deviation that of the population divided by the square root of the sample size. (Remember the *central limit theorem* of Chapter 4?) So we know, for example, that the sample mean has a 95 percent probability of falling within $1.96 \ \sigma/\sqrt{n}$ of the population mean. If we should find the mean weight of the coins in the pyx to be outside that range (either above or below), we could *reject* the

195

158

The Pyx Trial Today

The trial of the pyx is still held and annually about March the work begins. In the case of gold—for use as bullion abroad—its function is just the same as it was in the Middle Ages: the fineness of the gold is tested and publicly made known. Silver (now only Maundy Money) and cupro-nickel are also assayed but the results no longer have the old importance. For many years the trial was held at the Exchequer Office and the whole proceedings were carried through in a single day. Since 1870, however, it has always been held at Goldsmiths' Hall in the City of London and, due to the larger number of coins involved, it takes about eight weeks.

The object is to *try,* that is weigh and assay for purity, the gold, silver and cupro-nickel coinage of the United Kingdom. A jury of freemen of the Worshipful Company of Goldsmiths is summoned each year and in the presence of the Queen's Remembrancer is sworn to perform this work. The jury's duty is to consider samples taken from each batch of coins made, the samples being placed in bags, which are then locked in the pyx box and are brought to Goldsmiths' Hall on the day of the trial.

The Royal Mint provides one coin out of each *journey* or batch of coins. A *journey* is a weight of 720 ozs. Troy of minted coins, except in the case of gold when it is 2,000 coins. The sample coins are put into a bag labelled with the date of minting, the number of *journeys* minted, the denomination of the coin, the number of coins in the bags and their value.

The jury have first to count the contents of the pyx bags and check with the numbers given on the labels. One coin is taken at random from each bag and put aside for individual weighing and assaying. The remainder are weighed in bulk. The weight of the individual coins and that of the coins in bulk have to be within specified legal limits, known as the *remedy.*

Reprinted from *The Trial of the Pyx,* a report published by Westminster Abbey, London, 1968.

FIGURE 6.1

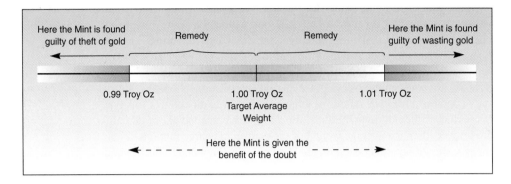

null hypotheis (the mint's claim of innocence), knowing that what we have observed had only a small probability of occurrence (a 5 percent chance) if the null hypothesis were indeed true. Of course, other cutoff probability values could be used, for example, 10 percent (use $z = 1.645$ instead of 1.96, as in constructing a 90 percent confidence interval) or 1 percent (use $z = 2.576$ as in constructing a 99 percent confidence interval).

In Figure 6.1, the remedy is 0.01 troy ounce. (For some probability of being wrong, say 5 percent, in this example $1.96\sigma/\sqrt{n} = 0.01$.) Within the range of $\pm 1.96\sigma/\sqrt{n}$ away from the target average weight, the mint is given the benefit of the doubt and the null hypothesis is not rejected. When the average weight of the coins in the pyx is outside the range, the null hypothesis is rejected and the master of the mint is in deep trouble.

Amazingly, the general concept underlying the procedure was understood by the Britons of the 12th century! The only difference between the way the trial was carried out then and the way it is carried out today is in the computation of the remedy. Stephen Stigler has shown that the remedy calculated in the old trial does not agree with modern theory that tells us that it should be proportional to $\sigma/\sqrt{n}$; instead it was computed as being proportional to σ/n. If that is all that we have learned in 800 years, the 12th-century Britons did quite well! For other reasons, Stigler hypothesizes that they may have purposely set the remedy larger than is done today because they wanted to give the master of the mint, as a respected servant of the crown, a greater benefit of the doubt.[2]

The modern procedure used to test hypotheses is somewhat simpler. In this chapter, we'll describe a *standardized* form of the test. Instead of computing a remedy in the units of the sample mean, we'll compute this region in standardized z-units. Then we'll compare a computed value of the associated standard normal random variable with this region. Using a 5 percent probability, we come up with a rejection region (which we'll discuss more fully in Section 6.2) for the null hypothesis that consists of just the points above and including 1.96 and below and including −1.96 (a simpler region to define and graph). This is shown in Figure 6.2.

---

[2]Ibid.

FIGURE 6.2

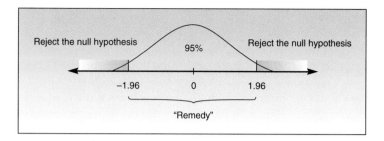

PROBLEMS

1. Why is it important in the trial of the pyx that the coins be selected at random?
2. If the trial of the pyx is not successful, that is, if the master of the mint is convicted, is he or she certainly guilty?
3. If the remedy used in the 12th-century ceremony is greater than modern statistical theory would call for, what effect does this have on the outcomes of the pyx trials?
4. If the standardized test is to be carried out using a 1 percent probability level, what should be the z-values defining the remedy?
5. If the standardized test is to be carried out using a 10 percent probability level, what should be the z-values defining the remedy?

6.2 STATISTICAL HYPOTHESIS TESTING

The modern theory of **statistical hypothesis testing** was developed in the early part of the 20th century. As we have already suggested, the procedure is almost identical to that of the trial of the pyx, which began in the Middle Ages.

In this section, we'll define the concepts and sketch the framework of statistical tests of hypotheses. First, we define the null and alternative hypotheses.

A **null hypothesis,** denoted by H_0, is an assertion about one or more population parameters. This is the assertion we hold as true until we have sufficient statistical evidence to conclude otherwise.

The **alternative hypothesis,** denoted by H_1, is the assertion of all situations *not* covered by the null hypothesis.[3]

---

[3]Some authors use the notation H_a or H_A for the alternative hypothesis.

Together, the null and the alternative hypotheses cover all possible values of the parameter or parameters in question. In the example of the trial of the pyx, the pair of hypotheses is as follows:

$$H_0: \mu = 1.00$$

$$H_1: \mu \neq 1.00$$

The null hypothesis here is the assertion that the population mean is equal to 1.00 troy ounce. The alternative hypothesis is the assertion that the population mean is not equal to 1.00 troy ounce. Only one of the two competing hypotheses may be true, and one of them must be true: Either the population mean is equal to 1.00 or it is equal to any of the infinitely many numbers other than 1.00.

Often, the null hypothesis represents a status quo situation or an existing belief—here the belief that the mint is operating as it should. We wish to *test* the null hypothesis and see whether we can reject it in favor of the alternative hypothesis. The hypothesis test is carried out using information obtained by random sampling, for example, using the random sample of coins in the pyx.

> A **test statistic** is a sample statistic computed from the data. The value of the test statistic is used in determining whether or not we may reject the null hypothesis.

The average weight of the coins in the pyx is a test statistic. We decide whether or not to reject the null hypothesis by following a rule called the *decision rule*.

> The **decision rule** of a statistical hypothesis test is a rule that specifies the conditions under which the null hypothesis may be rejected.

Does the average weight of a coin in the pyx fall outside of the remedy? If yes, we can use this statistical evidence to convict the master of the mint.

In the trial of the pyx, the mint is either innocent (H_0 is true) or guilty (H_1 is true). The jury does not know the real *state of nature* (the actual innocence or guilt of the mint). The jury must, however, make a decision as to whether or not they believe the mint guilty (*reject H_0*) or not guilty (*not reject H_0*).

Let's consider the consequences of the jury's decision. If the jury rejects the null hypothesis and declares the mint guilty when in reality it is guilty (the state of nature is H_1), then the jury has made a correct decision. But if the mint is actually innocent (the state of nature is H_0), then the jury has made an error. We will call this error a **Type I error.** Now consider the other possibility. Suppose the jury decides not to reject the null hypothesis. If the mint is innocent, a correct decision has been made. But if it is guilty, then the jury has failed to convict a guilty mint. We will call the error of failing to reject a false null hypothesis a **Type II error.** The four possible outcomes are shown in Table 6.1.

A moment's reflection will reveal that—within our legal system—a Type I error is considered more serious. "Innocent until proven guilty" means that we want to guard against the possibility of convicting an innocent person—more so than guarding

The State of Nature, the Decision, and the Two Possible Errors   **TABLE 6.1**

	State of Nature	
	H_0	H_1
H_0	Correct decision	Type II error
H_1	Type I error	Correct decision

(left axis label: Decision)

against letting a guilty person go free. What are the probabilities of committing the two types of errors?

The probability of committing a Type I error is denoted by **α**.

The probability of committing a Type II error is denoted by β.

While we would like both error probabilities, α and β, to be small, the preceding discussion implies that it is more important for us to control the level of α. The probability of committing a Type I error should be set to a predetermined small number. In criminal court cases, we would like α to be a very small number indeed.

We don't know the value of the population mean. If we decide to reject the null hypothesis that the average weight of a coin is 1.00 troy ounce when in actuality the population mean is 1.00, then we are committing a Type I error. If we fail to reject the null hypothesis when in reality the population mean is different from 1.00, then we are committing a Type II error. In either of the remaining two possibilities (rejecting the null hypothesis when the population mean is not 1.00, or *not* rejecting the null hypothesis when the population mean is equal to 1.00), we are making a correct decision. Since, as stated earlier, the null hypothesis is usually a statement of an existing belief, we would like to control α, the probability of committing a Type I error. Of course, we also hope not to commit a Type II error; that is, we hope to reject H_0 when H_0 is false. But if we can control only one of the two error probabilities, then it is α, the probability of a Type I error, that should be controlled.[4] In common statistical applications, α is usually set to less stringent standards than one would expect in the legal analogy ("beyond a reasonable doubt"). For us, Type I error probabilities will often be $\alpha = 0.05$ or $\alpha = 0.01$.

We can write α and β using the probability notation of Chapter 2. Both are conditional probabilities: α is the probability that, when we sample and obtain a value of the test statistic, we will end up rejecting the null hypothesis when the null

---

[4]As we will see, α is exactly the same error probability we used in Chapter 5 in the construction of confidence intervals.

hypothesis is actually true; similarly, β is the probability that we will end up not rejecting the null hypothesis when the null hypothesis is actually false.

$$\alpha = P(\text{Reject } H_0 \text{ when } H_0 \text{ is true})$$
$$\beta = P(\text{Not reject } H_0 \text{ when } H_0 \text{ is false})$$

A word of explanation is in order here. Usually, we will be presented with a null hypothesis, a statistical assertion, which we will try to *reject*. Before carrying out the actual test, we know the probability that we will make a Type I error. This probability, α, is preset to a small number, say 0.05. Knowing that we have a small probability of committing a Type I error (rejecting a true null hypothesis) makes our rejection of a null hypothesis a *strong conclusion*. Usually, the same cannot be said about not rejecting the null hypothesis. This is so because, unlike α, the probability β of failing to reject a null hypothesis when it should be rejected is usually not preset to a known small number. Thus, failing to reject the null hypothesis is usually a *weak conclusion* because we do not know the probability of failing to reject a false null hypothesis. When we reject the null hypothesis, we feel fairly confident that the hypothesis should indeed be rejected. When we fail to reject the null hypothesis, we feel that *we did not have enough evidence to reject the hypothesis.* Either the null hypothesis is indeed true or more evidence is needed to reject it.

The probability of committing a Type I error, α, is also called the level of **statistical significance.** In fact, statistical hypothesis testing is often called *significance testing*. A statistical result is considered significant at level α (e.g., at $\alpha = 0.05$) if it leads us to reject a given null hypothesis.

The decision rule of a statistical hypothesis test consists of a comparison of the computed value of the test statistic with the values defining the rejection and nonrejection regions. (The nonrejection region is the remedy of the pyx trial.) We reject the null hypothesis at the level of significance α if and only if the test statistic falls in the rejection region.

The **rejection region** of a statistical hypothesis test is the range of numbers that will lead us to reject the null hypothesis in case the test statistic falls within this range.

The rejection region, also called the **critical region,** is defined by the **critical points.** The rejection region is designed so that, before the sampling takes place, our test statistic will have a small probability α of falling within the rejection region if the null hypothesis is actually true.

The **nonrejection region** is the range of values (also determined by the critical points) that will lead us *not* to reject the null hypothesis if the test statistic should fall within this region.

The nonrejection region is designed so that, before the sampling takes place, our test statistic will have a probability $1 - \alpha$ of falling in the nonrejection region if the null hypothesis is true.

Before demonstrating how a test is conducted using a standardized test statistic and critical points, let's look at an example that links hypothesis tests with the *confidence intervals* of Chapter 5.

⊏⊑⊑
205

A company that delivers packages within a large metropolitan area claims that it takes an average of 28 minutes for a package to be delivered from your door to the destination. Suppose that you want to carry out a hypothesis test of this claim.

EXAMPLE 6.1

Consistent with our desire to assume the defendant innocent until proven guilty, we set the null and alternative hypotheses as follows:

SOLUTION

$$H_0: \mu = 28$$

$$H_1: \mu \neq 28$$

To conduct the test, we select a random sample of $n = 100$ deliveries. We record the delivery times and compute the sample mean $\bar{x} = 31.5$ minutes. The population standard deviation is known to be $\sigma = 5$. We will construct a 95 percent confidence interval for the mean of *all* delivery times of packages in the area (the population mean, μ), and we will use the normal distribution. We compute the following confidence interval for μ:

$$\bar{x} \pm z_{\alpha/2}\frac{\sigma}{\sqrt{n}} = 31.5 \pm 1.96\frac{5}{\sqrt{100}} = 31.5 \pm 0.98 = [30.52, 32.48]$$

Consistent with our interpretation of the meaning of confidence intervals, we can be 95 percent confident that the average time it takes a package to be delivered across town is anywhere from 30.52 to 32.48 minutes.

Let's take Example 6.1 one step further. If we are 95 percent confident that it takes, on average, anywhere from 30.52 to 32.48 minutes for a package to be delivered, then we are 95 percent confident that μ does not lie outside this range of values. Since μ under the null hypothesis is 28, a value outside the confidence interval, we may reject the null hypothesis. If the null hypothesis is actually true, then—before the sampling takes place—there is only a 0.05 probability that the constructed interval would not include the population mean. If we reject the null hypothesis whenever the population mean as stated in the null hypothesis lies outside the 95 percent confidence interval for μ, then we have at most a 0.05 probability of committing a Type I error. We are thus carrying out our test at the $\alpha = 0.05$ *level of significance*. Here we reject the null hypothesis in favor of the alternative hypothesis and conclude that we believe the average delivery time is *not* 28 minutes. Furthermore, since our confidence interval lies entirely *above* the hypothesized value of the population mean (*hypothesized* will mean "stated in the null hypothesis"), we may further conclude that, based on the information in our random sample, we believe that the average time it takes a package to be delivered in the metropolitan area is more than 28 minutes.

What have we learned from this example? We learned that we may use a confidence interval for a population parameter as our decision rule for a statistical hypothesis test. If we use a confidence interval based on a $(1 - \alpha)$ probability (obtained through the

FIGURE 6.3   Confidence Interval for μ and the Null-Hypothesized Population Mean, μ_0, in Example 6.1

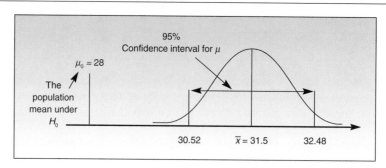

sampling distribution of our statistic), then our probability of a Type I error—rejecting a true null hypothesis—is at most α. This is demonstrated in Figure 6.3.

A confidence interval is useful when we are interested in giving a *range* of possible values that we believe the unknown parameter is likely to have. However, if we are interested only in testing a particular hypothesis about the value of a population parameter, we can carry out the test in an easier way by using a test statistic calculated from our data.

Recall that in Chapter 5 we centered the sampling distribution of our statistic $\overline{X}$ over its obtained value because the value of the parameter μ was unknown to us. We used the sampling distribution to establish bounds on either side of the obtained $\bar{x}$ such that we could be, say, 95 percent sure that the unknown parameter μ lies within these bounds.

Let's now use the same idea in testing hypotheses about the population mean, μ. If someone asserts that the population mean is equal to 28, then we'll center our sampling distribution of $\overline{X}$ over this hypothesized value of μ. We denote the null-hypothesized value of μ by μ_0. If the null hypothesis is correct, then $\mu_0 = 28$ *is* the mean of the sampling distribution of $\overline{X}$, and the distribution should be centered over that value. If μ_0 is indeed the population mean, then—when we sample from the population and obtain a value of the sample mean—this value, $\bar{x}$, should not fall too far away from μ_0. In fact, when H_0 is true, $\overline{X}$ will have a 0.95 probability of falling within the bounds:

$$\mu_0 \pm 1.96 \frac{\sigma}{\sqrt{n}}$$

Thus, if we want to take only a 0.05 chance of making a Type I error when we reject the null hypothesis, then we should reject the null hypothesis only if the sample mean $\overline{X}$ should fall outside the bounds given in the above equation. The range of values in the equation is the nonrejection region of the test, and the values outside this range are the rejection region of the test.

Our test statistic is $\overline{X}$. The obtained value of the test statistic, $\bar{x}$, is compared with the range of values above. If the test statistic falls outside the bounds, then we reject the null hypothesis at the $\alpha = 0.05$ level of significance.

In Example 6.1, the rejection region consists of all values outside the bounds: $\mu_0 \pm 1.96\sigma/\sqrt{n} = 28 \pm 1.96(5/\sqrt{100}) = [27.02, 28.98]$. Since $\bar{x} = 31.5$, a value outside this range, we reject the null hypothesis at the $\alpha = 0.05$ level of significance.

The Equivalence of a 95 Percent Confidence Interval for μ and a Hypothesis Test about μ at the 0.05 Level of Significance

FIGURE 6.4

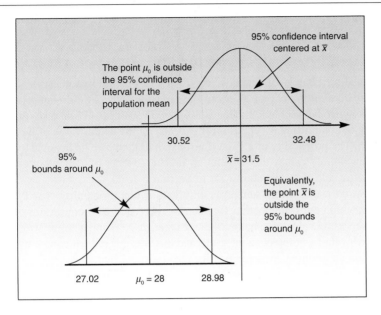

The Rejection Region, the Nonrejection Region, and the Actual Value of the Test Statistic, Falling in the Rejection Region (Example 6.1)

FIGURE 6.5

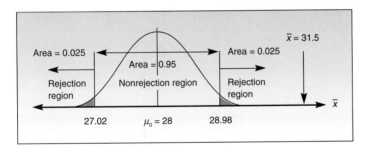

We see that when the level of significance is α, rejection of the null hypothesis occurs if and only if a confidence interval with confidence coefficient $(1 - \alpha)$ based on the same sample results does *not* include the hypothesized value of the parameter. Similarly, nonrejection of the null hypothesis at the significance level α occurs if and only if a confidence interval with confidence coefficient $(1 - \alpha)$ does contain the hypothesized value. This equivalence of (some) hypothesis tests and confidence intervals is demonstrated in Figure 6.4.

Figure 6.5 shows the nonrejection and rejection regions for the test $H_0: \mu = 28$ versus $H_1: \mu \neq 28$. The figure shows the resulting rejection of the null hypothesis due to the fact that the value of the test statistic $\bar{x}$ falls in the rejection region.

The critical points define the separation between the rejection and the nonrejection regions. In Example 6.1, these points were found to be 27.02 and 28.98, with the test statistic $\bar{x}$ in the rejection region (see Figure 6.5).

What about a Type II error? What is the probability that such an error will be made? This depends on what value (covered by the alternative hypothesis) the parameter actually has. Remember our hypothesis test:

$$H_0: \mu = 28$$

$$H_1: \mu \neq 28$$

If the population mean is 29.5, and we conclude that the null hypothesis $\mu = 28$ is true, then we are committing a Type II error. The same happens if we do not reject the null hypothesis when the population mean is 30.7, or 35.9, or 26.432, and so on. Thus, the probability of committing a Type II error *depends* on the actual value of the parameter in question. The probability of a Type II error should therefore be denoted by $\beta\,(\mu)$—a whole collection of values that depend on μ. The concept of β is more complicated than that of α, and we will return to it in Section 6.7.

For now, we will concern ourselves only with α, the probability of a Type I error. Recall that within our framework a Type I error is considered more serious than a Type II error. We set the value of α before the test takes place (similar to the way we set the confidence level of a confidence interval), and this sets our critical points, which define our rejection and nonrejection regions for the statistical hypothesis test.

The kind of test we have seen so far is called a **two-tailed test** because rejection of the null hypothesis is allowed either when we believe the parameter is *greater than* the value stated in the null hypothesis or when we believe it is *less than* the value stated in the null hypothesis. That is, rejection is allowed when the test statistic falls in either of the two tails of the sampling distribution.

As we have seen in Example 6.1,

A two-tailed test will result in rejection of the null hypothesis if and only if the null-hypothesized value of the parameter in question (the population mean) falls outside a confidence interval for the parameter based on the same information and with the same error level α. (For example, doing the test using $\alpha = 0.05$ will give the same result as a 95 percent confidence interval.)

In a two-tailed test, the null hypothesis is stated in terms of an equality. The format is $H_0: \mu = \mu_0$, where μ_0 is some stated number. The alternative hypothesis includes *all* other possible values of the population parameter μ. Logically, if the population parameter is not equal to μ_0, then the alternative hypothesis, $H_1: \mu \neq \mu_0$, is true. This is demonstrated in Figure 6.6.

You may raise some practical objections at this point. For example, if our null hypothesis is $H_0: \mu = 5.0$, you may ask what we should do if the true population mean is not exactly 5.0, but is equal to 4.99999. In such a case, strictly speaking, we should reject the null hypothesis, as it is not true that $\mu = 5.0$. However, 4.99999 is not very

FIGURE 6.6   The Value of μ under H_0 and the Values of μ under H_1

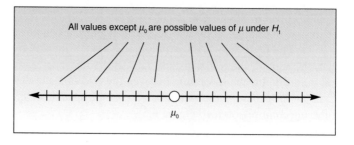

different from 5.0. Do we really want to detect such an infinitesimal deviation from the hypothesized value? The answer is, of course, no. Usually, we will end up rejecting null hypotheses only when the deviation from the hypothesized value is more substantial than in this hypothetical example. You should know, however, that as the sample size increases, we are better able to detect smaller and smaller deviations from any hypothesized value of the parameter.

Next, we'll discuss both the two-tailed and the more commonly used one-tailed tests in their easier, standardized form.

PROBLEMS

6. Explain the difference between the null and the alternative hypotheses.
7. What is the purpose of the test statistic?
8. Define the decision rule for a statistical test.
9. What are the two types of error in a statistical test? What are their respective probabilities?
10. Which of the two types of error is worse? Why?
11. Why is rejection of the null hypothesis a stronger conclusion than nonrejection?
12. A 95 percent confidence interval for the population mean is [17, 19]. If a test were carried out using $\alpha = 0.05$ of the null hypothesis that the population mean is equal to 20 versus the alternative that it is not, using the same data, what would have been the result?
13. What is a two-tailed test?

6.3 TESTS ABOUT THE POPULATION MEAN, μ, WHEN THE POPULATION STANDARD DEVIATION, σ, IS KNOWN

In all statistical hypothesis tests (not only in the two-tailed test), the equal sign (=) must be stated in the null hypothesis. In the one-tailed tests, we will have an *inequality* in the null hypothesis. This inequality, however, will always be a nonstrict inequality and will include the equal sign. This is so because when we carry out a hypothesis test, we center the sampling distribution of the test statistic over the value stated in the null hypothesis. This is the assumption of innocent until proven guilty—we *assume* that the mean is as stated in the null hypothesis by centering the distribution of $\overline{X}$ over the hypothesized value, μ_0. The equal sign in the null hypothesis is thus essential in all hypothesis tests.

Standardizing the Test

Recall that if we subtract from a normally distributed random variable its mean and then divide the result by the standard deviation, we get the *standard normal random variable, Z*. Therefore, assuming that the null hypothesis is true (innocent until proven guilty), the mean of $\overline{X}$ is μ_0. We subtract this mean from the obtained $\bar{x}$ and then divide the result by the standard deviation of $\overline{X}$, which is $\sigma/\sqrt{n}$. (When σ is not known, we divide by $s/\sqrt{n}$.) If the null hypothesis is true, the result (when the sample size is large) should be close to a value drawn from the standard normal distribution. If the null hypothesis is *not* true, the population mean, μ, is either larger than or smaller than μ_0. Hence, when H_0 is not true, the standardized test statistic will tend to be either too large or too small. The terms *too large* and *too small* will mean, respectively, above and below the $\alpha = 0.05$ probability bounds for the standard normal random variable (which are ± 1.96), or the $\alpha = 0.01$ probability bounds for the standard normal random variable (which are ± 2.576), or other bounds for some given significance level. Let's now define our terms in the context of a standardized, σ-known (or large-sample), two-tailed test for the population mean, μ.

FIGURE 6.7   The Test of Example 6.1 Carried Out on the z Scale, Where the Test Statistic Is Standardized

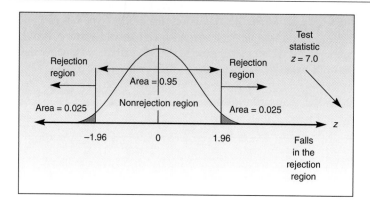

Figure 6.7 demonstrates the standardized form of the test for Example 6.1. A standardized test is equivalent to a test in the unstandardized units of the problem. Instead of comparing the sample mean, $\bar{x}$, with the computed critical points, we compare a standardized test statistic (a test statistic in z-units) with critical points of the standard normal distribution, which are ±1.96 for a test at $\alpha = 0.05$. The computed z statistic is:

$$z = \frac{\bar{x} - \mu_0}{\sigma/\sqrt{n}} = \frac{31.5 - 28}{5/\sqrt{100}} = 7.0$$

The test statistic value is clearly far in the right-hand rejection region, leading us to reject the null hypothesis that the population mean is equal to 28. Since rejection occurred on the right, the actual population mean is presumably larger than 28.

The elements of a two-tailed test for the population mean, μ:	
The null hypothesis:	$H_0: \mu = \mu_0$
The alternative hypothesis:	$H_1: \mu \neq \mu_0$
The significance level of the test:	α (often, $\alpha = 0.05$ or 0.01)
The test statistic:	$z = \dfrac{\bar{x} - \mu_0}{\sigma/\sqrt{n}}$ (assuming σ is known; otherwise, with large n, use $s/\sqrt{n}$ in the denominator)
The critical points:	These depend on α. They are the bounds $\pm z_{\alpha/2}$ that capture between them an area of $1 - \alpha$. (When $\alpha = 0.05$, the critical points are ±1.96; when $\alpha = 0.01$, the critical points are ±2.576. For other values of α, the critical points may be obtained from the standard normal table.)
The decision rule:	Reject the null hypothesis if either $z > z_{\alpha/2}$ or $z < -z_{\alpha/2}$, where $z_{\alpha/2}$ and $-z_{\alpha/2}$ are the two critical points.

EXAMPLE 6.2

Recently the airlines and their regulating agencies have been locked in a battle over baggage. The Air Transport Association, the airlines' trade group, believes that most passengers try to carry on board with them as much of their baggage as they possibly can, to avoid long delays at the baggage-claim carousels upon arrival. Consequently, airlines have been spending millions of dollars redesigning storage space in aircraft cabins. Federal regulators, on the other hand, worry about safety when too much baggage is stored in aircraft cabins. As part of a survey to determine the extent of required in-cabin storage capacity, a researcher needs to test the null hypothesis that the average weight of carry-on baggage per person is $\mu_0 = 12$ pounds (traditionally considered the average weight), versus the alternative hypothesis that the average weight is not 12 pounds. The study is undertaken to determine whether the average weight of carry-on baggage is indeed still 12 pounds or whether trends have changed carry-on baggage weight in either direction—either increased it to more than the traditional average of 12 pounds or decreased it to below that average. In other words, the researcher does not take sides as to the direction of possible change in passenger behavior, allowing for the possibility that *less* baggage may now be carried on board, to avoid delays at X-ray checking machines, for example. The analyst wants to test the null hypothesis at $\alpha = 0.05$.

Suppose that the analyst collects a random sample of 144 passengers traveling on different airlines and different routes (both of which are randomly selected) and finds that the average weight of carry-on baggage per passenger is $\bar{x} = 14.6$ pounds. The population standard deviation is known from past experience to be $\sigma = 7.8$. Carry out the statistical hypothesis test, and state your conclusion.

The null and alternative hypotheses are:

$$H_0: \mu = 12$$

$$H_1: \mu \neq 12$$

Using the standardized form of the test makes things easier. For example, the fact that we want to carry out the test at the $\alpha = 0.05$ level of significance means that we already know the critical points before computing the test statistic. The critical points for Z are ± 1.96. These are shown in Figure 6.8.

We now have an objective decision rule, known even before the sample is collected. We will reject the null hypothesis that the mean weight is 12 pounds if and only if our test statistic, Z, falls in the rejection region shown in Figure 6.8. Thus, if we should end

Determining the Rejection and Nonrejection Regions of the Test of Example 6.2

FIGURE 6.8

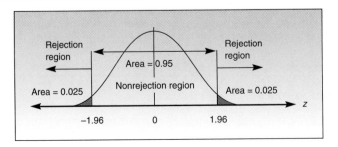

FIGURE 6.9   Conducting the Test of Example 6.2

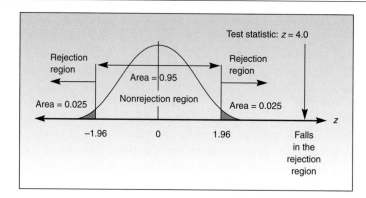

up rejecting the null hypothesis, we will be taking only a 0.05 chance of committing a Type I error. We now compute our test statistic:

$$z = \frac{\bar{x} - \mu_0}{\sigma/\sqrt{n}} = \frac{14.6 - 12}{7.8/\sqrt{144}} = 4.0$$

Since the computed value of the test statistic falls in the rejection region, our decision is to reject the null hypothesis. This is shown in Figure 6.9.

We reject the null hypothesis that the average weight of carry-on baggage per person is equal to 12 pounds in favor of the alternative that it is not equal to 12 pounds. Furthermore, since the rejection occurred in the right tail of the distribution, we may conclude that the average weight is probably higher than 12 pounds. (This conclusion should be made with some care. As we will see later, there is a very small probability that rejection on the right tail may occur when the value of the parameter is actually smaller than the value stated in the null hypothesis—in this example, if the mean is less than 12.)

One-Tailed Tests

The tails of a statistical hypothesis test are determined by the need for an action.

If action is to be taken if a parameter is greater than some value a, *then the alternative hypothesis is that the parameter is greater than* a, *and the test is a* **right-tailed test.**

For example, if a population mean is claimed to be 50 and we are concerned about the possibility that the mean may be greater than 50 (but do not care if the mean is below 50), then the hypotheses are:

$$H_0: \mu \leq 50$$

$$H_1: \mu > 50$$

which is a right-tailed test.

If action is to be taken only if the value of a parameter is believed to be less than a stated value a, *then the alternative hypothesis is that the parameter is less than* a, *and the test is a* **left-tailed test.**

For example, if the population mean is claimed to be 50 but action should only be taken if it is below 50, then the null and alternative hypotheses are:

$$H_0: \mu \geq 50$$

$$H_1: \mu < 50$$

which is a left-tailed test.

In a one-tailed test, the rejection region is either completely on the left (a left-tailed test) or completely on the right (a right-tailed test). Therefore, we have the following important property:

In a one-tailed test, the entire probability of a Type I error, α, is placed in the tail of rejection.

This changes the critical points of the tests. For example, in a right-tailed test for the population mean, with a known σ or a large sample size and $\alpha = 0.05$, there is only one critical point: $z = +1.645$. Why? The answer is shown in Figure 6.10.

The distribution of the statistic is normal, and the probability of a Type I error is 0.05. We place this probability *entirely* on the right-hand side of the distribution. We

---

The elements of a right-tailed test for the population mean, μ:

The null hypothesis:	$H_0: \mu \leq \mu_0$
The alternative hypothesis:	$H_1: \mu > \mu_0$
The significance level of the test:	α (often, $\alpha = 0.05$ or 0.01)
The test statistic:	$z = \dfrac{\bar{x} - \mu_0}{\sigma/\sqrt{n}}$ (assuming σ is known; otherwise, for large n, use $s/\sqrt{n}$ in the denominator)
The critical point:	This depends on α. It is the bound z_α that captures an area of α to its right. (When $\alpha = 0.05$, the critical point is 1.645; when $\alpha = 0.01$, the critical point is 2.326. For other values of α, the critical point may be obtained from the standard normal table.)
The decision rule:	Reject the null hypothesis if $z > z_\alpha$ where z_α is the critical point.

Critical Point and Rejection and Nonrejection Regions for a Right-Tailed Test at $\alpha = 0.05$ **FIGURE 6.10**

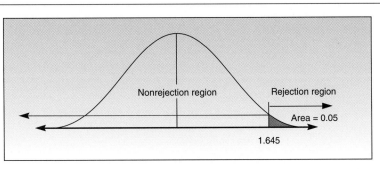

have to find the point a such that the area to its right is 0.05 (not 0.025, as in the case of a two-tailed test, where the sum of the areas in *both* tails is 0.05). From the standard normal table, we find that the point z_α that cuts off an area of 0.05 to its right is 1.645. (This is the same point we would use together with its negative, -1.645, in a two-tailed test at $\alpha = 0.10$. Here it is used alone.)

Example 6.3 demonstrates the use of a right-tailed test.

EXAMPLE 6.3

The Environmental Protection Agency (EPA) sets limits on the concentrations of pollutants emitted by various industries. Suppose that the upper allowable limit on the emission of vinyl chloride is set at an average of 55 parts per million (ppm) within a range of two miles around the plant emitting this chemical. To check compliance with this rule, the EPA collects a random sample of 100 readings at different times and dates within the two-mile range around the plant. The findings are that the sample average concentration is 60 ppm and the population standard deviation is known to be 20 ppm. Is there evidence to conclude that the plant in question is violating the law?

SOLUTION

Courtesy of CILCO.

The EPA is interested in determining whether or not the plant operators are violating the law. The EPA will be taking action against the plant operators only if there is enough evidence to conclude that the average concentration of vinyl chloride within the given range of two miles from the plant is *above* the allowed average of 55 ppm. Therefore, the test is a right-tailed test:

$$H_0: \mu \leq 55$$

$$H_1: \mu > 55$$

If a level of significance $\alpha = 0.01$ is desired, the critical point to be used is $+2.326$. The test statistic is the same as that used in the two-tailed test:

$$z = \frac{\bar{x} - \mu_0}{\sigma/\sqrt{n}} = \frac{60 - 55}{20/10} = 2.5$$

FIGURE 6.11

The One-Tailed Test in Example 6.3, Compared with a Two-Tailed Test

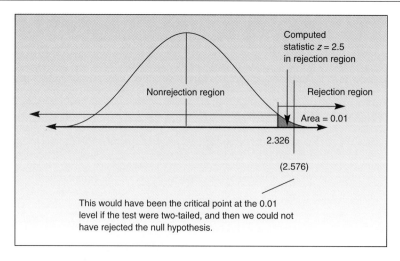

As can be seen from Figure 6.11, the value of the test statistic falls in the rejection region for $\alpha = 0.01$, and the EPA may therefore reject the null hypothesis (nonviolation of the emission control rule) in favor of the alternative hypothesis that the plant operators are indeed violating the law. By using $\alpha = 0.01$, the EPA is willing to take at most a 0.01 chance of concluding that the plant operators are guilty when in reality they are innocent.

What are the advantages of a one-tailed test? To answer this, refer back to Example 6.3 and imagine that the EPA had set up the test as a two-tailed test, with the understanding that action would be taken only if the rejection happened on the right-hand side. In such a case, the test would have had two critical points at the 0.01 level of significance, ± 2.576, and the obtained test statistic value, $z = 2.5$, would be found inside the nonrejection region for the test (although the null hypothesis would still be rejected if the EPA were willing to use an α of 0.05). Thus, we see that setting up the hypothesis test as a one-tailed test increases the *power* of the test (we will formally define power in Section 6.7) in the sense that this allows us to reject null hypotheses in more cases than would be possible if we used two tails. It is therefore very important to first determine whether a test should be one-tailed or two-tailed.

The elements of a left-tailed test for the population mean, μ:

The null hypothesis:	$H_0: \mu \geq \mu_0$
The alternative hypothesis:	$H_1: \mu < \mu_0$
The significance level of the test:	α (often, $\alpha = 0.05$ or 0.01)
The test statistic:	$z = \dfrac{\bar{x} - \mu_0}{\sigma/\sqrt{n}}$ (assuming σ is known; otherwise, for large n, use $s/\sqrt{n}$ in the denominator)
The critical point:	This depends on α. It is the bound z_α that captures an area of α to its left. (When $\alpha = 0.05$, the critical point is -1.645; when $\alpha = 0.01$, the critical point is -2.326. For other values of α, the critical point may be obtained from the standard normal table.)
The decision rule:	Reject the null hypothesis if $z < z_\alpha$ where z_α is the critical point.

Example 6.4 demonstrates the use of a left-tailed test.

EXAMPLE 6.4

A certain kind of packaged food bears the following statement on the package: "Average net weight 12 ounces." Suppose that a consumer group has received complaints from users of the product who believe that they were getting smaller quantities than the manufacturer states on the package. The consumer group therefore wants to test the hypothesis that the average net weight of the product in question was 12 ounces versus the alternative hypothesis that the packages were, on average, underfilled. A random sample of 225 packages of the food product is collected, and it is found that the average net weight in the sample is 11.8 ounces and the sample standard deviation is 7.5 ounces. Given these findings, is there evidence that the manufacturer is underfilling the packages?

SOLUTION

As in every problem, we first need to determine whether the situation calls for a one-tailed or a two-tailed test. In this particular example, the required test clearly is one-tailed and the tail of rejection is the left-tail. This is seen by answering the question: Under what conditions do we want to take an action? The action always determines the *alternative hypothesis*, and the null hypothesis then covers all other possibilities (always including the equal sign). Here it is easy to see that since the consumer group is the one carrying out the test, they would be interested in taking action against the manufacturer (complaint, lawsuit, etc.) only if they end up believing, to within a small probability of being wrong, that the manufacturer is underfilling the packages it sells. If the packages are believed to be overfilled or filled with exactly 12 ounces of food, then no action whatsoever will be taken. Since action is to be taken only in the underfilled case, the test is left-tailed. It may help to think of "less than" implying a left-tailed and "greater than" implying a right-tailed test. Our test is:

$$H_0: \mu \geq 12$$

$$H_1: \mu < 12$$

Now that we know what kind of test to perform, we can set up the distribution of the test statistic under the null hypothesis, noting that our sample size, $n = 225$, allows us to use the standard normal distribution. In the statement of the problem, we left out the required probability of a Type I error, α. This was done purposely, because in reality the choice of α will often be left to the statistician performing the test. Let's start by choosing a commonly used value, such as $\alpha = 0.05$. Figure 6.12 shows the distribution and the critical point $z_\alpha = -1.645$, which corresponds to $\alpha = 0.05$ and a left-tailed test using the normal distribution. The figure also shows the value of the test statistic, computed as:

$$z = \frac{\bar{x} - \mu_0}{s/\sqrt{n}} = \frac{11.8 - 12}{7.5/15} = -0.4$$

FIGURE 6.12   Nonrejection of the Null Hypothesis in Example 6.4

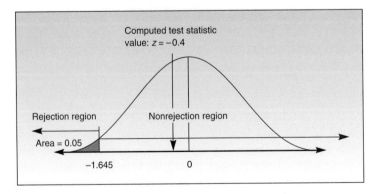

As we see, the value of the text statistic falls in the nonrejection region. We cannot reject the null hypothesis H_0: $\mu \geq 12$. There is insufficient evidence to conclude that the manufacturer is underfilling the packages. The fact that we accepted the null hypothesis does not mean that the manufacturer is *not* underfilling the packages; it merely means that, given our data, we cannot reject the null hypothesis without taking a large probability of being wrong. The probability of committing a Type I error is large because the value of our test statistic falls far inside the nonrejection region, and we would have to accept the null hypothesis even if we were to choose a much larger value of α than 0.05.

Table 6.2 gives the critical points for a test using the normal distribution in the cases of a one-tailed and a two-tailed test with commonly used values of α. Critical values for other levels of α are found in the standard normal table. Table 6.2 is only a quick-reference summary table provided for convenience.

The *p*-Value

Notice that rejection of the null hypothesis, if it occurs, happens at various points in the rejection region for, say, $\alpha = 0.05$. Is there a difference between our rejection of the null hypothesis because we got a test statistic value $z = 2.00$ and rejection of the null hypothesis because we got $z = 15.8$?

The answer is a very definite yes! Assuming the null hypothesis is true (the ever present assumption of innocence), it is much, much less likely for us to obtain a z of 15.8 than a z of 2.00. The former has an extremely small a priori probability of occurring if H_0 is true (the probability is a number close to zero), while the value $z = 2.00$ has a small—but not excessively so—a priori probability of occurring when H_0 is indeed true. (In a two-tailed test, this probability is actually just under 5 percent, or 0.0456 to be exact, as we will see later.)

Since the z value of 15.8 is so unlikely when the null hypothesis is true, and the value 2.00 is not nearly as extreme, the larger value casts a much stronger doubt on the veracity of the null hypothesis. Our rejection of the null hypothesis in this case should therefore be considered much more substantial. The a priori probability of the actual result, assuming H_0 is true, is called the *p-value*. The *p*-value is the smallest level of significance, α, at which a null hypothesis may be rejected using the obtained value of

Critical Points of Z for Selected Levels of Significance

TABLE 6.2

	Level of Significance α:		
	0.10	**0.05**	**0.01**
One-Tailed Test	+ or −1.28	+ or −1.645	+ or −2.326
Two-Tailed Test	+ and −1.645	+ and −1.96	+ and −2.576

Note: + or − means we use the positive value in the table for a right-tailed test and the negative value for a left-tailed test. In a two-tailed test, we use both the positive and the negative values.

the test statistic. Since this probability tells us how strongly we should feel about our rejection of the null hypothesis, it is an extremely important concept, which goes much further in its information content than does a simple rejection at a prespecified level of significance, α.

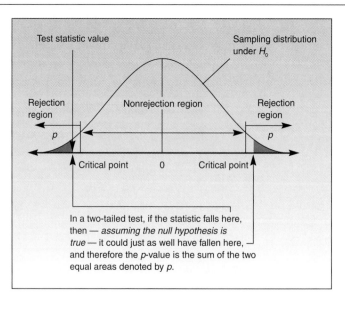

> The **p-value** is the probability of obtaining a result as extreme as, or even more extreme than, the result actually obtained when the null hypothesis is true.

As such, in a right-tailed test the p-value is the area to the *right* of the actually obtained test statistic value. In Example 6.3 it is the area to the right of $z = 2.5$, which is seen using the normal table to equal 0.0062.

In a left-tailed test, being more extreme means being to the left, so in a left-tailed test the p-value is the area to the left of the test statistic value. In Example 6.4, it is the area to the *left* of $z = -0.4$, which is 0.3446. Since this is such a large number, we do not reject the null hypothesis.

In a two-tailed test, we compute the area in the tail where the test statistic value occurs and then *double* that value. This is because in a two-tailed test, if the null hypothesis is true, then a value on the left is as likely to occur as one on the right. Since both would lead to rejection of the null, the tail area is doubled. This is shown in Figure 6.13.

Refer back to Figure 6.7, which shows the standardized test statistic for Example 6.1. The p-value is double the area to the right of $z = 7.00$. Our table ends before 7.00, so we cannot find the p-value, but we know it is a very small number. In Example 6.2, the p-value is double the area to the right of $z = 4.00$, which is equal to 0.00006.

FIGURE 6.13   The p-Value in a Two-Tailed Test

Test statistic value

Sampling distribution under H_0

Rejection region

Nonrejection region

Rejection region

p

p

Critical point   0   Critical point

In a two-tailed test, if the statistic falls here, then — *assuming the null hypothesis is true* — it could just as well have fallen here, and therefore the p-value is the sum of the two equal areas denoted by p.

To summarize:

1. In a right-tailed test, the p-value is the area to the right of the test statistic.
2. In a left-tailed test, the p-value is the area to the left of the test statistic.
3. In a two-tailed test, the p-value is twice the area to the right of a positive test statistic or to the left of a negative test statistic.

For a quick interpretation of reported p-values you may want to use the following rule:

For a given level of significance, α, reject the null hypothesis if and only if $\alpha \geq$ p-value.

In other words, a result is significant at level α if the associated p-value is less than or equal to α. This rule is useful when looking at tables of reported results, such as computer-generated test results. It is easy to scan the results, comparing p-values (often reported as "significance") with, say, the value 0.05. Any result with a p-value less than or equal to 0.05 will then be noted as significant at the $\alpha = 0.05$ level.

The results in Example 6.1, 6.2, and 6.3 are significant at $\alpha = 0.05$; the result of Example 6.4 is not. Remember, however, that a p-value contains much more information than just whether or not to reject the null hypothesis at a *given* level of significance, α. The p-value tells us how unusual our particular sample result is if we assume that the null hypothesis is true. This, in turn, gives us an indication as to how strongly we should feel about rejecting (or not rejecting) the null hypothesis in light of our evidence. The p-value is the attained significance level of the test. It stands on its own, and can be interpreted without a given level α.

Logically, if our obtained value of the test statistic is not very likely (i.e., has a small p-value), assuming that the null hypothesis is true, then we should reject the null hypothesis. Similarly, if the obtained value of the test statistic is relatively likely (i.e., has a high p-value, say, above 0.05 or 0.10), then we should not reject the null hypothesis because there is no convincing evidence against it.

The p-value acts as a "personalized" level of significance that goes with our value of the test statistic. It is the *exact α*, or *attained α*, attached to the test statistic value we have and at which we may reject the null hypothesis.

Reporting the p-value is therefore a more general way of reporting the results of a statistical hypothesis test: It leaves the choice of α to the person using the results rather than to the statistician performing the test. For example, suppose a test is carried out and we report that its p-value is 0.0002. This conveys much information to the reader. It tells him or her that the null hypothesis can be rejected at $\alpha = 0.01$ and that it can be rejected even at smaller values of α, such as 0.001. In fact, this p-value says that the null hypothesis can be rejected at values of α as small as $\alpha = 0.0002$. It also tells the receiver of the information that—assuming the null hypothesis is true—the probability of obtaining a test statistic value as extreme as the value obtained in the experiment, or more extreme than the obtained value, is 0.0002. Here, if one chooses to reject the null hypothesis given the obtained value of the test statistic, one is willing to take a 2-in-10,000 chance of committing a Type I error and thus "convicting an innocent person." Contrast the information content in the statement "p-value $= 0.0002$" with the very limited amount of information in the statement "we reject the null hypothesis at $\alpha = 0.05$," which could have been the reported result of the same test.

Suppose we have another case, where the p-value is 0.06. Reporting this value allows the reader to reject the null hypothesis if he or she is willing to use $\alpha = 0.06$. If the reader wants to use the $\alpha = 0.05$ level of significance, or anything smaller, then he or she will not be able to reject the null hypothesis. Reporting the p-value allows the reader to choose his or her own α and still know how to interpret the reported results of the experiment. The following rules of thumb are used by some statisticians as aids in interpreting *p*-values. (We note, however, that interpretation may change across applications of statistics and these guidelines are rough and informal.) A star system is often used in the research literature to convey *p*-value information. The number of stars or *ns* (not significant) appropriate for each statement are shown to its right.

1. When the *p*-value is *smaller than 0.01,* the result is called *very significant (**).*
2. When the *p*-value is *between 0.01 and 0.05,* the result is called *significant (*).*
3. When the *p*-value is *greater than 0.05,* the result is considered by most as *not significant (ns).*

Reporting and using the *p*-value eliminates the necessity of making standard conclusions that may be too restrictive. Reporting a *p*-value of 0.000003 implies that the result is very unlikely if the null hypothesis is true; therefore, we feel *strongly* about rejecting the null hypothesis. The *p*-value gets smaller as the test statistic falls further away in the tail of the distribution. Therefore, even if we are unable to compute the *p*-value exactly, we may have an idea about its size. For example, suppose that we compute the value of a test statistic as $z = 120.97$. In a right-tailed test, this means that the *p*-value is equal to the area under the standard normal curve to the right of the point 120.97. Without a computer (which may not even give accurate results for a value as large as 120.97), we are unable to compute the exact *p*-value. We do know, however, that this probability is very, very small. We may then say that the *p*-value, although unknown, is an extremely small number; hence, we reject the null hypothesis with much conviction.

The further away in the tail of the distribution our test statistic falls, the smaller is the p-*value and, hence, the more convinced we are that the null hypothesis is false and should be rejected.*

Conversely, the closer our test statistic is to the center of the sampling distribution, the larger is the *p*-value; hence, we may be more convinced that we do not have enough evidence to reject the null hypothesis.

If you understand the idea of a *p*-value, and use it, your statistical conclusions will be more meaningful and more convincing. Try to report the *p*-value when you can, and when you can't, try at least to make a statement about its relative magnitude ("small *p*-value," "very small *p*-value," etc.).

Computers are very useful in conducting statistical hypothesis tests and reporting *p*-values. In MINITAB, to conduct a test for the population mean when the population standard deviation is known, we use the command ZTEST. We follow this with the null-hypothesized value of the mean and the known value of the population standard deviation, σ. The default is a two-tailed test of the mean being the stated value. To specify a right-tailed test, we add a subcommand: ALTERNATIVE = 1. Remember that a subcommand can be specified only if the command preceding it is ended with a semicolon (;). The subcommand must end with a period (.). For a left-tailed test, we write the subcommand: ALTERNATIVE = −1. Here is an example:

```
MTB > SET C1
DATA > 15 14 13 14 15 16 16 14 17 19 13 17 16
DATA > END

MTB > ZTEST MU=17 SIGMA=2 C1;
SUBC > ALTERNATIVE= −1.
```

```
TEST OF MU =   17.000 VS MU L.T. 17.000
THE ASSUMED SIGMA = 2.00

             N    MEAN   STDEV   SE MEAN      Z   P VALUE
C1          13  15.308   1.750     0.555  -3.05   0.0012
```

This is a left-tailed test:

$$H_0: \mu \geq 17$$

$$H_1: \mu < 17$$

where $\sigma = 2$, using the data set that was input above. Notice that the test statistic value was -3.05 and the p-value 0.0012. Compare the p-value with what you obtain using the standard normal table.

PROBLEMS

14. What is the difference between a one-tailed test and a two-tailed test?
15. What is the p-value, and how is it computed?
16. The following are math SAT scores of a random sample of 20 students at a Los Angeles high school:

    485, 516, 600, 501, 618, 492, 521, 680, 552, 695, 577, 610, 652, 545, 560, 480, 640, 568, 620, 550

    The population standard deviation for all students at the school is believed to be 60. Use the MINITAB command ZTEST (or another computer package) to test the null hypothesis that the average math SAT score in this school is 540 versus the alternative that it is higher.

17. An automobile manufacturer substitutes a different engine in cars that were known to have an average miles per gallon (mpg) rating of 31.5 on the highway. The manufacturer wants to test whether the new engine changes the mpg rating of the automobile model. A random sample of 100 trial runs gives $\bar{x} = 29.8$ mpg. The population standard deviation is known to be 6.6 mpg. Using the 0.05 level of significance, is the average mpg rating on the highway for cars using the new engine different from the rating for cars using the old engine? What is the p-value?

18. A certain prescription medicine is supposed to contain an average of 247 parts per million (ppm) of a certain chemical. If the concentration is higher than 247 ppm, the drug may cause adverse side effects, and if the concentration is below 247 ppm, the drug may be ineffective. The manufacturer wants to check whether the average concentration in a large shipment is the required 247 ppm or not. A random sample of 60 portions is tested, and it is found that the sample mean is 250 ppm. The population standard deviation is known to be 12 ppm. Test the null hypothesis that the average concentration in the entire shipment is 247 ppm versus the alternative hypothesis that it is not 247 ppm using a level of significance $\alpha = 0.05$. Do the same using $\alpha = 0.01$. What is your conclusion? What is your decision about the shipment? If the shipment were guaranteed to contain an average concentration of 247 ppm, what would your decision be, based on the statistical hypothesis test? Explain. Find the p-value.

19. A metropolitan transit authority wants to determine whether there is any need for changes in the frequency of service over certain bus routes. The transit authority needs to know whether the frequency of service should increase, decrease, or remain the same. It is determined that if the average number of miles traveled by bus over the routes in question by all residents of a given area is about 5 per day, then no change will be necessary. If the average number of miles traveled per person per day is either above 5 or below 5, then changes in service may be necessary. The authority therefore wants to test the null hypothesis that the average number of miles traveled per person per day is 5.0 versus the alternative hypothesis that the average is not 5.0 miles. The required level of significance

for this test is $\alpha = 0.05$. A random sample of 120 residents of the area is taken, and it is found that the sample mean is 2.3 miles per resident per day and the sample standard deviation is 1.5 miles. Advise the authority on what should be done. Explain your recommendation. Could you state the same result at different levels of significance? Explain. Compute the p-value and interpret it.

20. A study was undertaken to determine customer satisfaction in Canadian automobile markets following certain changes in customer service. Suppose that before the changes, average customer satisfaction rating, on a scale of 0 to 100, was 77. A survey questionnaire was sent to a random sample of 350 residents who bought new cars after the changes in customer service were instituted, and the average satisfaction rating for this sample was found to be $\bar{x} = 84$; the sample standard deviation was found to be $s = 28$. Use an α of your choice, and determine whether there is statistical evidence of a change in customer satisfaction. If you determine that a change did occur, state whether you believe customer satisfaction has improved or deteriorated. Report your p-value.

21. An investment services company claims that the average annual return on stocks within a certain industry is 11.5 percent. An investor wants to test whether this claim is true and collects a random sample of 50 stocks in the industry of interest. He finds that the sample average annual return is 10.8 percent and the sample standard deviation is 3.4 percent. Does the investor have enough evidence to reject the investment company's claim? (Use $\alpha = 0.05$.) What is the p-value?

22. Tara Pearl founded a multimillion-dollar furniture business that specializes in making futons. Retail prices for the futons vary from outlet to outlet, and it is believed that the average price is $210 for a double futon.[5] To test this hypothesis, Tara's marketing director selects a random sample of 120 outlets and finds a mean price of $225 and a standard deviation of $82. Carry out the test using $\alpha = 0.05$ and also using $\alpha = 0.01$. What is the p-value?

23. Refer to the article "An Ideal Auto." Based on a random sample of 125 commuters, the number of filled car seats averages 5.1 and the population standard deviation is 2.0. Conduct a significance test. Find the p-value.

An Ideal Auto: 6.8 Plush Seats?

Under the Clean Air Act's formula, Manhattan has an average vehicle occupancy of about 6.8. This does not mean that there are 6.8 people in each car, but that the number of people arriving at work by bus, subway, train, bicycle and shoe, divided by the number of cars at worksites, comes to 6.8.

People who work in their homes are counted as arriving at the work site in zero vehicles. People who drive from home to the train station are counted as arriving by train, even though they may add substantial pollution because cars cause the most pollution in the first mile or two of driving.

Reprinted by permission from *The New York Times,* April 12, 1993, p. 82.

24. A new chemical process is introduced in the production of nickel-cadmium batteries. For batteries produced by the old process, it is known that the average life of a battery is 102.5 hours. To determine whether the new process affects the average life of the batteries, the manufacturer collects a random sample of 65 batteries produced by the new process and uses them until they run out. The sample mean life is found to be 107 hours, and the sample standard deviation is found to be 10 hours. Are these results significant at the $\alpha = 0.05$ level? Are they significant at the $\alpha = 0.01$ level? Explain. Draw your conclusion. What is the p-value?

---

[5]"Sweet Dreams," *Entrepreneurial Woman,* March 1991, p. 54.

25. Average soap consumption in a certain country is believed to be 2.5 bars per person per month. The standard deviation of the population is known to be $\sigma = 0.8$. While the standard deviation is not believed to have changed (and this may be substantiated by several studies), it is believed that the mean consumption may have changed either upward or downward. A survey is therefore undertaken to test the null hypothesis that average soap consumption is still 2.5 bars per person per month versus the alternative that it is not. A sample of size $n = 20$ is collected and gives $\bar{x} = 2.3$. The population is assumed to be normally distributed. Conduct the test and state your conclusion. Report your p-value.

26. A study of the quality of life of mentally ill people analyzed the responses of 729 mentally ill adults and found that the average score for satisfaction was 5.57 and the standard deviation was 1.61. (Responses were on a 0 to 10 scale.)[6] Use the information to test the null hypothesis that average population satisfaction is 5.00 versus the alternative that it is not. Find the p-value and interpret it.

27. According to a recent article, in 1988 motor vehicle theft had an expected punishment of 3.8 days in prison. This figure was based on all offenders that year. According to the same article, based on a random sample taken in 1990, the expected punishment for vehicle theft in that year was 1.5 days in prison.[7] If the sample size was 1,000 and the sample standard deviation was 1 day, test the hypothesis that the punishment for vehicle theft had decreased from 1988 to 1990. Report your p-value and comment on its meaning.

28. Imagine choosing a random sample of 15 lengths of wire from a large box where the average length of all pieces of wire is 60 millimeters and where the population standard deviation is 3 millimeters, and the population is normally distributed. Simulate the sampling results 20 times using MINITAB, and conduct tests of the null hypothesis that the population mean is 60, as follows:

```
RANDOM 15   C1-C20;
      NORMAL 60  3.
ZTEST MU=60   SIGMA=3 on C1-C20
```

   a. In how many of these tests did you, correctly, fail to reject the null hypothesis, using $\alpha = 0.05$?
   b. How many times did you, incorrectly, end up rejecting the null hypothesis that the population mean is 60? On the average, how many times out of 20 would you *expect* to make an incorrect decision when $\alpha = 0.05$?
   c. What are your obtained p-values of these tests?

29. Explain the accompanying advertisement for Vantin.

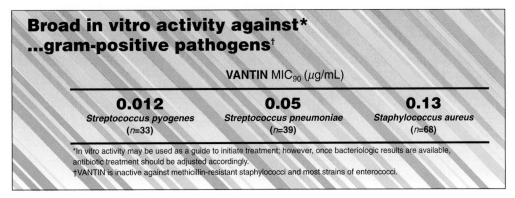

---

[6] F. Baker et al., "Social Support and Quality of Life," *Community Mental Health Journal* 28, no. 5 (October 1992), p. 58.

[7] "Ask Mr. Statistics," *Fortune,* March 8, 1993, pp. 139–40.

30. Explain the findings in the accompanying table, which gives results from the Bristol Social Adjustment Guide. What is the meaning of the asterisks?

Differences between High and Low Imitators' Mean Score Results on the Bristol Social Adjustment Guide

Measure	Significance
Withdrawal	0.92
Hostility to adults	0.92
Anxiety about adult interest	0.98
Unforthcoming	0.69
Emotional tension	0.66
Unconcerned about adult approval	0.51
Age	0.50
Environmental disadvantage	0.19
Backward	0.06
Shows hostility to other children	0.03^*
Depression/mood swings	0.05^*
Seeks approval from other children	0.04^*
Restlessness	0.002^{**}

Reprinted from B. Kniveton, "The Impact of a Child's Level of Social Adjustment on Imitation of Classroom Misbehavior," *Research in Education* 48 (1992), p. 81.

31. A certain brand of eggs claims that the eggs have reduced cholesterol content, with an average of only 2.5 percent cholesterol. A concerned health group wants to test whether the claim is true. The group believes that more cholesterol may be found, on the average, in the eggs. A random sample of 100 eggs reveals a sample average content of 5.2 percent cholesterol, and a sample standard deviation of 2.8 percent. Does the health group have cause for action? What is the *p*-value?

32. The following data are atomic charges reported in an article.[8] Use these data to test the null hypothesis that the average atomic charge in the population is zero versus the alternative that it is not zero. What is your *p*-value? Use a computer.

−0.512	−0.554	0.676	−0.589	−0.549
−0.512	−0.253	0.762	−0.265	−0.490
−0.534	−0.477	0.762	−0.265	−0.592
0.048	−0.498	0.744	−0.588	−0.592
0.135	−0.498	0.758	−0.262	−0.490
0.008	0.477	0.758	−0.263	−0.549
0.008	−0.253	0.744	−0.277	−0.544
0.135	−0.554	1.434	−0.277	−0.544
0.047	−0.534	1.434	−0.263	−0.262
0.165				

33. An advertisement for a Toshiba humidifier states: "Your bedroom may be 67 percent drier than the Sahara desert." The advertisement then states that the Sahara averages over 30 percent humidity. A competitor wants to prove that Toshiba's advertisement is incorrect

---

[8]O. Tamada, "Molecular Orbital Study of Atomic Charges in Modified Spinel," *Mineralogical Journal* 16, no. 2 (April 1992), p. 84.

and that the Sahara averages below 30 percent humidity. The competitor carries out a random set of 100 measurements of humidity in the Sahara desert and finds an average of 27 percent humidity and a standard deviation of 6 percent humidity. Are these findings significant? Explain.

34. Consider the following computer output for some hypothesis test. Do you accept or reject the null hypothesis? Explain.

MEAN	STDEV	SE MEAN	Z	P VALUE
7.262	5.310	0.819	-3.34	0.0009

35. Suppose that a random sample of 100 people treated with the protein described in the accompanying article showed an average reduction in body fat of 65 percent and a standard deviation of 20 percent. Use this information to test the null hypothesis that the average percentage of fat loss using such a diet would be 50 percent versus the alternative that it is greater than 50 percent. What is your conclusion? Report your *p*-value.

Protein Tied to Fat Craving; Weight-Control Hope Voiced

ASSOCIATED PRESS

MILWAUKEE—Researchers have found a brain protein that triggers a craving for fatty food and a second that blocks it, raising hopes for drugs that could curb weight gain without suppressing appetite.

Blocking the first protein or administering the second can cut body weight in animals by 50 percent, researchers reported.

It was the first time that scientists have identified substances that act on the appetite for fats, the researchers reported yesterday at the annual meeting of the North American Association for the Study of Obesity.

"We are extremely close to linking specific brain areas to specific appetites," said Sarah Leibowitz, a biologist at Rockefeller University in New York City. She identified a protein in the brain called galanin, and has shown that its level rises through the morning to stimulate an appetite for fats.

Reprinted by permission from the *Boston Globe,* October 20, 1993, p. 5.

36. Refer to the article on wind power on page 274. In a test run of a wind-power station, 80 trial runs reveal an average of 14.8 megawatts produced and a standard deviation of 0.6 megawatts. Is there statistical evidence to conclude that average output would be lower than 15 megawatts?

37. An analytical chemist wants to test the null hypothesis that the average purity of a laboratory chemical is 90 percent versus the alternative that it is not. The data—purity measurements, in percent, of a random sample of 40 batches of the chemical—are given below:

87, 69, 88, 91, 65, 67, 97, 88, 86, 68, 73, 79, 86, 78, 85

89, 91, 92, 84, 77, 81, 94, 91, 83, 77, 79, 68, 92, 87, 80

71, 69, 88, 89, 93, 82, 79, 87, 80, 94.

Use the MINITAB command ZTEST, or another computer package, to conduct the test. What is your *p*-value?

After Flirting with Dead Calm, Wind Power Gets Breath of Life in Northeast

New World Power in Limerock, Conn., has signed agreements for a 100-megawatt project in Chile and a 30-megawatt plant in Mexico. It went public last year.

In Maine, Endless Energy Corp. teamed up with another big California player this August, Zond Systems of Tehachapi, to build a wind farm that supplies 15 megawatts of power, beginning on Sugarloaf Mountain. Though much smaller than what US Windpower has in mind, it already has permits and power agreements.

But overall, US Windpower is far ahead of others in the marketplace.

Its biggest breakthrough, says Hap Ellis, vice president for marketing, has been the variable speed machine, which allows the company to run its turbines as fast as the wind blows the blades while still generating electricity at the standard utility-grid frequency of 60 cycles per second.

Reprinted by permission from Judy Tewes, October 24, 1993, p. A5.

6.4 TESTS ABOUT THE POPULATION MEAN, μ, WHEN THE POPULATION STANDARD DEVIATION, σ, IS UNKNOWN

When the population standard deviation, σ, is unknown—which is what most often happens in practice—and our sample size is not large enough to allow us to use the Z distribution, we must use the t distribution with $n - 1$ degrees of freedom. Remember that in doing so we are assuming that the population is at least roughly normally distributed.

The null and alternative hypotheses are exactly as they were in Section 6.3; the test statistic is as before, except that we use the sample standard deviation, s, instead of the unknown σ, and the distribution of the test statistic under H_0 is the t distribution with $n - 1$ degrees of freedom. The critical points of the test are now determined using the t distribution rather than Z. As a shortcut, determine the z value you *would* use if you could; find it in the last row of the t table, then look up to the row corresponding to your degrees of freedom.

The test statistic in tests for μ when σ is unknown:

$$t = \frac{\bar{x} - \mu_0}{s/\sqrt{n}}$$

When the population is normally distributed and the null hypothesis is true, the test statistic has a t distribution with $n - 1$ degrees of freedom.

For a given number of degrees of freedom, the t table lists only five values—corresponding to one-tail areas of 0.1, 0.05, 0.025, 0.01, and 0.005. Therefore, finding the p-value of a test that uses the t distribution cannot be done very precisely in most situations. If that's the case, then we have to see where our actual test statistic value falls as compared with these five tabulated values. This gives us bounds on the p-value.

It is important to practice using the t table in estimating the p-value of a test. Look at the t table, Table B.3 in Appendix B. Suppose that in a right-tailed test with $n = 20$ data points our test statistic value is 2.6. What is the p-value? We are in the row of $n - 1 = 19$ degrees of freedom. Where would the value 2.6 lie relative to the listed values in the table? Clearly it falls between 2.539 and 2.861. From the subscripts of t in the top of the table we see that the area in the tail to the right of the former value is 0.01 and the area to the right of the latter value is 0.005. Thus our p-value is a number between 0.005 and 0.01.

Suppose that in a left-tailed test for the mean with $n = 15$ data points our test statistic happens to be -1.8. What is the p-value? We are in the row corresponding to df $= n - 1 = 14$. By the symmetry of the t distributions about zero, we may put negative signs in front of all the listed values in the table, and the subscripted areas on the top of the table will apply as well. Now, -1.8 lies between -1.761 and -2.145; hence, our p-value is greater than 0.025 but smaller than 0.05.

Suppose that in a two-tailed test with 10 data points our test statistic value is 3.3. In the row corresponding to df $= 9$, we see that 3.3 lies somewhat to the right of the right-most value in that row, which is 3.250. The area in the right tail of the distribution to the right of our test statistic value, 3.3, is thus less than 0.005. Remember that in a two-tailed test the p-value is *twice* the area obtained in a table. Hence, in this case we conclude that our p-value is less than 0.01 (twice the tail area 0.005 from the subscript of t in the table).

As stated in the article about ex-pilot Norman Lyle Prouse, the federal limit on blood alcohol is 0.04 percent. Suppose that Mr. Prouse's reading of 0.13 percent is an average of five independent readings whose standard deviation was 0.05 percent. Is there statistical evidence that the average blood alcohol level of the pilot at the time he

EXAMPLE 6.5

Airline Gives Ex-Drunken Pilot a Second Chance

MINNEAPOLIS, Oct. 13 (AP)—Three years ago, Norman Lyle Prouse downed about a dozen rum and colas and took off in a Northwest Airlines jetliner, drunk at the controls.

This month, after a prison term and treatment for alcoholism, the former airline captain is returning to Northwest as a pilot instructor on the ground, in what he hopes will be a first step toward flying again.

Mr. Prouse's nearly 22-year career as a Northwest pilot came to a halt on March 8, 1990. After a night of heavy drinking, he flew a Boeing 727 with 58 passengers from Fargo, N.D., to Minneapolis-St. Paul.

High Blood-Alcohol Levels

The plane landed smoothly, but the authorities had been alerted by a bar patron who had seen the crew drinking. The three pilots were ordered to undergo sobriety tests, and all showed blood–alcohol levels higher than the Federal limit of .04 percent. Mr. Prouse's was the highest, at .13 percent; the first officer, Robert Kirchner, showed .06 percent, and the flight engineer, Joseph Balzer, showed .08 percent.

Northwest dismissed the three men, the Federal Aviation Administration revoked their licenses and in August 1990 they became the first pilots convicted of flying a commercial jet while intoxicated.

Reprinted by permission from *Associated Press*, October 14, 1993, p. B9.

FIGURE 6.14

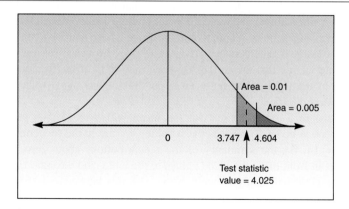

was tested (the "population" average, μ) was indeed above the federal limit, constituting cause for legal action against him?

SOLUTION

The null and alternative hypotheses are:

$$H_0: \mu \leq 0.04$$

$$H_1: \mu > 0.04$$

The test statistic value is:

$$t(4) = \frac{\bar{x} - \mu_0}{s/\sqrt{n}} = \frac{0.13 - 0.04}{0.05/\sqrt{5}} = 4.025$$

Figure 6.14 shows where this value lies as compared with the two values from the t table, 3.747 and 4.604, the former corresponding to a right-tail area of 0.01, the latter 0.005. The p-value is thus between these two probabilities. There is thus a less than 0.01 probability that five measurements of the pilot's blood alcohol level would have a mean of 0.13 percent and standard deviation 0.05 when his actual average blood alcohol level was at the federal maximum of 0.04 percent or less. There is cause for action based on these results.

EXAMPLE 6.6

According to the Japanese National Land Agency, average land prices in central Tokyo soared 49 percent in the first six months of 1993. Suppose that an international real estate investment company wants to test this claim versus the alternative that the average land price did not increase by 49 percent (that is, the null hypothesis is $H_0: \mu = 49$, and it is to be tested against the two-sided alternative $H_1: \mu \neq 49$). The company manages to find a random sample of 18 properties in central Tokyo for which the prices in both halves of 1993 are known. For each piece of property, the percentage of price increase is computed. Then the average percentage increase in price for the 18

Conducting the Hypothesis Test of Example 6.6

FIGURE 6.15

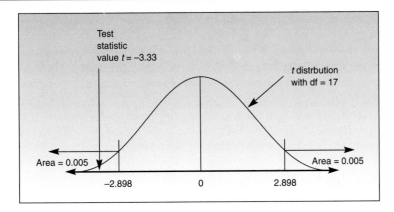

pieces of property in the sample is computed, along with the standard deviation of the percentage price increases. The calculated sample statistics are $\bar{x} = 38\%$ and $s = 14\%$. Given this information, conduct the statistical hypothesis test using a level of significance $\alpha = 0.01$.

Since the sample size is small ($n = 18$) and the population standard deviation is not known, we will use the t distribution with $n - 1 = 17$ degrees of freedom. We will implicitly assume that percentage increases in all property values in central Tokyo are at least roughly approximated by a normal distribution. (This is a reasonable assumption in this case, as some properties may have increased very much, others very little, and most may have had an increase of around the average percentage.) Since we have chosen to use the $\alpha = 0.01$ level of significance, we need to find the two critical points of the t distribution with 17 degrees of freedom that cut off an area of $\alpha/2 = 0.005$ in each tail. From the t table, Table B.3 in Appendix B, we find that the two values are ± 2.898. Our objective decision rule is therefore to reject the null hypothesis if and only if our test statistic t is either greater than 2.898 or less than -2.898. Computing the value of the t statistic, we find:

SOLUTION

$$t = \frac{\bar{x} - \mu_0}{s/\sqrt{n}} = \frac{38 - 49}{14/4.24} = -3.33$$

Since the test statistic falls in the rejection region, we reject the null hypothesis that the average percentage increase in price is 49 percent in favor of the alternative that the average price increase is not 49 percent. Furthermore, since the rejection took place on the left-hand tail of the t distribution, we are led to believe that the average percentage increase in property values in central Tokyo during the first half of 1993 was less than the 49 percent claimed by the Japanese National Land Agency. In rejecting their claim, we are taking less than a 0.01 chance of committing a Type I error. Figure 6.15 shows the t distribution and the estimation of the p-value as being less than 0.01 (twice 0.005).

Computers are very useful for computing p-values accurately, so that we do not have to estimate them using bounds from the t table. In fact, since complicated statistical tests are almost always done by computer, computer-generated p-values are very commonly reported. The p-value, as a single number, conveys so much information about the results of a statistical test that it is usually considered the most important number reported in a statistical test by computer. When looking at computer-produced statistical test results, always look first at the p-value or p-values when searching for a conclusion.

In MINITAB we conduct a t test by stating the command TTEST. We then specify the null hypothesized mean. We may write the subcommand ALTERNATIVE = 1 for a right-tailed test, and ALTERNATIVE = −1 for a left-tailed test.

EXAMPLE 6.7

If the ion accelerator described in the accompanying figure produces an average of more than 30 megawatts of ionization at a given time, the treatment may be harmful and should be stopped. Tests are therefore run regularly of

$$H_0: \mu \le 30 \text{ versus}$$
$$H_1: \mu > 30$$

A random sample of 20 runs of the machine revealed data as follows:

39, 35, 37, 28, 40, 25, 27, 24, 42, 45, 20, 38, 39, 25, 42, 26, 48, 51, 48, 41

These data are entered into column C1 of MINITAB. (Since we plan to use the t distribution, we must assume that the population is normally distributed.)

SOLUTION

Figure 6.16 shows the MINITAB command needed for executing the analysis. The subcommand ALTERNATIVE = 1 specifies that the test is a right-tailed test.[9] This comes in the row specifying the command: TTEST of MU = 30. The output is also shown in the figure. Since the p-value is 0.0045, as reported in the computer output, we may reject the null hypothesis; the process should be stopped. Check the reported p-value against the bounds you get using the t table.

FIGURE 6.16   MINITAB Program and Output for Example 6.7

```
MTB > TTEST OF MU = 30    ALTERNATIVE = 1 ON DATA IN C1
TEST OF  MU = 30.00 VS MU G.T. 30.00
```

	N	MEAN	STDEV	SE MEAN	T	P VALUE
C1	20	36.00	9.23	2.06	2.91	0.0045

[9]Notice that for a ZTEST specifying the population standard deviation, the ALTERNATIVE is stated as a subcommand.

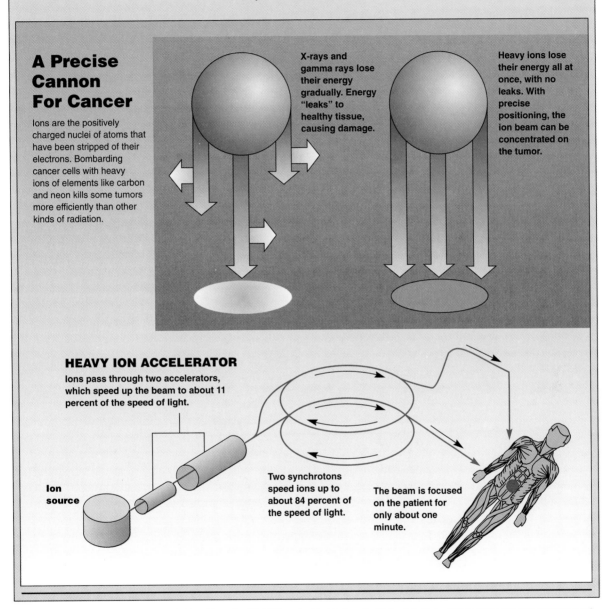

Japanese Project Aims to Harness Heavy Ions to Kill Malignant Cells
A Big Accelerator for Medical Treatment Nears Completion

By ANDREW POLLACK

A Precise Cannon For Cancer

Ions are the positively charged nuclei of atoms that have been stripped of their electrons. Bombarding cancer cells with heavy ions of elements like carbon and neon kills some tumors more efficiently than other kinds of radiation.

X-rays and gamma rays lose their energy gradually. Energy "leaks" to healthy tissue, causing damage.

Heavy ions lose their energy all at once, with no leaks. With precise positioning, the ion beam can be concentrated on the tumor.

HEAVY ION ACCELERATOR

Ions pass through two accelerators, which speed up the beam to about 11 percent of the speed of light.

Ion source

Two synchrotons speed ions up to about 84 percent of the speed of light.

The beam is focused on the patient for only about one minute.

Reprinted by permission from *The New York Times,* December 21, 1993, p. 3, with data from the National Institute of Radiological Sciences, Department of Accelerators (Japan).

PROBLEMS

N.R.C. Says 15 Reactors Need Testing

By MATTHEW L. WALD

The staff of the Nuclear Regulatory Commission has identified 15 nuclear reactors around the country whose reactor vessels have become so weakened by radiation that they will require careful analysis to determine if they are still safe.

The staff's report, affecting one in seven of the nation's dwindling fleet of reactors, is a sign that they are aging faster than their builders had anticipated and will take more effort to keep running.

The staff singled out the plants in a study that it began last year after a 32-year-old Massachusetts plant, Yankee Rowe, was retired because of questions about the strength of its reactor vessel.

The measure used by the commission is how much force is required to split a sample piece of metal when a wedge is dropped on a notch carved in the sample. When the vessels are new they are supposed to withstand a force of 75 foot-pounds, which would be the

equivalent of a 75-pound weight dropped from one foot. The 15 reactors now have metal that could be split with less than 50 foot-pounds of strength, the commission said, although some operators say that the number is above 50 foot-pounds for their plants. The basis of the disagreement is that steel has a grain somewhat like wood, and strength depends on whether the notch is carved parallel to the grain or perpendicular to it.

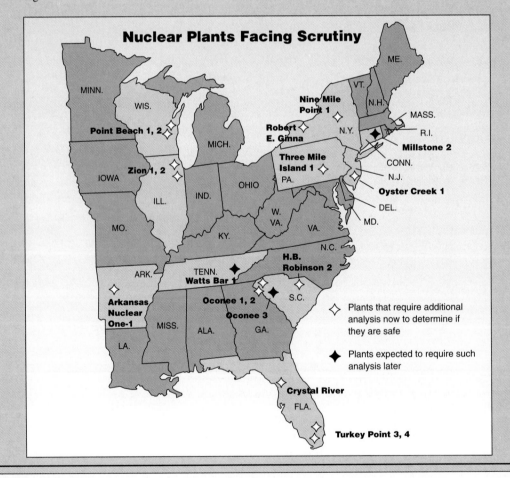

Nuclear Plants Facing Scrutiny

◇ Plants that require additional analysis now to determine if they are safe

◆ Plants expected to require such analysis later

Reprinted by permission from *The New York Times,* April 2, 1993, p. A21, with data from the Nuclear Regulatory Commission.

38. The Nuclear Regulatory Commission (NRC) wants to prove that for an aging reactor (see article), the average force that can be withstood is less than 50 foot-pounds. The NRC plans to use the 15 reactors as an assumed random sample. Measurements (in foot-pounds) obtained are as follows:

44, 41, 19, 53, 49, 77, 16, 63, 9, 40, 33, 47, 21, 48, 58

Using the 0.05 level of significance, does the NRC have cause for action? What is the p-value?

39. It is known that the average stay of tourists in Hong Kong hotels has been 3.4 nights.[10] A tourism industry analyst wanted to test whether recent changes in the nature of tourism to Hong Kong have changed this average. The analyst obtained the following random sample of the number of nights spent by tourists in Hong Kong hotels:

5, 4, 3, 2, 1, 1, 5, 7, 8, 4, 3, 3, 2, 5, 7, 1, 3, 1, 1, 5, 3, 4, 2, 2, 2, 6, 1, 7

Conduct the test using the 0.05 level of significance. Say something about the p-value.

40. The accompanying article gives data on the ozone layer. A random sample of 12 measurements shows a mean of 91 Dobson units and a standard deviation of 20. Is there evidence the population average in this area and time is below 100 Dobson units?

Antarctic Ozone Hits Record Low

A New Concern: Are Laws on Chemical Use Working?

By JOHN NOBLE WILFORD

The amount of ozone in the atmosphere over Antarctica reached a record low this month, government scientists reported yesterday. This was the lowest recorded ozone level anywhere in the world, raising new concerns that measures to restrict the use of ozone-destroying chemicals may not be strong enough.

The scientists said the area of intense ozone depletion, commonly called an ozone hole, was not quite as wide as it had been a year ago. But balloon-borne instruments showed that ozone, the atmospheric molecules that absorb harmful ultraviolet radiation from sunlight, was totally destroyed between the altitudes of 8.4 and 11.8 miles, creating an unusually thick void.

Ozone is measured in Dobson units, each of which represents the physical thickness of the ozone layer if it were compressed at the Earth's surface. About 300 Dobson units equal one-tenth of an inch. This is a usual value at the Antarctic in seasons when the ozone layer is not depleted.

In an announcement by NASA and the atmospheric administration, the satellite readings at the end of September and in early October showed values of less than 100 Dobson units. On Oct. 6, a surface instrument at the South Pole recorded 88 Dobson units. This is the lowest value ever measured, anywhere in the world, the agencies said.

Reprinted by permission from *The New York Times,* October 19, 1993, p. A23.

41. Look at the display labeled "War on Wounds: Many Kinds of Troops" on the next page. If the average concentration of autocrine in the surrounding tissue goes below 10,000 per gram of tissue, healing may be seriously slowed down. Five measurements are taken for a patient, yielding an average of 9,560 and a standard deviation of 2,380. Is there cause for concern about this patient? Say something about the p-value.

[10]"Hong Kong's Tourism: Stay a While, Spend a Little," *The Economist,* March 16, 1991, p. 72.

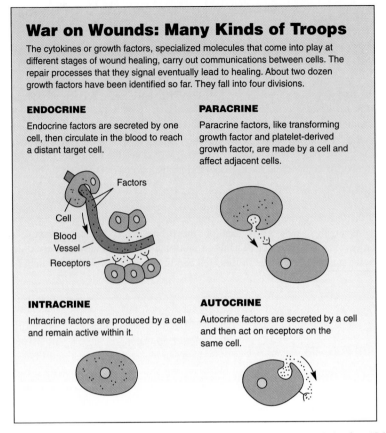

War on Wounds: Many Kinds of Troops

The cytokines or growth factors, specialized molecules that come into play at different stages of wound healing, carry out communications between cells. The repair processes that they signal eventually lead to healing. About two dozen growth factors have been identified so far. They fall into four divisions.

ENDOCRINE

Endocrine factors are secreted by one cell, then circulate in the blood to reach a distant target cell.

PARACRINE

Paracrine factors, like transforming growth factor and platelet-derived growth factor, are made by a cell and affect adjacent cells.

Factors
Cell
Blood Vessel
Receptors

INTRACRINE

Intracrine factors are produced by a cell and remain active within it.

AUTOCRINE

Autocrine factors are secreted by a cell and then act on receptors on the same cell.

Reprinted by permission from *The New York Times,* November 9, 1993, p. C14, with data from "Principles of Surgery" (McGraw-Hill).

42. Three airlines recently brought suit against a broker selling airline tickets that were earned by travelers through "frequent flyer" bonus programs. Prices for these bonus tickets, which are sold to the broker by people who earn them, are variable. The airlines claim that the average price for such tickets on the New York to Hawaii route is $1,250. The broker wants to prove that the average price is higher than claimed and, hence, that his profit margin upon resale is smaller than claimed by the airlines. An independent investigator is hired, and the investigator gathers a random sample of 18 tickets bought by the broker. The average price paid for tickets in the sample is $1,330 and the sample standard deviation is $120. Is there evidence to support the broker's claim? Use $\alpha = 0.01$. What is the p-value?

43. Read the article on the U.S. attempts to create a radiation weapon on the next page. If 15 measurements were made in Watrous, New Mexico, yielding a sample mean of 115 millirads and a standard deviation of 32 millirads, is it likely that the community was seriously harmed if a (population) average of over 60 millirads is considered extremely dangerous?

44. The following results of a hypothesis test are produced by the computer:

N	MEAN	STDEV	SE MEAN	T	P VALUE
15	17.283	0.397	0.102	7.65	0.0000

a. What are the degrees of freedom involved in this test?
b. Do you reject the null hypothesis?
c. Why did the computer report the p-value as 0.0000? Explain.

U.S. Spread Radioactive Fallout in Secret Cold War Weapon Tests

By KEITH SCHNEIDER

The United States deliberately released large amounts of radiation into the environment in the 1940's and early 1950's as part of a secret program aimed at developing a weapon that would kill enemy soldiers with radioactive fallout, according to a Congressional study made public yesterday.

The Government's extensive program to develop a radiation weapon came as Government scientists and medical specialists were insisting that radioactive particles falling to earth from open-air tests of atomic bombs posed little or no risk to civilians.

Efforts to design a device to turn radioactive fallout into a weapon took place at the Oak Ridge National Laboratory in Tennessee and the Dugway Proving Ground, the Army's testing site 90 miles west of Salt Lake City, according to a report by the General Accounting Office, the investigative arm of Congress.

In addition, the Congressional report documented tests conducted at the Los Alamos National Laboratory in northern New Mexico in which nuclear scientists had sought to develop methods for tracking radiation from atmospheric bomb blasts. In one such experiment on March 24, 1950, scientists exploded in the atmosphere near Los Alamos a conventional bomb containing metal with dangerous levels of radioactivity. They then measured particles of radiation in Watrous, a community 70 miles east of the laboratory.

12 Undisclosed Tests

After another similar blast five days later, Los Alamos scientists tracked a radioactive cloud of particles into the communities and desert west of the laboratory, although the G.A.O. was unable to determine how far the cloud had traveled.

All the tests released radiation at concentrations thousands of times higher than would be permitted by the Government today, but apparently far less than the amounts of radiation released into the atmosphere by the explosion in 1986 at the Chernobyl nuclear power plant in the Ukraine.

Reprinted by permission from *The New York Times,* December 16, 1993, p. A1.

---

45. Examine the data given in the article on air traffic errors on page 284. If the average yearly number of air traffic errors was 710 in the years prior to 1990, is there statistical evidence that air traffic is now more dangerous? (Assume the three years 1991, 1992, and 1993 represent a random sample of numbers of air traffic errors from the population of years beyond 1990, and that this population is normally distributed. What are the limitations of this situation?)

46. Dunhill Tailors of New York City claims that its average custom-tailored suit costs $1,925. A competitor, suspecting that Dunhill is trying to advertise its suits as more expensive than they really are in an attempt to build a high-class image, gets a random sample of 12 sales slips for Dunhill and finds $\bar{x} = \$1,645$ and $s = \$250$. Carry out the test at $\alpha = 0.05$. What is the *p*-value?

47. Executives at Gammon & Ninowski Media Investments, a top television station brokerage, believe that the current average price for an independent television station in the United States is $125 million. An analyst at the firm wants to check whether the executives' claim is true. The analyst has no prior suspicion that the claim is incorrect in any particular direction and collects a random sample of 25 independent TV stations around the country. The results are (in millions of dollars):

233, 128, 305, 57, 89, 45, 33, 190, 21, 322, 97, 103, 132, 200, 50, 48, 312, 252, 82, 212, 165, 134, 178, 212, 199

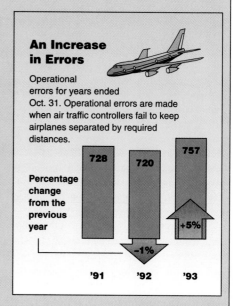

F.A.A. Finds More Errors by Air Traffic Controllers

By MARTIN TOLCHIN

WASHINGTON, Dec. 31—An increasing number of aircraft are flying too close to other planes because of errors made by air traffic controllers, the Federal Aviation Administration has reported.

The required distances of separation were breached 757 times in the 12-month period that ended last October, the agency found. That was a 5 percent increase in the number of such incidents, called "operational errors," compared with the preceding 12 months, when 720 were reported. "Operational errors" is the F.A.A.'s term for mistakes made by air traffic controllers who fail to keep aircraft separated by the required distance, which is, on average, three miles laterally or 1,000 feet vertically; sometimes both buffers are required and sometimes the distances are larger.

Some operational errors have been "relatively close," said Bill Jeffers, the F.A.A.'s deputy associate administrator for air traffic control. Dangerously close encounters between aircraft are called near-collisions.

Reprinted by permission from *The New York Times*, January 1, 1994, p. 6, with data from the Federal Aviation Administration.

Use a computer to test the hypothesis that the average station price nationwide is $125 million versus the alternative that it is not $125 million. Use a significance level of your choice. State the *p*-value.

6.5 LARGE-SAMPLE TESTS FOR THE POPULATION PROPORTION

As we know, when the sample size is large, the distribution of the sample proportion, $\hat{P}$, may be approximated by a normal distribution with mean p and standard deviation $\sqrt{pq/n}$. Therefore, for large samples (recall our rule of thumb: both np and nq greater than 5 or more), we can test hypotheses about the population proportion p in a similar manner to the way we constructed confidence intervals for p in Chapter 5. The test statistic we use is z, defined below. In computing the test statistic, we use p_0—the hypothesized value of p under the null hypothesis.

Large-sample test statistic for the population proportion, p:

$$z = \frac{\hat{p} - p_0}{\sqrt{p_0 q_0 / n}}$$

where $q_0 = 1 - p_0$.

We will demonstrate the use of this statistic in a two-tailed test for the population proportion in Example 6.8.

EXAMPLE 6.8

According to the Centers for Disease Control (reported in *The New York Times,* January 29, 1993, p. B1), 70 percent of American adults over 18 exercise regularly. Health officials in a major metropolitan area want to test whether this figure applies to their particular area. A random sample of 210 adults in this area reveals that 130 of them exercise regularly. Can the null hypothesis be rejected?

The Stock Market/John Henley, 1993.

SOLUTION

We want to test the null hypothesis H_0: $p = 0.70$ versus the alternative hypothesis H_1: $p \neq 0.70$. Our test statistic is z, and therefore, using $\alpha = 0.05$, our rejection region is defined by the two critical points ± 1.96. This is shown in Figure 6.17. We now compute the value of the test statistic using the calculated sample proportion $\hat{p} = x/n = 130/210 = 0.619$.

$$z = \frac{\hat{p} - p_0}{\sqrt{p_0 q_0 / n}} = \frac{0.619 - 0.70}{\sqrt{(0.7)(0.3)/210}} = -2.5614$$

As can be seen from Figure 6.17, the computed value of the test statistic falls in the left-hand rejection region; hence, at $\alpha = 0.05$ our result is significant. We may

FIGURE 6.17   Conducting the Hypothesis Test of Example 6.8

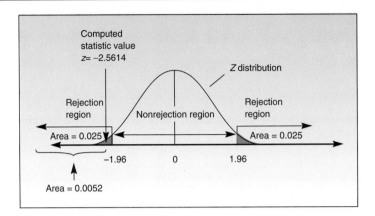

therefore state that based on our random sample, we believe that the percentage of adults in this area who exercise regularly is *lower* than 70 percent. The *p*-value is 0.0104.

EXAMPLE 6.9

"After looking at 1,349 hotels nationwide, we've found 13 that meet our standards." This statement by the Small Luxury Hotels Association, which appeared in advertisements of the association in national magazines, implies that the proportion of all hotels in the United States that meet the association's standards is $13/1{,}349 = 0.0096$. The management of a hotel that was denied acceptance to the association on the grounds that it did not meet the association's standards wanted to prove that the standards set by the association are not quite as stringent as claimed and that, in fact, the proportion of hotels in the United States that would qualify under present standards is higher than 0.0096. The management hired an independent research agency, which visited a random sample of 600 hotels nationwide and found that 7 of them satisfied the exact standards set by the association. Is there evidence to conclude that the population proportion of all hotels in the country satisfying the standards set by the Small Luxury Hotels Association is greater than 0.0096?

SOLUTION

Note that in order to use the normal approximation for the sample distribution of the sample proportion, $\hat{P}$, we must satisfy the rule that both np and nq be at least equal to 5. In this particular example, it is especially important to check this because under the null hypothesis, the population proportion, p, is very small ($p_0 = 0.0096$). This means that a relatively large sample size is needed. The condition is satisfied because $np_0 = 600(0.0096) = 5.76$. The other condition clearly holds.

Our test is a right-tailed test, because we are trying to prove that the proportion of hotels that satisfies the standards is greater than $p_0 = 0.0096$. The null and alternative hypotheses are $H_0\colon p \leq 0.0096$ and $H_1\colon p > 0.0096$. Again, no particular level of significance, α, is given, so we'll first compute the value of the test statistic in this case. We find the required value of the test statistic as follows:

$$z = \frac{\hat{p} - p_0}{\sqrt{p_0 q_0 / n}} = \frac{(7/600) - 0.0096}{\sqrt{(0.0096)(0.9904)/600}} = 0.519$$

Carrying Out the Test, Using $\alpha = 0.10$, in Example 6.9 **FIGURE 6.18**

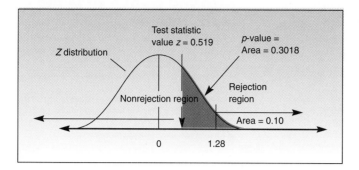

As can be seen, this result is not significant at any conventional level of significance because even if we were to take a 0.10 chance to a Type I error, our right-hand critical point would be equal to $+1.28$, and the value of the test statistic would still lie in the nonrejection region. The p-value is 0.3018, a relatively large number. This is shown in Figure 6.18.

PROBLEMS

48. It has been believed that one in four Americans relies on unconventional medicine. Recently, a study reported in the *New England Journal of Medicine* found that in a random sample of 1,539 adults, 554 of them have used some form of unconventional medicine.[11] Carry out the appropriate two-tailed hypothesis test. Find the p-value.

Newsweek Poll

Do you approve of the way Bill Clinton is handling his job as president?
51% Approve
32% Disapprove

For this *Newsweek* Poll, the Gallup Organization interviewed a national sample of 774 adults Jan. 28–29. Margin of error +/−4 percentage points. "Don't know" and other responses not shown. The NEWSWEEK POLL © 1993 by NEWSWEEK, Inc. Reprinted by permission from *Newsweek*, February 8, 1993, p. 22.

49. Look at the *Newsweek* Poll for President Clinton's approval rating. Test the null hypothesis that 50 percent of the public approved of the way President Clinton handled his job at the time of the survey versus the alternative that the percentage was not 50 percent. (Also comment on the reported margin of error.) What is the p-value?

50. In-vitro fertilization is a relatively new scientific method of forming embryos in the test tube to help couples who cannot conceive children. If in 200 trials 43 eggs were fertilized, test the null hypothesis that the method's success rate is at most 15 percent versus the alternative that it is greater than 15 percent. Assume a random sample. What is the p-value?

51. Refer to the display "Altering Genes to Make Pigs More Human" on page 288. Suppose that 10,000 attempts resulted in 78 successful incorporations of the human gene in pigs. Test the implied right-tailed hypotheses. What is the p-value?

---

[11]D. Eisenberg et al., "Unconventional Medicine in the United States" (Special Article), *The New England Journal of Medicine*, January 28, 1993, pp. 246–52.

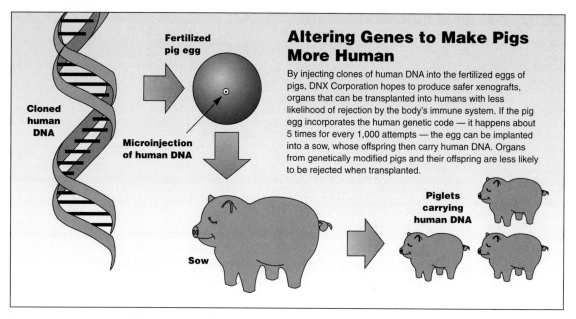

Altering Genes to Make Pigs More Human

By injecting clones of human DNA into the fertilized eggs of pigs, DNX Corporation hopes to produce safer xenografts, organs that can be transplanted into humans with less likelihood of rejection by the body's immune system. If the pig egg incorporates the human genetic code — it happens about 5 times for every 1,000 attempts — the egg can be implanted into a sow, whose offspring then carry human DNA. Organs from genetically modified pigs and their offspring are less likely to be rejected when transplanted.

Fertilized pig egg

Cloned human DNA

Microinjection of human DNA

Sow

Piglets carrying human DNA

Reprinted by permission from "U.S. Again Patenting 'Engineered' Animals," *The New York Times,* February 3, 1993, p. D5.

52. A news release by the IRS on March 11, 1993, stated that 90 percent of all taxpayers fill out their tax return forms correctly. A random sample of 1,500 returns revealed that 1,316 of them were correctly filled out. Test the null hypothesis that the statement by the IRS was correct versus the alternative that is was not, that is, that fewer people fill out tax forms correctly than the IRS claims. What is the p-value?

Ticked Off

Want to avoid Lyme disease–bearing deer ticks? Don't use Off!—take it *all* off. A Connecticut doctor surveyed 300 visitors to a local nudist camp in prime deer-tick breeding grounds. To his surprise, Dr. Henry Feder Jr. found that, far from being a tick snack bar, the nudists weren't bothered. Only those at the camp wearing clothes reported tick bites. Feder hypothesizes that the insects "prefer to do their biting under cover" of pants legs and socks. Though he has written to the *Journal of the American Medical Association* with his findings, Feder says he plans a more scientific study before he reaches any final conclusions.

From *Newsweek,* February 8, 1993, *Newsweek, Inc.* All rights reserved. Reprinted by permission.

53. Read the article "Ticked Off." Suppose that in the Lyme disease–plagued area in question, 10 percent of all clothed people who would sit or lie on the ground for a couple of hours would be infected by the tick-borne illness. Carry out the implied left-tailed hypothesis test about the rate of infection for nudists. Find the p-value.

54. One of the most ingenious advertising schemes in recent years is Gallo Winery's invention of Bartles and Jaymes, the two down-home entrepreneurs selling a wine cooler. Bartles & Jaymes wine coolers have been rapidly increasing in sales since their introduction, largely due to the believability of television commercials featuring the two elderly gentlemen in their suspenders sitting on the porch with bottles of the wine cooler. As part of marketing research efforts to design new commercials, it was of interest to find out what proportion of the general public knew that Bartles and Jaymes were not real people and that the wine cooler was actually made by Gallo. An assertion by advertising industry experts was that

no more than 10 percent of the public knew the truth. If it could be shown that more than 10 percent of the public knew that the wine cooler was made by Gallo, then the company would change the television commercials in some way; otherwise nothing would be changed. Thus, it was necessary to test the null hypothesis that the percentage of the public that was aware of the true maker of the wine cooler was less than or equal to 10 percent, versus the alternative that the population percentage was above 10 percent. A random sample of 800 people was selected, 95 of whom admitted they knew that the wine cooler was made by Gallo and that Bartles and Jaymes were fictional characters. Should Gallo change its commercials?

55. It has been believed that about 68 percent of incarcerated youths owned a revolver when arrested. A recent article aimed at showing that this figure underestimated the true percentage looked at a random sample of 835 criminally active youths and found that 601 of them had owned revolvers.[12] Carry out the appropriate test. Find the p-value.

56. It has been believed that two-thirds of all women work outside the home. Recently a Gallup poll (carried out December 9, 1992) of 506 randomly selected women found that 339 of them worked outside the home. Is there evidence that the percentage is actually higher than was believed?

57. The accompanying article on unrequited love contains some interesting data. Test the null hypothesis that 5 percent of all people have not suffered from unrequited love versus the alternative that the percentage is smaller. What is the p-value?

Pain of Unrequited Love Afflicts the Rejecter, Too

Science Studies Hurdles to Finding Reciprocal Passion

By DANIEL GOLEMAN

Since Young Werther died from it and Cyrano de Bergerac was so noble about it, unrequited love has been one of the great themes of literature and drama. Now, at last, unrequited love is getting systematic scrutiny from psychologists.

The first studies to look at the two sides of unrequited love—the would-be lover and the rejecter—show there is pain on both sides and, surprisingly, the rejecter often suffers just as much as the rejected.

And in studying the dynamics of love that goes unreturned, psychologists are gaining greater understanding of common hurdles in the sometimes tortuous route to finding a lasting love.

"We rarely hear about the agony of those who are the target of an unwanted love," said Dr. Roy Baumeister, a psychologist at Case Western Reserve University who has done much of the new research. "Literature and film almost always tell the story from the viewpoint of the rejected lover. But both rejecters and would-be lovers can end up feeling like victims."

The experience of unrequited love—not just a minor crush, but an intense, passionate yearning—is virtually universal at some point in life. Dr. Baumeister and Sara Wotman, a graduate student, found in a study of 155 men and women that only about 2 percent had never loved someone who spurned them, or found themselves the object of romantic passion they did not reciprocate. Their findings will be published later this year in *The Journal of Personality and Social Psychology*.

Reprinted by permission from *The New York Times*, February 9, 1993, p. C1.

---

[12]J. Wright et al., "Kids, Guns, and Killing Fields," *Society* 30, no. 1 (November–December 1992), pp. 84–89.

58. In a test run of the Eurostar train (see article), which European Passenger Services decided to view as a random sample, 650 passengers took the train from London to the Paris Gare du Nord station. Of these, 124 were business travelers. Test the null hypothesis that 18 percent of passengers in the future regular service of the Eurostar will be business travelers versus the alternative that the percentage will not be 18 percent. What is the *p*-value? What are the limitations of the inference here?

Train or Plane to Take Strain?

Steve Keenan Predicts that a Battle for Passengers Is About to Begin

The Eurostar train boarding at London Waterloo from next June for Paris and Brussels will be pulling 16 passenger coaches.

Those carriages will carry 210 seats in first class, 584 in second. There will be an hourly service for most of the day, and every 30 minutes in peak: the equivalent of two jumbo jets leaving Waterloo up to every half hour, bound for the two biggest business destinations on the Continent.

It is a staggering addition in capacity to routes on which airlines are already struggling to break even. However, it is hard to resist the argument from Eurostar—three hours, station-to-station from central London to Paris Gare du Nord, or an extra 15 minutes to the Gare Midi in Brussels.

European Passenger Services (EPS), the British Rail subsidiary jointly operating Eurostar with the French and Belgian railway companies, is confident. While it expects only 18 per cent of its customers to be on business, EPS also expects them to contribute 40 per cent of revenue. EPS also expects to win 40 per cent of its customers from the airlines.

Steve Keenan from *The London Times,* November 15, 1993, p. 8, © 1993 Times Newspapers Limited. Reprinted by permission.

59. Suppose that Canadian doctor Keith MacLeod (see article titled "Americans Filching Free Health Care in Canada") has 220 patients, 20 of whom are U.S. citizens. Also make the assumption that this physician's practice may be viewed as a random sample of all patients in Windsor, Ontario. Use this information to test the null hypothesis that 10 percent of people seeking medical help in Windsor are Americans versus the alternative that the percentage is lower than 10 percent. What is the *p*-value?

60. To test the hypothesis stated in the accompanying excerpt from a review of *Breaking the Maya Code,* a random sample of 100 glyphs was selected and 20 were found to be syllabic or phonetic. Conduct the two-tailed test. What is the *p*-value?

Review: Breaking the Maya Code, *by Michael D. Coe.*
Thames and Hudson, 1992

Mayan, too, was as pervasively syllabic and phonetic as Egyptian. Yucatec and Cholan would be as relevant as Coptic had been in Egypt, and the scholarly dictionaries of Maya tongues offered him workable read-ings that explained de Landa's few dozen badly misunderstood syllables. The glyphs do echo the spoken word. Of the 800 glyphs, about 150 are now known to be syllabic or phonetic. There are plenty of gaps still.

Reprinted by permission from *Scientific American,* April 1993, p. 126.

Americans Filching Free Health Care in Canada

By CLYDE H. FARNSWORTH

TORONTO, Dec. 19—Lacking a national health care system of their own, thousands of Americans are tapping into Canada's—illegally.

"It's not an epidemic in any one person's practice," said Keith MacLeod, an obstetrician in Windsor, Ontario, across from Detroit, "but I would estimate that from 12 to 20 of my patients at any one time are ineligible Americans. And I'm just one of 520 doctors in Windsor, 23,000 in Ontario."

Dr. MacLeod, former president of the Essex County Medical Society, delivers about 400 babies a year.

A report prepared for Ontario's Health Minister indicated that from August 1992 to February 1993, 60,000 medical claims had been made on behalf of patients who held American driver's licenses. The total number of improper claims in Ontario was estimated at 600,000.

Only legal residents qualify for free medical care in Canada, using plastic health cards for identification. Others are supposed to pay for medical services they may require, but many are submitting counterfeit, borrowed or fraudulently obtained cards.

Loopholes and the lack of stringent controls are costing the provincial health care system as much as $691 million a year, the Ontario report found.

"The ministry is open to the fraudulent use of health care in all programs," the report said. "Almost no analytical tools exist at this time, and lenient registration policies encourage abuse by non- and new residents."

Reprinted by permission from *The New York Times,* December 20, 1993, p. A1.

Using Lab Animals to Make Environmental Rules: Are Data Good Enough?

The Government first began experimenting with rodent studies in the early 1960's, and the program grew exponentially after the Nixon Administration announced the Government's "war on cancer" in 1971. Even with some known weaknesses, scientists enthusiastically embraced the animal studies as clear indicators of cancer risks.

Though there was no legal requirement to act on the studies results, a welter of laws did require Government agencies to protect the public from foods, drugs, household products, industrial chemicals and other substances that caused cancer. So Government officials responsible for protecting the public health accepted the data as justification for many new regulations in the 1970's.

By the mid-1980's, however, new research findings began to cast new doubts on the validity of the animal research. Government was no longer so quick to accept the results automatically in every case. But by then, dozens of substances had been ruled safe or dangerous based on the animal studies alone. It was suggested that two thirds of the substances that proved to be cancerous in the animal tests would present no cancer danger to humans at normal doses.

Dr. Maronpot swept his hand toward a long row of blue books stretching more than 10 feet along an upper shelf, reports on all 450 animal studies the Government has conducted over the last 30 years.

"It's an impressive product, not produced by anyone else in the world," he said. Still, Dr. Maronpot acknowledged, neither he nor anyone else at the institute knows how many of the tested substances that produced tumors or other harmful effects in animals—about half the total—might now be shown to be benign at normal levels.

Even more worrisome, perhaps, is the opposite question: How many substances that caused no harm to rodents might be

---

dangerous to humans? One chance finding demonstrated this problem.

"Arsenic is not a carcinogen in animal studies," said Dr. Joseph F. Fraumeni, director of epidemiology and bio-statistics at the National Cancer Institute. But several years ago, he recalled, a study of smelter workers exposed to high levels of arsenic in the air showed a high level of lung cancer.

From that, Dr. Olden's review committee concluded that the Government should no longer rely only on animal studies. They should be simply one part of a program of research also involving studies of population groups found to have been exposed to the substances without knowing of the possible risk, and laboratory analyses showing how the chemicals interact with cells.

That is easy to say, institute officials agree, but difficult and costly to do.

Reprinted by permission from *The New York Times,* March 23, 1993, p. A16.

61. Based on the article about laboratory animals on the previous page, $n = 450$, $\hat{p} = 0.50$. Test the null hypothesis that two-thirds of the time, a substance shown to be cancerous to animals would present no danger to humans at normal doses versus the alternative that the proportion is less than two-thirds. What is your p-value?

6.6 HOW HYPOTHESIS TESTING WORKS

In this section, we will look at the mechanics of hypothesis testing and gain an insight into what makes the procedure work. We will also make some general comments about hypothesis tests—aspects to consider when conducting a test, and some caveats.

Let's consider a hypothesis test for the mean of a population. Assume that the sample size n is large and the normal distribution applies. Assume also that the population standard deviation σ is known. (When it is not, use s in its place.) We have a hypothesized value for the mean, μ_0, and we compute from our sample the sample mean, $\bar{x}$. Now suppose that the test is two-tailed, or, if it is one-tailed, then assume that $\bar{x}$ is away from μ_0 in the direction of the rejection region ($\bar{x}$ larger than μ_0 in the right-tailed test and smaller than μ_0 in a left-tailed test). The first thing to consider is the distance between $\bar{x}$ and μ_0. This is the numerator of our test statistic:

$$z = \frac{\bar{x} - \mu_0}{\sigma/\sqrt{n}}$$

Suppose we test H_0: $\mu \leq 60$ versus H_1: $\mu > 60$, and $\bar{x}$ is equal to 70. The distance between $\bar{x}$ and μ_0 is $70 - 60 = 10$ units. This is the numerator in the above equation. Now suppose that you report your findings to a person who doesn't know anything about statistics. You say: "The claim was that the population mean is no greater than 60; I sampled n items and found that the sample mean was 70." The person to whom you are reporting may say: "Well, then, the population mean can't be 60, because your sample result is 10 units over the claimed mean!" When we do statistics, however, we make conclusions much more carefully and in a scientific manner. We must first ask ourselves: What is the *significance* of the distance of 10 units from $\bar{x}$ to μ_0? This is a question of *scale*. In the scale of the problem, do 10 units represent a large distance or a small one? The scale depends on two quantities: the population standard deviation and the sample size. Let's see why. Suppose the values of the population are the following: 0, −30, 400, 68, −298, 1,095, and so on. The variance in these values is large, and therefore a difference of 10 units between what you find from a sample and the true population mean may be relatively minor. (Again, the statistical significance depends also on the sample size.) On the other hand, a difference of 10 units is *enormous* if the population values are 69.9, 70.01, 69.8, 70.1, 69.99, 70.03, and so on. Here the population is very closely clustered around a mean of about 70, and the distance of 10 units is very significant: The claimed mean of 60 is indeed far away, in this scale, from the sample mean, 70. This is shown in Figure 6.19.

The dependence of the scale of measurement on the population standard deviation (a measure of the variation in the population) is taken into account in the equation of the test statistic by the fact that we divide the distance $\bar{x} - \mu_0$ by the standard deviation, σ.

The second factor that affects the scale of the problem is the sample size. This has to do with the fact that our sample mean, $\bar{x}$, should carry more weight (carry more importance) if it is based on a larger sample. If you invert the fraction in the denominator of the equation and multiply the numerator by $\sqrt{n}$, you see how the "importance" factor $\sqrt{n}$ acts as a weight that multiplies the difference between $\bar{x}$ and μ_0. The larger the sample size, the greater the importance given to the difference between our observed $\bar{x}$ and the hypothesized mean, μ_0.

Another way to see how our test statistic is a scaled distance is to note something we have known all along. The sample mean is a normally distributed random variable with mean μ and standard deviation equal to $\sigma/\sqrt{n}$. As σ decreases, or as n increases, the width of the normal curve centered at the true mean, μ, becomes smaller, and the sample mean $\overline{X}$ becomes "constrained" to fall within smaller distances from μ for any probability level. The sample mean should not be too far away from μ, when measured in standard deviations of the random variable $\overline{X}$. This is exactly what the test statistic z measures: *the distance from $\bar{x}$ to μ in standard deviations of the random variable $\overline{X}$.* This is done by dividing the actual distance between $\bar{x}$ and μ_0 by $\sigma/\sqrt{n}$. This gives us the standardized distance, z. The equation assumes that the true population mean is μ_0 (the assumption of innocence). If the distance between the sample result $\bar{x}$ and the hypothesized mean μ_0, measured in standard deviations of $\overline{X}$ (and, hence, scaled to the specifications of the problem), is a large distance, then we reject the null hypothesis. We know, for example, that if the population mean is indeed μ_0, then the standardized

A Distance of 10 Units in Two Different Settings   **FIGURE 6.19**

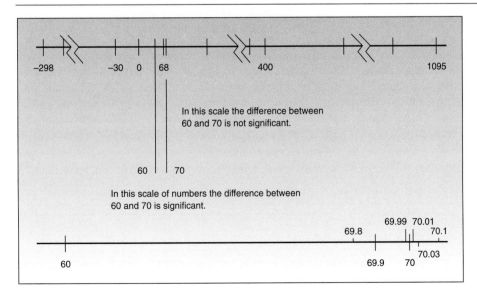

FIGURE 6.20   The Standardized Distance between $\bar{x}$ and μ_0 as Measured by the Test Statistic z

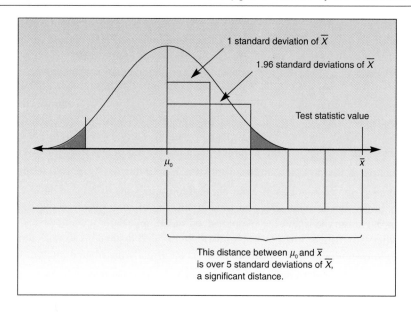

distance measured by Z has only a 0.05 chance of being greater than 1.96 or less than −1.96. Similar facts about the probability of the standardized distance Z are obtained by consulting the normal probability table. This is demonstrated in Figure 6.20.

We see that the standardized test statistic allows us to measure the distance between the value of the statistic and the hypothesized value of the parameter in terms of standard deviations of the statistic. We can now state a general formula (except for cases involving the population variance, where the mechanism is somewhat different) for a standardized test statistic.

$$\text{Test statistic} = \frac{\text{Estimate} - \text{Hypothesized parameter}}{\text{Standard deviation of estimator}}$$

The test statistic measures the distance between what is claimed under H_0 and the estimate obtained from the data, in terms of standard deviations of the estimator. Such a distance, in standard units (hence the statistical term *standard deviation*), may or may not be *significant*. When the standardized distance between estimate and hypothesized parameter value has a small (presampling) probability of occurrence, the result is statistically significant and the null hypothesis should be rejected. On the other hand, when the standardized distance between estimate and hypothesized parameter value has a relatively high (presampling) probability, the result is common (not significant) and the null hypothesis is not rejected.

Some Comments and Caveats

1. When you are conducting a right-tailed test and the estimate $\bar{x}$ is smaller than the hypothesized value of the parameter, μ_0, you do not need to compute the

value of the test statistic at all. You may immediately decide not to reject H_0. This is so because if $\bar{x}$ is less than μ_0, then the value of the test statistic will be negative, and in a right-tailed test we reject only for some (extreme) *positive* values of the test statistic. A similar argument holds for left-tailed tests.

2. If you are conducting a two-tailed test and you end up rejecting the null hypothesis in the left-hand rejection region, one of three things may be true. First, it is possible that, as you concluded, the parameter is smaller than the value stated in the null hypothesis. Second, it is possible that the null hypothesis is true (in which case you just observed an event that had a probability of occurrence equal to your p-value) and you committed a Type I error. The third possibility is that the null hypothesis is indeed false, as you concluded, but the true value of the parameter is actually greater than the value stated in the null hypothesis. This is an event with an extremely small probability in most situations, but it can happen. This is so because curves centered to the right of the null-hypothesized value still have tails extending to the left of the left-hand critical point. This is demonstrated in Figure 6.21. This possibility is of little practical importance, but you should understand it. Here you correctly reject the null hypothesis, but for the wrong reason.

3. **PRACTICAL SIGNIFICANCE.** A two-tailed test has the form H_0: $\mu = 10$ versus H_1: $\mu \neq 10$. Let's think about this logically. The mean could be equal to 10 exactly, or it could be equal to some other number. If the mean is not 10 but, say, 12, then we probably want to know about it and hope to reject the null hypothesis. Suppose, however, that the population mean is not 10, but 10.001. The fact that the mean differs so slightly from 10 means that we probably do not want to know about it. (Again, this must be considered within the scale of the problem, but suppose that for practical considerations 10 and 10.001 are no different to us.) In this case, we do not want to reject the null hypothesis.

A two-tailed hypothesis test will cause us to reject the null hypothesis if the sample size is large enough. As the sample size increases, the power of the test (explained in Section 6.7) increases, and this allows us to reject almost any null hypothesis in a two-tailed test. What is the meaning of this rejection? We must be very careful with our interpretations and be aware that, while we may not want to reject H_0 if the true mean is 10.001 instead of 10.000, a huge sample

The Possibility of Correctly Rejecting the Null Hypothesis, for the Wrong Reason

FIGURE 6.21

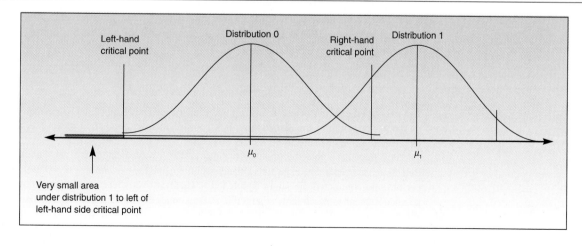

(say, 10,000) may make us reject this null hypothesis anyway. Therefore, if we reject H_0 and the p-value is very small, this means that we feel strongly about rejecting H_0. It does *not* always mean that the population mean is very far away from the value stated in the null hypothesis; it merely means that there is strong evidence to reject the claim that the mean is exactly 10. The true mean may be very different from 10, or it may be close to 10. In general, beware of inferences made on the basis of very large samples. While rejection of the null hypothesis may be very significant in the statistical sense, it may not be significant in a *practical* sense.

Statistical significance does not *necessarily imply practical significance.*

Of course the opposite is true as well. The true parameter value may be far from its value under the null hypothesis in practical terms and yet our test may fail to reject the null hypothesis.

4. The formulation of the null and alternative hypotheses should never be done *after* you have seen the data. This is called *data snooping,* and it is a bad thing. Why? We will demonstrate the answer with an example. Suppose that you are interested in estimating the proportion of consumers who use a certain product. You do not have any idea as to what the proportion may be, but you want to carry out a "hypothesis test." You collect a random sample of 100 consumers and, after looking at the data and noticing that 34 of the consumers indicated that they use the product, you decide to test the null hypothesis that the population proportion is $p = 0.34$. If you use the sample results just obtained, your "hypothesis test" will be meaningless and will never lead to rejection. Here is why:

$$H_0: p = 0.34$$

$$H_1: p \neq 0.34$$

$$z = \frac{\hat{p} - p_0}{\sqrt{p_0 q_0 / n}} = \frac{0.34 - 0.34}{\sqrt{p_0 q_0 / n}} = 0$$

The value we obtain for the test statistic, 0, lies right in the middle of the nonrejection region and thus cannot lead to rejection of the null hypothesis at any level α.

In this example, we used all the information in the sample in formulating the "hypothesis," but the same holds true, to a lesser degree, if we "snoop around" and look at just some of our data. Doing this, we lower our power of rejecting any null hypothesis, and we are also cheating. Data snooping is an example of lying with statistics.

PROBLEMS

62. In a two-tailed test of whether or not a population mean is equal to 52, rejection happened on the left-hand tail of the distribution. Is it possible that the population mean is actually larger than 52? Is it likely to be larger? Explain.

63. The null hypothesis that the population mean is 407.5 was rejected in a two-tailed test. Rejection occurred on the right tail of the distribution, and the p-value was 0.0000001. The person who contracted the study concluded: "There is overwhelming evidence that the population mean is much larger than 407.5." Do you agree with this conclusion? Explain.

64. What is wrong with examining your data before carrying out a statistical test? Discuss.

65. Does practical significance necessarily imply statistical significance? Does statistical significance imply practical significance? Explain; give some examples.
66. How does statistical significance relate to practical significance?
67. In a right-tailed test, where the null hypothesis is that the population mean is equal to 55 or less versus the alternative that it is greater than 55, we get a sample mean of 50. Can a statistical decision be made even though the sample size and the standard deviation are both unknown? Explain.
68. If the null-hypothesized population mean in a two-tailed test is 90, the sample mean is 170, and the sample standard deviation is 2, can we make a statistical decision?

6.7 THE PROBABILITY OF A TYPE II ERROR AND THE POWER OF THE TEST[13]

You may have wondered why, until now, we have concerned ourselves only with the probability of a Type I error, α—completely ignoring β, the probability of committing a Type II error. Part of the reason for this is that we have determined—consistent with our analogy of the legal system—that a Type I error is more serious than a Type II error, and therefore we should try to control the Type I error by setting its probability to a small level. That's why we have carried out our tests at given levels of significance without knowing our probabilities of committing errors of Type II.

Another reason we have ignored the Type II error probability for so long is that it is complicated to compute, because it depends on the particular value of the parameter, out of all the values covered by the alternative hypothesis. Recall that in the case of tests for μ, we said that β should actually be written $\beta(\mu)$ to indicate this dependence. We will now see why β depends on the value of the parameter. We will compute the probability β in certain situations, as well as the value $1 - \beta$, called the *power* of the test. Usually, our hypotheses are not simple but rather *composite* ones—covering a range of values, but as a simple example, consider the following hypothesis test:

$$H_0: \mu = 60$$

$$H_1: \mu = 65$$

The set consists of two *simple hypotheses:* The mean is equal either to 60 or to 65. Clearly, such a situation rarely arises in a real application because it is too restrictive. This type of hypothesis testing is useful, however, in helping us understand the Type II error and its probability.

Let's investigate the probability of a Type II error in this simple case. Suppose that the sample size is $n = 100$ and that the population standard deviation is known and given as $\sigma = 20$. In our investigation of the probability of a Type II error, we will find it convenient to consider the test in the scale of the problem itself, rather than in the standardized z scale. This particular simple test is a right-tailed test.

Now we'll determine the rejection region and the critical point for a given level of significance, $\alpha = 0.05$. We know that the value of Z for a right-tailed test at $\alpha = 0.05$ is $z = 1.645$. Transforming this value to the scale of the problem in order to determine the critical point, C, we have:

$$C = \mu_0 + 1.645\sigma/\sqrt{n} = 60 + 1.645(20/10) = 63.29$$

---

[13]In a short course, this section may be optional. However, some discussion of *power* should be provided.

FIGURE 6.22   Determining the Rejection Region and the Probability of a Type II Error for the Test

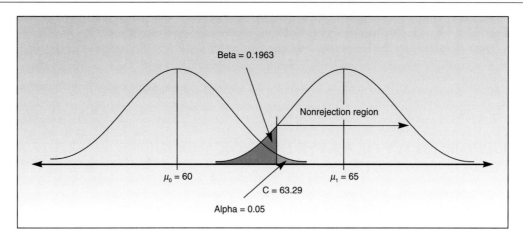

The critical point and the rejection region are shown in Figure 6.22.

The figure shows two curves. One is the curve corresponding to the sampling distribution of $\overline{X}$ if the null hypothesis is true; this is the curve centered over the mean $\mu = 60$. The area under this curve to the right of the critical point $C = 63.29$ is equal to 0.05. This is the probability that the sample mean $\overline{X}$ will fall above the critical point when the population mean is indeed equal to 60; it is the probability of a Type I error. Now, however, we are also interested in the probability of a Type II error. The true state of nature could be only one of two possibilities: Either the population mean is 60 (null hypothesis) or it is 65 (alternative hypothesis). The question here is: When is a Type II error committed? The answer: If the test statistic $\overline{X}$ falls to the left of the critical point C when the population mean is equal to 65. This is so because if the mean is 65 and the test statistic falls to the left of C, then the test statistic is in the nonrejection region of the null hypothesis. We then fail to reject a false null hypothesis.

What is the probability, β, of committing such an error?

The probability that $\overline{X}$ will fall to the left of point C *when the population mean is* $\mu = 65$ *is given by the area to the left of* C *under the normal curve centered over the mean 65.*

We have the following conditional probabilities.

Probability of Type I error:

$$\alpha = P(\overline{X} > C \mid \mu = \mu_0)$$

Probability of Type II error:

$$\beta = P(\overline{X} < C \mid \mu = \mu_1)$$

The probability α was preset to 0.05 by construction. The probability of the Type II error, β, is not determined by us; it is whatever the area to the left of C under the curve centered at 65 may be. Let's solve the equation to find the value of β:

$$\beta = P(\overline{X} < C \mid \mu = \mu_1) = P\left(\frac{\overline{X} - \mu_1}{\sigma/\sqrt{n}} < \frac{C - \mu_1}{\sigma/\sqrt{n}}\right)$$

$$= P\left(Z < \frac{63.29 - 65}{20/10}\right)$$

$$= P(Z < -0.855) = 0.1963$$

The situation is shown in Figure 6.22 with both curves and both areas. We conclude that carrying out this hypothesis test allows us a 0.05 chance of rejecting the null hypothesis when the null hypothesis is true, and a 0.1963 chance of failing to reject the null hypothesis when the null hypothesis is false and the actual population mean is 65.

We now define the *power* of a statistical hypothesis test.

The **power** of a statistical hypothesis test is the probability of rejecting the null hypothesis when the null hypothesis is false.

Thus, the power of a test is the probability of rejecting a null hypothesis that should indeed be rejected. Clearly, the probability of rejecting H_0 when H_0 is false is equal to 1 minus the probability of failing to reject H_0 when H_0 is false.

$$\text{Power} = 1 - \beta$$

The term *power* is quite appropriate. Considering the fact that we want to reject null hypotheses that are false, a test is *powerful* if it gives us a high probability of rejecting a false null hypothesis. Power is a measure of our ability to demonstrate that H_1, rather than H_0, is the correct hypothesis. In the preceding example, the power of the test is equal to $1 - \beta = 1 - 0.1963 = 0.8037$. We have a 0.8037 chance of rejecting the null hypothesis when the true mean is 65 rather than 60 as stated in the null hypothesis.

As we've already said, in most situations, the hypotheses to be tested are not simple but rather composite; that is, they include a range of values to be tested. Whatever the form of our test, the null hypothesis always contains a single number to be used in carrying out the test. (This is the value obtained from the equal sign in the null hypothesis.) For example, the null hypothesis $H_0: \mu \leq 60$ is tested as if it were $H_0: \mu = 60$ because 60 is the maximum of all values of μ under the null hypothesis. The assumption of innocence means we allow the mean to be as large as 60, and test to see if it is greater than this maximum value. In two-tailed tests, the null hypothesis contains a single value only. However, the alternative hypothesis (except in simple cases such as the one just considered) always contains many—in fact, infinitely many—possible values. If the null hypothesis is $H_0: \mu \leq 60$, then the alternative hypothesis is $H_1: \mu > 60$. How can we define the power of the test in such situations?

Again, power is the probability of rejecting a false null hypothesis. But in the example above, the null hypothesis can be false in an infinite number of ways: The population mean could be 61, or 67, or 72.8653, or 999.87, or *any* value above 60. What, then, is the probability of correctly rejecting the null hypothesis when the population mean is any number above 60? It should be easy to see that this probability depends on the actual value of the population mean. The larger the population mean,

FIGURE 6.23

Power at Different
Values of μ

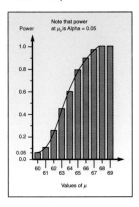

FIGURE 6.24

The Power Function
(for a One-Tailed Test)

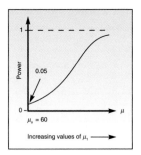

μ, the greater the probability that the sample mean, $\overline{X}$, will fall away from 60 and away from the critical point, C, determined by our level of significance, α.

Let's demonstrate the power of the hypothesis test H_0: $\mu \leq 60$ versus H_1: $\mu > 60$ for various values of μ covered by the alternative hypothesis. We assume that the sample size is $n = 100$, the population standard deviation is $\sigma = 20$, and $\alpha = 0.05$. This is shown in Figure 6.23. The actual power, evaluated at each of several selected values of μ, is given in Table 6.3. Note that the figure also shows the power at $\mu = \mu_0 = 60$. This is the probability of rejecting the null hypothesis when $\mu = \mu_0$—the probability of a Type I error, $\alpha = 0.05$.

Table 6.3 contains the value of β for each selected value of μ_1 and the corresponding power, $1 - \beta$. (The values selected for the table are integers, although the power is defined for any value μ_1, integer or otherwise.) We see that the power of a test depends on the possible values of the population parameter under the alternative hypothesis. Thus, both the probability of a Type II error, β, and its complement, the power, depend on the value of the mean. The dependence of the power on the value of the parameter is called the **power function.** The power function, also called the *power curve,* is shown for our particular test in Figure 6.24.

Note that the values in Table 6.3, which were used in the construction of the function in Figure 6.23, were calculated in exactly the same way we computed the power in the simple hypothesis test described earlier, where the alternative value of the mean was $\mu_1 = 65$. Since our present, composite, one-tailed test has the same critical point, sample size, and assumed standard deviation, the same formula applies, and we obtain the values of β by substituting in the equation each appropriate value for the population mean for which we want to compute β and the power. Note that the power as shown for $\mu_1 = 65$ in Table 6.3 conforms with the value we computed earlier in the simple hypothesis test.

The power of a statistical hypothesis test depends on the following factors:

1. The power depends on the distance between the value of the parameter under the null hypothesis and the *true value* of the parameter in question. *The greater this distance, the greater the power.*
2. The power depends on the population standard deviation. *The smaller the population standard deviation, the greater the power.*
3. The power depends on the sample size used. *The larger the sample, the greater the power.*
4. The power depends on the level of significance of the test. *The smaller the level of significance, α, the lower the power.*

TABLE 6.3 The Power of the Test at Selected Values of μ under H_1

Value of μ	β	Power = $1 - \beta$
61	0.8739	0.1261
62	0.7405	0.2595
63	0.5577	0.4423
64	0.3613	0.6387
65	0.1963	0.8037
66	0.0877	0.9123
67	0.0318	0.9682
68	0.0092	0.9908
69	0.0021	0.9979

Suppose that we look at the values of the power function shown in Table 6.3 and decide that it is important to be able to reject the null hypothesis if the true population mean is $\mu = 62$. The value $\mu = 62$ falls under the alternative hypothesis; however, for this particular value, the test is rather weak—the power is only 0.2595. That is, if the true population mean is 62, we only have a 0.2595 chance of rejecting the null hypothesis. As we see from the table and from Figure 6.23, the power increases quickly, and if the true mean is equal to 69, we have a 0.9979 chance of rejecting the null hypothesis. That still doesn't solve the problem of what to do if we want to have a high power—a high probability of rejecting the null hypothesis—if the true mean is $\mu = 62$. Looking at the four statements about power, we realize that we cannot control the first two factors. We can, however, use the next two properties. We can either reduce our demands on the level of significance, that is, increase α, or we can increase the sample size (the better solution). Increasing the sample size is a way of increasing the power of a test without sacrificing the level of significance, α (our protection against a Type I error).

Let's now use our equation to compute the power of the test at $\mu = 62$ if the sample size is increased to $n = 400$ (and the level of significance stays the same, $\alpha = 0.05$). Note that increasing the sample size changes the critical point as well. A larger sample size brings the critical point closer to the center of the distribution (the null-hypothesized value). First we solve for the new critical point. We have:

$$C = \mu_0 + 1.645\sigma/\sqrt{n} = 60 + 1.645(20/20) = 61.645$$

Now we compute the new power for $\mu = 62$. We have:

$$\beta = P\left(Z < \frac{C - \mu_1}{\sigma/\sqrt{n}}\right) = P\left(Z < \frac{61.645 - 62}{20/20}\right)$$

$$= P(Z < -0.355) = 0.3613$$

The power is $1 - \beta = 0.6387$. This is much higher than the power we had when the sample size was only 100.

What about two-tailed tests? The power curve for our test of the null hypothesis H_0: $\mu = \mu_0$ versus the two-tailed alternative H_1: $\mu \neq \mu_0$ is shown in Figure 6.25. Also shown, for comparison, is the power curve for a one-tailed test.

FIGURE 6.25

The Power Curve for a Two-Tailed Test and the Power Curve for a One-Tailed Test

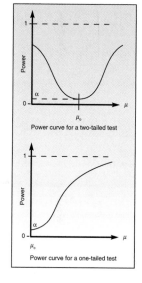

Power curve for a two-tailed test

Power curve for a one-tailed test

EXAMPLE 6.10

An advertisement for Saab states that *Car and Driver* magazine recently determined that the new Saab 9000 goes from 0 to 60 mph in 7.7 seconds, while *Road & Track* magazine tested this car at 7.6 seconds in going from 0 to 60 mph. Makers of a competing automobile feel that if the true average number of seconds it takes the Saab 9000 to reach 60 mph from zero is *above* the 7.6 figure claimed by *Road & Track,* and in particular, if it takes an average of 7.7 seconds, as determined by *Car and Driver,* the competitor could advertise its own car as a better model with respect to this performance aspect. Thus, the makers of the competing automobile want to test the Saab 9000 and carry out the following hypothesis test: H_0: $\mu \leq 7.6$ and H_1: $\mu > 7.6$. Since the competitor wants to have a high probability of rejecting the null hypothesis if μ is equal to the value 7.7, covered under the alternative hypothesis, the firm is interested in finding the power of the test evaluated at $\mu_1 = 7.7$.

Suppose that the sample size available for this test is $n = 150$ trials and that the population standard deviation (from previous driving tests) is believed to be $\sigma = 0.4$

Courtesy of SAAB Cars, Inc.

seconds. The required level of significance is $\alpha = 0.05$. What is the power? That is, if the true average time is 7.7 seconds, what is the probability of rejecting the null hypothesis that the average time is at most 7.6 seconds?

SOLUTION

First we find the critical point: $C = \mu_0 + 1.645\sigma/\sqrt{n} = 7.6 + 1.645(0.4/12.25) = 7.6537$. Now we find the power at $\mu = 7.7$.

$$\text{Power} = P(\overline{X} > C \mid \mu = 7.7) = P\left(Z > \frac{7.6537 - 7.7}{0.4/12.25}\right) = P(Z > -1.42) = 0.9222$$

The competitor thus has a 0.9222 chance of rejecting the null hypothesis when the true average time it takes the Saab 9000 to go from 0 to 60 mph is 7.7 seconds. Note that in our calculation we used the probability definition of power directly, without first computing β.

PROBLEMS

69. Recent near misses in the air, as well as several fatal accidents, have brought air traffic controllers under close scrutiny. As a result of a high-level inquiry into the accuracy of speed and distance determinations through radar sightings of airplanes, a statistical test was proposed to check the air traffic controllers' claim that a commercial jet's position can be determined, on the average, to within 110 feet in the usual range around airports in the United States. The proposed test was given as $H_0: \mu \leq 110$ versus the alternative $H_1: \mu > 110$. The test was to be carried out at the 0.05 level of significance using a random sample of 80 airplane sightings. The statistician designing the test wanted to determine the power of this test if the actual average distance at detection is 120 feet. An estimate of the standard deviation is 30 feet. Compute the power at $\mu_1 = 120$ feet.

70. McDonald's Corporation has been steadily moving into more countries around the world. Recent reports show that McDonald's has been interested in opening franchises in Poland. As part of its efforts to sell fast foods in Poland, McDonald's Corporation has been evaluating the potential of using Polish-grown potatoes not only in Poland but also for distribution in other European countries. The feasibility of such an option depends on the demand for french-fried potatoes in all of McDonald's European franchises. Company analysts believe that if the average weekly demand for fries per franchise per week is above 500 package units, it may be feasible to use Polish-grown potatoes. Thus, the analysts want to test the null hypothesis $H_0: \mu \leq 500$ versus the alternative hypothesis $H_1: \mu > 500$. The company has data on 100 weekly sales randomly obtained from franchises throughout the continent. The test is to be carried out at the 0.05 level of significance, and an estimate of the population variance is 2,500 (units squared). What is the power of the test if the true mean is 520 units per franchise per week?

71. The Polaroid Spectra® camera has an electronic device that makes complex focusing and exposure decisions in 50 thousandths of a second. Before each device is installed in a camera, it is tested by quality control inspectors. The device is linked to a simulator that runs a random sample of 80 situations and measures the sample average reaction time of the device. The statistical test is $H_0: \mu \leq 50$ (thousandths of a second) versus the alternative $H_1: \mu > 50$ (thousandths of a second). If the null hypothesis is accepted, the device is considered to have good quality and is installed in a camera; otherwise it is replaced. The test is carried out at the 0.01 level of significance, and the population standard deviation is $\sigma = 20$ (thousandths of a second). For quality control considerations, inspectors need to have a high power for this test, that is, a high probability of rejecting a faulty device, when the average speed of the device is 60 thousandths of a second. Find the power at this level of μ. Also compute the power at other levels, and sketch the power curve.

72. The management of a large manufacturing firm believes that the firm's market share is 45 percent. From time to time, a statistical hypothesis test is carried out to check whether the assertion is true. The test consists of gathering a random sample of 500 products sold nationally and finding what percentage of the sample constitutes brands made by the firm. Whenever the test is carried out, there is no suspicion as to the direction of a possible change in market share (i.e., increase or decrease); the company wants to detect any change at all. The tests are carried out at the $\alpha = 0.01$ level of significance. What is the probability of being able to statistically determine a true change in the market share of magnitude 5 percent in either direction? (That is, find the power at $p = 0.50$ or $p = 0.40$. Hint: Use the methods of this section in the case of sampling for proportions. You will have to derive the formulas needed for computing the power.)

73. Explain the importance of power in testing statistical hypotheses.

74. Why can't the power of a statistical test be given as a single number?

75. What are the drawbacks of a statistical test with low power?

76. Use MINITAB to generate 20 random samples of 100 units each from a normal population with mean 50 and standard deviation 10. Then use each sample in testing the null hypothesis that the population mean is 48 versus the alternative that it is not 48. Use the following commands:

```
RANDOM 100   C1-C20;
    NORMAL 50  10.
ZTEST MU = 48 SIGMA = 10 on C1-C20
```

   a. In how many of the 20 tests did you (correctly) end up rejecting the null hypothesis using $\alpha = 0.05$?
   b. What were your p-values?
   c. What did this simulation tell you about the power of the test at the point $\mu = 50$?
   d. Repeat the process simulating samples from populations with other values for the true population mean (49, 51, 52, etc.), always testing whether the mean is equal to 48, and keeping the same population standard deviation. Use your results to simulate the power function of the test.

SUMMARY

In this chapter we introduced the topic of **statistical hypothesis testing.** We saw how an analogy with legal procedures and the historical example of the trial of the pyx helped us determine two asymmetrical hypotheses: a **null hypothesis,** which enjoys the assumption of truth, and an alternative hypothesis, usually the one the researchers want to establish. We argued that in this framework the objective is to reject the null hypothesis, while allowing it the benefit of the doubt throughout the process. We defined two types of error: rejecting a true null hypothesis (Type I), and failing to reject a false null hypothesis (Type II). We showed why the first error may be considered worse than the second, and we defined its maximum probability as α, set to a small number: usually 0.05 or 0.01. We learned how to use the standardized sampling distribution of the sample mean based on the assumption that the null hypothesis is true in establishing a **rejection region** for a test for the population mean. We saw that the distribution we use is the standard normal when the population standard deviation is known, the t distribution when the population standard deviation is unknown. Similarly, we saw how large-sample tests for the population proportion may be carried out using the standard normal distribution and a similarly standardized test statistic. We defined the important concept of *p*-value: the actual, attained significance level of a statistical test. We learned how to compute it, report it, and use it as a measure of the statistical significance of the test results. We made a distinction, however, between **statistical significance** and **practical significance.** The **power** of a statistical hypothesis test was defined as one minus the probability of a Type II error.

KEY FORMULAS

Test statistic in tests for the population mean when σ is known:

$$z = \frac{\bar{x} - \mu_0}{\sigma/\sqrt{n}}$$

Test statistic in tests for the population mean when σ is unknown:

$$z = \frac{\bar{x} - \mu_0}{s/\sqrt{n}}$$

Large-sample test statistic in tests for the population proportion:

$$z = \frac{\hat{p} - p_0}{\sqrt{p_0 q_0/n}}$$

ADDITIONAL PROBLEMS

77. A CNN report in January 1994 claimed that 50 percent of all roller bladers get injured. A random sample of 849 roller bladers revealed that only 210 of them had been injured. Is there statistical evidence to reject the CNN report's credibility if the purpose of the study was to prove that CNN exaggerated upward the actual injury rate of this recreational activity? Incorporate a statement on the *p*-value in your conclusion.

78. A recent National Science Foundation (NSF) survey indicates that more than 20 percent of the staff in American research and development laboratories is foreign. Results of the study have been used for pushing legislation aimed at limiting the number of foreign workers in the United States. An organization of foreign-born scientists wants to prove that the NSF survey results do not reflect the true proportion of foreign workers in U.S. labs. The organization collects a random sample of 5,000 laboratory workers in all major labs in the country and finds that 876 are foreign. Can these results be used to prove that the NSF study overestimated the proportion of foreigners in American laboratories?

79. Corporate women are still struggling to break into senior management ranks, according to a study of senior corporate executives by a New York recruiter. In 1993, of 1,362 top executives surveyed by the firm, only 2 percent, or 29, were women. Assuming that the sample reported is a random sample, use the results to test the null hypothesis that the percentage of women in top management is 5 percent or more, versus the alternative hypothesis that the true percentage is less than 5 percent. If the test is to be carried out at $\alpha = 0.05$, what would be the power of the test if the true percentage of female top executives is 4 percent?

80. In the context of the article on mammography on the next page, write an essay explaining the concept of statistical significance, which is referred to repeatedly by the author of the article. What are the possible reasons for the lack of statistical significance of the findings in the article?

81. A rechargeable battery is claimed in television commercials to have the same output after an average of 25 recharges as it does originally. A competitor looks at a random sample of 105 batteries of this type and finds that the average number of recharges after which a battery retained its full power was 20.5 and the standard deviation was 8. Conduct the appropriate statistical test for the competitor. Make conclusions and state your *p*-value.

82. If the *p*-value is smaller than α, should the null hypothesis be rejected? Explain.

83. What is the difference between a result based on a *p*-value of 0.00000001 and a result based on a *p*-value of 0.0456? Explain.

Breast Cancer Debate: The Radiologists' Views

Some Radiologists Say Benefit Exists and Will Emerge in Time

By JANE E. BRODY

After the National Cancer Institute's decision to stop recommending regular mammograms for women in their 40's, many radiologists and other doctors who treat individual patients still are convinced by their clinical experience and intuition that the benefit of routine screening would be obvious if the proper studies were done.

The statisticians upon whose assessment public health policy is largely based want to see significant evidence that mammography offers lifesaving benefits to women younger than 50. Critics say the studies of screening mammography done so far have included too few women in their 40's and did not follow them long enough.

What is a woman to do if she is under 50, healthy and has no suspicious lumps or strong family history of breast cancer? Should she wait until menopause to begin annual mammography, or should she start at 40?

Even the cancer institute's director seems torn between how he would advise patients and what he would say to the public. As he told the National Cancer Advisory Board Subcommittee on Women's Health and Cancer: "What I would do as an individual is recommend annual mammograms. But I can't recommend it to the public because I don't have the facts."

The institute's decision to recommend routine mammography for women over 50 but not younger was prompted by a lack of statistically significant evidence that screening mammography can reduce deaths among younger women, as it has been proved to do among women in their 50's. Statistical significance is a way of determining that the finding was unlikely to have been a result of chance.

Yet many radiologists and some researchers who conducted the studies on which the institute's decision was based, insist that the existing evidence, though not statistically certain, indicates women of all ages can benefit from periodic mammograms. The very low-dose X-ray examinations can reveal breast cancers long before they can be felt. In some women, including those under 50, mammography finds cancers at a microscopic stage, when the cure rate is as high as 90 to 95 percent.

While the mortality reduction that Dr. Tabar observed was not statistically significant, because the actual numbers were so small, it was similar to the indications that routine mammography saves lives noted in five of seven other major studies of the effectiveness of mammography. Those studies found that screening mammography reduced breast cancer deaths among women initially 40 to 49 years old by 22 to 49 percent. Because the findings were not statistically significant, it is possible that the apparent benefit was a result of chance, and not any direct benefit of mammography.

Breast cancers, and breast cancer deaths, are far less common in women under 50 than in older women, although there is no dramatic change in breast tissue or cancer risk at any particular age. Only 22 percent of the cancers occur in women under 50. Contrary to belief, at any given stage of the disease, breast cancer is no more deadly in younger women than in older women. Dr. Tabar wrote, in the journal *Cancer,* "The prognosis of tumors diagnosed in women in their 40's is, if anything, better than that for older women." Younger women are more likely to survive the disease.

These observations mean that to detect a statistically significant lifesaving benefit among women in their 40's who undergo regular mammography, a much larger study is needed than any done so far.

Breast Cancer: Sorting Out the Numbers

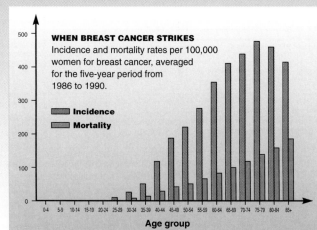

WHEN BREAST CANCER STRIKES
Incidence and mortality rates per 100,000 women for breast cancer, averaged for the five-year period from 1986 to 1990.

■ Incidence
■ Mortality

Age group

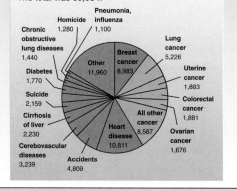

BREAST CANCER IN PERSPECTIVE
Breast cancer compared with 10 leading causes of death among women 35 to 55 years old in 1989. The total was 69,034.

Chronic obstructive lung diseases 1,440
Homicide 1,280
Pneumonia, influenza 1,100
Diabetes 1,770
Suicide 2,159
Cirrhosis of liver 2,230
Cerebovascular diseases 3,239
Accidents 4,809
Other 11,960
Breast cancer 8,983
Heart disease 10,811
All other cancer 8,587
Lung cancer 5,226
Uterine cancer 1,883
Colorectal cancer 1,881
Ovarian cancer 1,676

Reprinted by permission from *The New York Times,* December 14, 1993, p. C1, with data from the National Cancer Institute and the National Center for Health Statistics.

84. I want to test the null hypothesis that the population mean is equal to 50 versus the alternative that it is not equal to 50. My sample size is 10,000, my sample mean is 49.99, and my sample standard deviation is 0.1. What can I say about the p-value? Based on the result, can I say that there is evidence that the population mean is *far away* from 50? Explain. Think about the conclusion: Can I find here a general limitation of two-tailed tests? Other conclusions?

85. Suppose I rejected a null hypothesis, and my p-value was very small. What can you say about the probability that I committed a Type II error?

86. A social researcher looks at a random sample of 100 students and finds that 25 of them are nursing majors. Having seen the sample results, the researcher now wants to use these same data to test the hypothesis that 25 percent of all students in the population from which the sample was drawn are nursing majors. What is wrong here?

87. To win government approval, the developers of a cancer test such as the one described in the accompanying article must present statistical evidence that the test's success rate is greater than 60 percent. Use the information given in the article to conduct the appropriate test. Give the p-value and state your conclusions. Discuss the assumptions you must make, and whether they apply here.

Simple Test Found to Warn Those Prone to Rare Cancer

By NATALIE ANGIER

Scientists have developed a simple blood test to identify people with an inborn predisposition to one type of colon cancer, offering those in high-risk families the earliest possible warning if they have the trait and a large measure of relief if they do not.

The test detects mutant proteins that indicate whether a person has inherited the gene for familial adenomatous polyposis, a rare condition in which the lining of the colon sprouts hundreds or thousands of tiny wart-like polyps. If left untreated, the condition almost invariably leads to colon cancer, often when patients are in their 20's or 30's, and sometimes even younger.

The test is not foolproof. In the new study, the scientists identified mutant proteins in 54 of 62 patients known to be afflicted with familial polyposis, a success rate of 87 percent. But the researchers said the figure would be higher in many situations, particularly when there are others in the family with the trait. In such cases the results of the tests on the family members can be compared against each other. But Dr. Kinzer said people with no hereditary link who have unusually high numbers of polyps might also wish to be tested to see if they bear the trait.

Reprinted by permission from *The New York Times*, December 30, 1993, p. A12.

---

88. The Hunters Point project in New York (see display) was designed to provide apartments for an average price of $15,000. Apartments are to be sold on auction. In the preliminary stage, 12 apartments were auctioned, randomly distributed throughout the area of the project shown in the map. Their prices as determined in the auction were as follows (in thousands of dollars):

$$14.5, 12.8, 18.9, 16.4, 19.5, 16, 17.1, 14, 15.1, 18, 13.6, 17.7$$

Conduct a two-tailed test of the null hypothesis that the average price of an apartment in this project will be $15,000. Give an estimate of the p-value.

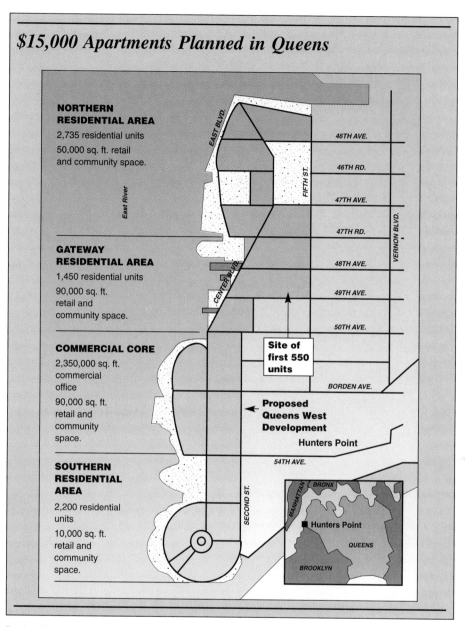

$15,000 Apartments Planned in Queens

NORTHERN RESIDENTIAL AREA

2,735 residential units

50,000 sq. ft. retail and community space.

GATEWAY RESIDENTIAL AREA

1,450 residential units

90,000 sq. ft. retail and community space.

COMMERCIAL CORE

2,350,000 sq. ft. commercial office

90,000 sq. ft. retail and community space.

SOUTHERN RESIDENTIAL AREA

2,200 residential units

10,000 sq. ft. retail and community space.

East River

EAST BLVD.

FIFTH ST.

CENTER BLVD.

SECOND ST.

46TH AVE.

46TH RD.

47TH AVE.

47TH RD.

48TH AVE.

49TH AVE.

50TH AVE.

VERNON BLVD.

BORDEN AVE.

54TH AVE.

Site of first 550 units

Proposed Queens West Development

Hunters Point

MANHATTAN BRONX

■ **Hunters Point**

QUEENS

BROOKLYN

Reprinted by permission from *The New York Times,* November 30, 1993, p. B7.

89. An article about women in business claims that 30 percent of all small businesses in the United States are owned by women.[14] An analyst wanted to test this claim without any prior intent to prove anything. The analyst looked at a random sample of 2,000 small businesses and found that 852 of them were owned by women. Conduct the test, and state your conclusion.

---

[14]"Selling to Uncle Sam," *Nation's Business,* March 1991, p. 29.

90. Use MINITAB to generate 100 random samples of size 50 each from a normal population with mean 100 and standard deviation 10. Use each sample to test the two-tailed hypothesis that the population mean is indeed 100. Use the 0.05 level of significance. In how many of the 50 samples did you (incorrectly) reject the null hypothesis? What are the p-values?

91. Use Table A.2 in Appendix A. Assume that each column represents a random sample of percentages of fatal injuries for the month for all occupations of these kinds. Test the null hypothesis that the average percentage of fatal injuries in July is less than or equal to 1/12 versus the alternative that it is greater than this number. Interpret your findings.

INTERVIEW WITH
Lee Wilkinson

"I said, "Can I call you back?" 'You can't call me back, this is the White House. I want to place an order.' It was amazing."

Lee Wilkinson is the founder and owner of Systat Inc., a leading publisher of microcomputer-based statistical analysis software based in Evanston, Illinois. His interest in statistics began as a graduate student in Yale where he had planned to study clinical psychology. His undergraduate work was at the Harvard Divinity School.

Aczel: How did you get into statistics and statistical computing?

Wilkinson: Actually, I started in psychology. My area was psychometrics. I was in graduate school at Yale, and my advisor was Bob Abelson, who is actually a Fellow of the ASA [American Statistical Association] and is a statistical psychologist. A lot of his research is in social psychology. I went over to the statistics department and took a course with John Hartigan—Clustering, and he inspired me in every way because not only was the clustering course interesting but he was really interested in statistical computing. John has been one of the pioneers in that area, and also had a point of view that impressed me because it was very data-oriented—not rigid, ideological, and he was very tolerant of alternative points of view. And he was a very good listener. From there I ended up spending almost all my time over in the

computer center. Yale computer center in the early 1970s was a very exciting place because unlike most computer centers, the person who ran it, that staff who ran it, let us near the machine. You remember a lot of places made users hand cards in at the window.

Aczel: Right, I was a student at Berkeley in those years. Were you an undergraduate or a graduate?

Wilkinson: I was at Yale as a graduate student.

Aczel: You were an undergrad at Harvard before that in the 60s.

Wilkinson: Yes, I went to Harvard Divinity School. My interest when I went to Yale was in going into clinical psychology and doing research and teaching in that area, but when I met Hartigan—things took a completely different path. My advisor also—we talked a lot of statistics. My advisor was Bob Abelson who, like Hartigan, had been at Princeton and had been influenced by John Tukey and the group, Tukey's thinking about exploratory data analysis. There was in that generation a solid group of Tukey disciples. Abelson and Tukey actually did a paper together on multidimensional scaling.

Aczel: Then you took a faculty position? In psychology?

Wilkinson: Yes, I taught in the department of psychology at the University of Illinois. I was the methodologist there. I was hired in the methods and measurement division to teach courses in statistics. Then I took a sabbatical some years later and during the sabbatical I built a microcomputer.

Aczel: By yourself?

Wilkinson: Yes, I had a friend who helped out. This was about 1978.

Aczel: There weren't any around.

Wilkinson: There were no commercial ones. The Apple hadn't been invented yet.

Aczel: The IBMs came much later.

Wilkinson: Yes, the IBMs were 82 . . 83 . . . something like that. . . 84 . . .

Aczel: How did you build it?

Wilkinson: It was a kit. There was something called a Cromemco and it was built here [San Francisco]—the manufacturers were here in California. It turned out to be the same computer that Jerry Pournelle in *Byte* magazine wrote all his columns on. The components were produced by Bill Godbout, one of the early hackers, manufactured right here in California. We put all the boards together, and soldered everything up.

Aczel: Was it like the Commodore that came with 16K, or did it have more?

Wilkinson: This actually was a very powerful machine. I still have it. When I finished with it, it actually had a megabyte of memory. Two large floppy drives, the big 8-inch floppy drives, with a megabyte each of storage space and lots and lots of serial and parallel ports and a printer and a plotter—I spent lots of money on this machine. I had been consulting at the University of Chicago and I spent all my consulting income on this machine. My wife was kind of aghast. Later, when I started Systat, it all made sense. But what happened was that very early after that machine was going, friends and I were communicating across machines and also we had our machines looking like terminals to the mainframe at the University, so we were able to use the mainframe as the downloading printer and do all our computations in BASIC, and eventually I got FORTRAN and LISP and a variety of languages running on the system. I just got in deeper and deeper. I had written at Yale a

fairly powerful general linear modeling program, so it did regression, analysis of variance, multivariate. . .

Aczel: How did you write it?

Wilkinson: I did that for my dissertation. I needed a repeated measures program and SAS didn't have one at the time—nobody did—multivariate repeated measures. So I ended up just writing one and it got bigger and bigger.

Aczel: And what you wrote is pretty much what went into the Systat package we have today?

Wilkinson: Originally, yes. I took that program, it was called REGM—multivariate regression—and basically chopped it into little pieces because I had to fit it on a 64K microcomputer with overlays. The segments that had to be in memory at any one time were so complicated that I had to take snips of paper and lay them out on the kitchen table, and it covered a large area. What it meant was that I had in 1984, when I came here to the ASA, a system for regression, analysis of variance, manova, and so on that was basically as powerful as the mainframe versions. Since I had grown it from the ground up and they were coming from the top down they were not able to get theirs to run on 64K machines.

Aczel: That's incredible! You wrote it at Yale on a mainframe but as you went to Chicago you were able to then transfer it to the machine that you built. . .

Wilkinson: Right. I had to tear the whole thing apart and that was the difference. It would not run on a microcomputer until I spent quite a bit of time devising the overlay structure, cutting the subroutines into very small subroutines.

I was able to develop a new code quite rapidly once I got the microcomputer. It's interesting that it's often quicker to write a new code than to try to revise the old code. So by understanding algorithms and then just sitting down and starting again, you have a tremendous advantage. That's something I faced in the last couple of years because Systat has gotten larger and larger and I began to bog down and I realized we had to rewrite the whole system.

Aczel: But why did you have to rewrite it?

Wilkinson: Because the architecture of the machines and operating systems changes so fast that nowadays a program does not stay current if the basis for it is more than about three years old. You basically have to revise your core code every three or four years. Now, not many companies are doing that because it takes a tremendous amount of energy and renewal, and so there are some that are doing it and there are some that have great difficulty doing it. They have such a large code base. It's just a law of nature that you have to rewrite every few years.

Aczel: It's an incredible story. People read something like this in the popular literature like *People* magazine or *Money* magazine about Bill Gates and Steven Jobs. What you did, I think, as a statistician, is at least as amazing.

Wilkinson: In one respect I think I know very prolific programmers, statistical programmers, around. I wouldn't begin to say I'm the most proficient or certainly the most knowledgeable, but I do think in terms of the amount of code I'm capable of producing that I'm pretty skillful. I produce a lot of code in just a short time.

Aczel: So, going back to your development in that area, when did that idea first strike you that you should start Systat Inc.? Was it when you came to this conference 10 years ago?

Wilkinson: Back in 83. I expected to sell it. At the conference, I showed it at a poster session. Then I expected I would sell copies for source code plus copies for a low price.

Aczel: For personal computers?

Wilkinson: Well, not PCs but CP/M systems because there were no PCs in 83. They were just coming out. But basically like the Apple II and CP/M systems.

Aczel: You were the first with that? You came with disks then, right? And you were the first with disks? Nobody else had disks?

Wilkinson: No, that year there were two of us showing that year. There was Neil Polimus from Statgraphics. His system was written in APL. He had it partially running, I believe. I forget . . . it was on some high level microcomputer. The following year he introduced the PC version of Statgraphics and that was our main competition at that point.

Aczel: So there you were at that poster session. What happened?

Wilkinson: There was a lot of interest. And two people who were doing a review for *Byte* magazine—

David Margenstein and Jim Carpenter from the Bureau of Labor Statistics—and they were just about finished. They were doing microcomputer packages and then they saw Systat and it sort of blew them away. They said give them a copy, and I sent them a copy and a little xeroxed manual and lo and behold April rolled around.

Aczel: Then you knew it would make it?

Wilkinson: At that point I knew it would make it. My brother-in-law did a very cautious business plan, and we assumed the company would disappear by the end of 84.

Aczel: That somebody would buy you?

Wilkinson: No, actually I assumed I would go out of business. I would sell several hundred copies of it and then go out of business because everybody who wanted it would have it. I thought there'd be maybe 50 or 100 or 200 people who'd want to use this thing.

Aczel: So it was like a limited time company. . .

Wilkinson: Yes, I'm by nature conservative anyway—not politically—but I mean in that sense I don't like to take big chances. April rolled around and I'd just put an ad in *Byte* magazine in March saying, Systat, but it didn't have much effect, but then the review came out and the phone just lit up. It's the April 84 issue—I'll never forget it. It came out March 20th or something and I got a panicked phone call from the bookkeeper. She said, "I don't know what to do. The phone is ringing off the hook. You've got to come back home. This is terrible. . ." It was a business phone down in the basement but it had call interrupt and I'd say "please hold" and I'd go to the other one and it was just unbelievable. So we started to package these things. I made a manual and got a looseleaf cover for it and I would trundle off to the post office and mail these packages.

Aczel: You were producing them in the basement?

Wilkinson: Yeah, at that point I was literally hand-assembling these little disks, copying them in a little computer. Every disk, I copied, stuck a label on, made the whole thing, and literally racing off to the post office. I wasn't using UPS yet. I'd walk up to the window with a stack about this high. The woman behind the window got to know me—she was very nice—and one day she asked me: "You're not insuring these. How much are you charging for these things?" $400. She just was absolutely flabbergasted because she had seen me coming with stacks of these things and she knew it was a computer program. Well needless to say, we started expanding. I moved out of the house. I almost broke down. I was hysterical. I called an old friend and I said, "I don't know what to do." He said, "I know what you do. You hire some people."

I literally had the White House calling. At one point this guy had this very deep voice and was saying "This is the White House," and I said, "Can I call you back?" "You can't call me back, this is the White House. I want to place an order." It was amazing. People sort of yelling at the other end of the phone, "I want my package now. Do you know who I am? I'm Jet Propulsion Labs," and that sort of thing. It was a very hot time. People had just gotten these microcomputers, were told they could actually do things with them, and then read that there was software that would actually do this. They were desperate to get off the mainframes. So, anyway, we grew then. We moved into an office in a bank and then we moved again to a larger office and then in 1987 we moved to where we are now, large offices in Evanston. A very nice building in Evanston. I now have 50 people in the company. It's been fun. One of the things is people. I've really learned about management and what not to do and how to get people who were skilled. I had to find and hire people who were good managers and every company you know goes through these stages that we've gone through. I think we've successfully managed to get past the entrepreneurial stage into a second stage where we're now working in teams.

Aczel: Programming in teams. . .

Wilkinson: Programming in teams. The marketing department's become very efficient. We've focused a lot on academic marketing. We have a very, very active documentation department. I have some terrific people in tech support. I've always had this free-floating kind of work environment. All of the offices are glass so you can look right through anybody's office out to the lake—Lake Michigan, and everybody walks around in socks. It's very pleasant, very casual. We've managed to move into more formal production plans without losing that informality.

Aczel: The Silicon Valley brain tank approach.

Wilkinson: Yes, very much.

CHAPTER

7

Making Comparisons

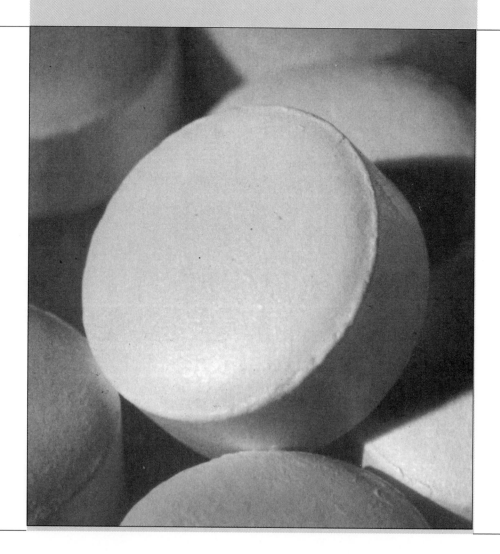

CHAPTER OUTLINE

7.1 Introduction 316

7.2 Paired-Observations
Comparisons 316

7.3 A Test for the Difference between Two
Population Means Using Independent
Random Samples 322

7.4 A Test for the Difference between Two
Population Means, Assuming Equal
Population Variances 330

7.5 A Large-Sample Test for the
Difference between Two Population
Proportions 336

The findings of the study had such potentially far-reaching implications that the Reuters news agency reported its advance knowledge of the information to its subscribing newspapers more than 24 hours before the article describing the study appeared in the January 28, 1988, issue of the *New England Journal of Medicine*. The large-scale study had begun in 1982. Its subjects were a group of 22,071 male physicians 40 to 84 years old living in the United States. Each physician was randomly assigned to take one of two kinds of pills: aspirin or beta carotene (as a placebo). Thus, 11,037 physicians were randomly assigned to take aspirin and 11,034 physicians were assigned to take the placebo. No physician knew which of the two kinds of pills he was taking. Each physician was to take his assigned pill every other day until the scheduled conclusion of the study in 1990. However, at a special meeting on December 18, 1987, the Data Monitoring Board of the Physicians' Health Study took the unusual step of recommending the early termination of the randomized aspirin experiment. At this point, the statistical results of the study had already exceeded everyone's expectations. It was shown, with a *p*-value of less than 0.00001, that a population (of American male physicians, at least) taking a single aspirin pill once every two days would have a significantly lower proportion of heart attacks than a similar population not taking aspirin. The study confirmed what doctors had suspected for years—that the wonder drug aspirin can actually help prevent heart attacks.[1]

---

[1] The editors of the *New England Journal of Medicine* barred the London-based Reuters from all new information from the *Journal* for a period of six months. For details of this study, see "Preliminary Report: Findings from the Aspirin Component of the Ongoing Physicians' Health Study," *New England Journal of Medicine,* January 28, 1988.

7.1 INTRODUCTION

The comparison of two populations with respect to some population parameter—the population mean, the population proportion, or the population variance—is the topic of this chapter. Testing hypotheses about population parameters in a single population, as was done in Chapter 6, is an important statistical undertaking. However, the true usefulness of statistics manifests itself in allowing us to make *comparisons*. Almost daily we compare products, services, investment opportunities, management styles, and so on. This chapter will show you how to conduct such comparisons in an objective and meaningful way. You'll learn how to find statistically significant differences between two populations. If you understood the methodology of hypothesis testing presented in Chapter 6 and the idea of a confidence interval from Chapter 5, you will find the extension to two populations straightforward and easy to understand.

In Section 7.2 we will see how a comparison may be made in the special case where the observations may be paired in some way. In Section 7.3 we will show how to conduct a test for the difference between two population means using independent random samples. In Section 7.4 we will present a test for the existence of a difference between two population means, assuming equal population variances. Then we will see how to compare two population proportions. Along the way, we will explain how to construct confidence intervals for the difference between two population parameters.

7.2 PAIRED-OBSERVATIONS COMPARISONS

249
205

In this section, we describe a method for conducting a *statistical hypothesis test* and constructing a *confidence interval* when our observations come from two populations and are paired in some way. What is the advantage of pairing our observations? Suppose that a taste test of two flavors is carried out. It seems intuitively plausible that if we let every person in our sample rate each one of the two flavors (with random choice of which flavor is tasted first), the resulting paired responses will convey more information about the taste difference than if we had used two different sets of people, with each group rating only one flavor. Statistically, when we use the same people for rating the two products, we tend to remove much of the *extraneous variation* in taste ratings—the variation in people, experimental conditions, and other such factors—and concentrate on the difference between the two flavors. Pairing the observations thus makes the experiment more precise.

Experimental Design

A paired-observations comparison is one example of the very important concept of **experimental design**. In looking for ways of reaching meaningful statistical conclusions, we need to design procedures that will honestly and accurately measure exactly what it is we are trying to measure. Experimental design is the science of designing such studies. When we pair observations for the purpose of conducting a statistical test, we are using a design called *blocking design*. Here, every pair is a block—it is an experimental unit made up of two elements that were paired because they are uniform in some sense. Pairing can greatly reduce experimental errors and improve our testing ability. A description of the formula we use in such experiments follows.

Test statistic for the paired-observations t test:

$$t = \frac{\bar{D} - \mu_{D_0}}{s_D / \sqrt{n}}$$

where $\bar{D}$ is the sample average difference between each pair of observations, s_D is the sample standard deviation of these differences, and the sample size, n, is the number of pairs of observations. The symbol μ_{D_0} is the population mean difference under the null hypothesis. When the null hypothesis is true and the population mean difference is μ_{D_0}, the statistic has a t distribution with $n - 1$ degrees of freedom.

Let's look at an example of a paired-observations t test.

Ultra Trim Quick wants to show that people who choose to follow its diet plan can lose, in a week, more weight than people who try another diet. Realizing that present weight, as well as age and gender, are important factors in how much weight can actually be lost, it is desirable to design a study that will correctly account for these variables and accurately measure (as much as possible) only the weight loss due to the difference between the two diets. Therefore, instead of simply taking two random groups of people and assigning one group to one diet and the other group to the other diet (a design we will use in the other sections in this chapter), we pair people according to age, gender, and weight. One block (pair) might be two men, both aged 41, both weighing 187 pounds. Then one of them is randomly assigned to the Ultra Trim Quick diet, the second man to the other diet. Similarly, a second block might be two women, both 38, each weighing 162 pounds. Suppose that the paired results— pounds lost in one week for all 12 blocks, for each diet—are as follows:

EXAMPLE 7.1

Ultra Trim Quick:	5	6	4	7	3	9	2	4	6	5	7	7
Other:	3	2	4	8	1	6	3	5	4	1	6	5
Difference:	2	4	0	−1	2	3	−1	−1	2	4	1	2

Conducting the test is rather easy, since we have already seen the procedure in Chapter 6. The novelty here is only in the concept of the experimental design—the pairing of observations.

SOLUTION

Ultra Trim Quick wants to test the following hypotheses:

$$H_0: \mu_D \leq 0$$

$$H_1: \mu_D > 0$$

where μ_D is the population average difference between the number of pounds lost per week for the two diets: Ultra Trim Quick − Other. Ultra Trim Quick wants to show that this difference is greater than zero, that is, that people lose more pounds on the Ultra Trim Quick diet.

We first compute the differences; then we find the mean and standard deviation and conduct a single-sample t test.

FIGURE 7.1   Carrying Out the Test of Example

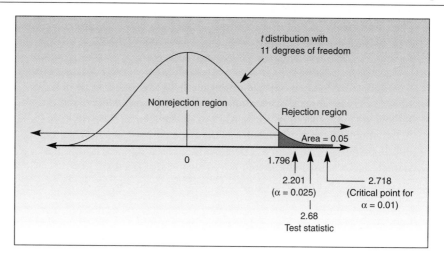

Analyzing the data, we find that their mean is $\overline{D} = 1.417$ and their standard deviation is $s_D = 1.832$. Since the sample size is small, $n = 12$, we use the t distribution with $n - 1 = 11$ degrees of freedom. (We assume a normal population of differences.) The null-hypothesized value of the population mean is $\mu_{D_0} = 0$. The value of our test statistic is obtained as follows:

$$t = \frac{\overline{D} - \mu_{D_0}}{s_D/\sqrt{n}} = \frac{1.417 - 0}{1.832/\sqrt{12}} = 2.68$$

As can be seen from Figure 7.1 and from consulting the t table (Table B.3 in Appendix B), our p-value is less than 0.025 but greater than 0.01. There is some statistical evidence that a person may lose more pounds on the Ultra Trim Quick diet than on the other diet.

Confidence Intervals

Just as we constructed confidence intervals in the single-population setting in Chapter 5, it is also possible to construct them for the average population difference, μ_D. We define a $(1 - \alpha)100\%$ confidence interval for the parameter μ_D as follows.

A $(1 - \alpha)100\%$ confidence interval for the mean difference μ_D:

$$\overline{D} \pm t_{\alpha/2}\frac{s_D}{\sqrt{n}}$$

where $t_{\alpha/2}$ is the value of the t distribution with $n - 1$ degrees of freedom that cuts off an area of $\alpha/2$ to its right. When the sample size, n, is large, we may use $z_{\alpha/2}$ instead.

In Example 7.1, we can construct a 95 percent confidence interval for the average difference in number of pounds lost for the two diets:

$$\bar{D} \pm t_{\alpha/2}\frac{s_D}{\sqrt{n}} = 1.417 \pm 2.201 \frac{1.832}{\sqrt{12}} = [0.253, 2.581] \text{ pounds}$$

We can therefore be 95 percent confident that, on the average, someone using the Ultra Trim Quick diet loses anywhere from 0.253 to 2.581 pounds more per week than someone using the competing diet.

PROBLEMS

1. What are the advantages of pairing observations in a statistical hypothesis test?
2. A study is undertaken to determine how consumers react to energy conservation efforts. A random group of 60 families is chosen. Each family's consumption of electricity is monitored in a period before and a period after they are offered certain discounts to reduce their energy consumption. Both periods are the same length. The difference in electric consumption between the periods is recorded for each family. Then the average difference in consumption and the standard deviation of the difference are computed. The results are $\bar{D} = 0.2$ kW (kilowatts) and $s_D = 1.0$ kW. At $\alpha = 0.01$, is there evidence to conclude that conservation efforts reduce consumption?
3. Recent studies indicate that in order to be globally competitive, firms must form global strategic partnerships. An investment banker wants to test whether the return on investment for international ventures is different from return on investment for similar domestic ventures. A sample of 12 firms that recently entered into ventures with foreign companies is available. For each firm, the return on investment for both the international venture (I) and a similar domestic venture (D) is given:

    D(%): 10 12 14 12 12   17   9 15   8.5 11 7 15
    I(%): 11 14 15 11 12.5 16 10 13 10.5 17 9 19

    Assuming that these firms represent a random sample from the population of all firms involved in global strategic partnerships, can the investment banker conclude that there are differences between average returns on domestic ventures and average returns on international ventures? Explain.
4. Two wines are compared in a taste test. The average difference in taste ratings is 0.5 and the standard deviation of the differences in taste ratings is 2.29. The sample consisted of 142 randomly selected individuals, each tasting both wines, the order determined at random. Are the results statistically significant? Explain.
5. For Problem 3, give a 95 percent confidence interval for the average difference in return on investment for domestic and international ventures. Compare the confidence interval with the conclusion you reached in Problem 3.
6. Assume that the regions listed in the display on page 320 about the northern spotted owl are randomly selected from a large population of such regions in Oregon, California, and Washington. Test the null hypothesis that the average difference between the number of Category 2 sites and Category 1 sites in the population of regions designed as a refuge for the northern spotted owl is 5 versus the alternative that the average difference is greater than 5. (Ignore the totals.)
7. In Problem 6, give a 99 percent confidence interval for the average difference between the number of Category 1 sites and Category 2 sites in the population from which the regions were sampled.
8. Refer to the table preceding Problem 58 in Chapter 4. The table "Mean Age at First Marriage" reports the average age at first marriage for both men and women in many

Photo Researchers, Inc.

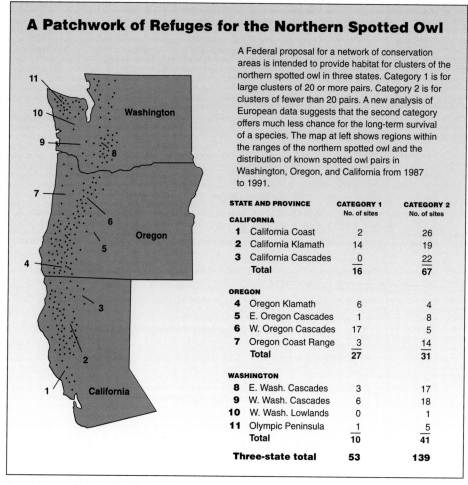

A Patchwork of Refuges for the Northern Spotted Owl

A Federal proposal for a network of conservation areas is intended to provide habitat for clusters of the northern spotted owl in three states. Category 1 is for large clusters of 20 or more pairs. Category 2 is for clusters of fewer than 20 pairs. A new analysis of European data suggests that the second category offers much less chance for the long-term survival of a species. The map at left shows regions within the ranges of the northern spotted owl and the distribution of known spotted owl pairs in Washington, Oregon, and California from 1987 to 1991.

STATE AND PROVINCE	CATEGORY 1 No. of sites	CATEGORY 2 No. of sites
CALIFORNIA		
1 California Coast	2	26
2 California Klamath	14	19
3 California Cascades	0	22
Total	16	67
OREGON		
4 Oregon Klamath	6	4
5 E. Oregon Cascades	1	8
6 W. Oregon Cascades	17	5
7 Oregon Coast Range	3	14
Total	27	31
WASHINGTON		
8 E. Wash. Cascades	3	17
9 W. Wash. Cascades	6	18
10 W. Wash. Lowlands	0	1
11 Olympic Peninsula	1	5
Total	10	41
Three-state total	**53**	**139**

Reprinted by permission from *The New York Times,* December 14, 1993, p. C1, with data from the Federal Environmental Management Assessment Team.

countries of the world. Using MINITAB, or another statistical computing package, do the following:

a. Compute the average difference in age at first marriage between men and women for all the countries in the table. This will be our population age difference at first marriage.

b. Select 50 random samples of 20 countries each and use each sample in computing a 95 percent confidence interval for the average age difference for the population of countries. In how many of your samples does your interval contain the actual mean computed in (*a*)?

c. What is the difference between the population parameter defined above and the average world difference in age at first marriage between men and women? (Hint: population.)

9. Using the entire population of countries in the table in Problem 58 in Chapter 4, generate a random sample of 25 countries and test the null hypothesis that the difference in "average annual change, 1950–85" in the mean age at first marriage for women and men is zero versus the alternative that it is not zero. Then compute the actual population parameter of interest. Is the parameter zero? Discuss your findings.

Sexes Equal on South Sea Isle

By JOHN NOBLE WILFORD

Anthropologists have long been attracted to South Pacific islands for more than the warm breezes and sparkling lagoons. They go to such places seeking in the simpler societies the rudiments of human community, sometimes glimpsing a rough-hewn harmony beyond the experience of more complex societies.

So it was for Dr. Maria Lepowsky, who in 1978 became the first and only anthropologist to live with the people of Sudest Island—or to use the local name, Vanatinai, meaning motherland. This is a sparsely populated island in the Louisiade Archipelago of Papua New Guinea, which extends southeast of New Guinea and separates the Solomon and the Coral seas. The people there are an anthropologist's dream because they have had only minimal contact with Western colonialism and missionaries.

The longer Dr. Lepowsky observed the people, sharing their daily lives for two years and learning their rituals and ideologies, the more she realized that life on Vanatinai was different from other societies in one fundamental respect. Men and women were living and working as virtual equals. This was, she concluded, a striking example of what anthropologists call a gender egalitarian culture, and perhaps the first one to be studied in detail by any anthropologist.

"It is not a place where men and women live in perfect harmony and where the privileges and burdens of both sexes are exactly equal," Dr. Lepowsky said, "but it comes close."

Close enough, she said, to challenge the position of some theorists in anthropology that male dominance is universal or somehow inherent in human cultures and that only its forms and intensity vary. The new findings suggest instead that the island's culture could serve as a model of what a sex-egalitarian society would be like and a counter-model to relations between the sexes and ideologies in nearly all other cultures.

On Vanatinai, Dr. Lepowsky reported, women figure prominently in the acquisition of ceremonial valuables like shell necklaces and greenstone ax blades. They may lead their own expeditions to other islands to trade for these goods. It is a "dramatic indicator of the sexually egalitarian nature of the culture," she said.

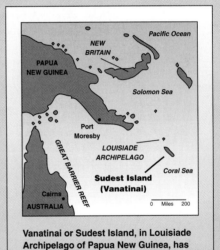

Vanatinai or Sudest Island, in Louisiade Archipelago of Papua New Guinea, has had minimal contact with the West.

Reprinted by permission from *The New York Times,* March 29, 1994, p. C1.

10. Read the article on the South Sea island of Vanatinai. Suppose that the worth, in some currency, of a random sample of ceremonial valuables each belonging to a man or a woman of the same family is assessed and the figures are as given below. Test for equality of assessed values for men's and women's ceremonial possessions on Vanatinai. What is the *p*-value, and what is its meaning in this context?

Family Number

	1	2	3	4	5	6	7	8	9	10
Man	28	17	19	45	29	51	18	53	72	65
Woman	30	13	22	40	30	47	26	50	69	71

11. Refer back to Problem 10. Give a 90 percent confidence interval for the average difference in value of ceremonial possessions of men and women on the island of Vanatinai.

12. Give some examples where pairing observations may not make sense.

13. What is experimental design? Discuss.

7.3 A TEST FOR THE DIFFERENCE BETWEEN TWO POPULATION MEANS, USING INDEPENDENT RANDOM SAMPLES

It may not always be possible to pair our observations from two populations and carry out a test on the differences. In cases where the paired-observations test of the previous section does not apply, we may still compare the means of two populations by drawing two independent *random samples* from the two populations of interest. The samples drawn from the populations do not have to be the same size. We denote the size of the sample from the first population by n_1 and the size of the sample from the second population by n_2.

Several different hypotheses may be tested. We may want to test whether or not the two population means are equal, with no prior intention to prove that one mean is greater than the other (a two-tailed test), or we may want to test whether the mean of one particular population is greater than the mean of the other population (a one-tailed test). Or we may wish to test whether or not the difference between the two population means is greater than (or less than) some particular number, D.

We will call the three basic hypothesis-testing situations, described above, situation I, situation II, and situation III. These situations are not exhaustive. Other possible tests are simple extensions of the three situations discussed here. The test we show, for example, may be a right-tailed test; the analogous left-tailed test is also possible, and you should have no problem carrying it out as well.

> Hypothesis-testing situation I:
>
> $$H_0: \mu_1 - \mu_2 = 0$$
>
> $$H_1: \mu_1 - \mu_2 \neq 0$$

Situation I is the most common test for the difference between two population means, μ_1 and μ_2. The *null hypothesis* states that the two means are equal (their difference is 0), while the two-tailed alternative states that the two population means are not equal.

> Hypothesis-testing situation II:
>
> $$H_0: \mu_1 - \mu_2 \leq 0$$
>
> $$H_1: \mu_1 - \mu_2 > 0$$

Situation II is another common set of hypotheses. The null hypothesis says that the mean of population 1 is equal to or smaller than the mean of population 2. The alternative hypothesis states that the mean of population 1 is greater than the mean of population 2. The alternative hypothesis is the one we try to prove. We want to prove that the mean of population 1 is greater; we choose the label 1 for the population we believe to have the higher mean. The test is right-tailed.

> Hypothesis-testing situation III:
>
> $$H_0: \mu_1 - \mu_2 \leq D$$
>
> $$H_1: \mu_1 - \mu_2 > D$$

Situation III is not very common. It is a special-purpose test where we want to prove that the mean of one population is greater than the mean of the other population by more than D units. (A two-tailed test of the hypothesis that the difference between μ_1 and μ_2 is equal to D is also possible.)

The test statistic in all three hypothesis-testing situations is the same. If we assume normal populations with known population variances, the statistic is a z statistic.

> A test statistic for the difference between two population means:
>
> $$z = \frac{(\bar{x}_1 - \bar{x}_2) - (\mu_1 - \mu_2)_0}{\sqrt{\dfrac{\sigma_1^2}{n_1} + \dfrac{\sigma_2^2}{n_2}}}$$
>
> The term $(\mu_1 - \mu_2)_0$ is the difference between μ_1 and μ_2 under the null hypothesis. It is equal to zero in situations I and II, and to the prespecified value D in situation III. The term in the denominator is the standard deviation of the difference between the two sample means. (It relies on the assumption that the two samples are independent.) If the population variances, σ_1^2 and σ_2^2, are unknown, they are replaced by the sample variances, s_1^2 and s_2^2. For large samples, the standard normal distribution may be assumed.

Note that when the sample sizes are small, using the sample variances in the square root in the denominator instead of the unknown population variances will not result in the statistic following an exact t *distribution*. In such cases, however, it is possible to estimate the distribution using an approximate t distribution with degrees of freedom that may not be a whole number. If the analysis is done by computer, the computer program will do this estimation of the t distribution correctly and will give the appropriate degrees of freedom. Doing this without a computer is difficult. In such cases you may use an approximate t distribution with degrees of freedom being the smallest of $n_1 - 1$ and $n_2 - 1$. This approximation is conservative in the sense that the obtained *p-value* will be equal to or larger than the actual *p*-value. It is highly recommended that a computer be used in such cases so that the more appropriate approximation by a t distribution will be applied.

Example 7.2 demonstrates the test for situation I.

214

266

EXAMPLE 7.2

Until a few years ago, the market for consumer credit was considered to be segmented. Higher-income, higher-spending people tended to be American Express cardholders, and lower-income, lower-spending people were usually Visa cardholders. In the last few years, Visa has intensified its efforts to break into the higher-income segments of the market by using magazine and television advertising to create a high-class image. Recently, a consulting firm was hired by Visa to determine whether or not average monthly charges on the American Express Gold Card are approximately equal to the average monthly charges on Preferred Visa. A random sample of 1,200 Preferred Visa cardholders was selected, and it was found that the sample average monthly charge was $\bar{x}_1 = \$452$ and the sample standard deviation was $s_1 = \$212$. An independent random sample of 800 Gold Card members revealed a sample mean $\bar{x}_2 = \$523$ and $s_2 = \$185$. (Anyone who held both the Gold Card and Preferred Visa was excluded from the study.) Is there evidence to conclude that the average monthly charge in the entire population of American Express Gold Card members is different from the average monthly charge in the entire population of Preferred Visa cardholders?

SOLUTION

Since we have no prior suspicion that either of the two populations may have a higher mean, the test is two-tailed. The null and alternative hypotheses are therefore as stated in situation I:

$$H_0: \mu_1 - \mu_2 = 0$$
$$H_1: \mu_1 - \mu_2 \neq 0$$

The value of our test statistic is computed as follows:

$$z = \frac{(452 - 523) - 0}{\sqrt{\dfrac{212^2}{1,200} + \dfrac{185^2}{800}}} = -7.926$$

This value of the z statistic falls in the left-hand rejection region for any commonly used α, and the p-value is very small. We conclude that there is a statistically significant difference in average monthly charges between Gold Card and Preferred Visa cardholders, with American Express Gold Card members spending more, on average, than their Preferred Visa counterparts. Note that this does not imply any *practical significance*. That is, while a difference in average spending in the two populations may exist, we cannot necessarily conclude that this difference is large. The test is illustrated in Figure 7.2.

FIGURE 7.2   Carrying Out the Test of Example 7.2

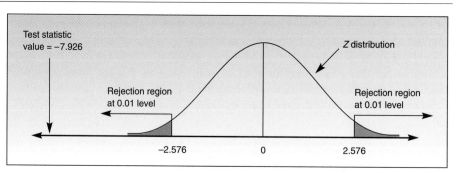

Confidence Intervals

Recall from Chapter 6 that there is a strong connection between hypothesis tests and confidence intervals. The same holds true in dealing with the difference between two population means.

A $(1 - \alpha)100\%$ confidence interval for the difference between two population means, $\mu_1 - \mu_2$, using independent random sampling:

$$(\bar{x}_1 - \bar{x}_2) \pm z_{\alpha/2} \sqrt{\frac{\sigma_1^2}{n_1} + \frac{\sigma_2^2}{n_2}}$$

If σ_1^2 and σ_2^2 are unknown, they are replaced by s_1^2 and s_2^2. When sample sizes are small, an approximate t distribution applies, as discussed earlier.

The equation should be intuitively clear. The bounds on the difference between the two population means are equal to the difference between the two sample means, plus or minus the z coefficient for $(1 - \alpha)100\%$ confidence times the standard deviation of the difference between the two sample means (which is the expression with the square root sign).

In the context of Example 7.2, a 95 percent confidence interval for the difference between the average monthly charge on the American Express Gold Card and the average monthly charge on the Preferred Visa Card is:

$$(523 - 452) \pm 1.96 \sqrt{\frac{212^2}{1,200} + \frac{185^2}{800}} = [53.44, 88.56]$$

The consulting firm may report to Visa it is 95 percent confident that the average American Express Gold Card monthly bill is anywhere from $53.44 to $88.56 higher than the average Preferred Visa bill.

In MINITAB, the command TWOSAMPLE will conduct the statistical analysis described in this section. As in the one-sample case of Chapter 6, ALTERNATIVE = -1 specifies a left-tailed test, ALTERNATIVE = 1 a right-tailed one, and no ALTERNATIVE (the default) results in a two-tailed test.

The following are teaching evaluation score *averages* (original scores are on a scale of 1 to 5, with 5 the best) for two university professors. Based on several separate classes for each professor, these are assumed to be two independent random samples of their courses. (The data, average scores for classes, are closely normally distributed by the central limit theorem; each class average is based on roughly 40 students.) Is there evidence that the first professor (C1) is better (gets generally higher scores in the population of all classes of this type that he or she might teach) than the second professor (C2)?

EXAMPLE 7.3

```
MTB > SET C1
DATA> 4.1 4.0 3.9 4.2 3.8 3.7 3.9 4.3
MTB > SET C2
```

SOLUTION

```
DATA> 3.5 3.6 3.1 3.7 4.0 3.1 2.8
DATA> END
MTB > TWOSAMPLE C1 C2;
SUBC> ALTERNATIVE = 1.

TWOSAMPLE T FOR C1 VS C2
      N     MEAN    STDEV  SE MEAN
C1 8       3.987   0.203   0.072
C2 7       3.400   0.416   0.16

95 PCT CI FOR MU C1 – MU C2: (0.189, 0.99)

TTEST MU C1 = MU C2 (VS GT): T = 3.40 P = 0.0047 DF = 8
```

As seen from the *p*-value, if the assumptions are correct, there is definitely strong statistical evidence that the answer to the above question is yes. Note that using MINITAB solved the problem of having to approximate the distribution using a *t* distribution. The computer reports the *p*-value based on all the information.

PROBLEMS

14. An important aspect in the area of earthquake intensity prediction is the comparison of tremors across various geological regions. In California, earthquake research focuses on two major metropolitan areas: San Francisco Bay and the Los Angeles Basin. Researchers at the University of California Earthquake Center want to compare these two areas, each of which is near an important geological fault line. The following data, assumed to be two independent random samples from the two areas (SF and LA, respectively), are the intensity, on the Richter scale, of tremors felt during several months in 1994:

    SF: 3.4, 1.2, 2.8, 3.3, 2.1, 1.0, 3.7, 2.8, 1.5, 1.7, 3.4, 4.0, 3.6, 1.7

    LA: 1.7, 4.2, 3.1, 3.2, 1.1, 1.2, 2.4, 1.0, 1.0, 0.8, 1.9, 3.7, 2.6, 0.9, 2.3, 3.0

    Is there evidence that the population average tremor strength is different in these two areas? If so, *how* different is it? How sure are you of your result? Use a computer.

15. Compare the two-sample procedures with the paired-sample ones of the previous section.

16. In a recent study, behavioral, cardiovascular, and self-report of cognitive and affective responses to two interpersonal challenges were examined among 20 men with a positive (FH+) and 20 with a negative (FH−) family history of hypertension (see table). Conduct a two-tailed test for differences in responses to male interactions and then to female interactions, for "positive nonverbal" FH+ and FH− men.

17. Using the table for Problem 16, conduct similar tests for the "positive verbal" variables.

18. Again referring to the table for Problem 16, conduct similar tests for the "negative nonverbal" variables.

19. Read the article "Sunscreen Lowers Risk of Melanoma." Assume an average of 3.5 keratoses for the 294 persons assigned the sunscreen and a standard deviation of 1.2. For the 294 people assigned the placebo, assume an average of 5.8 keratoses and a standard deviation of 1.4. Conduct a one-tailed test designed to prove that the sunscreen is effective in reducing the average number of sun-induced keratoses.

20. An archaeologist wants to test whether there is a difference in the average lead content of Roman glass made during the time of the Republic and that made later, during early Imperial times. Two samples of Roman glass, 12 pieces from each of the two periods, are available for analysis and are assumed to be independent random samples. The lead content, in percent, for each piece of glass is as follows:

Behavioral Measures for FH+ and FH− Subjects   Table for Problem 16

Measure	Male Interaction		Female Interaction	
	M	SD	M	SD
Positive nonverbal				
FH−	14.10	4.15	15.80	4.29
FH+	9.80	5.22	10.45	4.05
Positive verbal				
FH−	17.90	4.17	18.60	4.02
FH+	18.55	5.48	19.25	5.99
Negative nonverbal				
FH−	1.55	1.84	1.25	1.07
FH+	4.00	2.88	3.85	3.08
Negative verbal				
FH−	1.45	1.76	1.25	1.07
FH+	4.35	2.83	3.85	3.08

Reprinted from E. Semenchuk and K. Larkin, "Behavioral Responses to Interpersonal Challenges," *Health Psychology* 12, no. 5 (1993), p. 416.

Sunscreen Lowers Risk of Melanoma

Australian Study Offers First Proof That Blocking Rays Reduces Cancer Cells

BOSTON, Oct. 13 (AP)—Scientists have gathered the first direct proof that using a sunscreen helps prevent skin cancer.

Australian researchers have completed a summerlong experiment showing that people who used sunscreen before going outside cut their chances of developing the first signs of skin cancer.

The study was conducted on 588 men and women who were randomly assigned to use either a sunscreen with a sun-protection factor of 17 or a lookalike lotion without the blocker. The experiment was run from September 1991 through March 1992, one Australian summer.

Then they were checked for solar keratoses, small, wart-like growths that result from overexposure to the sun. These growths are forerunners of squamous-cell skin cancer, a common, usually harmless form of skin cancer.

Reprinted by permission from *The New York Times,* October 14, 1993, p. B9.

Republic: 1.2, 1.4, 1.3, 1.1, 1.2, 1.2, 1.7, 1.2, 1.4, 1.0, 1.3, 1.4

Imperial: 1.7, 1.1, 2.3, 1.8, 1.3, 1.2, 1.0, 1.9, 2.0, 2.9, 1.7, 2.5

Conduct the test and state your conclusion. Use MINITAB or another computing package.

21. Two tomato fertilizers are compared to see if one is better than the other. The weight measurements of two independent random samples of tomatoes grown using each of the two fertilizers (in ounces) are as follows:

Fertilizer A: 12, 11, 7, 13, 8, 9, 10

Fertilizer B: 13, 11, 10, 6, 7, 4, 10

Conduct the test. Use a computer.

22. Many companies that cater to teenagers have learned that young people respond to commercials that provide dance-beat music, adventure, and a fast pace, rather than words. In one test, a group of 128 teenagers was shown commercials featuring rock music, and the purchasing frequency of the advertised products over the following month was recorded as a single score for each person in the group. Then a group of 212 teenagers was shown commercials for the same products, but with the music replaced by verbal persuasion. The purchase frequency scores of this group were computed as well. The results for the music group were $\bar{x} = 23.5$ and $s = 12.2$, and the results for the verbal group were $\bar{x} = 18.0$ and $s = 10.5$. Assume that the two groups were randomly selected from the entire teenage consumer population. Using the $\alpha = 0.01$ level of significance, test the null hypothesis that both methods of advertising are equally effective versus the alternative hypothesis that they are not equally effective. If you conclude that one method is better, state which one it is, and explain how you reached your conclusion.

23. The Marcus Robert Real Estate Company wants to test whether the average sale price of residential properties in a certain size range in Bel Air, California, is approximately equal to the average sale price of residential properties of the same size range in Marin County, California. The company gathers data on a random sample of 32 properties in Bel Air and finds $\bar{x} = \$345,650$ and $s = \$48,500$. A random sample of 35 properties in Marin County gives $\bar{x} = \$289,440$ and $s = \$87,090$. Is the average sale price of all properties in both locations approximately equal or not? Explain.

24. The photography department of a glamour magazine needs to choose a camera. Of the two models the department is considering, one is made by Nikon and the other by Minolta. The department contracts with an agency to determine if one of the two models gets a higher average performance rating by professional photographers, or whether the average performance ratings of these two cameras are not statistically different. The agency asks 60 different professional photographers to rate one of the cameras (i.e., 30 photographers rate each model). The ratings are on a scale of 1 to 10 (with ten being the best). The average sample rating for Nikon is 8.5, and the sample standard deviation is 2.1. For the Minolta sample, the mean is 7.8, and the standard deviation is 1.8. Is there a difference between the average population ratings of the two cameras? If so, which one is rated higher?

25. Construct a 95 percent confidence interval for the average difference in Problem 21. Compare the result with the one you obtained in that problem.

26. Construct a 99 percent confidence interval for average difference using the data in Problem 22 and compare the results with those of the test conducted in that problem.

27. Construct a 90 percent confidence interval for the average difference using the information in Problem 23. Does the result accord with your conclusion in that problem?

28. A volunteer tries two methods of cracking equal-sized pieces of the numerical puzzle described in the article on the next page, and wants to find out whether the two methods are equally fast or whether one of them is faster than the other. The data, in computer seconds to crack the codes, are as follows:

Method A: 12, 14, 11, 17, 9, 10, 11, 16, 14, 14, 13, 10, 17, 10, 11, 15, 14

Method B: 10, 10, 12, 9, 15, 16, 8, 17, 9, 10, 11, 12, 9, 11, 13

Conduct the appropriate test and state your conclusions.

29. The table on the next page shows the actual snowfall in inches for the winter seasons of 1976–77 through 1993–94 in southern Vermont. A commonly held belief of the people in southern Vermont is that their climate began to change drastically starting in 1984–85, characterized by less snow. Assume that the data in the table are independent random samples, one from a population of possible snowfall characteristic of pre–1984–85 seasons, and one characteristic of 1984–85 and beyond. Use the data to test the Vermonters' hypothesis. Additionally, construct a bar chart for these data and a box plot. Describe what you are seeing in the graphs.

The Assault on 114,381,625,757,888,867,669,235,779, 976,146,612,010,218,296,721,242,362,562,561,842,935, 706,935,245,733,897,830,597,123,563,958,705,058,989, 075,147,599,290,026,879,543,541

By GINA KOLATA

Mathematicians say they are close to breaking a cryptographic stronghold that was not expected to fall for many years. The item is a 129-digit number that was first described in 1977 as proof of the security of a new public cryptographic system.

The number is known for short as RSA 129 after the initials of its inventors and its number of digits. The new coding system depended on very large numbers that were multiples of two primes, a prime being a number divisible only by itself and one.

The code could be cracked only by finding the component primes, one of the most mathematically difficult tasks imaginable. The inventors proposed RSA 129 as an example. Only they knew its component primes, and they asserted it would take others at least 40 quadrillion years to factor it, using the best methods and the fastest computers that were then available.

But over the years the number proposed as uncrackable simply became a challenge. Eight months ago, with the power of computers growing, cryptography enthusiasts proposed a cunning scheme to attack it. They would break the problem into millions of tiny pieces and then use volunteers recruited on the Internet, an international electronic mail system, to do the calculations on their computers, at night or in other fallow periods.

RSA 129 has not crumbled yet. But several factoring experts said that so many of the calculations have already been completed that they are confident the solution will emerge in a few weeks.

The inventors of RSA are Dr. Ronald Rivest of the Massachusetts Institute of Technology, Dr. Adi Shamir of the Weizmann Institute of Science in Rehovoth, Israel, and Dr. Leonard Adelman of the University of Southern California.

Reprinted by permission from *The New York Times,* March 22, 1994, p. C1.

Table for Problem 29

Season	Snowfall (inches)
1976–77	136
1977–78	138
1978–79	55
1979–80	70
1980–81	169
1981–82	106
1982–83	137
1983–84	121
1984–85	131
1985–86	155
1986–87	116.2
1987–88	145
1988–89	103
1989–90	137
1990–91	95
1991–92	95
1992–93	189
1993–94	160

7.4 A TEST FOR THE DIFFERENCE BETWEEN TWO POPULATION MEANS, ASSUMING EQUAL POPULATION VARIANCES

When the population variances σ_1^2 and σ_2^2 are assumed to be equal, there is another test that can be done for the difference between two population means. This test is especially useful in the case of small samples, because it allows us to conduct a test for the difference between two population means without having to worry about the degrees of freedom of the approximate t statistic. In addition to assuming equal (but usually unknown) population variances, we assume that the two populations of interest have, approximately, a normal distribution.

We will concentrate on the three situations we considered in the previous section:

Situation I: $H_0: \mu_1 - \mu_2 = 0$
                 $H_1: \mu_1 - \mu_2 \neq 0$

Situation II: $H_0: \mu_1 - \mu_2 \leq 0$
                 $H_1: \mu_1 - \mu_2 > 0$

Situation III: $H_0: \mu_1 - \mu_2 \leq D$
                 $H_1: \mu_1 - \mu_2 > D$

We note again that other tests are possible as well. While the tests in situations II and III are right-tailed tests, it is possible in each of these situations to conduct a similar left-tailed test. It is also possible to conduct a two-tailed test, in situation I, using a specified value D as the null-hypothesized difference between μ_1 and μ_2, where D is not necessarily zero. Before we describe the test statistic and its distribution, let's analyze the situation giving rise to our statistic.

We select two independent random samples: a random sample from population 1 and a random sample from population 2. We denote the sample sizes by n_1 and n_2, respectively. Once we compute the mean for each sample, $\bar{x}_1$ and $\bar{x}_2$, we lose one degree of freedom in each sample with respect to the computation of the sample standard deviations, s_1 and s_2. Figure 7.3 shows why, with two independent samples from which the two sample means are computed, the total number of degrees of freedom associated with deviations from the sample means is as follows:

$$\text{df} = (n_1 - 1) + (n_2 - 1) = n_1 + n_2 - 2$$

This parameter, df, is associated with the *pooled estimate* of the common population variance, defined as follows.

> A **pooled estimate** of the common population variance, based on a sample variance s_1^2 from a sample of size n_1 and a sample variance s_2^2 from a sample of size n_2, is given by:
>
> $$s_p^2 = \frac{(n_1 - 1)s_1^2 + (n_2 - 1)s_2^2}{n_1 + n_2 - 2}$$

We assume that the two population variances, σ_1^2 and σ_2^2, are equal. We denote this common variance of the two populations by σ^2. The two sample variances, s_1^2 and

The Degrees of Freedom Associated with Deviations from the Sample Means
of Two Independent Samples

FIGURE 7.3

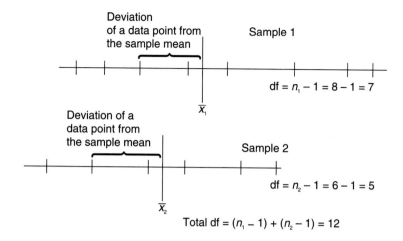

Total df $= (n_1 - 1) + (n_2 - 1) = 12$

s_2^2, are then estimates of the same quantity, σ^2. What we are looking for is a *pooled*
variance estimate, s_p^2. The variance estimator from sample 1 has $n_1 - 1$ degrees of
freedom, and the variance estimator from sample 2 has $n_2 - 1$ degrees of freedom. We
want to pool the two variance estimators S_1^2 and S_2^2 in such a way that the relative
weights given to the two estimates produced will be proportional to the degrees of
freedom of the estimators. (Obviously, an estimate based on a larger sample—larger
df—should be given proportionally more weight in computing the pooled variance
than an estimate based on smaller df.)

The equation should make sense to you. It gives a weighted average of two sample
variances, where the weights are the degrees of freedom upon which each estimate is
based, and the total weight—the denominator in the formula—is the sum of the
degrees of freedom associated with the two estimates: df $= (n_1 - 1) + (n_2 - 1) =$
$n_1 + n_2 - 2$.

The estimate of the standard error of $(\overline{X}_1 - \overline{X}_2)$ is given by:[2]

$$\sqrt{s_p^2\left(\frac{1}{n_1} + \frac{1}{n_2}\right)}$$

The expression in the above equation has the same function as $s/\sqrt{n}$ in the case of a
single-sample inference about a population mean. We are now ready to define the test
statistic.

[2]Our assumption of independent random sampling from the two populations implies that $\overline{X}_1$ and $\overline{X}_2$
are *independent* random variables. Thus, the variance of $(\overline{X}_1 - \overline{X}_2)$ is equal to $V(\overline{X}_1 - \overline{X}_2) = V(\overline{X}_1) +$
$V(\overline{X}_2) = (\sigma^2/n_1) + (\sigma^2/n_2) = \sigma^2[(1/n_2) + (1/n_2)]$, where σ^2 is the common population variance. This
common variance is estimated by s_p^2.

> Test statistic for the difference between two population means, assuming equal population variances:
>
> $$t = \frac{(\bar{x}_1 - \bar{x}_2) - (\mu_1 - \mu_2)_0}{\sqrt{s_p^2 \left(\dfrac{1}{n_1} + \dfrac{1}{n_2}\right)}}$$
>
> where $(\mu_1 - \mu_2)_0$ is the difference between the two population means under the null hypothesis (zero or some number, D). The degrees of freedom of the test statistic t are $n_1 + n_2 - 2$ (the degrees of freedom associated with s_p^2, the pooled estimate of the population variance).

For large samples, we may use the standard normal distribution instead of the t distribution with $n_1 + n_2 - 2$ degrees of freedom. Note that the test statistic in this equation follows the general form of a standardized test statistic given in Chapter 6: Test statistic = (Estimate − Hypothesized value of parameter)/(Standard error of estimator).

EXAMPLE 7.4

A recent article in the British journal *Lancet* reports that babies who were fed mother's milk tended to have a higher IQ than formula-fed babies. Suppose that two groups of babies are compared, a group fed mother's milk and a group fed formula. The resulting IQ scores are as follows:

Mother: 121, 105, 111, 119, 108, 101, 90, 131, 106, 112

Formula: 102, 110, 107, 98, 89, 103, 86, 117, 113, 87

Mother: mean = 110.4, s.d. = 11.4

Formula: mean = 101.2, s.d. = 11.03

Is there a significant difference between the two groups?

SOLUTION

Since the sample sizes are small, we need to use the t distribution with degrees of freedom $n_1 + n_2 - 2 = 10 + 10 - 2 = 18$. We make the assumption that the standard deviations of IQ scores for both populations are equal. Using the equation, the pooled variance of scores is:

$$s_p^2 = [9(11.4)^2 + 9(11.03)^2]/18 = 125.81$$

The estimated standard error of the difference between the two sample means is:

$$\sqrt{125.81(2/10)} = 5.02$$

Testing for a difference between the two population means, we get:

$$t = (110.4 - 101.2)/5.02 = 1.82$$

If this is carried out as a one-tailed test to the right (that is, we hypothesize before the experiment that mother's milk–fed babies may be more intelligent), then the critical point at the 5 percent level is 1.734 and the result is significant (although it is not significant at lower levels).

Confidence Intervals

As usual, it is possible to construct confidence intervals for the parameter in question—here, the difference between the two population means. The confidence interval for this parameter is based on the t distribution with $n_1 + n_2 - 2$ degrees of freedom (or z, used as an approximation when df is large).

A $(1 - \alpha)100\%$ confidence interval for $(\mu_1 - \mu_2)$, assuming equal population variances:

$$(\bar{x}_1 - \bar{x}_2) \pm t_{\alpha/2}\sqrt{s_p^2\left(\frac{1}{n_1} + \frac{1}{n_2}\right)}$$

The confidence interval in the above equation has the usual form: Estimate $\pm$ Distribution coefficient $\times$ Standard error of estimator.

In Example 7.4 let's compute a 95 percent confidence interval for the difference between the two means. The 95 percent confidence interval for $(\mu_1 - \mu_2)$ is:

$$(\bar{x}_1 - \bar{x}_2) \pm t_{0.025}\sqrt{s_p^2\left(\frac{1}{n_1} + \frac{1}{n_2}\right)} = 9.2 \pm 2.101 \times 5.02 = [-1.347, 19.74]$$

This range of values contains zero. This means that if our test had been two-tailed, it would have resulted in nonrejection of the null hypothesis of no difference in average IQ scores using $\alpha = 0.05$.

Note that in MINITAB, adding POOLED to the TWOSAMPLE command will perform the analysis described in this section.

---

30. See the article on Pooh and the president. Suppose the rating for the president is an average score for a random sample of eight children who watched him and their standard deviation[o]

PROBLEMS

Wherein a President Easily Outpolls Pooh

Say what you will about his economic plan, President Clinton is outperforming Winnie-the-Pooh.

In an ABC special broadcast on Saturday morning, the President answered questions from a group of children gathered in the White House. The broadcast, from 11:30 A.M. to 1:30 P.M., received an average rating of 6.4 in the nation's largest media markets, according to Nielsen Media Research.

The previous week, "The New Adventures of Winnie-the-Pooh," which usually runs on ABC from 11:30 A.M. to noon, got a 3.3 rating. And "Darkwing Duck," which usually runs from noon to 12:30 P.M., got a 2.7 rating.

Reprinted by permission from *The New York Times*, February 23, 1993, p. A18.

was 2.1. Suppose further that for Pooh the result is an average based on a random sample of 12 children who watched him and their standard deviation was 2.3. Assume equal population variances and test for a significant difference between average rating of all children this age for President Clinton versus Winnie-the-Pooh.

31. Ikarus, the Hungarian bus maker, has lost its important Soviet market and is reported on the verge of collapse. The company is now trying a new engine in its buses and has gathered the following random samples of miles-per-gallon figures for the old engine versus the new:

    Old engine: 8, 9, 7.5, 8.5, 6, 9, 9, 10, 7, 8.5, 6, 10, 9, 8, 9, 5, 9.5, 10, 8

    New engine: 10, 9, 9, 6, 9, 11, 11, 8, 9, 6.5, 7, 9, 10, 8, 9, 10, 9, 12, 11.5, 10, 7, 10, 8.5

    Is there evidence that the new engine is more economical than the old one? (Assume equal population variances.)

32. *Air Transport World* recently named the Dutch airline KLM "Airline of the Year." One measure of the airline's excellent management is its research effort in developing new routes and improving service on existing routes. The airline wanted to test the profitability of a certain transatlantic flight route and offered daily flights from Europe to the United States over a period of six weeks on the new proposed route. Then, over a period of nine weeks, daily flights were offered from Europe to an alternative airport in the United States. Weekly profitability data for the two samples were collected, under the assumption that these may be viewed as independent random samples of weekly profits from the two populations with equal variances. (One population is flights to the proposed airport, and the other population is flights to an alternative airport.) For the proposed route, $\bar{x} = \$96,540$ per week and $s = \$12,522$. For the alternative route, $\bar{x} = \$85,991$ and $s = \$19,548$. Test the hypothesis that the proposed route is more profitable than the alternative route. Use a significance level of your choice.

33. Authors T. Peters and R. Waterman state in their book *In Search of Excellence* that the giant advertising firm of Ogilvy and Mather is more concerned with customer satisfaction than it is with company profits.[3] Suppose a test is carried out to determine whether or not new management decisions at Ogilvy and Mather increase average customer satisfaction. To test this claim, a random sample of company clients is polled before an important management decision, and customer satisfaction is measured for this sample on a scale of 1 to 10. Then, some time after the management decision has been made and a new company policy implemented, another sample of clients is polled, and their satisfaction scores are computed (on the same scale). Letting the subscript b denote "before the new decision" and the subscript a denote "after the new decision," the results of the surveys are $n_b = 18$, $\bar{x}_b = 7.4$, $s_b = 1.3$; $n_a = 23$, $\bar{x}_a = 8.2$, $s_a = 2.4$. Using these data, do you believe that customer satisfaction is increased, on the average, after the new management decision and the resulting new company policy? Assume normal populations with equal variances.

34. Zim Container Service ships containers across both the Atlantic and the Pacific oceans. In 1993, the company had to decide whether to include Kingston, Jamaica, as a port of call instead of Savannah, Georgia, where the company's ships had been calling for some time. Zim's general manager of operations decided to use statistics to determine which of the two ports of call—Savannah or Kingston—would have a greater demand for shipped containers, or whether the average number of containers demanded at the two ports might

---

[3] T. Peters and R. Waterman, *In Search of Excellence* (New York: Harper & Row, 1982).

be approximately equal (in which case the decision would be based on the cost of service). The manager had data on the last 17 calls at Savannah. The data, in number of containers unloaded per call, were 9, 6, 7, 7, 8, 7, 9, 4, 7, 6, 6, 5, 7, 5, 9, 8, 6. Then, as a test run, company ships unloaded containers at Kingston during 9 arrivals at that port. The data (number of containers unloaded at each trip) were 10, 5, 8, 8, 9, 10, 4, 9, 11. Use these data to conduct the test, and advise Zim's general manager of operations as to which port should be the regular port of call, or whether the ports will have approximately equal demand for containers.[4] Assume normal populations with equal variances.

35. The U.S. Census Bureau's *Current Industrial Report on Confectionery* (1994) indicates that the average per capita consumption of candy in the United States was about 19 pounds per person per year for 1992–94 and about 17 pounds per person per year for 1989–91. Suppose that an industry analyst wants to test the claim that average candy consumption in the United States has increased by an average of over 2 pounds per person per year. The analyst is able to obtain a random sample of 30 data items for 1989–91, and finds a sample mean of 16.54 pounds per person per year and a sample standard deviation of 5.3 pounds. A similar random sample of consumer data consisting of 30 items for 1992–94 gives a sample mean of 21.2 pounds per person per year and a sample standard deviation of 3.5 pounds. Using these data, do you believe that Americans increased their candy consumption by over 2 pounds per person per year from one period to the other?

36. For Problem 35, give a 95 percent confidence interval for the increase in average candy consumption between the two time periods in question.

37. The power of supercomputers derives from the idea of parallel processing. Engineers at Cray Research are interested in determining whether one of two parallel processing designs produces faster average computing time, or whether the two designs are equally fast. The following are the results, in seconds, of independent random computation times using the two designs.

Design 1	Design 2
2.1, 2.2, 1.9, 2.0, 1.8, 2.4, 2.0, 1.7, 2.3, 2.8, 1.9, 3.0, 2.5, 1.8, 2.2	2.6, 2.5, 2.0, 2.1, 2.6, 3.0, 2.3, 2.0, 2.4, 2.8, 3.1, 2.7, 2.6

Assume that the two populations of computing time are normally distributed and that the two population variances are equal. Is there evidence that one parallel processing design allows for faster average computation than the other? Use a computer.

38. Construct a 95 percent confidence interval for the difference in Problem 37.

39. The senior vice president for marketing at Westin Hotels believes that the company's recent advertising of the Westin Plaza in New York has increased the average occupancy rate at that hotel by at least 5 percent. To test the hypothesis, a random sample of daily occupancy rates (in percentages) before the advertising is collected. A similar random sample of daily occupancy rates is collected after the advertising took place. The data are as follows:

Before Advertising (%)	After Advertising (%)
86, 92, 83, 88, 79, 81, 90 76, 80, 91, 85, 89, 77, 91 83	88, 94, 97, 99, 89, 93, 92 98, 89, 90, 97, 91, 87, 80 88, 96

---

[4] Data provided courtesy of Zim Navigation Company, Ltd.

Assume normally distributed populations of occupancy rates with equal population variances. Test the vice president's hypothesis. Use a computer.

7.5 A LARGE-SAMPLE TEST FOR THE DIFFERENCE BETWEEN TWO POPULATION PROPORTIONS

When sample sizes are large enough that the distributions of the sample proportions $\hat{P}_1$ and $\hat{P}_2$ are both approximated well by a normal distribution, the difference between the two sample proportions is also approximately normally distributed, and this gives rise to a test for equality of two population proportions based on the standard normal distribution. It is also possible to construct confidence intervals for the difference between the two population proportions. Assuming the sample sizes are large and assuming independent random sampling from the two populations, the following are possible hypotheses:

Situation I: $H_0: p_1 - p_2 = 0$
$H_1: p_1 - p_2 \neq 0$

Situation II: $H_0: p_1 - p_2 \leq 0$
$H_1: p_1 - p_2 > 0$

Situation III: $H_0: p_1 - p_2 \leq D$
$H_1: p_1 - p_2 > D$

where D is some number other than 0. (These situations are similar to the ones discussed in the previous two sections; other tests are also possible.)

For tests about the difference between two population proportions, there are two test statistics. One statistic is appropriate when the null hypothesis is that the difference between the two population proportions is equal to (or greater than or equal to, or less than or equal to) zero. This is the case, for example, in situations I and II. The other test statistic is appropriate when the null-hypothesized difference is some number, D, different from zero. This is the case, for example, in situation III (or in a two-tailed test, situation I, with D replacing 0).

Test statistic for the difference between two population proportions where the null-hypothesized difference is zero:

$$z = \frac{(\hat{p}_1 - \hat{p}_2) - 0}{\sqrt{\hat{p}(1 - \hat{p})\left(\frac{1}{n_1} + \frac{1}{n_2}\right)}}$$

where $\hat{p}_1 = x_1/n_1$ is the sample proportion in sample 1 and $\hat{p}_2 = x_2/n_2$ is the sample proportion in sample 2. The symbol $\hat{p}$ stands for the *combined sample proportion in both samples,* considered as a single sample. That is,

$$\hat{p} = \frac{x_1 + x_2}{n_1 + n_2}$$

Note that the 0 in the numerator of the equation is the null-hypothesized difference between the two population proportions; we retain it only for conceptual reasons—to maintain the form of our test statistic: (Estimate − Hypothesized value of the parameter)/(Standard error of the estimator). When we carry out computations using the equation, we will, of course, ignore subtracting zero. Under the null hypothesis that the difference between the two population proportions is zero, both sample proportions $\hat{p}_1$ and $\hat{p}_2$ are estimates of the same quantity, and therefore—assuming, as always, that the null hypothesis is true—we pool the two estimates when computing the estimated standard error of the difference between the two sample proportions: the denominator of the equation.

When the null hypothesis is that the difference between the two population proportions is a number other than zero, we cannot assume that $\hat{p}_1$ and $\hat{p}_2$ are estimates of the same population proportion (because the null-hypothesized difference between the two population proportions is $D \neq 0$); in such cases we cannot pool the two estimates when computing the estimated standard error of the difference between the two sample proportions. In such cases, we use the following test statistic.

Test statistic for the difference between two population proportions when the null-hypothesized difference between the two proportions is some number, D, other than zero:

$$z = \frac{(\hat{p}_1 - \hat{p}_2) - D}{\sqrt{\dfrac{\hat{p}_1(1 - \hat{p}_1)}{n_1} + \dfrac{\hat{p}_2(1 - \hat{p}_2)}{n_2}}}$$

Examples 7.5 and 7.6 demonstrate the use of the test statistics presented in this section.

EXAMPLE 7.5

Finance incentives by the major automakers are reducing banks' share of the market for automobile loans. In 1984, banks made about 53 percent of all car loans, and in 1994, the banks' share was only 43 percent. Suppose that the 1984 data are based on a random sample of 100 car loans, of which 53 were bank loans, and that the 1994 data are also based on a random sample of 100 loans, 43 of which were bank loans. Carry out a two-tailed test of the equality of banks' share of the car loan market in 1984 and in 1994.

SOLUTION

Our hypotheses are those described as situation I, a two-tailed test of the equality of two population proportions. We have H_0: $p_1 - p_2 = 0$ and H_1: $p_1 - p_2 \neq 0$. The null-hypothesized difference between the two population proportions is zero. First we calculate $\hat{p}$, the combined sample proportion:

$$\hat{p} = \frac{x_1 + x_2}{n_1 + n_2} = \frac{53 + 43}{100 + 100} = 0.48$$

We also have: $1 - \hat{p} = 0.52$.

We now compute the value of the test statistic:

$$z = \frac{\hat{p}_1 - \hat{p}_2}{\sqrt{\hat{p}(1-\hat{p})\left(\frac{1}{n_1}+\frac{1}{n_2}\right)}} = \frac{0.53 - 0.43}{\sqrt{(0.48)(0.52)(0.01 + 0.01)}} = 1.415$$

This result is not statistically significant. We conclude that there is no evidence of a change in proportion of bank car loans from 1984 to 1994.

Confidence Intervals

When constructing confidence intervals for the difference between two population proportions, we do not use the pooled estimate because we do not assume that the two proportions are equal. The estimated standard error of the difference between the two sample proportions, to be used in the confidence interval, is the denominator in the previous boxed equation.

> A large-sample $(1 - \alpha)100\%$ confidence interval for the difference between two population proportions:
>
> $$(\hat{p}_1 - \hat{p}_2) \pm z_{\alpha/2} \sqrt{\frac{\hat{p}_1(1-\hat{p}_1)}{n_1} + \frac{\hat{p}_2(1-\hat{p}_2)}{n_2}}$$

In the context of Example 7.5, let's now construct a 95 percent confidence interval for the difference between the proportion of bank car loans in 1984 and 1994:

$$(0.53 - 0.43) \pm 1.96 \sqrt{\frac{(0.53)(0.47)}{100} + \frac{(0.43)(0.57)}{100}} = [-0.038, 0.238]$$

This confidence interval contains zero, which agrees with our nonrejection of the null hypothesis of no difference.

EXAMPLE 7.6

The Integrated Studies Program (ISP) at the Massachusetts Institute of Technology (MIT) is a nationally acclaimed educational program aimed at teaching college freshmen humanities, social sciences, and hands-on skills in a way that integrates these with the more traditional engineering and science disciplines emphasized at engineering schools. As part of a program review effort in 1994, administrators wanted to prove that students who enroll in the ISP for a full year are more successful in coping with the intensive MIT curriculum and that their retention rate at the institute, measured as the percentage who graduate within at most six years, is at least 10 percent higher than that of students who do not enroll in the program.

A tracking study found that out of 198 ISP students, 189 graduated within the time limit; out of a random sample of 210 non-ISP students, 158 graduated within the time limit. These data are shown in the matrix below.

SOLUTION

	ISP	**Non-ISP**
Graduated	189	158
Not Graduated	9	52
Total:	198	210

Our null and alternative hypotheses are:

$$H_0: p_{ISP} - p_N \leq 0.10$$
$$H_1: p_{ISP} - p_N > 0.10$$

Computing the test statistic, we find:

$$z = \frac{(\hat{p}_{ISP} - \hat{p}_N) - 0.10}{\sqrt{\dfrac{\hat{p}_{ISP}(1 - \hat{p}_{ISP})}{n_{ISP}} + \dfrac{\hat{p}_N(1 - \hat{p}_N)}{n_N}}} = \frac{0.9545 - 0.7524}{\sqrt{\dfrac{(0.9545)(0.0455)}{198} + \dfrac{(0.7524)(0.2476)}{210}}} = 3.07$$

This result is statistically significant and establishes the administrators' claim.

PROBLEMS

40. Airline mergers cause many problems for the airline industry. One variable often quoted as a measure of an airline's efficiency is the percentage of on-time departures. Following the merger of Republic Airlines with Northwest Airlines, the percentage of on-time departures for Northwest planes declined from approximately 85 percent to about 68 percent. Suppose that these percentages are based on two random samples of flights: a sample of 100 flights over a period of two months before the merger, of which 85 are found to have departed on time; and a sample of 100 flights over a period of two months after the merger, 68 of which are found to have departed on time. Based on these data, do you believe that Northwest's on-time percentage declined during the period following its merger with Republic?

41. A physicians' group is interested in testing to determine whether more people in small towns choose a physician by word of mouth in comparison with people in large metropolitan areas. A random sample of 1,000 people in small towns reveals that 850 have chosen their physicians by word of mouth; a random sample of 2,500 people living in large metropolitan areas reveals that 1,950 of them have chosen a physician by word of mouth. Conduct a one-tailed test aimed at proving that the percentage of popular recommendation of physicians is larger in small towns than in large metropolitan areas. Use $\alpha = 0.01$.

42. The EPA looked at a random sample of 1,200 children living in areas where the water main pipes are known to contain lead and found that these children had an average blood lead level of 10 micrograms per deciliter (mpd) and a standard deviation of 3 mpd. In a random sample of 5,100 children from among those living in areas where no lead is known to be present in water main pipes, the sample mean was 6 mpd and the standard deviation was 2 mpd. Test for significance and explain your findings. (For more information on lead testing see the *EPA Journal.*)

Benefits of Broccoli Confirmed as Chemical Blocks Tumors

By NATALIE ANGIER

WASHINGTON, April 11—Yes, broccoli is as good for you as it's chopped up to be. Likewise for cauliflower, brussels sprouts, cabbage and other crunchy, tastily bitter members of the cruciferous family of vegetables.

Offering new proof for the protective benefits of chemicals found in many plant foods, scientists have shown that a compound isolated from broccoli, called sulforaphane, blocks the growth of tumors in rats treated with a cancer-causing toxin.

In the latest experiments, the scientists injected 29 rats with either a low dose or a high dose of a synthetic version of sulforaphane, and then followed up with a shot of dimethyl benzanthracene (DMBA), a toxin known to cause mammary tumors. For comparison, 25 rats were injected with DMBA without the benefit of a sulforaphane pretreatment.

The results were striking. While 68 percent of the group that did not receive sulforaphane came down with breast cancer, only 26 percent of the rodents given sulforaphane contracted cancer. What is more, sulforaphane delayed the onset of the cancer, and kept the number and size of any resulting tumors comparatively small.

Reprinted by permission from *The New York Times,* April 12, 1994, p. C11.

---

43. Conduct the statistical test called for by the data reported in the article on broccoli and state your conclusions. What assumptions are you making? Are these assumptions satisfied?

44. The accompanying article on women's bones includes data on women buried in a London church crypt. Suppose that 5 of the buried women had what we would define as weak bones and that 49 of the present-day women had weak bones. Test for significance.

Women's Bones Appear to Have Become Weaker

LONDON—A study of old human bones, unearthed after two centuries in a church crypt, offers a possible clue to the mystery of why so many elderly women today suffer hip fractures associated with osteoporosis.

Millions of people develop osteoporosis, a condition in which bones lose calcium and become thinner and more susceptible to fractures in the spine, hip and wrist. It primarily affects women after menopause when their estrogen levels decline.

Previous studies have shown that osteoporotic hip fractures are much more common today than would be expected, even with increased life expectancy. In Britain, the incidence of hip fractures in women and men has doubled in the last 30 years. Fractures have increased in the United States and Canada as well. One possible explanation is that modern women's bones are weaker than those of their ancestors.

During restoration of Christ Church Spitalfields in the East End of London, scientists from the Wynn Institute for Metabolic Research, Britain's Natural History Museum and University College London, examined the thigh bones of 87 women buried in the crypt from 1729 to 1852. After comparing their findings with bone-density measurements of 294 present-day women, the scientists found that the older bones were indeed stronger than contemporary ones.

Reprinted by permission from *The New York Times,* March 16, 1993, p. C4.

45. To test the benefits of a mediterranean diet to heart patients, a hospital in Lyons, France, instituted two nutritional programs. One program, 358 patients who have had one heart attack, featured a traditional low-fat diet customarily given to heart patients. The other program, 412 patients who have had one heart attack, featured a mediterranean diet, including fruit, olive oil, and bread. When the two-year programs ended in 1994, it was found that 17 of the patients in the traditional-diet program had suffered a second ^{o}heart attack. Of the patients on the mediterranean diet, only 4 had a second heart attack. Are the results significant? Conduct a test aimed at showing that the mediterranean diet reduces the percentage of a second heart attack in the population of people who had suffered one attack by at least 1 percent as compared with a traditional cure diet. Also give a 95 percent confidence interval for the difference between the two population proportions.

46. Study the data in the article and graphs on TV program choice on page 342. Assuming two independent random samples, conduct a test aimed at showing that the population proportion of women who watch current affairs programs is at least 10 percent higher than the population proportion of women who watch sports programs. Also construct a 95 percent confidence interval for the population difference between these two proportions. Also construct a 90 percent confidence interval for the difference in proportion of people classified high education between those who watch soap operas and those who watch current affairs.

47. Study the data in the display on cheetahs on page 343. Assuming 2 random samples of 100,000 genes, test for significance of the difference in proportions of genes with more than one version between cheetahs and humans. Also give a 90 percent confidence interval for the difference between the two proportions.

48. Refer to the article on the study of leukemia risk and day care. Assume that of the 136 children with leukemia, 45 were in day care, the rest not. Assume that of 150 children who did not have leukemia 70 were in day care, 80 not. Construct a two-by-two table showing these results and the column and row totals. Using the table, conduct a test of the null hypothesis of equal leukemia proportions for day care and non–day care children versus the two-tailed alternative. Use $\alpha = 0.05$. State your conclusions.

49. The accompanying table shows data from a study on AIDS risk factors and women. For the "Not traditional" sample under the "Tell men about contraception" variable, conduct a test for equality of proportions between the main partner and client groups.

Frequencies on the Analytic Variables for Partner and Client Subsamples

Variables	Main Partner (N = 268)		Client (N = 109)	
	Count	%	Count	%
Decision to use condom				
Self	55	(20)	61	(56)
Both	123	(46)	20	(18)
Partner	17	(6)	9	(8)
Never discuss	73	(27)	19	(17)
God and equality				
Not traditional	144	(54)	57	(52)
Traditional	124	(46)	52	(48)
Tell men about contraception				
Not traditional	190	(71)	76	(70)
Traditional	78	(29)	33	(30)

Reprinted from Marie Withers Osmond et al., "The Multiple Jeopardy of Race, Class, and Gender for AIDS Risk among Women," *Gender & Society* 7, no. 1 (March 1993), pp. 99–120.

50. Repeat Problem 49 for the "Self" sample under the "Decision to use condom" variable.

51. Repeat Problem 49 for the "Traditional" sample under the "God and equality" variable.

Class, Gender and Good Taste

The moral hierarchy of programme types . . . is relatively independent of the preferences of individuals in the sense that although different individuals say they watch different programmes, their discourses can be interpreted within the same collective moral hierarchy. Although there is inter-individual variation, this is confined within the limits of the said hierarchy. No one, for instance, explains why they are interested in current affairs programmes, and no one has excuses for not watching a certain fictional serial.

Let us begin by taking a closer look at gender differences. They are here analysed by counting the proportions of women among those who in the interview, in one way or another, reported that they watch different programme types. In that way, since there were 60 women and 39 men among the 99 interviewees, the percentage numbers themselves are not that important. It is rather the gendered preference order of the programme types achieved that way.

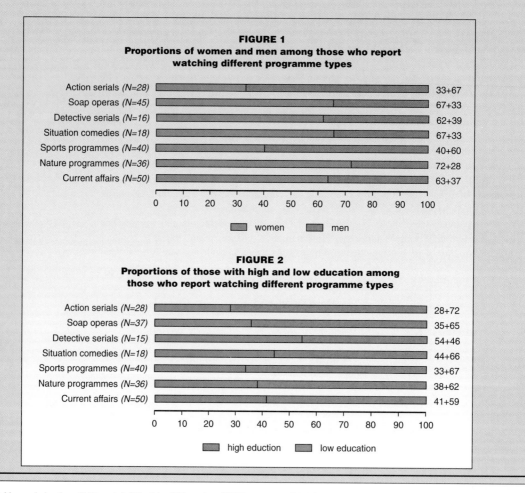

FIGURE 1
Proportions of women and men among those who report watching different programme types

Action serials *(N=28)*	33+67
Soap operas *(N=45)*	67+33
Detective serials *(N=16)*	62+39
Situation comedies *(N=18)*	67+33
Sports programmes *(N=40)*	40+60
Nature programmes *(N=36)*	72+28
Current affairs *(N=50)*	63+37

women ▢ men ▢

FIGURE 2
Proportions of those with high and low education among those who report watching different programme types

Action serials *(N=28)*	28+72
Soap operas *(N=37)*	35+65
Detective serials *(N=15)*	54+46
Situation comedies *(N=18)*	44+66
Sports programmes *(N=40)*	33+67
Nature programmes *(N=36)*	38+62
Current affairs *(N=50)*	41+59

high eduction ▢ low education ▢

Reprinted with permission from P. Alasuutari, "The Moral Hierarchy of TV Programmers," *Media, Culture, and Society* 14 (1992), by permission of Sagi Publications Ltd.

Cheetahs Appear Vigorous despite Inbreeding

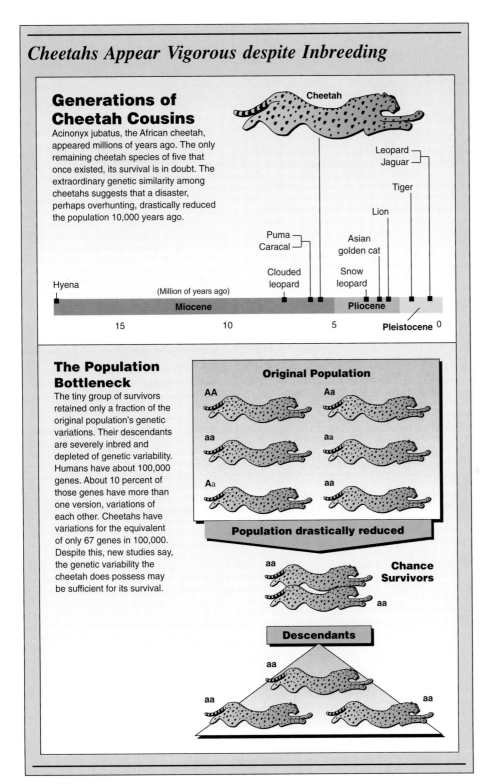

Generations of Cheetah Cousins

Acinonyx jubatus, the African cheetah, appeared millions of years ago. The only remaining cheetah species of five that once existed, its survival is in doubt. The extraordinary genetic similarity among cheetahs suggests that a disaster, perhaps overhunting, drastically reduced the population 10,000 years ago.

Cheetah

Leopard
Jaguar

Tiger

Lion

Puma
Caracal

Asian
golden cat

Clouded
leopard

Snow
leopard

Hyena

(Million of years ago)

Miocene

Pliocene

Pleistocene 0

15

10

5

The Population Bottleneck

The tiny group of survivors retained only a fraction of the original population's genetic variations. Their descendants are severely inbred and depleted of genetic variability. Humans have about 100,000 genes. About 10 percent of those genes have more than one version, variations of each other. Cheetahs have variations for the equivalent of only 67 genes in 100,000. Despite this, new studies say, the genetic variability the cheetah does possess may be sufficient for its survival.

Original Population

AA

Aa

aa

aa

Aa

aa

Population drastically reduced

aa

Chance Survivors

aa

Descendants

aa

aa

aa

Reprinted by permission from *The New York Times*, November 10, 1992, p. C13.

Leukemia Risk Lower in Day Care, Study Says

By LAWRENCE K. ALTMAN

Children who attended a day care center for at least three months before the age of 2 were found to have a lower risk of developing childhood leukemia than children who were not placed in such an environment, a study to be reported Saturday in the British Medical Journal says.

In the study of 136 children with childhood leukemia and a slightly larger number of healthy children, conducted in Greece, children who attended a day care center before the age of 2 had a 70 percent lower risk for childhood leukemia than children who had never been in day care. Children who had been in day care at any age had a 30 percent lower risk. The statistical association was "startlingly strong," the authors of the study said.

Reprinted by permission from *The New York Times*, September 24, 1993.

SUMMARY

In this chapter we learned how to carry out a statistical comparison of two population parameters. We learned about **experimental design** and saw how it is used in designing good test procedures by **pairing observations**—forming **blocks** of two observations each—when comparing two population means. We saw how when pairing is not possible, we may still test for the existence of a difference between two population means using separate, independent random samples from the two populations of interest. We learned how to conduct such tests both when we may assume a common population variance and when this assumption cannot be made. We also learned how to carry out similar tests for the difference between two population proportions. In each situation, confidence intervals for the difference between two population parameters are also possible.

KEY FORMULAS

A paired-observations test statistic for the difference between two poulation means:

$$t = \frac{\bar{D} - \mu_{D_0}}{s_D / \sqrt{n}}$$

A test statistic for the difference between two population means:

$$z = \frac{(\bar{x}_1 - \bar{x}_2) - (\mu_1 - \mu_2)_0}{\sqrt{\dfrac{\sigma_1^2}{n_1} + \dfrac{\sigma_2^2}{n_2}}}$$

A test statistic for the difference between two population means assuming equal population variance:

$$t = \frac{(\bar{x}_1 - \bar{x}_2) - (\mu_1 - \mu_2)_0}{\sqrt{s_p^2 \left(\dfrac{1}{n_1} + \dfrac{1}{n_2}\right)}} \quad \text{where } s_p^2 = \frac{(n_1 - 1)s_1^2 + (n_2 - 1)s_2^2}{n_1 + n_2 - 2}$$

A test statistic for zero difference between two population proportions:

$$z = \frac{(\hat{p}_1 - \hat{p}_2) - 0}{\sqrt{\hat{p}(1 - \hat{p})\left(\dfrac{1}{n_1} + \dfrac{1}{n_2}\right)}} \quad \text{where } \hat{p} = \frac{x_1 + x_2}{n_1 + n_2}$$

Problems 52–56 refer to the display "Index of Leading Cultural Indicators." Conduct the implied hypothesis tests given the sample sizes and standard deviations in each problem.

*ADDITIONAL
PROBLEMS*

52. Average daily TV viewing:

   1960: $n_1 = 1,500$ $s_1 = 2.01$
   1992: $n_2 = 1,200$ $s_2 = 2.05$

53. Average SAT scores:

   1960: $n_1 = 2,800$ $s_1 = 280$
   1992: $n_2 = 3,500$ $s_2 = 300$

54. Births to unwed mothers:

   1960: $n_1 = 6,220$
   1992: $n_2 = 8,300$

55. Teen suicide:

   $n_1 = n_2 = 100,000$

56. Violent crimes:

   $n_1 = n_2 = 10,000$

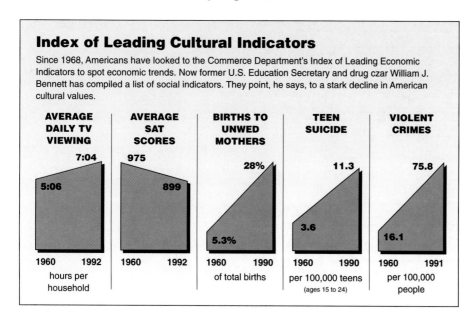

Index of Leading Cultural Indicators

Since 1968, Americans have looked to the Commerce Department's Index of Leading Economic Indicators to spot economic trends. Now former U.S. Education Secretary and drug czar William J. Bennett has compiled a list of social indicators. They point, he says, to a stark decline in American cultural values.

AVERAGE DAILY TV VIEWING	AVERAGE SAT SCORES	BIRTHS TO UNWED MOTHERS	TEEN SUICIDE	VIOLENT CRIMES
7:04	975	28%	11.3	75.8
5:06	899	5.3%	3.6	16.1
1960 1992	1960 1992	1960 1990	1960 1990	1960 1991
hours per household		of total births	per 100,000 teens (ages 15 to 24)	per 100,000 people

Reprinted by permission from *Time* magazine, March 29, 1993, p. 18, with data from the Heritage Foundation and Empower America.

Problems 57–59 refer to the display "As Girls Grow, Math Skills Decline."

57. Assume:

$$n_{Girl} = 1,000 \qquad n_{Boy} = 1,200$$

Using the "Less Enjoyment" bar graph, test for equality of proportions at the high school level.

As Girls Grow, Math Skills Decline

Though girls and boys enter eighth grade with about the same interest and achievement level, girls soon fall behind. Among college-bound students, girls score lower on the S.A.T. and achievement tests, and when they reach college, only one-quarter of young women major in math or science-related subjects.

Less Enjoyment
Percentage of students at each level who say they like math.

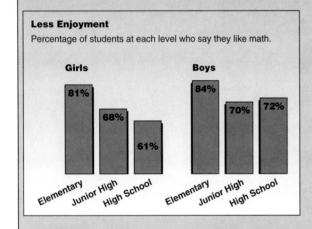

Lower Achievement
Percentage of students in each grade at each math achievement level.

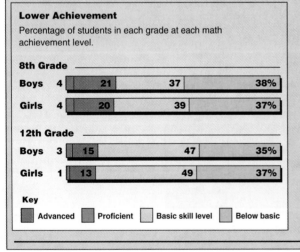

Lower Test Scores
College-bound students
Average scores on the math portion of the Scholastic Aptitude Test, 1969–1992.

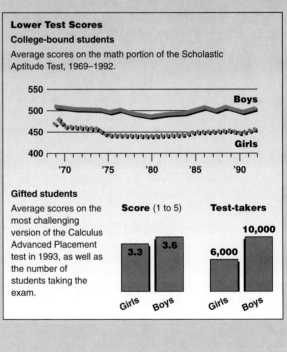

Gifted students
Average scores on the most challenging version of the Calculus Advanced Placement test in 1993, as well as the number of students taking the exam.

Reprinted by permission from *The New York Times,* November 24, 1993, p. B8, with data from the American Association of University Women, the Department of Education, the College Board, and the Educational Testing Service.

58. Using the "Lower Achievement" bar graph, test for equality of proportions among boys and girls for "Proficient" at the 8th grade level. Assume:

$$n_{Boy} = n_{Girl} = 800$$

59. Using the "Lower Test Scores" graphs, conduct a test for equality of means for gifted students. Assume:

$$s_{Boy} = 1.2 \text{ and } s_{Girl} = 1.1$$

Also assume independent random samples and equal population variances.

Big Grocery Chain Reaches Landmark Sex-Bias Accord

By JANE GROSS

OAKLAND, Calif., Dec. 16—In a sex discrimination settlement that could alter personnel practices throughout the grocery industry, Lucky Stores Inc. has agreed to pay nearly $75 million in damages to women in Northern California who were denied promotion opportunities and to invest $20 million in affirmative action programs for female employees.

The women said they were segregated into jobs working the cash registers and in store departments like bakeries or delicatessens that were not as high-level as the grocery and produce sections, where the men worked.

New Personnel Policies

Since carrying out some of the new personnel policies on a trial basis—including the posting and bidding for jobs that were previously filled at management discretion without women knowing they were available—Lucky's has filled 58 percent of its entry-level management jobs with women. Ten years ago, when the first of several women at Lucky's filed complaints with the Federal Equal Employment Opportunities Commission, 12 percent of those entry-level management jobs was held by women.

Reprinted by permission from *The New York Times,* December 17, 1993, p. A1.

---

60. Read the article on the sex-bias settlement against Lucky Stores. Assume that the results for the present and 10 years ago are based on two independent random samples of 1,000 employees each. Test for significance. Also give a 95 percent confidence interval for the difference between the proportion now and the proportion 10 years ago of women entry-level managers.

61. An article on the global labor market reports that 39.5 percent of the work force in the United States is under age 34, while for Russia the figure is 42.9 percent.[5] Assuming that these percentages are based on a random sample of 1,000 workers in each country, conduct a test aimed at showing that, in the two populations, the difference between the proportion of the work force under age 34 is at least one percent higher in Russia than it is in the United States. Use $\alpha = 0.05$.

---

[5] W. B. Johnston, "Global Work Force 2000: The New World Labor Market," *Harvard Business Review,* March–April 1991, p. 120.

62. Interpret the following computer output:

TWOSAMPLE T FOR C1 VS C2

	N	MEAN	STDEV	SE MEAN
C1	16	23.94	1.91	0.48
C2	17	24.82	2.88	0.70

95 PCT CI FOR MU C1 − MU C2: (−2.62, 0.85)

TTEST MU C1 = MU C2 (VS GT): T= −1.05   P=0.85 DF= 27

63. Interpret the following computer output:

TWOSAMPLE T FOR C1 VS C2

	N	MEAN	STDEV	SE MEAN
C1	16	34.63	2.13	0.53
C2	17	24.82	2.88	0.70

95 PCT CI FOR MU C1 − MU C2: (8.00, 11.61)

TTEST MU C1 = MU C2 (VS NE): T= 11.07 P=0.0000 DF= 31

POOLED STDEV = 2.54

Risk of Cerebral Palsy Found Higher for Twins

By the ASSOCIATED PRESS

Twins are almost 12 times as likely as single-birth babies to develop cerebral palsy, a serious brain disorder diagnosed in about 5,000 children each year, a new study found.

An examination of the health records of 155,000 children in four counties in Northern California showed that cerebral palsy occurred 12 times per 1,000 twin pregnancies and only about 1.1 times per 1,000 single pregnancies, the journal *Pediatrics* reported in the December issue.

Among children whose twin died before birth, the risk of cerebral palsy was 108 times the risk among children from single pregnancies, the study said.

Dr. Karin Nelson, a neurologist at the National Institute of Neurological Disorders and Stroke and a co-author of the study, said that the increased risk of cerebral palsy among twins was important to society because multiple births were becoming more and more common.

Reprinted by permission from *The New York Times*, December 14, 1993, p. C3.

64. Read the article "Risk of Cerebral Palsy Found Higher for Twins." Assume the two sample sizes are indeed 1,000 each, and that samples were randomly selected. Are the results significant? Explain.

65. Refer to the accompanying table with data on physicians and drug companies. Test the null hypothesis that, in the entire population from which these samples were randomly selected, the proportion of physicians who prescribe drug-company medicines is equal for those who share meals with drug-company staff and those who do not. Interpret your findings.

	Number of Physicians Who Prescribed Drug-Company Medicines	Number of Physicians Who Did Not Prescribe Drug-Company Medicines
Number of physicians who shared meals with drug-company staff	19	27
Number of physicians who never shared meals with drug-company staff	17	42

M. Chren and S. Landefeld, "Physicians' Behavior and Their Interactions with Drug Companies," *Journal of the American Medical Association* 271, no. 9 (March 2, 1994), pp. 684–89.

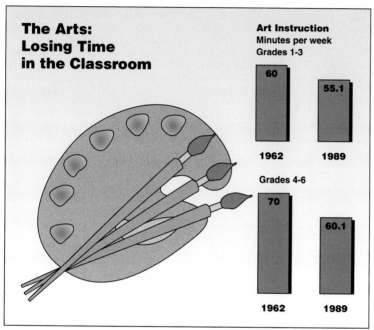

Reprinted by permission from *The New York Times,* February 3, 1993, p. A19.

66. Refer to the display on arts in the classroom. Assuming all sample sizes are 1,000 and all standard deviations are 5 minutes, are there differences in the average amount of time spent teaching the arts from 1962 to 1989?

67. Use Table A.2 in Appendix A. Assume that the data columns in the table of fatal occupational injuries constitute independent random samples with equal variances. Test the null hypothesis that the average percentage of fatal occupational injuries for the population of high-risk occupations is equal in January and in May versus the alternative that it is not.

INTERVIEW WITH
Toni Falvo

"I think the most challenging problem is knowing how to use statistical data *and* 'gut' feelings in making station decisions."

Toni Falvo holds the position of Research Director at the ABC affiliate in Chicago—WLS-TV. She has worked in the broadcast industry for fourteen years and during her career has made numerous contributions to the growth of WLS—now the city's most watched station.

Her role as the station's chief researcher and data analyst experience at a major ABC affiliate lends an insider's perspective on the varied uses of statistics and decision-making in the television industry. Decision-making and revenue generation within the broadcast industry are heavily dependent on gathering and correctly interpreting data, in many cases within very short time periods. Ms. Falvo's role in this process at the nerve center of the station demands both statistical analysis and intuition.

Aczel: Please tell us something about your background in statistics.

Falvo: I took the required basic business statistics course, but nothing beyond that.

Aczel: In what ways is decision-making at the station dependent on statistical data?

Falvo: As a television station, we are evaluated by our ratings performance and position in the market. From Sales to News, decisions can be based on what "the numbers" are telling us. In Sales, how we price and sell our product is based on the strength of our ratings performance and our competitor's. In News, we may make changes and adjustments to content and talent based on those same ratings. In Programming, ratings and audience composition play a key role in decisions.

Aczel: What are the primary sources of your data? Is it generated by your department, externally, etc?

Falvo: The A.C. Nielsen company is our source for ratings data. We also use various qualitative sources, such as Scarborough Research (market specific consumer, media and retail information). We periodically conduct our own research studies via phone surveys or focus group studies.

Aczel: How do you use and interpret Nielsen data?

Falvo: Since Chicago is a metered market we receive "overnight" ratings every day. We can immediately see how well a program is doing. We track programs daily and conduct month-to-month and year-to-year comparisons of regularly scheduled programs. We look for positive stories and trends to develop sales pieces to sell our station. We can also see how well our news or local programming is doing, or if there is a competitive concern. When you are looking at a particular program you want to see if it maintains its audience throughout the show, any audience drop-off may signal a problem with content or talent. How well a show maintains its lead-in or increases audience from its lead-in is another way to interpret Nielsen information. If you see a drop-off in your audience and a gain in your competitors we can conduct a study to see where our audience is going minute-by-minute to see if our competitor is gaining audience at our expense. There is also a variety of Nielsen studies to analyze ratings by household income, education, employment characteristics to help define your audience composition.

Aczel: How are the Nielsen viewers selected? How are they motivated?

Falvo: Stations subscribe to the Nielsen service, but we do not have anything to do with the selection process. There are very strict and specific rules regarding the Nielsen sample that must be in place to guarantee the integrity of the sample.

Nielsen follows a very specific selection process for the metered sample. Their goal is to select and maintain a representative sample of television households. They use census data and geographical data to select and recruit sample households. Once a household is selected all televisions are metered and they may remain in the sample for up to five years. There is turnover of the sample every month due to planned and unplanned turnover. You have people moving, wanting out of the survey and other reasons.

People are more willing to be a part of the metered sample than the diary sample because of the amount of work involved. With a meter there is virtually nothing the home has to do but watch TV. With a diary the home must write in what they watch and when. All Nielsen sample participants are given a small compensation.

Aczel: What types of statistical research, polling, etc., does the station generate or sponsor?

Falvo: We will conduct, at least once a year a very detailed research study. We work in conjunction with a vendor that we have selected, to conduct either a telephone survey or a focus group survey. In fact, this year we tried a new technique with our focus groups. The focus group participants were given a hand-held device that (enabled the) to key in positive and negative responses to the video (being tested). This proved to be helpful at the time of the focus group. People seemed more focused and precise with their comments since they had a visual reference and recorded their own responses. It also gave us information to analyze after the groups. All the respondents were recorded individually and we were able to analyze the data based on different variables.

Aczel: Do you use any inferential statistics in your work? If so, what and how?

Falvo: We project audience ratings performance for all programming on the station. Audience projections are based on past performances of a particular program to project it's future ratings performance, or if a show has not run before, you would look for a similar show to base your projection on. You are also looking at the time of day a show is running and what the HUTS (households using television) are during that time period. You may also look at the competitive situation to see what available audience you can expect to (obtain).

Aczel: Do you use statistics in evaluating economic data?

Falvo: We use economic data in two ways. First, in a sales application we would analyze the economic status of a household that watches a particular program. That information becomes a sales tool. For example, if you have "X" amount of households with an income of $75,000+ and they watch your station at a particular time more than any other station this becomes a way to sell your station.

We also use economic data in analyzing the Nielsen sample. Based on U.S. Census statistics we want to make sure that the Nielsen sample accurately reflects the economic composition of the market.

Aczel: Do you design public opinion polls in a statistical way? How do you interpret and report the findings?

Falvo: As I mentioned earlier, we do at least one major study per year. We may do a similar study each year so we can track results year-to-year. Or we may have a specific objective and design a study around that.

We do not conduct public opinion polls, our studies are based on a questionnaire of specific questions that we conduct over the phone. We decide what base we will need, and try to model our recruitment process to mirror the Nielsen sample. Which means we want to get the right ethnic and demographic mix so the sample is now skewed. We randomly select homes in the geographic area we are targeting and select the people by age and sex that we want included. A telephone survey should not last more than 20 minutes and should be a mix of close-ended (selection of responses) and open-ended, where you would encourage individual responses (which (is) more time consuming).

Once you have collected all the data you can make a variety of assumptions based on the age and sex of your viewers and what the main pur-

pose of the survey was. If you are trying to find out what your viewers feel about your station or a competitor's you can get a clear picture based on a survey that has an adequate base of people and is set up correctly.

Depending on what the study was on, all the departments that this study was (conducted) for would meet together and discuss the results. Many time decisions are influenced by what is (discovered) in these studies, but never solely based on these results.

Aczel: From your perspective, what are the most challenging, statistically-related problems the station deals with?

Falvo: I think the most challenging problem is knowing how to use statistical data *and* "gut" feelings in making station decisions. I think there is a skill or a sense that you get, especially after being in the television industry for a period of time, that you know when something will work. [After you look at the facts, the ratings, the demographics etc. you also have to have a "gut" feeling that this is the right decision you should make. I don't think you can have one without the other. Statistical facts alone and "gut" feelings alone will not work.]

Aczel: Given that advertising rates are essentially governed by the Nielsen ratings, does the station or network ever attempt to "verify" their accuracy by conducting your own surveys? If so, have any statistically relevant discrepancies surfaced? What happens in those cases?

Falvo: There is the EMRC (Electronic Media Rating Council) that accredits the Nielsen service. They are the "watchdog" of the ratings service. If problems are uncovered they are brought to Nielsen's attention.

Aczel: At your station, how would you statistically determine the saturation point of certain types of programming, e.g. local news, game shows, talk shows, etc?

Falvo: Statistically there is no real measure of "saturation." Your measure is your viewers, once they stop watching you know you need to make a change. And you will see that (loss) if you are trending and comparing your programming month-to-month and year-to-year. Also you see trends in particular types of programs. For example, at one time there were numerous game shows in daytime TV, those game shows have now been replaced with talk shows and the cycle goes on. In primetime you see the same cycles, police shows, medical shows and variety shows, they start with a big audience appetite for a particular type of show and after there are enough of them their appeal starts to wane.

CHAPTER

8   Making Extended Comparisons

Courtesy of the Bancroft Library.

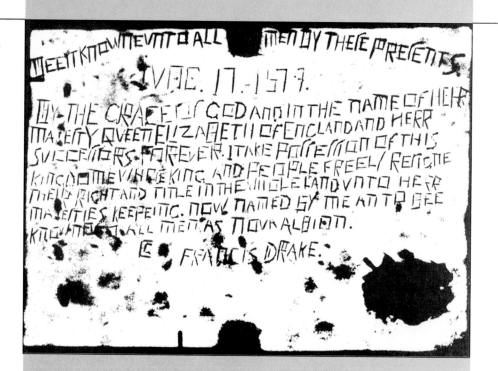

8.1 Introduction   358

8.2 The Hypothesis Test of Analysis of Variance (ANOVA)   359

8.3 The Theory and the Computations of ANOVA   365

8.4 Contingency Table Analysis: A Chi-Square Test for Independence   384

B*EE IT KNOWNE UNTO ALL MEN BY THESE PRESENTS*
IUNE.17.1579.
BY THE GRACE OF GOD AND IN THE NAME OF HERR
MAIESTY QUEEN ELIZABETH OF ENGLAND AND HERR
SUCCESSORS FOREVER. I TAKE POSSESSION OF THIS
KINGDOME WHOSE KING AND PEOPLE FREELY RESIGNE
THEIR RIGHT AND TITLE IN THE WHOLE LAND UNTO HERR
MAIESTIES KEEPING. NOW NAMED BY ME AN TO BEE
KNOWNE UNTO ALL MEN AS NOVA ALBION.

G   *FRANCIS DRAKE.*

In the summer of 1936, while hiking in a hilly area overlooking Point San Quentin and the San Francisco Bay, a young man named Beryle Shinn chanced upon a brass plate bearing the above inscription. Mr. Shinn put the plate in the trunk of his car, where it remained for a few years. Later, by chance, the plate came to the attention of Professor Herbert E. Bolton at the Department of History at the University of California, Berkeley.

In a logbook kept during his circumnavigation of the earth, Sir Francis Drake wrote of entering a sheltered area in 1579 on what is now the northern California coast to refit his ship. He also mentioned leaving a brass plate attached to a post to record the event. Thus, upon being presented the Shinn plate, Professor Bolton declared: "One of the world's long-lost historical treasures apparently has been found!"

Although questions about the authenticity of the plate were immediately raised by many scholars—mainly because of the curious forms of many of the letters, and the writing style, which is different from known Elizabethan styles—the plate was nonetheless pronounced genuine and put on permanent display at the Bankroft Library of the University of California.

Contentions that the plate was the work of a modern forger continued to be expressed, and these led the Bankroft Library to order tests of the metallurgic structure of the brass. Finally, in 1976, several tiny holes were drilled in the plate, and a sample of brass particles was sent to the Research Laboratory of Archaeology at Oxford University for analysis. There the sample of brass particles from the plate was statistically compared with two random samples of brass: a sample of brasses made in the 20th century and a sample of English and Continental brasses created between 1540 and 1720. Analysis of the average zinc content of the sample from the discovered plate and that of the two other samples led to the conclusion that the average zinc content of the plate was equal to the average zinc content of modern brasses and very different from the average zinc content of brasses made in the 16th to 18th centuries.[1]

The results of the analysis led Dr. R. Hedges of Oxford University to conclude: "I would regard it as quite unreasonable to continue to believe in the authenticity of the plate." Thus, what was thought to be an ancient artifact was shown to be an ingenious modern forgery. The scientific studies left unanswered the questions of who made the plate, and why.

8.1 INTRODUCTION

The statistical method of comparing the means of several populations, such as the mean zinc content of the three brasses relevant to the study of the plate, is the **analysis of variance.** The method is often referred to by its acronym, **ANOVA.** Analysis of variance is the first of two advanced statistical techniques to be discussed in this chapter. (The second is contingency table analysis.) What is analysis of variance?

> Analysis of variance (ANOVA) is a statistical method for determining the existence of differences among several population means.

The name of the technique may be misleading. While the aim of ANOVA is to detect differences among several population *means,* the technique requires the analysis of different forms of *variance* associated with the random samples under study—hence the name *analysis of variance.*

46

51

The original ideas of analysis of variance were developed by the English statistician Sir Ronald A. Fisher during the first part of this century. Much of the early work in this area dealt with agricultural experiments where crops were given different "treatments," such as different kinds of fertilizers. The researchers wanted to determine

---

[1]Several tests of the brass plate were conducted. See "The Plate of Brass Reexamined," a report by the Bankroft Library (1979).

whether all treatments under study were equally effective or whether some treatments were better than others. "Better" referred to those treatments that would produce crops of greater average weight. This question, of choosing the best treatment, is answerable by the analysis of variance and follow-up work. Since the original investigations involved different **treatments,** the term remained, and we use it interchangeably with *populations* even when no actual treatment is administered. Thus, for example, if we compare the mean incomes in four different communities, we may refer to the four populations as four different *treatments.*

8.2 THE HYPOTHESIS TEST OF ANALYSIS OF VARIANCE (ANOVA)

As with all tests of hypotheses, we follow certain steps in testing for the equality of several population means:

1. State the *null hypothesis* and the alternative hypothesis.
2. Determine the *rejection region* based on the distribution of a test statistic when the null hypothesis is true.
3. Compute the value of the test statistic and compare it with the rejection region for a given α.
4. Compute a *p-value,* and make a decision.

The hypothesis test of analysis of variance:

$$H_0: \mu_1 = \mu_2 = \mu_3 = \cdots = \mu_r$$
$$H_1: \text{Not all } \mu_i \ (i = 1, \ldots, r) \text{ are equal}$$

Let's say there are r populations, or treatments, under study. We draw an independent *random sample* from each of the r populations. The size of the sample from population i $(i = 1, \ldots, r)$ is n_i, and the total sample size is:

$$n = n_1 + n_2 + \cdots + n_r$$

From the r samples we compute several different quantities, and these lead to a computed value of a test statistic that follows a known distribution when the null hypothesis is true. From the value of the statistic and the critical point for a given level of significance (or the p-value), we are able to make a determination of whether or not we believe that the r population means are equal.

Usually, the number of compared means, r, is greater than 2. Why greater than 2? If r is equal to 2, then the above test is just a test for equality of two population means, the two-sample t test discussed in Chapter 7, and would give the same results. In this chapter, we are interested in investigating whether or not *several* population means may be considered equal. This is a test of a *joint hypothesis* about the equality of several population parameters.

But why can't we just use the two-sample t tests repeatedly? Suppose we are comparing $r = 5$ treatments. That means that there are 10 possible comparisons. (Count them!) It should be possible to make all 10 separate comparisons. However, if we use, say, $\alpha = 0.05$ for each test, then the probability of committing a Type I error in any particular test (deciding that the two population means are not equal when indeed they are equal) is 0.05. If each of the 10 tests has a 0.05 probability of a Type I error, what is the probability of a Type I error if we state "not all the means are equal" (i.e., rejecting H_0 in the *joint test* above)? The answer to this question is not known![2]

If we need to compare more than two population means and we want to remain in control of the probability of committing a Type I error, we need to conduct a joint test, which is what analysis of variance provides.

The reason for ANOVA's widespread applicability is that there are many situations where we need to compare more than two populations simultaneously. Even in cases where we need to compare only two treatments, say, test the relative effectiveness of two different prescription drugs, our actual test may require the use of a third treatment: a control treatment, or placebo.

We now present the assumptions that must be satisfied so that we can use the analysis of variance procedure in testing our hypotheses.

The required assumptions of ANOVA:

1. We assume *independent random sampling* from each of the r populations.
2. We assume that the r populations under study are *normally distributed*, with means μ_i that may or may not be equal, but with *equal variances, σ^2.*

Suppose, for example, that we are comparing three populations and want to determine whether or not the three population means μ_1, μ_2, and μ_3 are equal. We draw separate random samples from each of the three populations under study, and we assume that the three populations are distributed as shown in Figure 8.1.

These model assumptions are necessary for the test statistic used in analysis of variance to possess the known distribution when the null hypothesis is true. If the

FIGURE 8.1   Three Normally Distributed Populations with Different Means but with Equal Variance

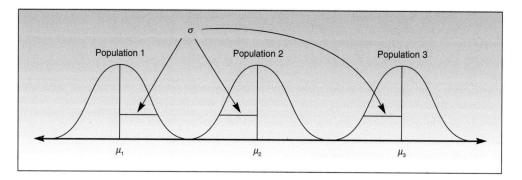

---

[2]The problem is further complicated because we cannot assume independence of the 10 tests, and therefore we cannot use a probability computation for independent events. The sample statistics used in the 10 tests are not independent since two such possible statistics are $(\overline{X}_1 - \overline{X}_2)$ and $(\overline{X}_2 - \overline{X}_3)$. Because both statistics contain a common term, $\overline{X}_2$, they are not independent of each other.

populations are not exactly normally distributed, but have distributions that are close to normal, the method still yields good results.

The Test Statistic and the F Distribution

When the null hypothesis is true, the test statistic of analysis of variance follows an F distribution. The F distribution has two kinds of degrees of freedom: those for the numerator and those for the denominator.

In the analysis of variance, the numerator degrees of freedom are $r - 1$, and the denominator degrees of freedom are $n - r$.

$$\text{ANOVA test statistic} = F_{(r-1, n-r)}$$

Figure 8.2 shows the F distribution with 3 and 15 degrees of freedom, which would be appropriate for a test of the equality of four population means using a total sample size of 19. Also shown is the critical point for $\alpha = 0.05$, found in Table B.4, Appendix B. The critical point is 3.29. The test is carried out as a right-tailed test. Unlike the normal and the t distributions, the F distributions are not symmetric—they are all right-skewed.

Figure 8.3 shows several F distributions, with different numbers of degrees of freedom for numerator and denominator. As can be seen from the figure, as the number of degrees of freedom of numerator and denominator increase, the skewness lessens and the distribution looks more and more like a normal distribution. (The F distributions *converge* to the normal.) Table 8.1 is a reproduction of part of the F table, Table B.4 in Appendix B. It shows how the critical point for $\alpha = 0.05$ is found for our particular F distribution with 3 degrees of freedom for the numerator and 15 degrees of freedom for the denominator.

Analysis of variance is an involved technique, and the required computations are difficult and time-consuming to carry out by hand. Consequently, computers are indispensable in most situations involving analysis of variance, and we will make extensive use of the computer in this chapter. For now, let's assume that a computer is available to us, and that it provides us with the value of the test statistic, as well as the p-value. (The computations are a topic in themselves and will be presented in Section 8.3.)

The Null Distribution of the ANOVA Test Statistic for $r = 4$ Populations and a Total Sample Size $n = 19$, the F Distribution with 3 and 15 Degrees of Freedom

FIGURE 8.2

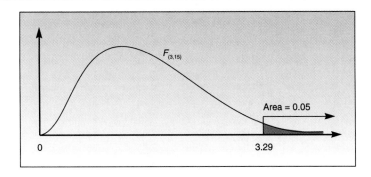

FIGURE 8.3 Several *F* Distributions, with Different Degrees of Freedom

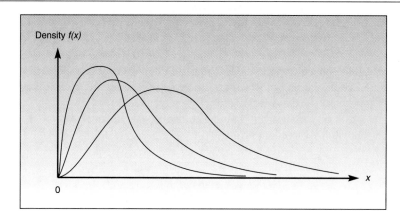

TABLE 8.1 Critical Points Cutting Off a Right-Tail Area of 0.05 for Selected *F* Distributions

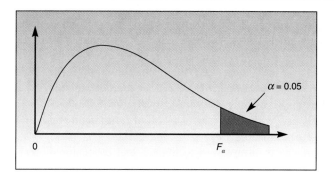

	k₁	\multicolumn{9}{c}{**Degrees of Freedom of the Numerator**}								
k₂		1	2	3 ↓	4	5	6	7	8	9
Degrees of	1	161.4	199.5	215.7	224.6	230.2	234.0	236.8	238.9	240.5
Freedom	2	18.51	19.00	19.16	19.25	19.30	19.33	19.35	19.37	19.38
of the	3	10.13	9.55	9.28	9.12	9.01	8.94	8.89	8.85	8.81
Denominator	4	7.71	6.94	6.59	6.39	6.26	6.16	6.09	6.04	6.00
	5	6.61	5.79	5.41	5.19	5.05	4.95	4.88	4.82	4.77
	6	5.99	5.14	4.76	4.53	4.39	4.28	4.21	4.15	4.10
	7	5.59	4.74	4.35	4.12	3.97	3.87	3.79	3.73	3.68
	8	5.32	4.46	4.07	3.84	3.69	3.58	3.50	3.44	3.39
	9	5.12	4.26	3.86	3.63	3.48	3.37	3.29	3.23	3.18
	10	4.96	4.10	3.71	3.48	3.33	3.22	3.14	3.07	3.02
	11	4.84	3.98	3.59	3.36	3.20	3.09	3.01	2.95	2.90
	12	4.75	3.89	3.49	3.26	3.11	3.00	2.91	2.85	2.80
	13	4.67	3.81	3.41	3.18	3.03	2.92	2.83	2.77	2.71
	14	4.60	3.74	3.34	3.11	2.96	2.85	2.76	2.70	2.65
→	15	4.54	3.68	3.29	3.06	2.90	2.79	2.71	2.64	2.59

Major roasters and distributors of coffee in the United States have long felt the great uncertainty in the price of coffee beans. Over the course of one year, for example, coffee futures prices went from a low of $1.40 per pound up to $2.50 and then down to $2.03. The main reason for such wild fluctuations in price, which strongly affect the performance of coffee distributors, is the constant danger of drought in Brazil. Since Brazil produces 30 percent of the world's coffee, the market for coffee beans is very sensitive to the annual rumors of impending drought.

Recently a domestic coffee distributor decided to avert the problem altogether by eliminating Brazilian coffee from all blends the company distributes. Before taking such action, the distributor wanted to minimize the chances of suffering losses in sales volume. Therefore, the distributor hired a marketing research firm to conduct a statistical test of consumers' taste preferences. The research firm made arrangements with several large restaurants to serve randomly chosen groups of their customers different kinds of after-dinner coffee. Three kinds of coffee were served: a group of 21 randomly chosen customers were served pure Brazilian coffee; another group of 20 randomly chosen customers were served pure Columbian coffee; and a third group of 22 randomly chosen customers were served pure African-grown coffee.

To prevent a response bias, the people in this experiment were not told the kind of coffee they were being served. The coffee was listed as a "house blend."

Suppose that data for the three groups were consumers' ratings of the coffee on a scale of 0 to 100 and that certain computations were carried out with these data leading to $F = 2.02$ as the value of the ANOVA test statistic. Is there evidence to conclude that any of the three kinds of coffee leads to an average consumer rating different from the other two kinds?

EXAMPLE 8.1

SOLUTION

The null and alternative hypotheses here are:

$$H_0: \mu_1 = \mu_2 = \mu_3$$

$$H_1: \text{Not all three } \mu_i \text{ are equal}$$

The null hypothesis states that average consumer responses to each of the three kinds of coffee are equal. The alternative hypothesis says that not all three population means are equal. What are the possibilities covered under the alternative hypothesis? Some of the possible relationships among the relative magnitudes of any three real numbers μ_1, μ_2, and μ_3 are shown in Figure 8.4.

As you can see from Figure 8.4, the alternative hypothesis comprises different possibilities—it includes all the cases where *not all* three means are equal. Thus, if we reject the null hypothesis, all we know is that there is statistical evidence to conclude that not all three population means are equal. However, we do not know in what way the means are different. Therefore, if we reject the null hypothesis, we need to conduct further analysis to determine which population means are different from one another.

We have a null hypothesis and an alternative hypothesis. We also assume that the conditions required for ANOVA are met; that is, we assume that the three populations of consumer responses are (approximately) normally distributed with equal population variance. Now we need to conduct the test.

Since there are three populations, or treatments, under study, the degrees of freedom for the numerator are $r - 1 = 3 - 1 = 2$. Since the total sample size is: $n = n_1 + n_2 + n_3 = 21 + 20 + 22 = 63$, we find that the degrees of freedom for the denominator are $n - r = 63 - 3 = 60$. Thus, when the null hypothesis is true, our test statistic has an F distribution with 2 and 60 degrees of freedom: $F_{(2,60)}$. From Table B.4 in Appendix B,

FIGURE 8.4

Some of the Possible Relationships among the Relative Magnitudes
of the Three Population Means μ_1, μ_2, and μ_3

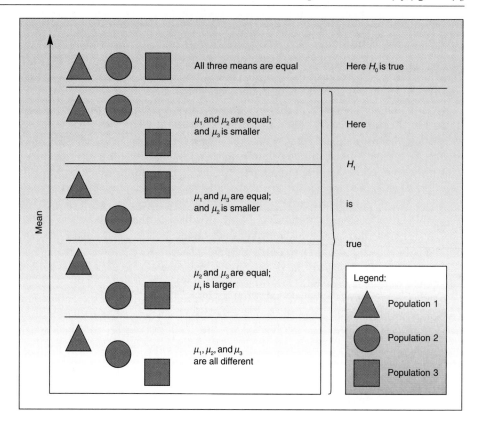

FIGURE 8.5

Carrying Out the Test of Example 8.1

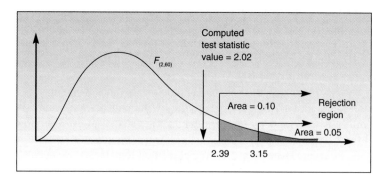

we find that the right-tail critical point at $\alpha = 0.05$ for an F distribution with 2 and 60 degrees of freedom is $C = 3.15$. Since the computed value of the test statistic is equal to 2.02, we may conclude that at the 0.05 level of significance the null hypothesis that all three population means are equal cannot be rejected. Since the critical point for $\alpha = 0.10$ is 2.39, we find that the p-value is even greater than 0.10.

Based on our data, there is no evidence that consumers tend to prefer the Brazilian coffee to the other two brands. The distributor may substitute one of the other brands for the price-unstable Brazilian coffee. (Note, however, that we usually prefer to make conclusions based on the *rejection* of a null hypothesis because nonrejection is often considered a weak conclusion.) The results of our test are shown in Figure 8.5

In this section, we have seen the basic elements of the hypothesis test underlying analysis of variance: the null and alternative hypotheses, the required assumptions, the test statistic, and the decision rule. We have not, however, seen how the test statistic is computed from the data or the reasoning behind its computation. The theory and the computations of ANOVA are explained in Section 8.3.

PROBLEMS

1. Four populations are compared by analysis of variance. What are the possible relations among the four population means covered under the null and alternative hypotheses?
2. What are the assumptions of ANOVA?
3. Three methods of training managers are to be tested for relative effectiveness. The management training institution proposes to test the effectiveness of the three methods by comparing two methods at a time, using a paired t test. Explain why this is a poor procedure.
4. In an analysis of variance comparing the output of five plants, data sets of 25 observations per plant are analyzed. The computed F statistic value is 3.6. Do you believe that there are differences in average output among the five plants? What is the approximate p-value? Explain.
5. A real estate development firm wants to test whether there are differences in the average price of a lot of a given size in the center of each of four cities: Philadelphia, New York, Washington, and Baltimore. Random samples of 52 lots in Philadelphia, 38 lots in New York, 43 lots in Washington, and 47 lots in Baltimore lead to a computed test statistic value of 12.53. Do you believe that average lot prices are equal in the four cities? How confident are you of your conclusion? Explain.
6. A recent article reported the results of an analysis of variance of injury severity among several types of athletes.[3] The reported results were $F(7,90) = 1.68$, $p > 0.1$.
   a. How many groups of athletes were compared?
   b. What was the total sample size?
   c. What is the conclusion?
7. A study of Japanese children's respect for fathers, mothers, and teachers reported an ANOVA F-statistic value of 380.04.[4] The degrees of freedom are 2 and 1,675. Are respect levels for all three categories of adults equal on average? Explain.

8.3 THE THEORY AND THE COMPUTATIONS OF ANOVA

Recall that the purpose of analysis of variance is to detect differences among several population means based on evidence provided by random samples from these populations. How can this be done? We want to compare r population means. We use

---

[3]T. Petrie, "The Moderating Effect of Social Support and Playing Status on the Life Stress–Injury Relationship," *Journal of Applied Sport Psychology* 5 (1993), pp. 1–16.

[4]L. Mann et al., "Measuring Respect for Parents and Teachers in Japan," *Journal of Cross-Cultural Psychology* 25, no. 1 (March 1994), pp. 133–45.

r random samples, one sample from each population. Each random sample has its own mean. The mean of the sample from population i will be denoted by $\bar{x}_i$. We may also compute the mean of all data points in the study, regardless of which population they come from. The mean of all the data points (when all data points are considered a single set) is called the **grand mean** and is denoted by $\bar{\bar{x}}$.

In Example 8.1, $r = 3$, $n_1 = 21$, $n_2 = 20$, $n_3 = 22$, and $n = n_1 + n_2 + n_3 = 63$. The third data point (person) in the group of 21 people who consumed Brazilian coffee is denoted by $x_{1,3}$.

We will now state the main principle behind the analysis of variance.

> If the r population means are different (that is, at least two of the population means are *not* equal), then it is likely that the variation of the data points about their respective sample means, $\bar{x}_i$, will be *small* when compared with the variation of the r sample means about the grand mean, $\bar{\bar{x}}$.

To demonstrate this principle, we'll use three hypothetical populations, which we'll call the triangles, the squares, and the circles. Table 8.2 gives the values of the sample points from the three populations. For demonstration purposes, we'll use very small samples. In real situations, the sample sizes should be much larger. The data given in Table 8.2 are shown in Figure 8.6. The figure also shows the deviations of the data points from their sample means and the deviations of the sample means from the grand mean.

Look carefully at Figure 8.6. Note that the *average* distance (in absolute value) of data points from their respective group means (that is, the average distance, in absolute value, of a triangle from the mean of the triangles, $\bar{x}_1$, and similarly for the squares and the circles) is *relatively small* compared to the average distance (in absolute value) of

TABLE 8.2   Data and the Various Sample Means for Triangles, Squares, and Circles

Treatment (i)	Sample Point (j)	Value ($x_{i,j}$)
$i = 1$ (Triangles)	1	4
	2	5
	3	7
	4	8
Mean of triangles		6
$i = 2$ (Squares)	1	10
	2	11
	3	12
	4	13
Mean of squares		11.5
$i = 3$ (Circles)	1	1
	2	2
	3	3
Mean of circles		2
Grand mean of all data points		**6.909**

The Deviations of the Triangles, Squares, and Circles from Their Sample Means
and the Deviations of the Sample Means from the Grand Mean   **FIGURE 8.6**

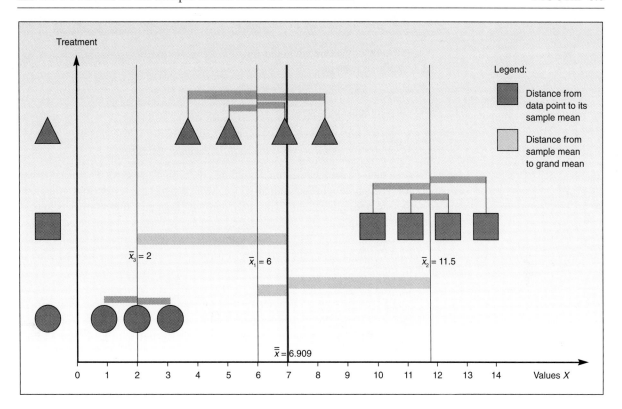

the three sample means from the grand mean. If you're not convinced of this, note that there are only three distances of sample means to the grand mean (in the computation, each distance is weighted by the actual number of points in the group), and that only one of them, the smallest distance—that of $\bar{x}_1$ to $\bar{\bar{x}}$—is of the relative magnitude of the distances between the data points and their respective sample means. The two other distances are much greater; hence, the average distance of the sample means from the grand mean is greater than the average distance of all data points to their respective sample means.

The *average* deviation from a mean is zero. We therefore talk about the average absolute deviation—actually, we will use the average *squared* deviation—to prevent the deviations from canceling out. This should remind you of the definition of the sample variance in Chapter 1.

Now let's define some terms that will make our discussion simpler.

> An **error deviation** is the difference between a data point and its sample mean. Errors are denoted by e.

Thus, all the distances from the data points to their sample means in Figure 8.6 are errors. (Some are positive, and some are negative.) These distances are called errors because they are unexplained by the fact that a given data point belongs to population *i*. The errors are assumed to be due to natural variation, or pure randomness, within the sample from population *i*.

> A **treatment deviation**, denoted by *t*, is the deviation of a sample mean from the grand mean.

The ANOVA principle thus says:

> *When the population means are not equal, the "average" error deviation is relatively small compared to the "average" treatment deviation.*

Again, if we actually averaged all the deviations, we would get zero. Therefore, when we apply the principle computationally, we will square the error and treatment deviations before averaging them. This way, we will maintain the relative (squared) magnitudes of these quantities. The averaging process is further complicated because we have to average based on degrees of freedom. (Recall that degrees of freedom were used in the definition of a sample variance.) For now, let the term *average* be used in a simplified, intuitive sense.

FIGURE 8.7   The Samples of Triangles, Squares, and Circles and Their Respective Populations (Normal Populations with Equal Variance but with Different Means)

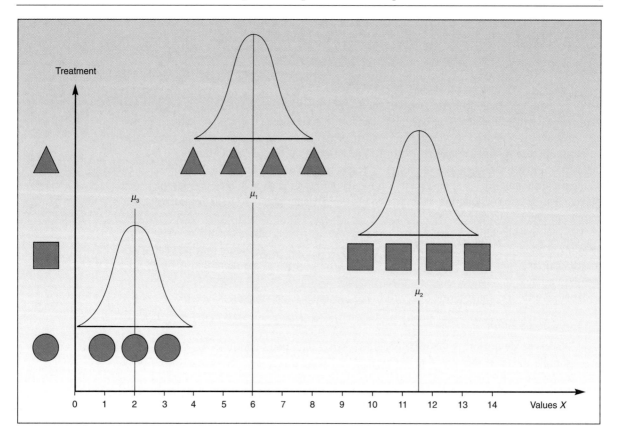

Since we noted that the average error deviation in our triangle-square-circle example looks small relative to the average treatment deviation, let's see what the populations that brought about our three samples look like. Figure 8.7 shows the three populations, assumed to be normally distributed with equal variance. (This can be seen from the equal width of the three normal curves.) The figure also shows that the three population means are not equal.

Figure 8.8, in contrast, shows three samples of triangles, squares, and circles in which the average error deviation is of about the same magnitude as the average treatment deviation. As can be seen from the superimposed normal populations from which the samples have arisen in this case, the three population means, μ_1, μ_2, and μ_3, are all equal. Compare the two figures to convince yourself of the ANOVA principle.

The Sum of Squares Principle

We have seen how, when the population means are different, the error deviations in the data are small when compared with the treatment deviations. We made general statements about the average (squared) error being small when compared with the average (squared) treatment deviation. The error deviations measure how close the data *within* each group are to their respective group means. The treatment deviations

Samples of Triangles, Squares, and Circles Where the Average Error Deviation Is
Not Smaller than the Average Treatment Deviation   **FIGURE 8.8**

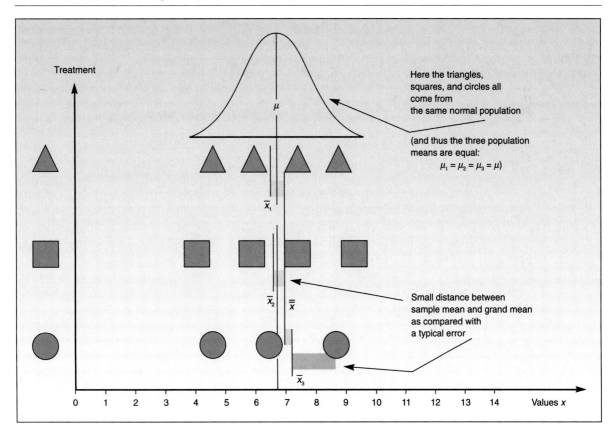

measure the distances *between* the various groups. It therefore seems intuitively plausible (as seen from Figures 8.6, 8.7, 8.8) that when these two kinds of deviations are of about equal magnitude, the population means are about equal. Why? Because when the average error is about equal to the average treatment deviation, then the treatment deviation may itself be viewed as just another error. That is, the treatment deviation in this case is due to pure chance rather than to any real difference among the population means. In other words, when the average *t* is of the same magnitude as the average *e*, then both are estimates of the internal variation within the data and carry no information about a difference between any two groups—about a difference in population means.

We will now make everything quantifiable, using the **sum of squares principle.** We start by returning to Figure 8.6, looking at a particular data point, and analyzing distances associated with the data point. We choose the fourth data point from the sample of squares (population 2). This data point is $x_{2,4} = 13$. (Verify this from Table 8.2.) A magnified section of Figure 8.6, the section surrounding this particular data point, is shown in Figure 8.9.

Figure 8.9 shows that the total deviation is equal to the treatment deviation plus the error deviation. This is true for *any* point in our data set (even when some of the numbers are negative).

The **total deviation** of a data point, denoted by Tot, is the deviation of the data point from the grand mean.

For any data point:

$$\text{Tot} = t + e$$

In words:

$$\text{Total deviation} = \text{Treatment deviation} + \text{Error deviation}$$

In the case of our chosen data point, $x_{2,4}$, we have:

$$t_2 + e_{2,4} = 4.591 + 1.5 = 6.091 = \text{Tot}_{2,4}$$

This works for every data point in our data set. In Figure 8.9, the *total* deviation of a data point from the grand mean is thus partitioned into a deviation due to *treatment* and a deviation due to *error*. The deviation due to treatment differences is the *between-treatments* deviation, while the deviation due to error is the *within-treatment* deviation.

We have considered only one point, $x_{2,4}$. In order to determine whether or not the error deviations are small when compared with the treatment deviations, we need to aggregate over all our data. We square each term above in our equation Tot = $t + e$. Although for any *given* data point the equality no longer holds after the squaring, the equality does hold again after *summing* all three terms over all our data. For example, while $5 = 3 + 2$, when we square the quantities the equality no longer holds: $25 \neq 9 + 4$. However, when all squared quantities in the data are aggregated, equality does hold for the totals.

The sum of squares principle can be stated as follows.

The Total Deviation as the Sum of the Treatment Deviation and the Error Deviation
for a Particular Data Point   **FIGURE 8.9**

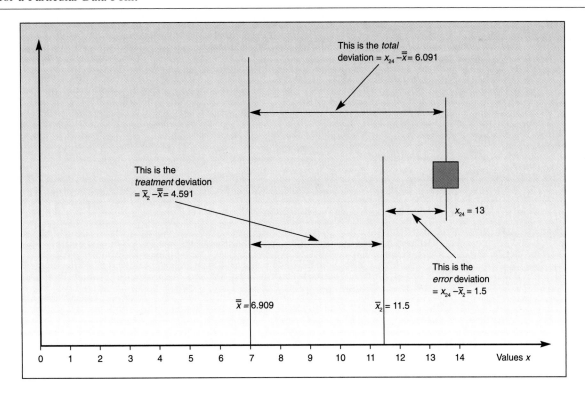

The **total sum of squares (SSTO)** is the sum of the two terms: the **sum
of squares for treatment (SSTR)** and the **sum of squares for error
(SSE):**

$$SSTO = SSTR + SSE$$

The sum of squares principle partitions the total sum of squares within the data, SSTO,
into a part due to treatment effect, SSTR, and a part due to errors, SSE. The squared
deviations of the treatment means from the grand mean are *counted for every data
point*.

Sums of squares measure variation within the data. SSTO is the total amount of
variation within the data set. SSTR is that part of the variation within the data that is
due to differences among the groups, and SSE is that part of the variation within the
data that is due to error—the part that cannot be explained by differences among the
groups. Therefore, SSTR is sometimes called the sum of squares *between* (variation
among the groups), and SSE is called the sum of squares *within* (within-group
variation). SSTR is also called the *explained variation* (because it is the part of the
total variation that can be explained by the fact that the data points belong to several
different groups). SSE is, then, called the *unexplained variation*. The partition of the
sum of squares in analysis of variance is shown in Figure 8.10.

FIGURE 8.10 Partition of the Total Sum of Squares into Treatment and Error Parts

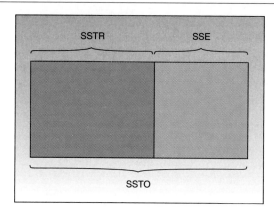

Breaking down the sum of squares is not enough, however. If we want to determine whether or not the errors are small as compared with the treatment part, we need to find the *average* (squared) error deviation and the *average* (squared) treatment deviation. Averaging, in the context of variances, is achieved by dividing by the appropriate number of degrees of freedom associated with each sum of squares.

The Degrees of Freedom

The **degrees of freedom** are the number of data points that are "free to move," that is, the number of elements in the data set minus the number of restrictions. A restriction on a data set is a quantity already computed from the entire data set under consideration; thus, knowledge of this quantity makes one data point fixed and reduces by one the effective number of data points that are free to move. This is why, when computing a variance, knowledge of the sample mean reduces the degrees of freedom of the sample variance to $n - 1$. What are the degrees of freedom in the context of analysis of variance?

Consider the total sum of squares, SSTO. In computing this sum of squares, we use the entire data set and information about *one* quantity computed from the data: the grand mean (because, by definition, SSTO is the sum of the squared deviations of all data points from the grand mean). Since we have a total of n data points and one restriction, the degrees of freedom are $n - 1$.

The number of degrees of freedom associated with SSTO is $n - 1$.

The sum of squares for treatment, SSTR, is computed from the deviations of r sample means from the grand mean. The r sample means are considered r independent data points, and the grand mean (which can be considered as having been computed from the r sample means) thus reduces the degrees of freedom by one.

The number of degrees of freedom associated with SSTR is $r - 1$.

The sum of squares for error, SSE, is computed from the deviations of a total of n data points ($n = n_1 + n_2 + \cdots + n_r$) from r different sample means. Since each of the sample means acts as a restriction on the data set, the degrees of freedom for error are $n - r$. This can be seen another way: There are r groups with n_i data points in group i. Thus, each group, with its own sample mean acting as a restriction, has degrees of freedom equal to $n_i - 1$. The total number of degrees of freedom for error is the sum of the degrees of freedom in the r groups: df $= (n_1 - 1) + (n_2 - 1) + \cdots (n_r - 1) = n - r$.

The number of degrees of freedom associated with SSE is $n - r$.

An important principle in analysis of variance is that the degrees of freedom of the three components are *additive* in the same way that the sums of squares are additive.

---

df(total) = df(treatment) + df(error)

This can easily be verified by noting the following: $n - 1 = (r - 1) + (n - r)$—the r drops out. We are now ready to compute the average squared deviation due to treatment and the average squared deviation due to error.

The Mean Squares

In finding the average squared deviations due to treatment and to error, we divide each sum of squares by its degrees of freedom. We call the two resulting averages **mean square treatment (MSTR)** and **mean square error (MSE),** respectively.

$$MSTR = \frac{SSTR}{r - 1}$$

$$MSE = \frac{SSE}{n - r}$$

The Expected Values of the Statistics MSTR and MSE under the Null Hypothesis

When the null hypothesis of ANOVA is true, all r population means are equal, and in this case there are *no treatment effects*. In such a case, the average squared deviation due to treatment is just another realization of an average squared error.

When the null hypothesis of ANOVA is true and all r population means are equal, MSTR and MSE both estimate the common population variance, σ^2.

If, on the other hand, the null hypothesis is not true and differences do exist among the r population means, then *MSTR will tend to be larger than MSE.*

The *F* Statistic

The preceding discussion suggests that the ratio of MSTR to MSE is a good indicator of whether or not the r population means are equal. If the r population means are equal, then MSTR/MSE would tend to be close to 1.00. Remember that both MSTR and MSE are sample statistics derived from our data. As such, MSTR and MSE will have some randomness associated with them, and they are not likely to exactly equal their expected values. Thus, when the null hypothesis is true, MSTR/MSE will vary around the value 1.00. When the r population means are not all equal, the ratio MSTR/MSE will tend to be greater than 1.00 because the expected value of MSTR will be larger than the expected value of MSE. How large is "large enough" for us to reject the null hypothesis?

This is where statistical inference comes in. We want to determine whether the difference between our observed value of MSTR/MSE and the number 1.00 is due just to chance variation, or whether MSTR/MSE is *significantly* greater than 1.00— implying that the population means are not all equal. We will make the determination with the aid of the F distribution.

Under the assumptions of ANOVA, the ratio MSTR/MSE possesses an F distribution with $r - 1$ degrees of freedom for the numerator and $n - r$ degrees of freedom for the denominator when the null hypothesis is true. Hence, the test statistic in analysis of variance is:

$$F_{(r-1, n-r)} = \frac{\text{MSTR}}{\text{MSE}}$$

The ANOVA Table

Table 8.3 shows the data for our triangles, squares, and circles. In addition, the table shows the deviations from the group means, and their squares. From these quantities, we find the sum of squares and mean squares.

As the last row of the table shows, the sum of all the deviations of the data points from their group means is zero, as expected. The sum of the *squared* deviations from the sample means (which is SSE) is 17.

Now we want to compute the sum of squares for treatment. Recall from Table 8.2 that $\bar{\bar{x}} = 6.909$. For every data point, we count the squared distance of the sample mean from the grand mean and then add over all our data:

$$\text{SSTR} = 4(6 - 6.909)^2 + 4(11.5 - 6.909)^2 + 3(2 - 6.909)^2$$

$$= 159.9$$

Next we compute the mean squares:

$$\text{MSTR} = \frac{\text{SSTR}}{r - 1} = \frac{159.9}{2} = 79.95$$

Computations for Triangles, Squares, and Circles **TABLE 8.3**

Treatment (i)	j	Value ($x_{i,j}$)	($x_{i,j} - \bar{x}_i$)	($x_{i,j} - \bar{x}_j$)^2
Triangle	1	4	$4 - 6 = -2$	$(-2)^2 = 4$
Triangle	2	5	$5 - 6 = -1$	$(-1)^2 = 1$
Triangle	3	7	$7 - 6 = 1$	$(1)^2 = 1$
Triangle	4	8	$8 - 6 = 2$	$(2)^2 = 4$
Square	1	10	$10 - 11.5 = -1.5$	$(-1.5)^2 = 2.25$
Square	2	11	$11 - 11.5 = -0.5$	$(-0.5)^2 = 0.25$
Square	3	12	$12 - 11.5 = 0.5$	$(0.5)^2 = 0.25$
Square	4	13	$13 - 11.5 = 1.5$	$(1.5)^2 = 2.25$
Circle	1	1	$1 - 2 = -1$	$(-1)^2 = 1$
Circle	2	2	$2 - 2 = 0$	$(0)^2 = 0$
Circle	3	3	$3 - 2 = 1$	$(1)^2 = 1$
			Sum $= 0$	Sum $= 17$

$$MSE = \frac{SSE}{n - r} = \frac{17}{8} = 2.125$$

And the value of the F statistic:

$$F_{(2,8)} = \frac{MSTR}{MSE} = \frac{79.95}{2.125} = 37.62$$

We are finally in a position to conduct the ANOVA hypothesis test to determine whether or not the means of the three populations are equal. From Table B.4 in Appendix B, we find that the critical point at $\alpha = 0.01$ for the F distribution with 2 degrees of freedom for the numerator and 8 degrees of freedom for the denominator is 8.65. We can therefore reject the null hypothesis. Since 37.62 is much greater than 8.65, the p-value is much smaller than 0.01. This is shown in Figure 8.11.

Note that, as usual, we must exercise caution in the interpretation of results based on such small samples. As we noted earlier, in real situations we use large data sets, and the computations are usually done by computer.[5]

An essential tool for reporting the results of an analysis of variance is the ANOVA table, which lists the sources of variation: treatment, error, and total, as well as the sums of squares, the degrees of freedom, the mean squares, and the F ratio. The table format simplifies the analysis and the interpretation of the results. The structure of the ANOVA table is based on the fact that both the sums of squares and the degrees of freedom are additive. Table 8.4 shows the ANOVA results computed above for the triangles, squares, and circles example.

Note that the entries in the second and third columns, sum of squares and degrees of freedom, are both additive. The entries in the fourth column, mean square, are

---

[5]If you must carry out ANOVA computations by hand, there are computational formulas for the sums of squares that may be used:

$$SSTO = \sum_i \sum_j (x_{i,j})^2 - \left(\sum_i \sum_j x_{i,j}\right)^2 / n$$
$$SSTR = \sum_i \left(\left(\sum_j x_{i,j}\right)^2 / n_i\right) - \left(\sum_i \sum_j x_{i,j}\right)^2 / n$$

and we obtain SSE by subtraction: SSE = SSTO − SSTR.

FIGURE 8.11   Rejecting the Null Hypothesis in the Triangles, Squares, and Circles Example

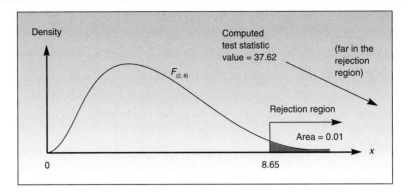

TABLE 8.4   ANOVA Table

Source of Variation	Sum of Squares	Degrees of Freedom	Mean Square	F Ratio	p-Value
Treatment	SSTR = 159.9	$r - 1 = 2$	$MSTR = \dfrac{SSTR}{r-1}$ $= 79.95$	$F = \dfrac{MSTR}{MSE}$ $= 37.62$	0.000
Error	SSE = 17.0	$n - r = 8$	$MSE = \dfrac{SSE}{n-r}$ $= 2.125$		
Total	SSTO = 176.9	$n - 1 = 10$			

obtained by dividing the appropriate sums of squares by their degrees of freedom. We do not define a mean square total, which is why there is no entry in that particular position in the table. The main objective of our analysis is the fifth column: the *F* ratio, which is computed as the ratio of the two entries in the previous column. The last column shows the *p*-value associated with the *F* ratio. It's the area under the *F* density to the right of the obtained value 37.62. Example 8.2 demonstrates the use of the ANOVA table.

EXAMPLE 8.2

Club Med has over 30 major resorts worldwide, from Tahiti to Switzerland. Many of the beach resorts are in the Caribbean, and at one point the club wanted to test whether or not the resorts on Guadeloupe, Martinique, Eleuthera, Paradise Island, and St. Lucia were all equally well liked by vacationing club members. The analysis was to be based on a survey questionnaire filled out by a random sample of 40 respondents in each of the resorts. From every returned questionnaire, a general satisfaction score, on a scale of 0 to 100, was computed. Analysis of the survey results yielded the statistics given in Table 8.5.

A computer program was used to calculate the sums of squared deviations from the sample means and from the grand mean. Given the values of SSTO and SSE, construct an ANOVA table and conduct the hypothesis test.

First we construct an ANOVA table and fill in the information we have: SSTO = 112,564; SSE = 98,356; $n = 200$; and $r = 5$. This has been done in Table 8.6. We now compute SSTR as the difference between SSTO and SSE and enter it in the appropriate place in the table. We then divide SSTR and SSE by their respective degrees of freedom to get MSTR and MSE. Finally, we divide MSTR by MSE to get the F ratio. All these quantities are entered in the ANOVA table. The result is the complete ANOVA table for the study, Table 8.7.

Table 8.7 contains all the pertinent information for this study. We are now ready to conduct the hypothesis test. The null hypothesis is that the average vacationer satisfaction for each of the five resorts is equal:

$$H_0: \mu_1 = \mu_2 = \mu_3 = \mu_4 = \mu_5$$

The alternative hypothesis is that, on the average, vacationer satisfaction is not equal among the five resorts:

$$H_1: \text{Not all } \mu_i \ (i = 1, \dots, 5) \text{ are equal}$$

As shown in Table 8.7, the test statistic value is $F_{(4, 195)} = 7.04$. As often happens, the exact number of degrees of freedom we need does not appear in the table in the

SOLUTION

Courtesy of Club Med.

Club Med Survey Results

TABLE 8.5

Resort (i)	Mean Response ($\bar{x}_i$)
1. Guadeloupe	89
2. Martinique	75
3. Eleuthera	73
4. Paradise Island	91
5. St. Lucia	85
SSTO = 112,564	SSE = 98,356

Preliminary ANOVA Table for Club Med Example

TABLE 8.6

Source of Variation	Sum of Squares	Degrees of Freedom	Mean Square	F Ratio
Treatment	SSTR =	$r - 1 = 4$	MSTR =	$F =$
Error	SSE = 98,356	$n - r = 195$	MSE =	
Total	SSTO = 112,564	$n - 1 = 199$		

ANOVA Table for Club Med Example

TABLE 8.7

Source of Variation	Sum of Squares	Degrees of Freedom	Mean Square	F Ratio	p-Value
Treatment	SSTR = 14,208	$r - 1 = 4$	MSTR = 3,552	$F = 7.04$	0.000
Error	SSE = 98,356	$n - r = 195$	MSE = 504.4		
Total	SSTO = 112,564	$n - 1 = 199$			

FIGURE 8.12   Club Med Test

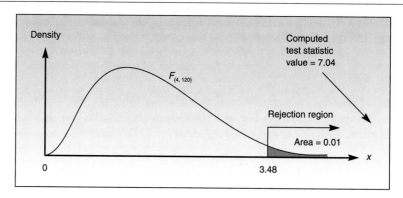

appendix. We use the nearest entry, which is the critical point for F with 4 degrees of freedom for the numerator and 120 degrees of freedom for the denominator. The critical point for $\alpha = 0.01$ is $C = 3.48$. The test is illustrated in Figure 8.12.

Since the computed test statistic value falls in the rejection region for $\alpha = 0.01$, we reject the null hypothesis and note that the p-value is smaller than 0.01. We may conclude that, based on the survey results and our assumptions, it is likely that the five resorts studied are not equal in terms of average vacationer satisfaction. Which resorts are more satisfying than others? This question may be answered by further analysis. MINITAB provides basic further analysis by reporting separate confidence intervals for the population means based on the pooled sample variance. More sophisticated analyses to follow ANOVA are also possible.

EXAMPLE 8.3

A university admissions officer wants to test whether differences exist in the average SAT mathematics scores of graduates of three high schools (here denoted A, B, and C). The scores of three groups of students, assumed to be independent random samples from these three schools, are given below. Use MINITAB to carry out the analysis.

SOLUTION

We will use the MINITAB command AOVONEWAY, which stands for "analysis of variance, one way." (A two-way analysis would include two sets of different treatments.) The command is followed by the names of the columns where data are stored.

```
MTB > SET C1
DATA> 640 655 590 610 625 585 570 620 480 500 630
DATA> SET C2
DATA> 635 670 590 630 610 700 620 650 690 710 640 705
DATA> SET C3
DATA> 570 545 600 658 590 600 612 608 550 587 637 620
DATA> END
MTB > AOVONEWAY C1 C2 C3

ANALYSIS OF VARIANCE
SOURCE      DF         SS         MS          F           p
FACTOR       2      27986      13993       7.32       0.002
```

```
ERROR       32    61187    1912
TOTAL       34    89174
                                  INDIVIDUAL 95 PCT CI'S FOR MEAN
                                  BASED ON POOLED STDEV
   LEVEL     N     MEAN   STDEV  ---------+---------+---------+-------
   C1       11   591.36   56.08   (-------*-------)
   C2       12   654.17   40.16                    (------*------)
   C3       12   598.08   33.02     (------*------)
                                  ---------+---------+---------+-------
POOLED STDEV =    43.73            595      630      665
```

The output includes the complete ANOVA table for this problem, with degrees of freedom, sums of squares, mean squares, F ratio, and p-value. Notice that MINITAB uses the term FACTOR for what we call in the text *treatment*. This terminology is quite common.

We see that the p-value is 0.002. The results are statistically significant. We have evidence, based on these small samples, that the three high schools are not all the same with respect to their students' average SAT math scores.

How different are the three high schools with respect to this variable? From the display of the 95 percent confidence intervals for average SAT math scores for the three schools we notice that high school B (in C2) has a confidence interval that is greater than, and does *not overlap,* those of high schools A (in C1) and C (in C3). These two have overlapping confidence intervals. Without using a more complicated follow-up technique, we may state that high school B is probably better than the other two (as measured by the average SAT math scores of its students), and there is probably no significant difference between the other two schools.

More on Experimental Design

Analysis of variance is strongly related with *experimental design.* For example, if we want to test the relative effectiveness of three drugs for reducing blood pressure, it is very important to design our experiment in a fair and meaningful way. If we assign our first 15 patients to take drug A, the next 15 to take drug B, and the last 15 to take drug C, we may introduce a bias. For instance, patients may have signed up for the test in an order relative to the severity of their problem. To avoid this and other biases, we must assign patients to treatments in a random way. When our data are collected, rather than produced in a randomized experiment, we need to check that sources of possible bias have been controlled.

Recall the *blocking* design, presented in Chapter 7, where we used blocks of two observations each in a paired-difference test. Using a blocking design in analysis of variance would allow us to reduce experimental error when comparing r populations.

Randomized Complete Block Design

Recall the Club Med example, Example 8.2, where we were interested in determining possible differences in average ratings among the five resorts. Suppose that Club Med can get information about its vacationers' age, sex, marital status, socioeconomic level, and so on, and can then randomly assign vacationers to the different resorts. The club could form five-member groups such that the vacationers within each group are similar to each other in age, sex, marital status, and so on. Each group of five vacationers is a block. Once the blocks are formed, one member from each block is randomly assigned to one of the five resorts (Guadeloupe, Martinique, Eleuthera, Paradise Island, or St. Lucia). Thus, the vacationers sent to each resort will comprise a mixture of ages, of males and females, of married and single people, of different socioeconomic levels, and so on. The vacationers within each block, however, will be more or less homogeneous.

The vacationers' ratings of the resorts are then analyzed using an ANOVA that utilizes the blocking structure. Since the members of each block are similar to each other (and different from members of other blocks), we expect them to react to similar conditions in similar ways. This brings about a *reduction in the experimental errors.* Why? If we cannot block, it is possible, for example, that the sample of people we get for Eleuthera will happen to be wealthier (or predominantly married, predominantly male, or whatever) and will tend to react less favorably to a resort in a developing country than a more balanced sample would react. In such a case, we will have greater experimental error. If, on the other hand, we can send one member of each homogeneous group to each of the resorts and then compare the responses of the block as a whole, we would be more likely to reduce the errors (that is, to find real differences among the resorts and not among the people). When all members of every block are randomly assigned to all treatments, such as in this example, our design is called the **randomized complete block design.**

Figure 8.13 shows the formation of blocks in the case of Club Med. We assume that the club is able—for the purpose of a specific study—to randomly assign vacationers to resorts.

FIGURE 8.13   Blocking in the Club Med Example

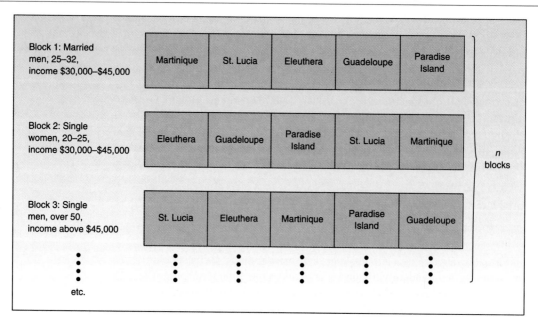

8. Define *treatment* and *error.*

9. Explain why trying to compute a simple average of all error deviations and of all treatment deviations will not lead to any results.

10. Explain how the total deviation is partitioned into the treatment deviation and the error deviation.

11. Explain the sum of squares principle.

12. Where do errors come from? That is, what do you think are their sources?

13. If, in an analysis of variance, you find that MSTR is greater than MSE, why can't you immediately reject the null hypothesis without determining the F ratio and its distribution? Explain.

14. What is the main principle behind analysis of variance?

15. Explain how information about the variance components in a data set can lead to conclusions about population means.

16. Explain the meaning of the terms *within, between, unexplained, explained,* and the context in which these terms arise.

17. By the sum of squares principle, SSE and SSTR are additive, and their sum is SSTO. Does such a relation exist between MSE and MSTR? Explain.

18. The Fidelity Overseas mutual fund consists of roughly equal proportions of Japanese and European stocks. The percentage of the fund invested in any individual country varies according to prevailing rates of return on stocks in different countries. At the end of October 1993, the fund manager was considering the possibility of shifting the proportions invested in French, Dutch, and Italian stocks. This change would be made if it could be statistically substantiated that differences in average annualized rates of return during the period ending in October existed among stocks from the three countries. Random samples of 50 French stocks, 32 Dutch stocks, and 28 Italian stocks were collected, and the annualized rate of return for each stock over the period under study was computed. Then an analysis of variance was carried out, which produced the following results: SSE = 22,399.8 and SSTO = 32,156.1. Based on these results, should the manager shift the proportions of the fund invested in the three countries? How confident are you of your answer? Construct a complete ANOVA table for this problem.

19. A study is undertaken to determine whether differences exist in average consumer quality ratings of the following brands of color television sets: Magnavox, General Electric, Panasonic, Zenith, Sears, Philco, Sylvania, and RCA. For each brand, 100 randomly chosen consumer responses are available, and from these the following quantities are computed: SSTR = 45,210; SSTO = 92,340. Construct an ANOVA table for this study, and test the null hypothesis that all eight brands have equal average consumer quality ratings versus the alternative hypothesis that they do not.

20. The Gulfstream Aerospace Company produced three different prototypes as candidates for mass production as the company's newest large-cabin business jet, the Gulfstream V. Each of the three prototypes has slightly different features, which may bring about differences in performance. Therefore, as part of the decision-making process as to which model to produce, company engineers are interested in determining whether or not the three proposed models have about the same average flight range. Each of the models is assigned a random choice of 10 flight routes and departure times, and the flight range on a full standard fuel tank is measured. (The planes carry additional fuel on the test flights, to allow them to land safely at certain destination points.) Range data for the three prototypes, in nautical miles (measured to the nearest 10 miles), are as follows:

   Do all three prototypes have the same average range? Use MINITAB, or another computer package, to construct an ANOVA table, and carry out the test. Explain your results.

21. The following is a partial computer output of an analysis of variance based on randomly chosen rents in four cities. Complete the table. Do you believe that the average rent is equal in the four cities studied? Explain.

Prototype A	Prototype B	Prototype C
4,420	4,230	4,110
4,540	4,220	4,090
4,380	4,100	4,070
4,550	4,300	4,160
4,210	4,420	4,230
4,330	4,110	4,120
4,400	4,230	4,000
4,340	4,280	4,200
4,390	4,090	4,150
4,510	4,320	4,220

```
MTB > AOVONEWAY C1, C2, C3, C4

ANALYSIS OF VARIANCE
SOURCE     DF           SS      MS      F
FACTOR      3        37402
ERROR      44       311303
TOTAL      47       348706
```

22. Interpret the following computer output:

```
ANALYSIS OF VARIANCE ON SALES
SOURCE     DF        SS      MS        F        P
STORE       2    1017.33  508.67   156.78    0.000
ERROR      15      48.67    3.24
TOTAL      17    1066.00
                                 INDIVIDUAL 95 PCT CI'S FOR MEAN
                                 BASED ON POOLED STDEV
LEVEL    N     MEAN   STDEV   -+---------+---------+---------+----
    1    6   53.667   1.862              (-*--)
    2    6   67.000   1.673                               (--*-)
    3    6   49.333   1.862          (-*---)
                                 -+---------+---------+---------+----
POOLED STDEV = 1.801             48.0   54.0   60.0   66.0
```

23. The following are student test score results for students being taught the same material but with three different teaching methods:

Method A	Method B	Method C	Method A	Method B	Method C
21	27	18	27	32	18
20	28	17	29	35	18
22	22	19	19	37	22
25	29	24	20	28	21
24	32	20	23	27	21
19	37	17	18	32	17
26	33	19	27	31	21
18	34	22	22	26	20
24	28	20	23	35	19
25	29	21	24	29	18
25	29	24	20	34	23

Are the three methods equally effective? Use $\alpha = 0.01$.

24. Complete the ANOVA table showing reading readiness data, and state your conclusions. Are there statistically significant differences among schools with respect to reading readiness? What can you say about the *p*-value?

Reading Readiness Score Data from Preschools

Source of Variation	Sum of Squares	Degrees of Freedom	Mean Square	*F* Ratio
Between groups (Preschools)	50.284	3		
Within groups (Preschools)	7214.462	116		
Total	7652.262	119		

Reprinted from "The Impact of Preschool Programs and Home Factors on the Reading Readiness of First Grade Learners," *Journal of Social and Behavioral Sciences* 36, no. 1 (Winter 1990–91), p. 32.

25. Four different television commercials are rated on a scale of 0 to 100. One commercial features a product package shot, another features a cast, a third features a celebrity, and the fourth commercial has a jingle. The advertising company testing these commercials wants to know if they are all equally liked, on average, by the population of viewers, or whether some are liked better than others. Refer to the accompanying table, and conduct the test, using $\alpha = 0.05$.

Consumer Recall Scores for Four Different Commercials

Product Package Shot	Cast	Celebrity	Jingle
45	19	61	81
40	29	71	89
52	36	68	68
36	28	59	75
56	49	65	86
57	27	58	82
35	27	49	65
61	20	43	69
49	29	62	64
40	41	52	68
50	47	47	82
58	42	41	80
59	46	48	68
62	38	47	63
65	25	39	65
63	26	44	82
39	52	62	70
38	31	65	81
46	39	61	59
47	38	59	87
48	37	70	82
46	46	68	82
47	52	58	84
65	49	69	76
62	41	*	56
*	40	*	*

26. Five different drugs used to treat AIDS patients are compared for effectiveness. An analysis of variance results in an error sum of squares 1,200 and a total sum of squares 2,650. The total sample size is 159. Are all five treatments equally effective?

27. A college student wanted to rate his fellow students' satisfaction with their accomodations; he chose to compare living in the dormitory, sharing an apartment, and sharing a house. The following scores, on a 0 to 100 scale, were given by random samples of students living in these arrangements.

Dorm: 45, 30, 0, 78, 55, 39, 15, 50, 25, 12, 45, 65, 40
Apt.: 50, 65, 30, 32, 10, 80, 45, 75, 60, 50, 55, 70, 50
House: 85, 75, 75, 80, 50, 65, 78, 90, 69, 70, 88, 70, 85, 60

Conduct the test. Who is happiest? Provide statistical evidence.

28. The following are color brightness measurements for three brands of film:

Kodak: 32, 34, 31, 30, 37, 28, 28, 27, 30, 32, 26, 29, 27, 30, 31
Fuji: 43, 41, 44, 50, 47, 32, 32, 36, 35, 34, 32, 38, 38, 40, 36
Agfa: 23, 24, 25, 21, 26, 25, 27, 26, 22, 25, 27, 30, 25, 25, 27

Assuming random sampling, are there statistically significant differences? Use $\alpha = 0.05$.

29. Explain the advantages of blocking in analysis of variance.

8.4 CONTINGENCY TABLE ANALYSIS: A CHI-SQUARE TEST FOR INDEPENDENCE

Recall the important concept of *independence of events,* which we discussed in Chapter 2. Two events, A and B, are independent if the probability of their joint occurrence is equal to the product of their marginal (i.e., separate) probabilities. This was stated as follows:

A and B are independent if P(A ∩ B) = P(A)P(B)

In this section, we will develop a statistical test that will help us determine whether or not two classification criteria, such as gender and job performance, are independent of each other. The technique will make use of **contingency tables:** tables with cells corresponding to cross-classifications of attributes or events. The basis for our analysis will be the property of independent events just stated.

Contingency tables may have several rows and several columns. The rows correspond to levels of one classification category, and the columns correspond to another. We will denote the number of rows by r, and the number of columns by c. The total sample size is n, as before. The count of the elements in cell (i, j), that is, the cell in row i and column j (where $i = 1, 2, \ldots, r$ and $j = 1, 2, \ldots, c$) is denoted by 0_{ij}. The total count for row i is R_i, and the total count for column j is C_j. The general form of a contingency table is shown in Figure 8.14. The table is demonstrated for $r = 5$ and $c = 6$. Note that n is also the sum of all r row totals and the sum of all c column totals.

The null and alternative hypotheses in the test for independence are as follows:

H_0: The two classification variables are independent of each other
H_1: The two classification variables are not independent

Now we need to find the *expected* cell counts, E_{ij}. Here is where we use the assumption that the two classification variables are independent. Remember that the

The Layout of a Contingency Table

FIGURE 8.14

Second Classification Category	First Classification Category						
	1	2	3	4	5	6	Total
1	O_{11}	O_{12}	O_{13}	O_{14}	O_{15}	O_{16}	R_1
2	O_{21}	O_{22}	O_{23}	O_{24}	O_{25}	O_{26}	R_2
3	O_{31}	O_{32}	O_{33}	O_{34}	O_{35}	O_{36}	R_3
4	O_{41}	O_{42}	O_{43}	O_{44}	O_{45}	O_{46}	R_4
5	O_{51}	O_{52}	O_{53}	O_{54}	O_{55}	O_{56}	R_5
Total	C_1	C_2	C_3	C_4	C_5	C_6	n

philosophy of hypothesis testing is to assume that H_0 is true and to use this assumption in determining the distribution of the test statistic. Then we try to show that the result is unlikely under H_0 and thus reject the null hypothesis.

Assuming that the two classification variables are independent, let's derive the expected counts in all cells. The expected number of items in a cell is equal to the sample size times the probability of the occurrence of the event signified by the particular cell. Look at a particular cell in row i and column j. In the context of an $r \times c$ contingency table, the probability associated with cell (i, j) is the probability of occurrence of event i *and* event j. Thus, the expected count in cell (i, j) is $E_{ij} = nP(i \cap j)$. If we assume independence of the two classification variables, then event i and event j are independent events, and, by the law of independence of events, $P(i \cap j) = P(i)P(j)$.

From the row totals, we can estimate the probability of event i as R_i/n. Similarly, we estimate the probability of event j by C_j/n. Substituting these estimates of the marginal probabilities, we get the following expression for the expected count in cell (i, j): $E_{ij} = n(R_i/n)(C_j/n) = R_iC_j/n$.

The expected count in cell (i, j):

$$E_{ij} = \frac{R_iC_j}{n}$$

This equation allows us to compute the expected cell counts. These, along with the observed cell counts, are used in computing the value of a *chi-square* statistic, which leads us to a decision about the null hypothesis of independence.

Chi-square test statistic for independence:

$$\chi^2 = \underset{\text{all cells}}{\overset{\text{Sum}}{\text{over}}} \left[\frac{(O_{ij} - E_{ij})^2}{E_{ij}} \right]$$

The statistic is equal to the *squared difference between the observed count and the expected count in each cell, divided by the expected count, summed over all cells.* The degrees of freedom of the chi-square statistic are df = $(r - 1)(c - 1)$.

One important note:

> *Use of the chi-square statistic assumes that all, or at least most, of the expected cell counts are at least 5.*

The following subsection explains the chi-square distribution and gives an example of how the chi-square statistic is used to test for independence.

The Chi-Square Distribution

Chi (pronounced $k\bar{\imath}$) is a Greek letter denoted by χ. Hence, we denote the chi-square distribution by χ^2.

> The **chi-square distribution** is the probability distribution of the sum of several independent, squared standard normal random variables.

As a sum of squares, the chi-square random variable cannot be negative and is therefore bounded on the left by zero. The resulting distribution is skewed to the right. The chi-square distribution, like the *t* distribution, has associated with it a degrees of freedom parameter, df. Unlike the *t* and the normal distributions, however, the chi-square distribution is *not* symmetric. It is similar in shape to the *F* distribution. Figure 8.15 shows several chi-square distributions with different numbers of degrees of freedom.

> *The mean of a chi-square distribution is equal to the degrees of freedom parameter, df. The variance of a chi-square distribution is equal to twice the number of degrees of freedom.*

Note in Figure 8.15 that as df increases, the chi-square distribution looks more and more like a normal distribution. In fact, as df increases, the chi-square distribution approaches a normal distribution with mean df and variance 2(df).

FIGURE 8.15   Several Chi-Square Distributions with Different Values of the df Parameter

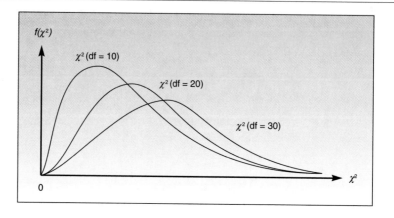

Values and Probabilities of Chi-Square Distributions **TABLE 8.8**

df	\multicolumn{10}{c}{Area in Right Tail}
df	.995	.990	.975	.950	.900	.100	.050	.025	.010	.005
1	$.0^4393$	$.0^3157$	$.0^3982$	$.0^2393$	0.158	2.71	3.84	5.02	6.63	7.88
2	.0100	.0201	.0506	.103	.211	4.61	5.99	7.38	9.21	10.6
3	.0717	.115	.216	.352	.584	6.25	7.81	9.35	11.3	12.8
4	.207	.297	.484	.711	1.06	7.78	9.49	11.1	13.3	14.9
5	.412	.554	.831	1.15	1.61	9.24	11.1	12.8	15.1	16.7
6	.676	.872	1.24	1.64	2.20	10.6	12.6	14.4	16.8	18.5
7	.989	1.24	1.69	2.17	2.83	12.0	14.1	16.0	18.5	20.3
8	1.34	1.65	2.18	2.73	3.49	13.4	15.5	17.5	20.1	22.0
9	1.73	2.09	2.70	3.33	4.17	14.7	16.9	19.0	21.7	23.6
10	2.16	2.56	3.25	3.94	4.87	16.0	18.3	20.5	23.2	25.2
11	2.60	3.05	3.82	4.57	5.58	17.3	19.7	21.9	24.7	26.8
12	3.07	3.57	4.40	5.23	6.30	18.5	21.0	23.3	26.2	28.3
13	3.57	4.11	5.01	5.89	7.04	19.8	22.4	24.7	27.7	29.8
14	4.07	4.66	5.63	6.57	7.79	21.1	23.7	26.1	29.1	31.3
15	4.60	5.23	6.26	7.26	8.55	22.3	25.0	27.5	30.6	32.8
16	5.14	5.81	6.91	7.96	9.31	23.5	26.3	28.8	32.0	34.3
17	5.70	6.41	7.56	8.67	10.1	24.8	27.6	30.2	33.4	35.7
18	6.26	7.01	8.23	9.39	10.9	26.0	28.9	31.5	34.8	37.2
19	6.84	7.63	8.91	10.1	11.7	27.2	30.1	32.9	36.2	38.6
20	7.43	8.26	9.59	10.9	12.4	28.4	31.4	34.2	37.6	40.0
21	8.03	8.90	10.3	11.6	13.2	29.6	32.7	35.5	38.9	41.4
22	8.64	9.54	11.0	12.3	14.0	30.8	33.9	36.8	40.3	42.8
23	9.26	10.2	11.7	13.1	14.8	32.0	35.2	38.1	41.6	44.2
24	9.89	10.9	12.4	13.8	15.7	33.2	36.4	39.4	43.0	45.6
25	10.5	11.5	13.1	14.6	16.5	34.4	37.7	40.6	44.3	46.9
26	11.2	12.2	13.8	15.4	17.3	35.6	38.9	41.9	45.6	48.3
27	11.8	12.9	14.6	16.2	18.1	36.7	40.1	43.2	47.0	49.6
28	12.5	13.6	15.3	16.9	18.9	37.9	41.3	44.5	48.3	51.0
29	13.1	14.3	16.0	17.7	19.8	39.1	42.6	45.7	49.6	52.3
30	13.8	15.0	16.8	18.5	20.6	40.3	43.8	47.0	50.9	53.7

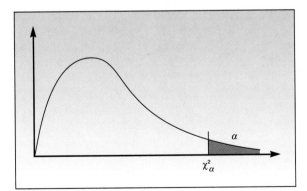

Table B.5 in Appendix B gives values of the chi-square distribution with different degrees of freedom, for given tail probabilities. An abbreviated version of the table (with numbers rounded up to fewer decimals) is shown as Table 8.8. As an example, to help you read the table, the probability that a chi-square random variable with 30 degrees of freedom will exceed 47.0 is 0.025.

Example 8.4 shows how the test for independence is carried out using the chi-square distribution.

EXAMPLE 8.4

An article in *Business Week* reports profits and losses of firms by industry. A random sample of 100 firms is selected, and for each firm in the sample, we record whether the company made money or lost money, and whether or not the firm is a service company. The data are summarized in the 2×2 contingency table, Table 8.9. Using the information in the table, determine whether or not you believe that the two events "the company made a profit this year" and "the company is in the service industry" are independent.

SOLUTION

Table 8.9 is the table of observed counts. We now use its marginal totals, R_1, R_2, C_1, and C_2, as well as the sample size, n, in creating a table of expected counts:

$$E_{11} = R_1 C_1/n = (60)(48)/100 = 28.8$$

$$E_{12} = R_1 C_2/n = (60)(52)/100 = 31.2$$

$$E_{21} = R_2 C_1/n = (40)(48)/100 = 19.2$$

$$E_{22} = R_2 C_2/n = (40)(52)/100 = 20.8$$

Next we arrange these values in a table of expected counts, Table 8.10. Using the values shown in the table, we can compute the chi-square test statistic:

$$\chi^2 = \frac{(42 - 28.8)^2}{28.8} + \frac{(18 - 31.2)^2}{31.2} + \frac{(6 - 19.2)^2}{19.2} + \frac{(34 - 20.8)^2}{20.8} = 29.09$$

To conduct the test, we compare the computed value of the statistic with critical points of the chi-square distribution with $(r - 1)(c - 1) = (2 - 1)(2 - 1) = 1$ degree of

TABLE 8.9   Contingency Table of Profit/Loss versus Industry Type

	Industry Type		
	Service	**Nonservice**	**Total**
Profit	42	18	60
Loss	6	34	40
Total	48	52	100

TABLE 8.10   Expected Counts (with the Observed Counts Shown in Parentheses) for Example 8.4

	Service	**Nonservice**
Profit	28.8	31.2
	(42)	(18)
Loss	19.2	20.8
	(6)	(34)

freedom. From Table B.5 in Appendix B, we find that the critical point for $\alpha = 0.01$ is 6.63, and, since our computed value of the χ^2 statistic is much greater than the critical point, we reject the null hypothesis and conclude that the two qualities, profit/loss and industry type, are probably not independent.

A Chi-Square Test for Equality of Proportions

Contingency tables and the chi-square statistic are also useful in another kind of analysis. Sometimes we are interested in whether the proportion of some characteristic is equal in several populations. An insurance company, for example, may be interested in finding out whether the proportion of people who submit claims for automobile accidents is about the same for the three age groups: 25 and under, 26 to 49, and 50 and over. In a sense, the question of whether the proportions are equal is a question of whether the three age populations are *homogeneous* with respect to accident claims. Therefore, tests of equality of proportions across several populations are also called tests of *homogeneity*.

The analysis is carried out in exactly the same way as in the previous application. We arrange the data in cells corresponding to population-characteristic combinations, and for each cell, we compute the expected count based on its row and column totals. The chi-square statistic is computed exactly as before. Two things are different in this analysis. First, we identify our populations of interest *before* the analysis and sample directly from these populations. Contrast this with the previous application, where we sampled from *one* population and then cross-classified according to two criteria. Second, because we identify populations and sample from them directly, the sizes of the samples are *fixed*. This is called *a chi-square analysis with fixed marginal totals*. This fact, however, does not affect the analysis.

We will demonstrate the analysis with the insurance company example just mentioned. The null and alternative hypotheses are as follows:

H_0: The proportion of claims is the same for all three age groups (i.e., the age groups are homogeneous with respect to claim proportions)

H_1: The proportion of claims is not the same across age groups (the age groups are not homogeneous)

Suppose that random samples, selected from company records for the three age categories, are classified according to *claim* versus *no claim* and counted. The data are presented in Table 8.11.

To carry out the test, we first calculate the expected counts in all the cells, as before. The expected count in each cell is equal to the row total times the column total, divided by the total sample size (the pooled sample size from all populations). The reason for the formula in this new context is that if the proportion of items in the class of interest (here, the proportion of people who submit a claim) is equal across all populations, as stated in the null hypothesis, then *pooling* this proportion across populations gives us the expected proportion in the cells for the class. Thus, the expected proportion in the claim class is estimated by the total in the claim class divided by the grand total, or $R_1/n = 135/300 = 0.45$. If we multiply this pooled

TABLE 8.11   Data for the Insurance Company Example

	Age Group			
	25 and Under	**26 to 49**	**50 and Over**	**Total**
Claim	40	35	60	135
No Claim	60	65	40	165
Total	100	100	100	300

TABLE 8.12   The Expected Counts for the Insurance Company Example

	25 and Under	**26 to 49**	**50 and Over**	**Total**
Claim	45	45	45	135
No Claim	55	55	55	165
Total	100	100	100	300

proportion by the total number in the sample from the population of interest (say, the sample of people 25 and under), this should give us the *expected* count in the cell *claim–25 and under.* We get $E_{11} = C_1(R_1/n) = (C_1 R_1)/n$. This is exactly as prescribed in the test for independence. Here we get $E_{11} = (100)(0.45) = 45$. This is the expected count under the null hypothesis. We compute the expected counts for all other cells in the table in a similar manner. Table 8.12 is the table of expected counts in this example.

Note that since we have used equal sample sizes (100 from each age population), the expected count is equal in all cells corresponding to the same class. The proportions are expected to be equal under the null hypothesis. Since these proportions are multiplied by the same sample size, the counts are also equal.

We are now ready to compute the value of the chi-square test statistic:

$$\chi^2 = \underset{\text{all cells}}{\overset{\text{Sum}}{\text{over}}} \frac{(O - E)^2}{E} = \frac{(40 - 45)^2}{45} + \frac{(35 - 45)^2}{45} + \frac{(60 - 45)^2}{45}$$
$$+ \frac{(60 - 55)^2}{55} + \frac{(65 - 55)^2}{55} + \frac{(40 - 55)^2}{55} = 14.14$$

The degrees of freedom are obtained as before. We have three rows and two columns, so the degrees of freedom are $(3 - 1)(2 - 1) = 2$.

Comparing the computed value of the statistic with critical points of the chi-square distribution with two degrees of freedom, we find that the null hypothesis may be rejected and that the *p*-value is less than 0.01. (Check this, using the table in the appendix.) We conclude that the proportions of people who submit claims to the insurance company are not the same across the age groups studied.

In general, when we compare *c* populations (or *r* populations, if they are arranged as the rows of the table rather than the columns), the hypotheses may be written as follows:

$$H_0: p_1 = p_2 = \cdots = p_c$$

$$H_1: \text{Not all } p_i, i = 1, \ldots, c, \text{ are equal}$$

where p_i ($i = 1, \ldots, c$) is the proportion in population i of the characteristic of interest.

The chi-square test for equality of proportions may also be applied to *several* proportions within each population. That is, instead of just testing for the proportion of *claim* versus *no claim,* we could be testing a more general hypothesis about the proportions of *several* different types of claims: no claim, claim under $1,000, claim of $1,000 to $5,000, and claim over $5,000. Here the null hypothesis would be that the proportion of each type of claim is equal across all populations. (This does not mean that the proportions of all types of claims are equal within a population.) The alternative hypothesis would be that not all proportions are equal across all populations under study. The analysis is done using an $r \times c$ contingency table (instead of the 2 $\times$ c table we used in the preceding example). The test statistic is the same, and the degrees of freedom are as before: $(r - 1)(c - 1)$. This situation is demonstrated, using MINITAB, in Example 8.5.

EXAMPLE 8.5

Let's assume that there were 10 "no claims" for people 25 and under, 12 claims under $1,000 in this age group, 16 claims between $1,000 and $5,000, and 29 claims over $5,000. Similarly, for the 26- to 49-year-old group these numbers were 4, 29, 18, and 70. Finally, for the 50 and over group the numbers were 23, 40, 45, and 17. Assuming random sampling, is there statistical evidence that the proportions of the different claim-amount categories are different among the three age groups?

SOLUTION

In MINITAB, the command is very simple: CHISQUARE. The counts, arranged in a table of rows and columns, are then read either column by column or row by row into the usual MINITAB columns (C1, C2, C3, etc.):

```
MTB > SET C1              DATA> SET C3
DATA> 10 12 16 29         DATA> 23 40 45 17
DATA> SET C2              DATA> END
DATA> 4 29 18 70          MTB > CHISQUARE C1 C2 C3
```

Expected counts are printed below observed counts:

	C1	C2	C3	Total
1	10	4	23	37
	7.92	14.30	14.78	
2	12	29	40	81
	17.34	31.31	32.35	
3	16	18	45	79
	16.91	30.54	31.55	
4	29	70	17	116
	24.83	44.84	46.33	
Total	67	121	125	313

ChiSq = 0.546 + 7.422 + 4.577 +
        1.644 + 0.171 + 1.810 +
        0.049 + 5.149 + 5.734 +
        0.700 + 14.112 + 18.564 = 60.479

df = 6

Comparing this value of the test statistic with tabulated values for a chi-square random variable with six degrees of freedom, we find that the *p*-value is much less than 0.005. Thus, there is strong statistical evidence that the proportions in question are not equal across age groups.

PROBLEMS

30. Refer to the table "Who Smokes?" If all the data are based on independent random samples of 1,000 people in each category, test for the appropriate significance of differences.

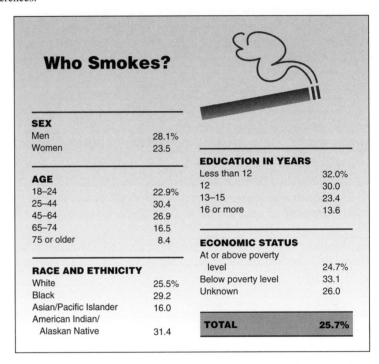

Who Smokes?

SEX		**EDUCATION IN YEARS**	
Men	28.1%	Less than 12	32.0%
Women	23.5	12	30.0
		13–15	23.4
AGE		16 or more	13.6
18–24	22.9%		
25–44	30.4	**ECONOMIC STATUS**	
45–64	26.9	At or above poverty	
65–74	16.5	level	24.7%
75 or older	8.4	Below poverty level	33.1
		Unknown	26.0
RACE AND ETHNICITY			
White	25.5%	**TOTAL**	**25.7%**
Black	29.2		
Asian/Pacific Islander	16.0		
American Indian/			
Alaskan Native	31.4		

Reprinted by permission from "Decline in Smoking Levels Off," *The New York Times,* April 2, 1993, p. A21, with data from the Centers for Disease Control Office of Smoking and Health.

The Hidden Price of Saving Lives

Why Air Bags Can Mean Higher Medical Bills

By WARREN BROWN
WASHINGTON POST STAFF WRITER

The Insurance Institute for Highway Safety, which conducts auto-safety research for the nation's property and casualty insurers, says its study of 12,000 head-on collisions between 1985 and 1991 shows that air bags reduced fatalities by 31 percent when drivers did not wear seat belts. When the drivers of cars with air bags also were belted, fatalities declined 51 percent.

Reprinted by permission from the *Washington Post National Weekly Edition,* March 29–April 4, 1993, p. 22.

31. Refer to "The Hidden Price of Saving Lives." Assume two independent random samples of 6,000 accidents each, and assume that the sample with only airbags had 500 fatalities, while the sample with both air bags and seat belts had 255 fatalities. Construct a contingency table and test for significance.

32. Look at the snooze-button survey results for Bentley College. A similar survey at competing Babson College, based on a random sample of 100 students, revealed the following percentages of number of times the snooze button is pressed. Once: 1 percent; twice: 12 percent; three times: 25 percent; four times: 22 percent; five times: 15 percent; six or more times: 25 percent. (In neither sample did anyone press the button *zero* times and just wake up.)

Is there statistical evidence of a difference in the proportions between the two colleges? Can you say that one college's students sleep longer? Explain.

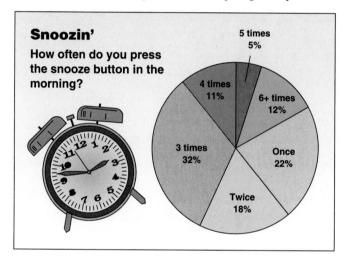

Source: Marketing Department, *The Vanguard,* Bentley College, April 12, 1994, p. 1; based on a survey of 100 students.

Problems 33–35 refer to the Mount Snow Ski Survey on page 394.

33. Out of a total of 1,538 respondents, 1,129 were alpine skiers, 313 were snowboarders, 89 were both, and the rest were nonskiers. Of the alpine skiers, 105 described themselves as super skiers, 631 as advanced intermediates, 202 as upper intermediates, 78 as lower intermediate, and the rest as beginners. Of the snowboarders, 80 were super skiers, 109 advanced intermediate, 16 upper intermediate, 19 lower intermediate, and the rest beginners. Among those who were both alpine skiers and snowboarders, all but 5 called themselves super skiers; the remaining 5 were upper intermediates. Conduct a chi-square analysis of these data and state your conclusions. Be careful with assumptions. Make conclusions only where you can make them.

34. Of all the survey respondents, 905 were married with children, 358 were married without children, 208 were single (without children), and 67 were single with children at home. Of the married people with children, 864 said they will return to Mount Snow this season, 22 said they will not return, and the rest said maybe. Of the people married without children, 152 said they will return, 60 said they will not, and the rest said maybe. Of the single people, 29 said they will return, 155 said they will not, the rest maybe. And for the single with children, 16 said they will return and the rest not.

What kind of a ski resort is Mount Snow? (Who are their most loyal customers?) Provide statistical support for your conclusions.

35. In the survey, when the data for Questions 2 and 6 were condensed, the following emerged: Of people with children, 40 percent skied 10 or fewer days each season, the rest 11 or more days. Of people without children, 25 percent skied 10 or fewer days each season, the rest 11 or more days. Conduct the appropriate statistical test and state your conclusions.

MOUNT SNOW / HAYSTACK ENTRY FORM
Forms must be filled out completely in order to be eligible for drawing.

Name:_____

Address:_____

City:_____

State/Province:_____ Zip:_____

Phone #:_____ Date of Birth:___/___/___

1. Are you: (1) ___ Male   (2) ___ Female

2. Are you: (1) ___ married with children at home
                   (2) ___ married without children at home
                   (3) ___ single
                   (4) ___ single with children at home

3. If you have children 18 years or younger living at home, what is the birthdate of your youngest child: ___/___/___

4. Are you a(n):   (1) ___ alpine skier
                   (2) ___ snowboard skier
                   (3) ___ both

5. Skiing ability: (1) ___ Never skied
                   (2) ___ Beginner
                   (3) ___ Lower intermediate
                   (4) ___ Upper intermediate
                   (5) ___ Advanced intermediate
                   (6) ___ Super skier

6. How many days do you ski each season at all ski areas?
   (1) ___ N/A   (5) ___ 11–15 days
   (2) ___ 1–3 days   (6) ___ 16–20 days
   (3) ___ 4–5 days   (7) ___ 21+ days
   (4) ___ 6–10 days

Date:_____

7. How many ski trips will you make to Mount Snow this season?
   (1) ___ 1   (3) ___ 3   (5) ___ 5–9
   (2) ___ 2   (4) ___ 4   (6) ___ 10+

8. Will you return to Mount Snow this season?
   (1) ___ Yes   (3) ___ Maybe
   (2) ___ No

9. Have you ever been to Mount Snow before this trip?
   (1) ___ Yes   (2) ___ No

10. Do you plan to purchase a 4x4 sport utility vehicle within 6 months to a year?
   (1) ___ Yes   (2) ___ No

11. Does your schedule allow you to ski Monday through Friday non-holiday?
   (1) ___ Yes   (2) ___ No

12. How many days do you ski at Mount Snow each season?
   (1) ___ N/A   (5) ___ 11–15 days
   (2) ___ 1–3 days   (6) ___ 16–20 days
   (3) ___ 4–5 days   (7) ___ 21+ days
   (4) ___ 6–10 days

13. I have skied (check all that apply):
   (1) ___ N/A   (3) ___ Between X-mas and mid-March
   (2) ___ Before X-mas   (4) ___ After mid-March

14. Are you currently attending college?
   (1) ___ Yes   (2) ___ No

15. I participate in the following activities (check all that apply):
   (1) ___ Golf   (3) ___ Tennis
   (2) ___ Bicycling   (4) ___ Mountain biking

Mount Snow
SOUTHERN VERMONT
NO ONE ELSE IS CLOSE.

Problems 36–41 refer to the table showing results of the Rogaine study.

36. Is there statistical evidence that Rogaine may be associated with respiratory, gastrointestinal, or cardiovascular disorders (when compared with the placebo)?

37. Is there statistical evidence that Rogaine may be associated with neurological, special senses, or psychiatric problems?

38. Is there statistical evidence that Rogaine may be associated with hematological, dermatological, or endocrinal disorders?

39. Is there statistical evidence that Rogaine may be associated with allergies, or urinary or genital tract disorders?

40. Conduct a complete single test for all body system problems for comparison of Rogaine's side effects as compared with the placebo.

41. What assumptions did you have to make when analyzing the Rogaine study results? State the assumptions as well as all your conclusions. Should Rogaine be marketed? Explain.

42. Discuss and compare the chi-square tests described in Section 8.3. When are the tests appropriate? What assumptions are necessary? How should conclusions be framed?

Source: Data provided courtesy of the Upjohn Company.

SUMMARY

In this chapter we introduced two important methods for testing for differences among several population parameters. First, **analysis of variance (ANOVA)** was presented as the method of carrying out a joint test for the equality of several population means. We saw that an **ANOVA table,** listing the sources of variation: **treatment, error,** and **total;** their **sums of squares, degrees of freedom, mean squares,** and **F-ratio,** is an integral part of the analysis. We saw that the sum-of-squares is a key concept in the analysis of variance. We introduced the **F distribution** and saw how it applies in ANOVA to test the null hypothesis of equal population means. Then we presented the **chi-square distribution** and its uses in testing for **differences among several population proportions.** We also saw how this statistic can be used in testing for the **independence** of two categorical variables.

KEY FORMULAS

The *F*-statistic in ANOVA:

$$F_{(r-1,\, n-r)} = \frac{\text{MSTR}}{\text{MSE}}$$

The chi-square test statistic:

$$\chi^2_{(r-1)\times(c-1)} = \underset{\text{cells}}{\overset{\text{Sum}}{\text{over all}}} \frac{(O - E)^2}{E}$$

ADDITIONAL PROBLEMS

When the Season Changes, So Do the Tourists

Characteristics of Colorado tourists, 1991

Summer vacationers		Winter (ski) vacationers
37	Age of head of household	37
68%	Married	55%
17%	Never married	30%
45%	Have children	36%
$41,500	Median income	$57,700
44.9%	White collar	81.6%
50.2%	College graduate	73.5%
66% arrive by car	Travel	66% arrive by air
797 miles	Median distance traveled	891 miles
5.3 days	Average stay	5.8 days
$44	Daily amount spent per person	$83

Reprinted by permission from *The New York Times,* April 13, 1993, p. B1, with data from the Colorado Tourism Board.

Problems 43–47 refer to the table on Colorado's tourism trade. Assume the table is based on two independent random samples of 1,500 tourists each.

43. If in both samples the average age of head of household is exactly 37, discuss what this would do to the various components of an ANOVA on the two groups of tourists aimed at testing for equality of average age. What would be the conclusion? Compare this with what would happen if a two-sample *t*-test were to be carried out to test the same hypotheses.

44. Is there a statistically significant difference in the percentage of married and never married tourists in summer versus winter?

45. Compare statistically the two types of tourists with respect to their proportions arriving by air versus by car. Are the results statistically significant? Explain.

46. Are there statistically significant differences between the two types of tourists with respect to the proportion of white-collar workers?

47. Are there statistically significant differences in the percentages of college graduates between winter and summer tourists?

48. Read the article on students' excuses on page 398. Assume the following contingency table for excuses in schools in the United States, France, and Britain:

	Dead Grandparent	Car Problem	Animal Trauma	Crime Victim
United States	158	187	12	65
France	22	90	239	4
Britain	220	45	8	125

Conduct the appropriate test and state your conclusions.

49. The following are the results of a statistical comparison to choose the best cruise line based on passenger ratings:

```
ANALYSIS OF VARIANCE
SOURCE     DF      SS      MS      F
FACTOR     3
ERROR      63      580
TOTAL      66      2825
```

Conduct the test, and explain your results.

50. In testing the effectiveness of a drug, it is usually compared with no treatment and with a placebo. The following data, in days to disappearance of symptoms, are reported. Conduct the test.

Drug: 2, 2, 3, 2, 5, 4, 6, 12, 1, 2, 11, 5, 3

Placebo: 3, 4, 6, 4, 1, 7, 4, 6, 5, 15, 12, 4, 10

Nothing: 8, 7, 9, 13, 23, 10, 9, 5, 19, 15, 9, 17, 7

51. Construct a 95 percent confidence interval for the average number of days to disappearance of symptoms for the "drug," "placebo," and "nothing" groups in Problem 50. Use MINITAB.

52. Three ads are being tested for effectiveness. The data are as follows:

A: 12, 17, 34, 11, 5, 42, 18, 27, 2, 37, 50, 32, 12, 27, 21, 10, 4, 33, 63, 22, 41, 19, 28, 29, 8

B: 44, 32, 28, 30, 22, 12, 3, 12, 42, 13, 27, 54, 56, 32, 37, 28, 22, 22, 24, 9, 20, 4, 13, 42, 67

Beyond "Dog Ate My Homework"

Douglas A. Bernstein, a psychology professor at the University of Illinois, recently asked faculty members on an electronic-mail network for "the most unusual, bizarre and amazing student excuses" they had ever heard. He got dozens of replies.

"As you read," he told readers of *APS Observer,* the journal of the American Psychological Society, "keep in mind that many of these excuses (and most that weren't funny enough to include) actually turn out to be true. As one contributor put it, 'It is easy to forget that our students have lives outside of class, and their lives are as chaotic as ours.' "

Here is a sampling:

GRANDPARENTAL DEATH This old favorite needs no description, but one professor's class established what must be a world's record when 14 out of 250 students reported their grandmothers' dying just before final exams. In another class, one student said he could not take the midterm because his grandmother had died. When the instructor expressed condolences a week later, the student replied, "Oh, don't worry. She was terminal, but she's feeling much better now."

AUTOMOBILE PROBLEMS "I had an accident, the police impounded my car, and my paper is in the glove compartment."

ANIMAL TRAUMA "I can't be at the exam because my cat is having kittens and I'm her coach."

"I can't take the test because my dog is having a Caesarean section."

CRIME VICTIMIZATION "I need to take the final early because the husband of the woman I am seeing is threatening to kill me."

"I missed the final because when I went to the convenience store yesterday, it was robbed and the robber locked me and the clerk in the basement until this morning."

OTHER "I'm too happy to give my presentation tomorrow." (The contributor noted, "This was easily fixed.")

"My paper is late because I lost a pair of eyeballs and I couldn't do anything else till I found them." (The student worked in an eye bank.)

"I'm too depressed to take the exam; I just found my girlfriend in bed with another man."

On a note slipped under a laboratory door before an experiment: "I am unable to come to lab because I don't have time."

Finally, there is the excuse given by two students who, after sitting next to each other during an exam, were asked why their answer sheets contained identical responses to different versions of the test: "We studied together."

Reprinted by permission from *The New York Times,* April 28, 1993, p. A19.

C: 32, 33, 21, 12, 15, 14, 55, 67, 72, 1, 44, 60, 36, 38, 49, 66, 89, 63,
   23, 6, 9, 56, 28, 39, 59

Is one ad more effective than the others? Explain. What is the *p*-value?

53. Explain the difference between ANOVA and the chi-square tests in this chapter.

54. Mr. Gonen Nissim, coffee roaster at Beans Coffee Boutique in Brookline, Massachusetts, wants to know whether the average time it takes to roast three kinds of coffee (Kenyan, Colombian, Costa Rican) to a particular roast level (Full City) is equal. The time to roast, in minutes, for 10 random batches of the three coffees are given below.

Kenya: 17.1 16.5 15.4 16.1 14.5 17.5 15.2 16.2 15.3 15.4

Colombia: 17.3 17.5 19.1 18.2 16.5 19.1 17.0 15.2 19.0 16.0

Costa Rica: 18.0 17.0 19.4 18.0 19.0 16.5 16.3 15.5 17.4 16.3

Conduct the test using the 0.05 level of significance.

55. A recent article reported an analysis of variance of listening scores for riddles of four kinds.[6] The reported results were: $F(3, 81) = 94.63$. Are the results statistically significant?

---

[6] H. Marmurek and M. Rossi, "Processing of Ambiguous Words," *Journal of Genetic Psychology* 154, no. 4 (1993), pp. 475–86.

INTERVIEW WITH
Nancy Kirkendall

"I personally think it is likely that there will be another energy crisis. After all, the world is highly dependent on energy and there are many ways an energy crisis could occur."

Nancy Kirkendall is Chief Statistician in the Energy Information Agency that is part of the Department of Energy. She was trained in engineering at Ohio State and received her PhD from George Washington University in statistics and worked at Bellcom as a programmer and at Mitre Corporation as a systems engineer.

Aczel: Where did you go to school and how did you get interested in statistics?

Kirkendall: My first degree was a Bachelor of Science in mathematics from Ohio State University. After I graduated I got a summer job at North American Aviation, as a computer programmer in an Analog Hybrid Simulation Group; however, my title with this job was engineer, which I thought was rather amusing. At that time I didn't know anything about computers, however, and they hired people with a background in mathematics because they knew that with that background you could pick up programming easily. And, as it turned out, they were right; I found the programming easy. In those days, the computers we used were DDP 2400 computers that had Fortran 2 on it and we used machine language as well. So statistics were involved in the job, even though I was there as just a programmer.

I got my Master's degree at Ohio State and as part of that curriculum I took my first statistics courses, which I thoroughly enjoyed—probably because I thought they were more practical than the mathematics courses I had been taking.

Aczel: When you graduated, you went . . .

Kirkendall: After graduation my husband went into the Army and was transferred to Washington, D.C. I got a job at Bellcom, which was the smallest company in the Bell system. Bellcom was organized to work on space projects for NASA, and I was hired as an applications programmer, under the condition that I would go ahead to graduate school and pursue a PhD. I was thrilled because it was what I wanted to do anyhow.

My new boss suggested that I should go to George Washington University and major in statistics. He evidently knew that it was a good, convenient program in statistics, and he thought that I would like it. At Bellcom I started moving from programming into applied statistics. My husband's company, however, bid on a job in Vietnam, and when they got that contract, they wanted very much for some of the people who had worked on the job to go. My husband said that he would go if they hired me. As a result, we spent a year in Saigon, working as senior operations research analysts.

One of the interesting projects we worked on there was a survey used to determine the quality of life and the extent of Viet Cong influence in the country. This was done on a large management information computer system. The input was from questionnaires that were filled out by people who were living in the hamlets and villages scattered around the country.

We came back to Washington, and I spent a year completing my PhD in statistics. After that I went to work for the Mitre Corporation working on projects for the Federal Aviation Administration [FAA] concerning separation standards for aircraft and air routes. Separation standards are basically rules for how far apart routes should be spaced or how far apart aircraft should fly to keep collision risk at a minimum.

Finally, 15 years ago I came to the Energy Information Administration [EIA]. The person who hired me is the same one I work for now, Dr. Yvonne Bishop, who is a well-known statistician. This was my first job as a real statistician, and I was thrilled.

Aczel: What role does statistics play in the department's business?

Kirkendall: Well, I can certainly talk about the role of statistics in the Energy Information Administration's business, because EIA is the statistical agency for the Department of Energy. We collect and publish data from energy producers, suppliers, and consumers. We operate forecasting models which use either econometric models or linear programming methods. And, in addition, we are currently advising the Department of Energy on implementing customer satisfaction surveys. This is something a little bit new for us, and it has been an interesting exercise. And that's one that I have been particularly involved with as well.

Aczel: Where does the EIA get data on such things as the usage rate for different kinds of energy?

Kirkendall: EIA conducts large-scale sample surveys to estimate energy use in homes, in commercial buildings, and in manufacturing establishments. Those are three separate surveys. We collect actual billing data for the sampled units from the utilities that provide the energy—from the electric power companies or, if natural gas is used, from the gas company that supplies it. We get good solid data on the actual energy used in either the home or the commercial building.

Those surveys collected a lot of information that could be used for explaining why certain buildings consume more or less energy than other buildings of the same type; that is, we collected information about the number of appliances they have and operate, the age of the refrigerators, whether they've replaced their windows, whether their hot water heater is wrapped, whether they have special lighting in commercial buildings or in manufacturing establishments—almost anything that might be a descriptor or might be related to energy savings.

Aczel: Does the department keep track of foreign sources and supplies? How and where does that data come from?

Kirkendall: We collect information on energy imports into the country by country of origin. Those data are actually from the U.S. Customs Service. We also have a stocks-at-sea report for petroleum that is in tankers at sea and that is a more long-range prediction of imports. The United States is a member of the International Energy Agency, and we therefore have to provide a lot of energy information to the IEA, as do all other member countries. As part of that affiliation, then, we have access to the data from the other member countries. We also collect some foreign data directly from other countries that are not members of the IEA. We do not go out and interview companies that work inside other countries. We rely on the foreign government to provide information to us.

Aczel: The Department does assessments of future energy needs for the United States. How are those done and what tools do statisticians and forecasters use in the process?

Kirkendall: EIA has two modeling systems—a short-term modeling system that predicts about a year and a half into the future and the National Energy Modeling System (NEMS), which predicts about thirty years into the future. The short-term modeling system uses some statistical time series methods and some econometric models. These tend to be smaller scale models—they could probably all be put on a PC (although I believe that some of them are still on a mainframe, just for the sake of convenience).

The National Energy Modeling System is a fairly new effort and that relies on econometric models and linear programming methods. And the modulars for NEMS are gigantic programs, some of which take hours to run, even on the mainframe computers.

Aczel: Do those models get revised? What do you see as the next major changes to the forecasting system?

Kirkendall: The short-term models are the older set of models and they are continually revised and improved. As people have worked on them they will notice that they haven't done a good job forecasting in the past and they will revise them and make improvements.

The National Energy Modeling System is new and, in fact, it's probably not even complete as yet. They have been under a serious time crunch to put this system together and I'm sure as they've done it, they know where there are weak places in the model and they will be continually working to keep that one up-to-date as well.

Aczel: The energy industry, or at least the oil industry, has been down and out the last few years. Does your Department or do these models predict a comeback?

Kirkendall: I can't speak for the Department and I have not looked that closely at specific model forecasts. However, I would think that the models most likely predict a slow increase in the price of oil. Our supplies of oil are finite and we do not yet have replacement technology securely lined up, so it would make sense that the oil industry would experience a comeback, unless there is some grand, inexpensive environmentally friendly replacement fuel or technology discovered before then. However, we don't see any of them on the horizon just yet.

Aczel: In the 1970s and 1980s the oil crisis was projected to stick around. What happened to it? How did good or bad forecasting or data analysis affect the results?

Kirkendall: Well, I'm not sure exactly what happened. I guess economists would probably tell you the supply of oil exceeded demand by a fair amount, but in early 1986, the price of oil plummeted. As a result U.S. production fell by 4 percent per month and it was no longer economically feasible for companies to operate marginal wells. There was little incentive to explore for new wells. In the Energy Information Administration, our official data series did not show this 4 percent drop. Because we get data on crude oil production too late to include in our monthly publications, we used forecasts that were just extrapolations. And these estimates were not revised for a full year because of our publication policy. These estimates worked real well up until early 1986 because crude oil production was relatively constant, nobody thought there would be a drop like that.

In 1986, however, the oil industry started complaining to Congress about their economic woes. Congress observed that EIA data did not show that there was a problem and our administrator was required to go testify to Congress. In response, of course, we were required to change our forecasting methods for crude oil production. We now make use of auxiliary data and we have changed our revision policies. Now our monthly crude production data accurately show changes when they occur.

I would like to point out that this is really a common problem in making forecasts. The methods that make the strongest use of past data will do a great job when there is no change; it will make a nice, smooth forecast. Unfortunately, these methods will tend not to show a significant change when there is one.

Other methods make strong use of the most current but perhaps slightly less accurate data. These estimates will have more variation when there's no change but will show a major change when there is one. So what you really need in a system like this is to make use of both systems and then use some expert judgment to decide whether the change you see in the data is a real change or just an exceptional noisy data point.

Aczel: In the future is there likely to be an energy crisis again? Why or why not?

Kirkendall: I personally think it is likely that there will be another energy crisis. After all, the world is highly dependent on energy and there are many ways an energy crisis could occur. However, it is unlikely that anybody will actually forecast either the timing or the impact of the next energy crisis.

Aczel: Do politics and policy and data and statistics ever clash or work against each other? How do you resolve such differences?

Kirkendall: Politics, policy, data, and statistics frequently work against each other. This is one reason why the Energy Information Administration was established as an independent part of the department. In 1978, when we were established, Congress realized that data and statistics should not be influenced by politics.

In fact, though, this is related to current discussions of paradigms and paradigm shifts. They say that many researchers did not see data or evidence that doesn't conform to their perception of how things are supposed to work. And so, this is perhaps the same example, people in politics want to see things turn out a certain way and perhaps they just dismiss data that doesn't go along with that expectation. At least within the Department of Energy, since the statistical agency is separate from the policy making department, you can be reasonably assured of honest statistics.

Aczel: If you could talk to students about statistics and how they are likely to need or use statistical thinking, what would you say?

Kirkendall: Statistical thinking is more important than people realize. We are surrounded by numbers, measures of things, and opinions. All of these are statistics. They are reported to us by the media, by politicians, and by people who are trying to sell us stuff. Statistical thinking is critical to making sense of all these numbers.

Aczel: What are your views about the state of quantitative and statistical literacy in the United States these days? How would you advise students to view statistics and mathematics in their education?

Kirkendall: Actually I have a lot of hope for the state of quantitative literacy. The American Statistical Association has initiated an active program working with schoolteachers at various levels to help them learn about projects and teaching methods to help kids appreciate statistics. In addition, the Department of Energy also has encouraged scientific education in many of its programs. There are students and teachers who are able to spend time at the National Labs working on projects. DOE sponsors an annual science bowl for high schools.

I would advise students to take advantage of any of these programs they can find and to take math and statistics, and really any of the sciences; I think they will find them very useful in the future. I would also advise students to look for practical applications of statistics—to read books describing real problems, to read statistics in the newspapers, some of the descriptions where they throw lots of numbers at you, to see if they make sense, to see if you can find errors in the reasoning.

Is There a Relationship?

CHAPTER OUTLINE

9.1 Introduction   408

9.2 Statistical Models   409

9.3 The Simple Linear Regression Model   410

9.4 Estimation: The Method of Least Squares   414

9.5 Error Variance and the Standard Errors of Regression Estimators   421

9.6 How Good Is the Regression?   431

9.7 Residual Analysis and Checking for Model Inadequacies   437

9.8 Use of the Regression Model for Prediction   444

9.9 Correlation   448

I n 1855, a 33-year-old Englishman settled down to a life of leisure in London after several years of travel throughout Europe and Africa. The boredom brought about by a comfortable life induced him to write, and his first book was, naturally, *The Art of Travel*. As his intellectual curiosity grew, he shifted his interests to science, and many years later published a paper on heredity, *Natural Inheritance* (1889). He reported his discovery that sizes of seeds of sweet pea plants appeared to "revert," or "regress," to the mean size in successive generations. He also reported results of a study of the relationship between heights of fathers and the heights of their sons. A straight line was fit to the data pairs: height of father versus height of son. Here, too, he found a "regression to mediocrity": The heights of the sons represented a movement away from their fathers, toward the average height. The man was Sir Francis Galton, cousin of Charles Darwin. We credit him with the idea of statistical regression.

9.1 INTRODUCTION

While most applications of regression analysis have nothing to do with the "regression to the mean" discovered by Galton, the term **regression** remains. It now refers to the statistical technique of modeling the relationship between variables. In this chapter we'll use a *simple linear regression* model, which shows the *straight-line relationship* between two variables: a **dependent variable,** denoted by Y, and an **independent variable,** denoted by X. If we were to model the relationship between the dependent variable Y and a set of several independent variables, or if the assumed relationship between Y and X is curved, which requires the use of other terms in the model, we would use a technique called *multiple regression* (discussed in advanced books on statistics).

Figure 9.1 is a general example of simple linear regression: fitting a straight line to describe the relationship between two variables, X and Y. The points on the graph are randomly chosen observations of the two variables, X and Y, and the straight line describes the general *movement* in the data—an increase in Y corresponding to an increase in X. An inverse straight-line relationship is also possible, consisting of a general decrease in Y as X increases. (In such cases, the slope of the line is negative.)

Regression analysis is one of the most important and widely used statistical techniques and has applications in many areas. An education researcher may be interested in the relationship between students' high school grade point averages (GPAs) and their GPAs in their freshman year in college. Such a relationship may be well approximated by a straight line. Students with a low high school GPA may tend to have a low GPA in the first year of college; high GPA in high school may be followed by high GPA in the freshman year—although not always. Such a *general* increasing trend is shown in Figure 9.1. The X variable (the horizontal direction) in Figure 9.1 would thus stand for the high school GPA; the Y variable (the vertical direction) would stand for the freshman GPA.

Although in reality our sample may consist of all available information on the two variables under study, we always assume that our data set constitutes a random sample of observations from a population of possible pairs of values of X and Y. The data pairs, values of X paired with corresponding values of Y, are the points shown in a

FIGURE 9.1   Simple Linear Regression

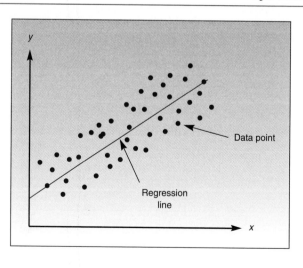

sketch of the data (such as Figure 9.1). A sketch of data on two variables is called a **scatter plot.** In addition to the scatter plot, Figure 9.1 shows the straight line believed to best indicate the general trend of increasing freshman GPA with increasing high school GPA. In simple linear regression, our data are bivariate *quantitative* measurements. Note that in chi-square tests for independence in Chapter 8, we use bivariate *qualitative* data.

Other common examples of the use of simple linear regression are the modeling of the relationship between job performance (the dependent variable Y) and extent of training (the independent variable X); the relationship between returns on a stock (Y) and the riskiness of the stock (X); and the relationship between degree of cure (Y) and the concentration of the active ingredient in a medicine (X).

9.2 STATISTICAL MODELS

A **statistical model** is a set of mathematical formulas and assumptions that describe a real-world situation. We would like our model to explain as much as possible about the process underlying our data. However, due to the uncertainty inherent in all real-world situations, our model will probably not explain everything, and we will always have some remaining errors. The errors are due to unknown outside factors that affect the process generating our data.

A good statistical model is *parsimonious,* which means that it uses as few mathematical terms as possible to describe the real situation. The model captures the systematic behavior of the data, leaving out the factors that are nonsystematic and cannot be foreseen or predicted—the errors. The idea of a good statistical model is illustrated in Figure 9.2. The errors constitute the random component in the model. In a sense, the statistical model breaks down the data into a nonrandom, systematic component (which can be described by a formula) and a purely random component.

How do we deal with the errors? This is where probability theory comes in. Since our model, we hope, captures much that is systematic in the data, the remaining random errors are probably due to a large number of minor factors that we cannot trace. We assume that the random errors, denoted by ϵ, have a *normal distribution.* If we have a properly constructed model, the resulting observed errors will have an average of zero (although few, if any, will actually equal zero), and they should also be *independent* of each other. We note that the assumption of a normal distribution of the errors is not absolutely necessary in the regression model. The assumption is made so that we can carry out statistical hypothesis tests using the F and t distributions. The only necessary assumption is that the errors ϵ have mean zero and a constant variance

157

92

A Statistical Model   **FIGURE 9.2**

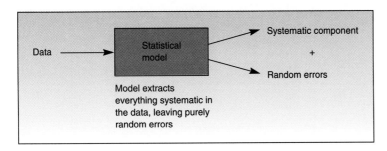

σ^2 and that they be uncorrelated with each other (correlation is defined in Section 9.9). In Section 9.3, we describe the simple linear regression model, but first we'll present a general model-building methodology.

The first step in building a model is to propose a particular type of model to describe a given situation. For example, we may propose a simple linear regression model for describing the relationship between two variables. Then we estimate the model parameters from the random sample of data we have. The next step is to consider the observed errors resulting from the fit of the model to the data. These observed errors, called **residuals,** represent the information in the data not explained by the model. For example, in the *ANOVA* model discussed in Chapter 8, the within-group variability (leading to SSE and MSE) is due to the residuals. If the residuals are found to contain some nonrandom, systematic component, we reevaluate our proposed model and, if possible, adjust it to incorporate the systematic component found in the residuals; or we may have to discard the model and try another. When we believe that model residuals contain nothing more than pure randomness, we use the model for its intended purpose: *prediction* of a variable, *control* of a variable, or the *explanation* of the relationships among variables. Figure 9.3 shows the usual steps in building a statistical model.

9.3 THE SIMPLE LINEAR REGRESSION MODEL

Recall from algebra that the equation of a straight line is $Y = MX + B$ where B is the Y-intercept and M is the slope of the line. For example, for a while in 1994, the U.S. dollar was trading at exactly 100 Japanese yen. A trader at the Bank of Tokyo gets 500 yen as a flat fee for every transaction. What straight line describes the amount, in yen, received for exchange of U.S. dollars? Answer: the line is $Y = 500 + 100X$. If \$17 are exchanged, the trader receives from the customer $500 + 100(17) = 2200$ yen. In **simple linear regression,** we model the relationship between two variables, X and Y, as

FIGURE 9.3   Steps in Building a Statistical Model

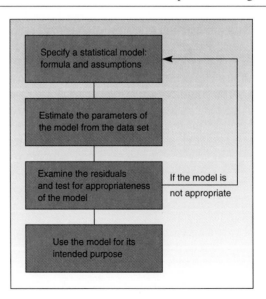

The Simple Linear Regression Model   **FIGURE 9.4**

$$Y = \beta_0 + \beta_1 X + \epsilon$$

random error

nonrandom component:
straight line

a straight line. Therefore, our model must contain two parameters: an intercept parameter and a slope parameter. The usual notation for the **population intercept** is β_0, and the notation for the **population slope** is β_1. If we include the error term, ϵ, we get the population simple linear regression model.

> The population simple linear regression model:
>
> $$Y = \beta_0 + \beta_1 X + \epsilon$$
>
> where Y is the dependent variable, the variable we wish to explain or predict; X is the independent variable, also called the *predictor* variable; and ϵ is the error term, the only random component in the model and thus the only source of randomness in Y.

The model parameters are as follows.

> β_0 is the Y-intercept of the straight line given by $Y = \beta_0 + \beta_1 X$.
> β_1 is the slope of the line $Y = \beta_0 + \beta_1 X$.

The simple linear regression model consists of two components: a nonrandom component, which is the line itself, and a purely random component—the error term, ϵ. This is shown in Figure 9.4. The nonrandom part of the model, the straight line, is the equation for the *mean of Y, given X*. Thus, if the model is correct, the *average* value of Y for a given value of X falls right *on* the regression line.

Our model says that each value of Y consists of the average Y for the given value of X (this is a point on the straight line), plus a random error. As X increases, the average population value of Y also increases (or decreases), assuming a positive (or negative) slope of the line. The *actual* population value of Y is equal to the average Y conditional on X, plus a random error, ϵ. We thus have, for a given value of X:

$$Y = \text{Average } Y \text{ for given } X + \text{Error}$$

Figure 9.5 illustrates the population regression model.

The model relies on the following assumptions:

1.   The relationship between X and Y is a straight-line relationship.
2.   The values of the independent variable X are assumed to be fixed (not random); the only randomness in the values of Y comes from the error term, ϵ.
3.   The errors, ϵ, are normally distributed with mean 0 and a constant variance σ^2.

FIGURE 9.5   The Population Regression Line

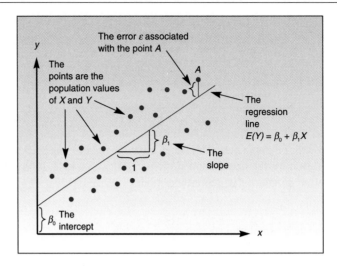

FIGURE 9.6   The Distributional Assumptions of the Linear Regression Model

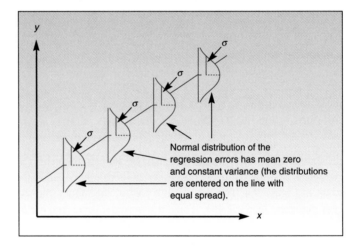

The errors are uncorrelated (not related) with each other in successive observations.[1]

Figure 9.6 shows the distributional assumptions of the errors of the simple linear regression model. The population regression errors are normally distributed about the population regression line, with mean zero and equal variance. (The errors are equally

---

[1]The idea of statistical *correlation* will be discussed in detail in Section 9.9. In the case of the regression errors, we assume that successive errors ϵ_1, ϵ_2, ϵ_3, . . . are uncorrelated; there is no trend, no joint movement in successive errors. Incidentally, the assumption of noncorrelation together with the assumption of a normal distribution of the errors implies the assumption that the errors are independent of each other. Independence implies noncorrelation, but noncorrelation does not imply independence, except in the case of a normal distribution. (This is a technical point.)

Some Possible Relationships between X and Y **FIGURE 9.7**

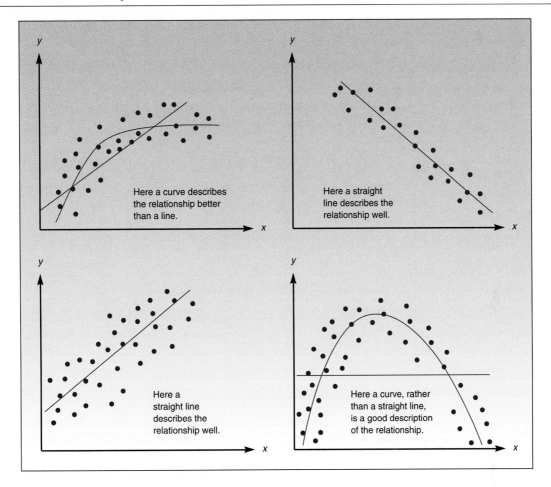

spread about the regression line; the error variance does not increase or decrease as X increases.)

The simple linear regression model applies only if the true relationship between the two variables X and Y is a straight-line relationship. If the relationship is curved (*curvilinear*), then we need to use more involved methods. Figure 9.7 shows four possible relationships between two variables. Some are straight-line relationships that can be modeled by simple linear regression, and others are not.

So far, we have described the population model, that is, the assumed true relationship between the two variables X and Y. Our interest is focused on this unknown population relationship, and we want to *estimate* it using sample information. We obtain a random sample of paired observations on the two variables, and we estimate the regression model parameters, β_0 and β_1, from this sample. This is done by the *method of least squares,* which is discussed in Section 9.4.

---

1. What is a statistical model?   *PROBLEMS*
2. What are the steps of statistical model building?

3. What are the assumptions of the simple linear regression model?
4. Define the parameters of the simple linear regression model.
5. What is the mean of Y, given X?
6. What are the uses of a regression model?
7. What is the purpose and meaning of the error term in regression?
8. Give examples of situations where you believe a straight-line relationship exists between two variables. What would be the uses of a regression model in each of these situations?
9. For the line $Y = 5 + 8X$, what is the slope? What is the Y-intercept?
10. For the line $Y = 12.5X$, what are the slope and Y-intercept?

9.4 ESTIMATION: THE METHOD OF LEAST SQUARES

We want to find good estimates of the regression parameters, β_0 and β_1. A method that will help us is the **method of least squares.**

> The least-squares estimators:
>
> $$b_0 \xrightarrow{\text{estimates}} \beta_0$$
>
> $$b_0 \xrightarrow{\text{estimates}} \beta_1$$

> The estimated regression relationship:
>
> $$Y = b_0 + b_1X + e$$
>
> where b_0 estimates β_0, b_1 estimates β_1, and e stands for the observed errors—the residuals from fitting the straight line $(b_0 + b_1X)$ to the data set of n points.

In terms of the data, the estimated regression relationship can be written with the subscript i to signify each particular data point:

$$y_i = b_0 + b_1x_i + e_i$$

where $i = 1, 2, \ldots, n$. Then, e_1 is the first residual, the distance from the first data point to the fitted regression line; e_2 is the distance from the second data point to the line; and so on to e_n, the nth error. The errors, e_i, are viewed as estimates of the true population errors, ϵ_i. The equation of the regression line itself is as follows.

> The regression line:
>
> $$\hat{Y} = b_0 + b_1X$$
>
> where $\hat{Y}$ (pronounced "Y hat") is the Y value *lying on the fitted regression line* for a given X.

Thus, $\hat{y}_1$ is the fitted value corresponding to x_1, that is, the value of y_1 without the error e_1, and so on for all $i = 1, 2, \ldots, n$. The fitted value $\hat{Y}$ is also called the *predicted value of Y* because, if we do not know the actual value of *Y*, it is the value we would predict for a given value of *X* using the estimated regression line. The predicted *Y* is the average *Y* for a given *X*. Since the average error is zero, this equation has no error term.

At last we come to the principle of least squares, which gives us the regression parameter estimates. Consider the data set shown in part (*a*) of Figure 9.8. In parts (*b*), (*c*), and (*d*) of the figure, we show different lines passing through the data set and the resulting errors, e_i.

Note that the regression line proposed in part (*b*) of Figure 9.8 results in very large errors. The errors away from the line of part (*c*) are smaller than the ones of part (*b*), but the errors resulting from using the line proposed in part (*d*) are by far the smallest. The line in part (*d*) seems to move with the data and *minimize* the resulting errors. This should convince you that the line that best describes the trend in the data is the line that lies "inside" the set of points; since some of the points lie above the fitted line and some below the line, some errors will be positive and others will be negative. If we want to minimize all the errors (both positive and negative ones), we may want to minimize the *sum of the squared errors* (SSE, as in ANOVA). Thus, we want to find the *least-squares* line—the line that minimizes SSE.

A Data Set of *X* and *Y* Pairs, and Different Proposed Straight Lines to Describe the Data   **FIGURE 9.8**

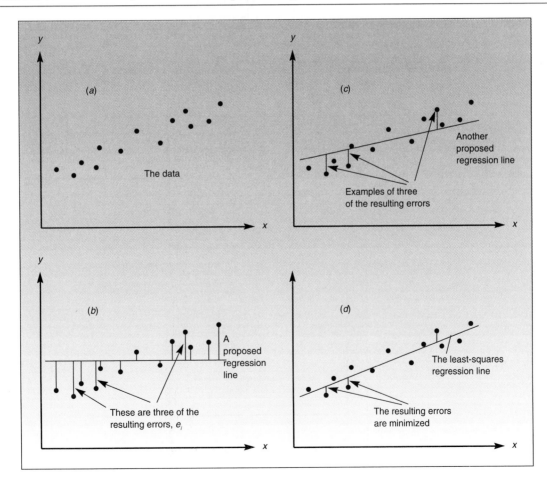

Figure 9.9 shows how the errors lead to the calculation of SSE. Note that least squares is not the only method of fitting lines to data. There are other methods, such as minimizing the sum of the absolute errors. The method of least squares, however, is the one most commonly used when estimating a regression relationship.

> The sum of squares for error in regression:
>
> $$SSE = \text{Sum of all squared errors}, (y - \hat{y})^2$$

The least-squares estimators b_0 and b_1, which minimize the sum of the squared errors, are calculated as follows.

> Least-squares regression estimators:
> Slope:
>
> $$b_1 = \frac{\text{Sum of } xy - \dfrac{(\text{Sum of } x)(\text{Sum of } y)}{n}}{\text{Sum of } x^2 - \dfrac{(\text{Sum of } x)^2}{n}}$$
>
> Intercept:
>
> $$b_0 = \bar{y} - b_1 \bar{x}$$

The formula for the estimate of the intercept makes use of the fact that the *least-squares line always passes through the point* $(\bar{x}, \bar{y})$, the intersection of the mean of X and the mean of Y.

It is important to remember that the obtained estimates, b_0 and b_1, of the regression relationship are just realizations of *estimators* of the true regression parameters, β_0 and

FIGURE 9.9   The Regression Errors Leading to SSE

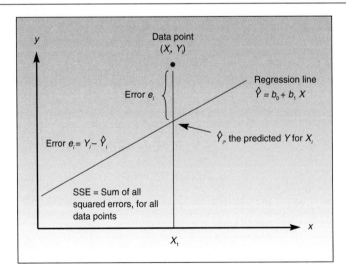

β_1. As always, our estimators have standard errors. The estimates can be used, along with the assumption of normality, in the construction of confidence intervals for, and the conducting of hypothesis tests about, the true regression parameters β_0 and β_1. This will be done in Section 9.5.

Example 9.1 demonstrates the process of estimating the parameters of a simple linear regression model.

EXAMPLE 9.1

As part of a comprehensive research effort undertaken by a New York marketing research firm on behalf of American Express, a study was conducted to determine the relationship between travel and charges on the American Express card. The research firm selected a random sample of 25 cardholders from the American Express computer file and recorded their total charges over a specified period of time. The firm also sent the selected cardholders a questionnaire on the total number of miles they traveled during the same period. The data for this study are given in Table 9.1. Figure 9.10 is a scatter plot of the data.

SOLUTION

As can be seen from the figure, it seems likely that a straight line will describe the trend of increase in dollar amount charged with increase in number of miles traveled. The least-squares line that fits these data is shown in Figure 9.11. Table 9.2 shows the computations necessary to obtain the least-squares regression line in Figure 9.11.

From Table 9.2 we get:

$$\text{Sum of } xy = 390{,}185{,}014$$
$$\text{Sum of } x = 79{,}448$$
$$\text{Sum of } y = 106{,}605$$
$$\text{Sum of } x^2 = 293{,}426{,}946$$

Therefore,

$$b_1 = \frac{51{,}402{,}852.40}{40{,}947{,}557.84} = 1.255333776$$

Used by permission of American Express Company.

American Express Study Data

TABLE 9.1

Miles	Dollars	Miles	Dollars
1,211	1,802	3,209	4,692
1,345	2,405	3,466	4,244
1,422	2,005	3,643	5,298
1,687	2,511	3,852	4,801
1,849	2,332	4,033	5,147
2,026	2,305	4,267	5,738
2,133	3,016	4,498	6,420
2,253	3,385	4,533	6,059
2,400	3,090	4,804	6,426
2,468	3,694	5,090	6,321
2,699	3,371	5,233	7,026
2,806	3,998	5,439	6,964
3,082	3,555		

FIGURE 9.10   The Data for the American Express Study

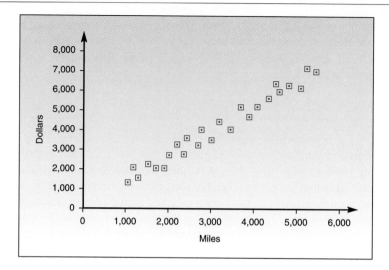

FIGURE 9.11   Least-Squares Line for the American Express Study

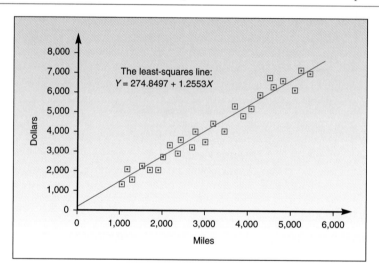

TABLE 9.2   The Computations Required for the American Express Study

Miles(X)	Dollars(Y)	X^2	Y^2	XY
1,211	1,802	1,466,521	3,247,204	2,182,222
1,345	2,405	1,809,025	5,784,025	3,234,725
1,422	2,005	2,022,084	4,020,025	2,851,110
1,687	2,511	2,845,969	6,305,121	4,236,057
1,849	2,332	3,418,801	5,438,224	4,311,868

Continued

TABLE 9.2

Miles(X)	Dollars(Y)	X^2	Y^2	XY
2,026	2,305	4,104,676	5,313,025	4,669,930
2,133	3,016	4,549,689	9,096,256	6,433,128
2,253	3,385	5,076,009	11,458,225	7,626,405
2,400	3,090	5,760,000	9,548,100	7,416,000
2,468	3,694	6,091,024	13,645,636	9,116,792
2,699	3,371	7,284,601	11,363,641	9,098,329
2,806	3,998	7,873,636	15,984,004	11,218,388
3,082	3,555	9,498,724	12,638,025	10,956,510
3,209	4,692	10,297,681	22,014,864	15,056,628
3,466	4,244	12,013,156	18,011,536	14,709,704
3,643	5,298	13,271,449	28,068,804	19,300,614
3,852	4,801	14,837,904	23,049,601	18,493,452
4,033	5,147	16,265,089	26,491,609	20,757,851
4,267	5,738	18,207,289	32,924,644	24,484,046
4,498	6,420	20,232,004	41,216,400	28,877,160
4,533	6,059	20,548,089	36,711,481	27,465,447
4,804	6,426	23,078,416	41,293,476	30,870,504
5,090	6,321	25,908,100	39,955,041	32,173,890
5,233	7,026	27,384,289	49,364,676	36,767,058
5,439	6,964	29,582,721	48,497,296	37,877,196
79,448	106,605	293,426,946	521,440,939	390,185,014

and

$$b_0 = \bar{y} - b_1\bar{x} = \frac{106,605}{25} - (1.255333776)\frac{79,448}{25} = 274.8496866$$

It is important to carry out as many significant digits as you can in these computations. Here we carried out the computations by hand, for demonstration purposes. Usually, all computations are done by computer or by calculator. There are many hand calculators with a built-in routine for simple linear regression. From now on, we will present only the computed results, the least-squares estimates. The estimated least-squares relationship is (reporting estimates to the second significant decimal):

$$Y = 274.85 + 1.26X + e$$

The equation of the line itself, that is, the predicted value of Y for a given X, is:

$$\hat{Y} = 274.85 + 1.26X$$

11. Explain the advantages of the least-squares procedure for fitting lines to data. Explain how the procedure works.

12. A banking analyst is interested in developing a prediction regression of the prime lending rate using the federal funds rate for the same week as the independent variable. A random sample of 15 weekly observations on both variables is obtained. Data are as follows:

PROBLEMS

Federal Funds Rate	Prime Lending Rate
6.23	7.75
6.87	8.50
5.54	6.85
5.90	6.78
6.45	8.00
6.55	7.80
5.75	7.05
6.00	7.35
6.20	7.30
6.70	8.00
7.00	8.69
7.23	8.86
5.30	6.25
6.35	7.90
7.15	8.96

Estimate the slope and intercept parameters of the simple linear regression model of the prime rate based on the federal funds rate.

13. The following data, provided by Herminio Agustin of Beans Coffee Boutique of Chestnut Hill, Massachusetts, are the roast time, in minutes, for nine batches of coffee (X) and the resulting shrinkage, in percent, of the coffee weight (Y). Estimate the regression relationship between these two variables.

X	Y
12	15
13	14
14.5	16
15	16
16.2	18
17	19
19.1	21
20	23
23.5	25

14. Read the article "Why Birds and Bees, Too, Like Good Looks." The degree of symmetry (X), on a scale of 0 to 100, versus the number of suitors (Y) for an insect species are given below. Estimate the regression relationship between the two variables.

X	Y
58	1
67	2
70	2
75	3
81	3
90	4
98	5

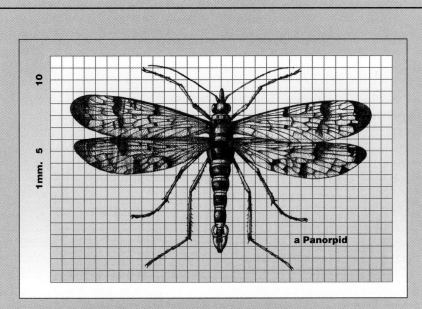

a Panorpid

Why Birds and Bees, Too, Like Good Looks

By NATALIE ANGIER

Beauty is only skin deep. How sweet that old chestnut is, equally comforting to the unbeautiful, who know they have so much beyond physical appearance to offer the world, and the beautiful, who, after years of being pursued for their prettiness, really do want to be loved for their inner selves.

The only problem with the cliché, say evolutionary biologists, is that it may not be true. In the view of a growing number of researchers who study why animals are attracted to each other, a beautiful face and figure may be alluring not for whimsical esthetic reasons, but because outward beauty is a reasonably reliable indicator of underlying quality. These biologists have gathered evidence from studies of species as diverse as zebra finches, scorpion flies, elk and

human beings that creatures appraise the overall worthiness of a potential mate by looking for at least one classic benchmark of beauty: symmetry.

By this theory, the choosier partner in a pair—usually though not always the female—seeks in a suitor the maximum possible balance between the left and right halves of the body. She looks for signs of exquisite harmony, checking that the left wing is the same length and shape as the right, for example, or that the lips extend out in mirror-image curves from the center of the face. In searching for symmetry, she gains essential clues to the state of the male's health, the vigor of his immune system, the ability of his genes to have withstood the tribulations of the environment as he was growing up.

Reprinted by permission from *The New York Times,* February 8, 1994, p. C1.

9.5 ERROR VARIANCE AND THE STANDARD ERRORS OF REGRESSION ESTIMATORS

Recall that σ^2 is the variance of the population regression errors, ϵ, and that this variance is assumed to be constant for all values of X in the range under study. The error variance is an important parameter in the context of regression analysis because it is a measure of the spread of the population elements about the regression line. Generally, the smaller the error variance, the more closely the population elements follow the regression line. The error variance is the variance of the dependent variable Y as "seen" by an eye looking in the direction of the regression line. (The error variance is not the variance of Y.) These properties are demonstrated in Figure 9.12.

The figure shows two regression lines. The top regression line in the figure has a larger error variance than the bottom regression line. The error variance for each regression is the vertical variation in the data points as seen by the eye located at the base of the line, looking *in the direction of the regression line*. The variance of Y, on the other hand, is the variation in the Y values regardless of the regression line. That is, the variance of Y for each of the two data sets in the figure is the variation in the data as seen by an eye looking in a direction parallel to the X-axis. Note also that the spread of the data is constant along the regression lines. This is in accordance with our assumption of equal error variance for all X.

Since σ^2 is usually unknown, we need to estimate it from our data. An estimator of σ^2, denoted by s^2, is the mean square error (MSE) of the regression. As you will soon see, sums of squares and mean squares in the context of regression analysis are very similar to those of ANOVA, presented in Chapter 8. The degrees of freedom for error in the context of simple linear regression are $n - 2$ because we have n data points, from which two parameters, β_0 and β_1, are estimated. (Thus, two restrictions are imposed on the n points, leaving df $= n - 2$.) The sum of squares for error (SSE) in regression analysis is defined as the sum of squared deviations of the data values Y from the fitted values $\hat{Y}$.

$$\text{df (error)} = n - 2$$
$$\text{SSE} = \text{Sum of } (y - \hat{y})^2$$

An estimator of σ^2, denoted by s^2, is:

$$\text{MSE} = \frac{\text{SSE}}{n - 2}$$

FIGURE 9.12   Two Examples of Regression Lines Showing the Error Variance

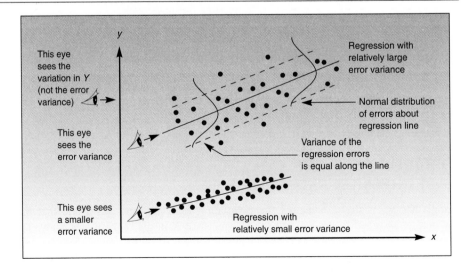

In Example 9.1, the sum of squares for error is:

$$SSE = 2,328,161.2$$

Although usually obtained by computer, it is possible to compute the SSE above by subtracting from each y value its predicted value using the regression equation. It may also be computed as shown in Section 9.6. The mean square error is:

$$MSE = \frac{SSE}{n-2} = \frac{2,328,161.2}{23} = 101,224.4$$

An estimate of the standard deviation of the regression errors, σ, is s, which is the square root of MSE. The estimate $s = \sqrt{MSE}$ of the standard deviation of the regression errors is sometimes referred to as *standard error of estimate*. In Example 9.1, we have:

$$s = \sqrt{MSE} = \sqrt{101,224.4} = 318.1578225$$

The computation of SSE and MSE for Example 9.1 is demonstrated in Figure 9.13.

The standard deviation of the regression errors, σ, and its estimate, s, play an important role in the process of estimation of the values of the regression parameters, β_0 and β_1. This is so because σ is part of the expressions for the standard deviations of both parameter estimators. The standard errors, defined next, give us an idea of the accuracy of the least-squares estimates, b_0 and b_1. *The standard error of b_1 is especially important because it is used in a test for the existence of a linear relationship between X and Y.*

The estimated standard error of b_0 is:

$$s(b_0) = \frac{s\sqrt{\text{Sum of } x^2}}{\sqrt{n(\text{Sum of } x^2) - (\text{Sum of } x)^2}}$$

The estimated standard error of b_1 is:

Computing SSE and MSE in the American Express Study   **FIGURE 9.13**

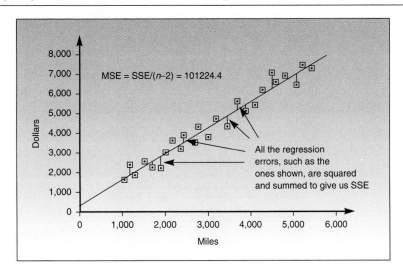

$$s(b_1) = \frac{s}{\sqrt{\text{Sum of } x^2 - (\text{Sum of } x)^2/n}}$$

Formulas like the ones above are nice to know, but you should not worry too much about having to use them.

Regression analysis is usually done by computer, and the computer output will include the values of the standard errors of the regression estimates.

The following subsection shows how the regression parameter estimates and their standard errors can be used in the construction of confidence intervals for the true regression parameters, β_0 and β_1.

Confidence Intervals for the Regression Parameters

Confidence intervals for the true regression parameters, β_0 and β_1, are easy to compute.

A $(1 - \alpha)$ 100% confidence interval for β_0:

$$b_0 \pm t_{(\alpha/2, n-2)} s(b_0)$$

where $s(b_0)$ is as given above.

A $(1 - \alpha)$ 100% confidence interval for β_1:

$$b_1 \pm t_{(\alpha/2, n-2)} s(b_1)$$

where $s(b_1)$ is as given above.

Let's construct 95 percent confidence intervals for β_0 and β_1 in the American Express study from Example 9.1:

$$s(b_0) = 318.16 \frac{\sqrt{293,426,946}}{\sqrt{(25)(40,947,557.84)}} = 170.338$$

where the various quantities were computed earlier, including the sum of X^2, which is found at the bottom of Table 9.2.

A 95 percent confidence interval for β_0 is:

$$b_0 \pm t_{(\alpha/2, n-2)} s(b_0) = 274.85 \pm 2.069(170.338) = [-77.58, 627.28]$$

where the value 2.069 is obtained from the Table B.3, Appendix B, for $1 - \alpha = 0.95$ and 23 degrees of freedom. We may be 95 percent confident that the true regression intercept is anywhere from -77.58 to 627.28.

$$s(b_1) = \frac{318.16}{\sqrt{40,947,557.84}} = 0.04972$$

A 95 percent confidence interval for β_1 is:

$$b_1 \pm t_{(\alpha/2, n-2)} s(b_1) = 1.25533 \pm 2.069(0.04972) = [1.15246, 1.35820]$$

Interpretation of the Slope Estimation for Example 9.1 **FIGURE 9.14**

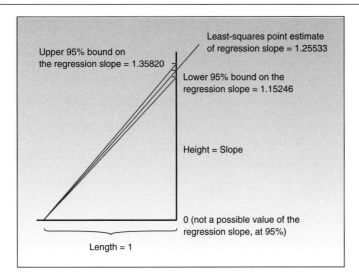

From the confidence interval given above, we may be 95 percent confident that the *true* slope of the (*population*) regression line is anywhere from 1.15246 to 1.3582. This range of values is far from zero, so we may be quite confident that the true regression slope is not zero. This conclusion is very important, as we will see in following sections. Figure 9.14 demonstrates the meaning of the confidence interval.

A Hypothesis Test for the Existence of a Linear Regression Relationship

When there is no linear relationship between X and Y, the population regression slope, β_1, is equal to zero. Why? Either one of the following two situations holds true:

1. The variable Y is *constant* for all values of X. For example, $Y = 457.33$ for all X. This is shown in part (*a*) of Figure 9.15. If Y is constant for all values of X, the slope of Y with respect to X, parameter β_1 is identically zero; there is no linear relationship between the two variables.
2. The two variables are *uncorrelated*. (The concept of correlation will be formally introduced in Section 9.9.) When the correlation between X and Y is zero, as X increases Y may increase, decrease, or remain constant. There is no *systematic* increase or decrease in the values of Y as X increases. This case is shown in part (*b*) of Figure 9.15. As can be seen in the figure, data from this process are not "moving" in any pattern; thus, there is no direction for the line to follow. Since there is no direction, the slope of the line is, again, zero.[2]

In cases other than these, there is at least *some* linear relationship between the two variables X and Y; the slope of the line in all such cases would be either positive or negative, but not zero. Therefore, *the most important statistical test in simple linear*

---

[2]Note, however, that the relationship between the two variables may be *curved,* with no linear correlation. We will see this when we discuss correlation. In such cases, the slope may also be zero.

FIGURE 9.15   The Two Possibilities Where the Population Regression Slope is Zero

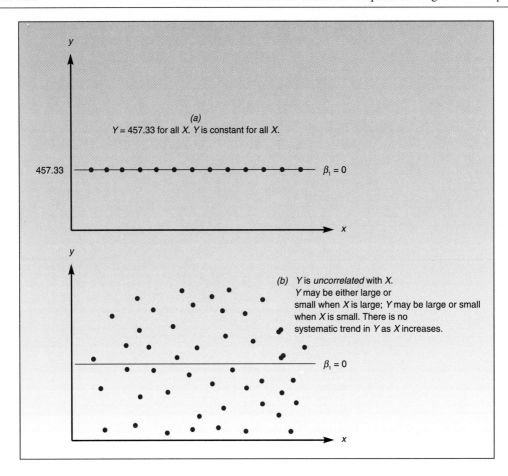

regression is the test of whether or not the slope parameter, β_1, *is equal to zero.* If we conclude in any particular case that the true regression slope is equal to zero, this means that there is no linear relationship between the two variables: Either the dependent variable is constant, or—more commonly—the two variables are not linearly related. We thus have the following test for determining the existence of a linear relationship between two variables, X and Y.

A hypothesis test for the existence of a **linear relationship** between X and Y:

$$H_0: \beta_1 = 0$$

$$H_1: \beta_1 \neq 0$$

This test is, of course, two-tailed. Either the true regression slope is equal to zero, or it is not. If it is equal to zero, there is no linear relationship between the two variables; if the slope is not equal to zero, then it is either positive or negative (the two tails of rejection), in which case there is a linear relationship between the two variables. The test statistic for determining the rejection or nonrejection of the null hypothesis is shown below. Given the assumption of normality of the regression errors, the test statistic possesses the t distribution with $n - 2$ degrees of freedom.

Test statistic for the existence of a linear relationship between X and Y:

$$t_{(n-2)} = \frac{b_1}{s(b_1)}$$

where b_1 is the least-squares estimate of the regression slope and $s(b_1)$ is the standard error of b_1. When the null hypothesis is true, the statistic has a t distribution with $n - 2$ degrees of freedom.

This test statistic is a special version of a general test statistic:

$$t_{(n-2)} = \frac{b_1 - \beta_{10}}{s(b_1)}$$

where β_{10} is the value of β_1 under the null hypothesis. This statistic follows the format: (Estimate − Hypothesized parameter value)/(Standard error of estimator). Since, in the test for existence of a linear relationship, the hypothesized value of β_1 is zero, we have the simplified version of the test statistic. One advantage of the simple form of our test statistic is that it allows us to conduct the test very quickly. Computer output for regression analysis usually contains a table similar to Table 9.3.

The estimate associated with X (or whatever name the user may have given to the independent variable in the computer program) is b_1. The standard error associated with X is $s(b_1)$. To conduct the test, all you need to do is divide b_1 by $S(b_1)$. In the example of Table 9.3, 4.88/0.1 = 48.8. The answer is reported in the table as the t ratio. The t ratio can now be compared with critical points of the t distribution with $n - 2$ degrees of freedom. Suppose that the sample size used was 200. Then the critical points for $\alpha = 0.05$ are ± 1.96, and since 48.8 > 1.96, we conclude that there is evidence of a linear relationship between X and Y in this hypothetical example. Actually, the p-value is very small. Most computer programs will also report the p-value in an extra column on the right.

What about the first row in the table? The test suggested here is a test of whether or not the *intercept,* β_0 (this is the "constant"), is equal to zero. The test statistic is the same, but with subscripts 0 instead of 1. This test, although suggested by the output of computer routines, is usually not a meaningful test and should generally be avoided.

To conduct the hypothesis test for the existence of a linear relationship between miles traveled and amount charged on the American Express card in Example 9.1, our hypotheses are $H_0: \beta_1 = 0$ and $H_1: \beta_1 \neq 0$. Recall that for the American Express study, $b_1 = 1.25533$ and $s(b_1) = 0.04972$.

An Example of a Part of the Computer Output for Regression   **TABLE 9.3**

Variable	Estimate	Standard Error	t Ratio
Constant	5.22	0.5	10.44
X	4.88	0.1	48.80

FIGURE 9.16   Test for a Linear Relationship for Example 9.1

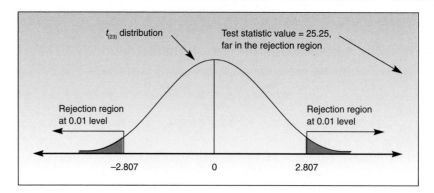

We now compute the test statistic:

$$t = \frac{b_1}{s(b_1)} = \frac{1.25533}{0.04972} = 25.25$$

From the magnitude of the computed value of the statistic, we know that there is statistical evidence of a linear relationship between the variables, because 25.25 is certainly greater than any critical point of a t distribution with 23 degrees of freedom. Figure 9.16 illustrates the test. The critical points of t with 23 degrees of freedom and $\alpha = 0.01$ are obtained from Appendix B, Table B.3.

Example 9.2 shows how to conduct a regression analysis using a computer.

EXAMPLE 9.2

A study was undertaken to assess the effects of noise on hearing loss. The independent variable was maximal noise level, in decibels, to which an individual was exposed regularly (seven hours a day, for a period of 10 years), and the dependent variable was hearing loss. Hearing loss was expressed as a percentage, compared with normal hearing.

The data below constitute a random sample of 12 patients:

	Decibel (C1)	Loss (C2)
1	65	15
2	70	10
3	72	20
4	80	18
5	86	25
6	90	20
7	112	36
8	125	40
9	130	49
10	145	52
11	108	32
12	100	35

Decibel Level	Example
30	Quiet library, soft whispers
40	Living room, refrigerator, bedroom away from traffic
50	Light traffic, normal conversation, quiet office
60	Air conditioner at 20 feet, sewing machine
70	Vacuum cleaner, hair dryer, noisy restaurant
80	Average city traffic, garbage disposals, alarm clock at two feet

The Following Noises Can Be Dangerous under Constant Exposure:

90	Subway, motorcycle, truck traffic, lawn mower
100	Garbage truck, chain saw, pneumatic drill
120	Rock band concert in front of speakers, thunderclap
140	Gunshot blast, jet plane
180	Rocket launching pad

Source: Information courtesy of the Deafness Research Foundation.

Figure 9.17 shows the data, and Figure 9.18 presents the regression analysis for percentage of hearing loss versus level of regular exposure in decibels. The analysis is done using the statistical computing package MINITAB, although very similar results would be obtained using any other package. Explain the figures.

From the MINITAB display we see that the estimated slope, b_1, is 0.50439, and the standard error of the estimate is $s(b_1) = 0.04234$. Thus the t-ratio, our test statistic for the existence of a linear relationship between the two variables, is 11.91. This is very significant, as we see from the reported p-value of "0.000." (This means that the p-value is less than 0.001.) There is very strong evidence for a linear relationship between noise level and hearing loss.

SOLUTION

Data for Hearing Loss Study

FIGURE 9.17

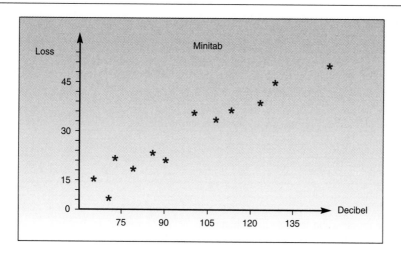

FIGURE 9.18   MINITAB Program and Output for Example 9.2

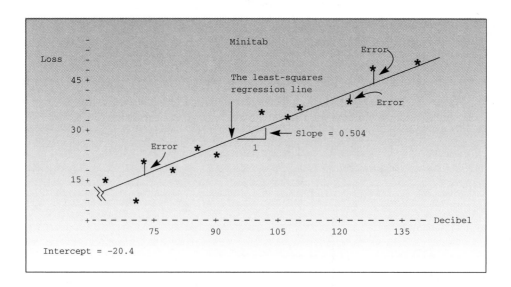

```
                              MINITAB
        MTB > PLOT  C2*C1
        MTB > REGRESS C2 on 1 C1

        The regression equation is
        Loss = - 20.4 + 0.504 Decibel

        Predictor    Coef       Stdev    t-ratio        p
        Constant   -20.391      4.304      -4.74     0.000
        Decibel     0.50439     0.04234    11.91     0.000

        s = 3.628      R-sq = 93.4%    R-sq(adj) = 92.8%

        Analysis of Variance

        SOURCE      DF       SS         MS         F        p
        Regression   1     1867.1     1867.1    141.89   0.000
        Error       10      131.6       13.2
        Total       11     1998.7
```

```
       -                     Minitab
       -                                        Error
 Loss  -                                         *          *
       -
   45 +             The least-squares
       -            regression line
       -
       -                          *      *
       -                            *  *        *    Error
   30 +                        ┌────          Slope = 0.504
       -              Error   *│  1
       -                    *  │
       -            *    *
   15 + *
       -  ╲╱
       -       *
       +- - - -+- - - -+- - - -+- - - -+- - - -+- - Decibel
            75      90     105     120     135

Intercept = -20.4
```

From the analysis of variance table also reported in the output, we find that the mean square error (MSE: in column MS, row Error) is 13.2. The square root of the MSE, as we know, is reported above the table as $s = 3.628$. This is a measure of the variation of the data points about the regression line.

PROBLEMS   15. Test for the existence of a linear relationship between the federal funds rate and the prime lending rate using the data in Problem 12.

16. Test for the existence of a linear relationship between coffee bean shrinkage and the duration of roasting using the data in Problem 13.

17. Test for the existence of a linear relationship between degree of symmetry and number of suitors using the data in Problem 14. Give a 90 percent confidence interval for the true regression slope.
18. What is measured by the mean square error (MSE) in regression analysis?
19. A regression analysis of fuel efficiency (X) versus sales (Y) of different types of corporate aircraft includes the following results: $b_1 = 2.435$, $s(b_1) = 1.567$, $n = 12$. Do you believe there is a linear relationship between sales of corporate aircraft and the aircraft's fuel efficiency?
20. A management recruiter wants to estimate a linear regression relationship between an executive's experience and the salary the executive may expect to earn after placement with an employer. From data on 28 executives, which are assumed a random sample from the population of executives that the recruiter places, the following regression results are obtained: $b_1 = 5.49$, $s(b_1) = 1.21$. Is there a linear relationship between the experience and the salary of executives placed by the recruiter?
21. The data in the accompanying table (provided by courtesy of Professor Charles Hadlock and the Center for Excellence in Teaching at Bentley College) are actual teaching evaluation scores for several instructors. Each data point is the average class response to three separate questions. Each individual student response is on a scale of 1 to 5 (5 indicating strongest answer). The class averages are continuous, and roughly normally distributed. Question 10 is the overall instructor rating by students, while Question 2 is the perception of course difficulty by the students, and Question 8 is course interest. Conduct a linear regression analysis of overall instructor rating versus interest, and a separate regression analysis of overall instructor rating on perceived difficulty. Draw your conclusions.

Q10	Q2	Q8	Q10	Q2	Q8
4.83	3.83	4.50	4.04	3.40	3.65
3.56	3.33	3.67	4.80	3.10	4.43
4.85	2.75	4.60	4.79	3.38	4.51
4.69	3.31	4.00	4.76	3.36	4.42
4.32	3.58	3.79	4.66	3.31	4.41
4.32	2.89	3.79	4.48	2.97	4.27
4.17	3.00	3.67	4.29	3.92	3.63
3.79	2.74	3.32	4.26	3.22	4.00
5.00	2.50	4.75	4.17	3.60	3.50
4.52	2.52	3.68	3.73	2.77	3.38
4.50	2.33	3.67	4.69	3.46	3.85
4.47	2.67	3.91	4.29	4.00	3.50
4.37	2.30	3.33	4.12	3.46	3.27
4.32	2.79	3.82	4.11	3.48	3.56
4.10	2.66	3.21	3.90	4.05	3.62
4.10	2.87	3.33	3.57	4.15	2.90
3.79	2.83	3.25	3.89	3.88	3.90
4.80	3.33	4.33	3.76	3.12	3.50
4.80	3.40	4.60	3.71	4.00	3.71
4.36	3.93	4.07	3.63	3.15	3.00
4.40	2.87	4.13	3.45	3.20	2.96
4.26	3.26	3.70	3.16	3.66	2.89

9.6 HOW GOOD IS THE REGRESSION?

Once we have determined that a linear relationship exists between the two variables, the question is: How strong is the relationship? If the relationship is a strong one, prediction of the dependent variable can be relatively accurate, and other conclusions drawn from the analysis may be given a high degree of confidence.

We have already seen one measure of the regression fit: the mean square error. The MSE is an estimate of the variance of the true regression errors and is a measure of the variation of the data about the regression line. The MSE, however, depends on the nature of the data, and what may be a large error variation in one situation may not be considered large in another. What we need, therefore, is a *relative* measure of the degree of variation of the data about the regression line. Such a measure allows us to compare the fits of different models.

The relative measure we are looking for is a measure that compares the variation of Y about the regression line with the variation of Y without a regression line. This should remind you of analysis of variance, and we will soon see the relation of ANOVA to regression analysis. It turns out that the relative measure of regression fit we are looking for is the coefficient of determination, denoted by r^2.

> The **coefficient of determination, r^2,** is a descriptive measure of the strength of the regression relationship, a measure of how well the regression line fits the data.[3]

Let's see how the coefficient of determination is obtained directly from a decomposition of the variation in Y into a component due to error and a component due to the regression. Figure 9.19 shows the least-squares line that was fitted to a data set. One of the data points, (X, Y), is highlighted. For this data point, the figure shows

FIGURE 9.19   The Three Deviations Associated with a Data Point

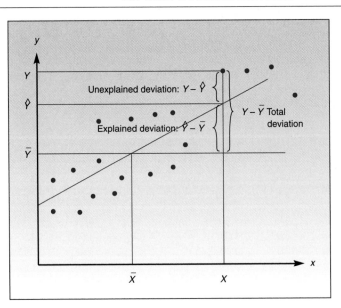

_____

[3] r^2 is the square of the sample correlation coefficient, discussed later.

three kinds of deviations: the deviation of y from its mean $(y - \bar{y})$, the deviation of y from its predicted value using the regression $(y - \hat{y})$, and the deviation of the regression-predicted value of y from the mean of y, which is $(\hat{y} - \bar{y})$. Note that the least-squares line passes through the point $(\bar{x}, \bar{y})$.

We will now follow exactly the same mathematical derivation we used in Chapter 8 when we derived the ANOVA relationships. There we looked at the deviation of a data point from its respective group mean—the error; here the error is the deviation of a data point from its regression-predicted value. In ANOVA, we also looked at the total deviation, the deviation of a data point from the grand mean; here we have the deviation of the data point from the mean of Y. Finally, in ANOVA we also considered the treatment deviation, the deviation of the group mean from the grand mean; here we have the **regression deviation:** the deviation of the predicted value from the mean of Y.

The error is also called the *unexplained deviation* because it is a deviation that cannot be explained by the regression relationship; the regression deviation is also called the *explained deviation* because it is that part of the deviation of a data point from the mean that can be explained by the regression relationship between X and Y. We *explain* why the Y value of a particular data point is above the mean of Y by the fact that its X component happens to be above the mean of X and by the fact that X and Y are linearly (and positively) related. As can be seen from Figure 9.19, and by simple arithmetic, we have the following equation.

$$\begin{matrix} \text{Total} \\ \text{deviation} \end{matrix} = \begin{matrix} \text{Unexplained} \\ \text{deviation (Error)} \end{matrix} + \begin{matrix} \text{Explained} \\ \text{deviation (Regression)} \end{matrix}$$

As in the analysis of variance, we square all three deviations for each one of our data points, and sum over all n points. Here, again, we are left with the following important relationship for the sums of squares.

$$\text{SST} = \text{SSE} + \text{SSR}$$
$$\begin{matrix} \text{(Total sum} \\ \text{of squares)} \end{matrix} = \begin{matrix} \text{(Sum of} \\ \text{squares for error)} \end{matrix} + \begin{matrix} \text{(Sum of} \\ \text{squares for regression)} \end{matrix}$$

The term *SSR* is also called the *explained variation* due to the model we selected: It is the part of the variation in Y that is explained by the relationship of Y with the explanatory variable X. Similarly, SSE is the *unexplained variation,* due to error. The sum of the two is the *total variation* in Y.

We define the coefficient of determination as the sum of squares due to the regression divided by the total sum of squares. Also, since SSE and SSR add up to SST, the coefficient of determination is equal to 1 minus SSE/SST.

$$r^2 = \frac{\text{SSR}}{\text{SST}} = 1 - \frac{\text{SSE}}{\text{SST}}$$

The coefficient of determination can be interpreted as *the proportion of the variation in Y that is explained by the regression relationship of Y with X. The r^2 can be* anywhere from 0 to 1. This is in accordance with the interpretation of r^2 as the *percentage of the variation in Y explained by the regression.* The coefficient is a measure of how closely the regression line fits the data; it is a measure of how much the variation in the values of Y is reduced once we regress Y on the variable X. When $r^2 = 1$, we know that 100 percent of the variation in Y is explained by X. This means that the data all lie right on the regression line, and there are no resulting errors.

At the other extreme is the case where the regression line explains nothing. Here the errors account for everything, and SSR is zero. In this case, $r^2 = 0$. When this happens, there is no linear relationship between X and Y, and the true regression slope is probably zero. (We say "probably" because r^2 is only an estimator, given to chance variation; it could be estimating a nonzero population parameter.) Between the two cases, $r^2 = 0$ and $r^2 = 1$, are values of r^2 that give an indication of the *relative fit* of the regression model to the data. *The higher the* r^2, *the better the fit and the higher our confidence in the regression.* Be wary, however, of situations where the reported r^2 is exceptionally high, such as 0.99 or 0.999. In such cases, something may be wrong.

How high should the coefficient of determination be before we can conclude that a regression model fits the data well enough to use the regression with confidence? There is no clear-cut answer to this question. The answer depends on the intended use of the regression model. If we intend to use the regression for *prediction,* the higher the r^2, the more accurate our predictions will be.

An r^2 value of 0.9 or above is very good, a value above 0.8 is good, and a value of 0.6 or above may be satisfactory in some applications, although we must be aware of the fact that in such cases errors in prediction may be relatively high. When the r^2 value is 0.5 or below, the regression explains only 50 percent or less of the variation in the data; therefore, predictions may be poor. If we are interested only in understanding the relationship between the two variables, lower values of r^2 may be acceptable, as long as we realize that the model does not explain much.

Figure 9.20 shows several regressions and their corresponding r^2 values. If you think of the total sum of squared deviations as being in a box, then the r^2 is the proportion of the box that is filled with the explained sum of squares, the remaining part being the squared errors. This is shown for each regression in the figure.

Going back to the data from Example 9.1, we may obtain the following quantities from the data:

$$SST = \text{Sum of } y^2 - (\text{Sum of } y)^2/n = 66,855,898$$

$$SSR = b_1 [\text{Sum of } xy - (\text{Sum of } x)(\text{Sum of } y)/n] = 64,527,736.8$$

and

$$SSE = SST - SSR = 2,328,161.2$$

(The SSE was computed when we found the MSE for this example.) We now compute r^2 as follows:

$$r^2 = \frac{SSR}{SST} = \frac{64,527,736.8}{66,855,898} = 0.96518$$

The r^2 in this example is very high. The interpretation is that over 96.5 percent of the variation in charges on the American Express card can be explained by the relationship between charges on the card and extent of travel (miles). The coefficient of determination, r^2, is always reported in a prominent place in regression computer output.

The Value of the Coefficient of Determination in Different Regressions   **FIGURE 9.20**

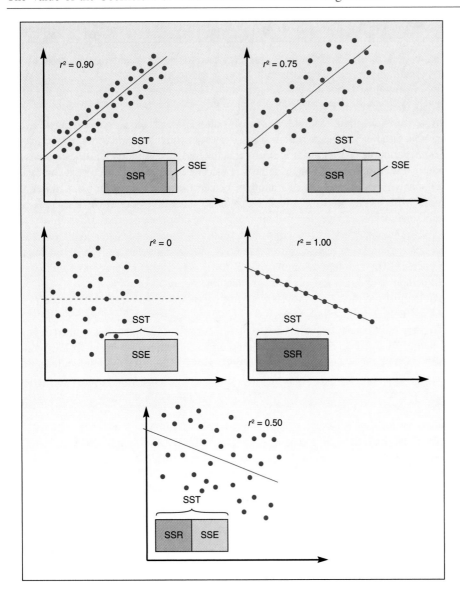

For Example 9.2, we see from Figure 9.18 that the coefficient of determination is $R - sq = 93.4\%$, which is excellent. This means that 93.4 percent of the variation in the hearing loss variable can be explained by long-term exposure to noise. Note also from the figure that the sum of squares for regression is 1,867.1 and the total sum of squares is 1,998.7. Dividing these two numbers, of course, gives us the same result.

Analysis of Variance Table and an *F* Test of the Regression Model

▲ We know from our discussion of the *t* test for the existence of a linear relationship that the degrees of freedom for error in simple linear regression are $n - 2$. For the regression, we have one degree of freedom because there is one independent *X*

optional

variable in the regression. The total degrees of freedom are $n - 1$ because here we consider only the mean of Y, to which one degree of freedom is lost. These are similar to the degrees of freedom for ANOVA in Chapter 8. Mean squares are obtained, as usual, by dividing the sums of squares by their corresponding degrees of freedom. This gives us the mean square regression, MSR, and mean square error (which we encountered earlier), MSE. Further dividing MSR by MSE gives us an F ratio with degrees of freedom 1 and $n - 2$. All these can be put in an ANOVA table for regression. This has been done in Table 9.4.

In regression, there are three sources of variation (see again Figure 9.19): *regression*—the explained variation; *error*—the unexplained variation; and their sum, the *total* variation. We know how to obtain the sums of squares and the degrees of freedom, and from them the mean squares. Dividing the mean square regression by the mean square error should give us another measure of the accuracy of our regression because MSR is the average squared explained deviation and MSE is the average squared error (where averaging is done using the appropriate degrees of freedom). The ratio of the two has an F distribution with 1 and $n - 2$ degrees of freedom *when there is no linear regression relationship between* X *and* Y. This suggests an F test for the existence of a linear relationship between X and Y. *In simple linear regression, this test is equivalent to the two-tailed* t *test for the slope being zero.*

We'll illustrate the analysis with data from Example 9.1. The ANOVA results are given in Table 9.5.

To carry out the test for the existence of a linear relationship between miles traveled and dollars charged on the card, we compare the computed F ratio of 637.47 with a critical point of the F distribution with 1 degree of freedom for the numerator and 23 degrees of freedom for the denominator. Using $\alpha = 0.01$, the critical point from Table B.4, Appendix B, is found to be 7.88. Clearly, the computed value is far in the rejection region, and the p-value is very small. We conclude, again, that there is strong evidence of a linear relationship between the two variables.

An F distribution with 1 degree of freedom for the numerator and k degrees of freedom for the denominator is the *square* of a t distribution with k degrees of freedom. In Example 9.1, our computed F statistic value is 637.47, which is the square of our obtained t statistic, 25.25 (to within rounding error). The same relationship

TABLE 9.4   ANOVA Table for Regression

Source of Variation	Sum of Squares	Degrees of Freedom	Mean Square	F Ratio
Regression	SSR	1	$MSR = \dfrac{SSR}{1}$	$F_{(1,n-2)} = \dfrac{MSR}{MSE}$
Error	SSE	$n - 2$		
Total	SST	$n - 1$	$MSE = \dfrac{SSE}{n-2}$	

TABLE 9.5   ANOVA Table for American Express Example

Source of Variation	Sum of Squares	Degrees of Freedom	Mean Square	F Ratio
Regression	64,527,736.8	1	64,527,736.8	637.47
Error	2,328,161.2	23	101,224.4	
Total	66,855,898.0	24		

holds for the critical points: For $\alpha = 0.01$, we have a critical point for $F_{(1,23)}$ equal to 7.88, and the (right-hand) critical point of a two-tailed test at $\alpha = 0.01$ for t with 23 degrees of freedom is $2.807 = \sqrt{7.88}$.

Consider the ANOVA table produced by the computer for Example 9.2. At this point, you should be able to compare the reported F value with the t ratio for the variable "decibel," and interpret the p-value.

---

22. An article in the *Strategic Management Journal* attempts to explain changes in the gross margin for plastics in Western Europe using an independent variable that measures the balance between buyer power and seller power. Four different simple linear regression models are reported, along with their r^2 values: model 1, $r^2 = 0.56$; model 2, $r^2 = 0.67$; model 3, $r^2 = 0.39$; model 4, $r^2 = 0.46$. Comment on the explanatory power of each model.

23. Results of a study reported in the *Financial Analysts Journal* include a simple linear regression analysis of firms' pension funding (Y) versus profitability (X). The regression coefficient of determination is reported to be $r^2 = 0.02$. (The sample size used is 515.)
    a. Would you use the regression model to predict a firm's pension funding?
    b. Does the model explain much of the variation in firms' pension funding on the basis of profitability?
    c. Do you believe these regression results are worth reporting? Explain.

24. What percentage of the variation in Y is explained by the regression in Problem 12?

25. What is the r^2 in the regression of Problem 13? Interpret its meaning.

26. What is the r^2 in the regression of Problem 14?

27. In a regression, SSE $= 2,850$ and SST $= 7,965$. What is the r^2?

28. Conduct the F test for the existence of a linear relationship between the two variables in Problem 12.

29. Carry out an F test for a linear relationship in Problem 13. Compare your results with those of the t test.

30. Repeat Problem 25 for the regression of Question 10 on Question 2 in Problem 21.

31. What is the coefficient of determination for the regression of Question 10 on Question 8 in Problem 21? Explain the meaning of your finding.

32. Explain the meaning and interpretation of the coefficient of determination.

33. If the coefficient of determination is 100 percent, what does this mean? When would you expect this to happen with real data?

34. If a statistical test for the existence of a linear relationship between two variables does not lead to rejection of the null hypothesis of no relationship, would you be surprised to find that the coefficient of determination was high?

35. What are the uses of an ANOVA table for regression?

9.7 RESIDUAL ANALYSIS AND CHECKING FOR MODEL INADEQUACIES

Recall our discussion of statistical models in Section 9.2. We said that a good statistical model accounts for the systematic movement in the process, leaving out a series of uncorrelated, purely random errors, ϵ, which are assumed to be normally distributed with mean zero and a constant variance, σ^2. In Figure 9.3, we saw a general methodology for statistical model building, consisting of model identification, estimation, tests of validity, and, finally, use of the model. We are now at the third stage of the analysis of a simple linear regression model: examining the residuals and testing the validity of the model.

Analysis of the residuals could reveal whether or not the assumption of normally distributed errors holds. In addition, the analysis could reveal whether the variance of the errors is indeed constant, that is, whether the spread of the data around the regression line is uniform. The analysis could also indicate whether or not there are any missing variables that should have been included in our model (leading to a *multiple regression* equation). The analysis may reveal whether the order of data collection (e.g., time of observation) has any effect on the data and whether the order should have been incorporated as a variable in the model. Finally, analysis of the residuals may determine whether or not the assumption that the errors are uncorrelated is satisfied.

A Check for the Equality of Variance of the Errors

A graph of the regression errors, the residuals, versus the independent variable X, or versus the predicted value $\hat{Y}$, will reveal whether or not the variance of the errors is constant. The variance of the residuals is indicated by the width of the scatter plot of the residuals as X increases. If the width of the scatter plot of the residuals either increases or decreases as X increases, then the assumption of constant variance is not met. This problem is called **heteroscedasticity.** When heteroscedasticity exists, we can't use the ordinary least-squares method for estimating the regression and should use a more complex method called *generalized least squares.* Figure 9.21 shows how a plot of the residuals versus X or $\hat{Y}$ looks in the case of heteroscedasticity. Figure 9.22 shows a residual plot in a good regression, with no heteroscedasticity.

Testing for Missing Variables

Figure 9.22 also shows how the residuals should look when plotted against time (or the order in which data are collected). There should be no trend in the residuals when plotted versus time. A linear trend in the residuals as plotted versus time is shown in Figure 9.23.

FIGURE 9.21   **FIGURE 9.22**
Residual Plots Indicating Heteroscedasticity and No Heteroscedasticity

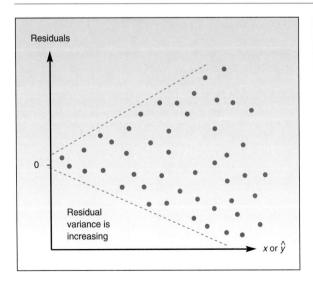

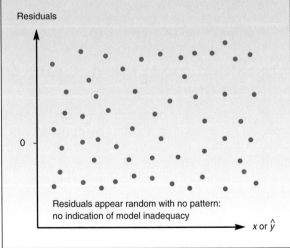

FIGURE 9.23   **FIGURE 9.24**

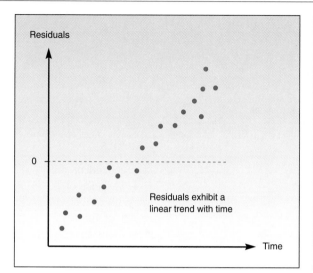

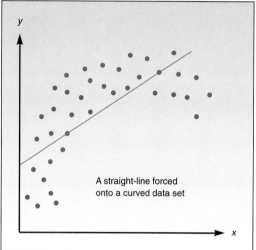

If the residuals exhibit a pattern when plotted versus time, then time should be incorporated as an explanatory variable in the model in addition to X. The same is true for any other variable against which we may plot the residuals: If there is any trend in the plot, the variable should be included in our model along with X. Incorporating additional variables leads to a multiple regression model.

Detecting a Curvilinear Relationship between Y and X

If the relationship between X and Y is curved, "forcing" a straight line to fit the data will result in a poor fit. This is shown in Figure 9.24. In this case, the residuals are at first large and negative, then decrease, become positive, and then again become negative. The residuals are not random and independent; they show curvature. This pattern appears in a plot of the residuals versus X, shown in Figure 9.25.

The situation can sometimes be corrected by adding the variable X^2 to the model. This also entails the techniques of multiple regression analysis. We note that, in cases where we have repeated Y observations at some levels of X, there is a statistical test for model lack of fit such as that shown in Figure 9.24. The test entails decomposing the sum of squares for error into a component due to lack of fit and a component due to pure error. This gives rise to an F test for lack of fit. This test is described in advanced texts. Note, however, that examination of the residuals is an excellent tool for detecting such model deficiencies, and this simple technique does not require the special data format needed for the formal statistical test.

Detecting Deviations from the Normal Distribution Assumption

When trying to detect deviations from the normal distribution assumption by use of residual plots, we need to plot the regression residuals on a special "probability paper" scale. Such paper is available in stationery stores. However, if we run the regression on a computer, we may include a command that will produce a residual plot on the

FIGURE 9.25
The Resulting Pattern of the Residuals When a Straight Line Is Forced
to Fit a Curved Data Set

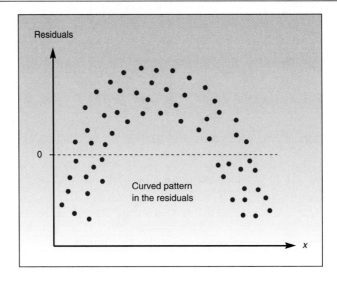

FIGURE 9.26   Approximately Normally Distributed Residuals as Plotted by the NSCORES Command

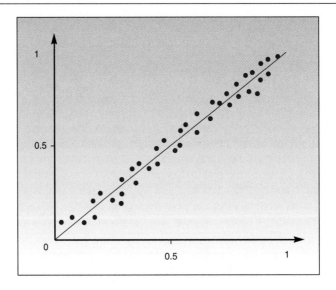

required scale. Using the MINITAB command NSCORES, we can check the normality of the residuals. Serious deviations from a straight line indicate that the residuals are not approximately normally distributed. An example of approximately normally distributed residuals is shown in Figure 9.26.

It may also be useful to plot *standardized* residuals, that is, the residuals divided by their (sample) standard deviation. (The mean of the residuals should be zero, so we need not subtract it when forming the standardized residuals.) The computer will standardize the residuals upon request. A histogram of the residuals should look similar to a normal curve. In MINITAB, we use the subcommand RESIDUALS, followed by a column number. Then the PLOT command will display the residuals against a specified variable. This is shown below for Example 9.1.

```
MTB > REGRESS 'DOLLARS' on 1 predictor 'MILES';
SUBC> RESIDS in C5
MTB > NAME C1 'MILES', C5 'RESIDS'
MTB > PLOT 'RESIDS' vs 'MILES'
```

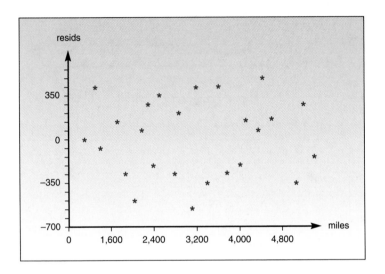

As can be seen, the residuals exhibit no pattern. The subcommand TRESID could be used to produce standardized residuals.

36. For each of the three accompanying plots of regression residuals versus X, state if there is any indication of model inadequacy, and if so, which inadequacy.

PROBLEMS

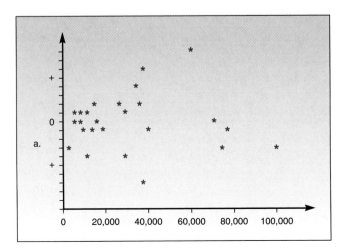

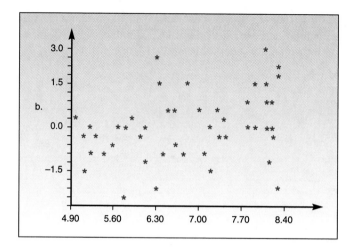

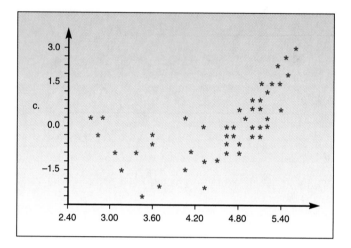

37. In the two accompanying plots of the residuals versus time of observation, state if there is evidence of model inadequacy. How would you correct any inadequacy?

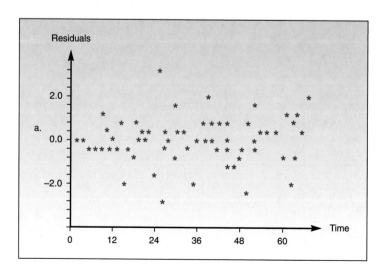

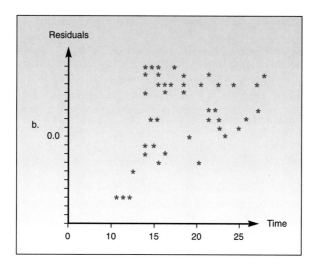

38. Is there any indication of model inadequacy in the two accompanying plots of residuals on a normal probability scale?

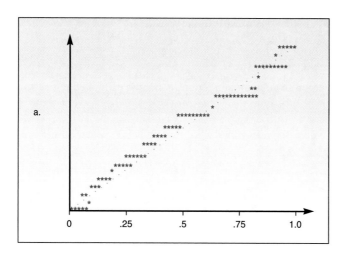

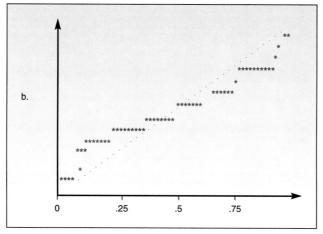

^o39.   Produce residual plots for the regression of Problem 12. Is there any apparent model inadequacy?

40.   Repeat Problem 39 for the regression of Problem 13.

41.   Repeat Problem 39 for the regression of Problem 14.

42.   Produce residual plots for the regressions in Problem 21.

43.   Use residual plots to check for model adequacy in Example 9.2. Use the RESIDUALS subcommand in MINITAB. Also use the NSCORES command to check for normality.

44.   Check for normality of the errors in the two regressions in Problem 21.

9.8 USE OF THE REGRESSION MODEL FOR PREDICTION

As mentioned in the introduction to this chapter, a regression model has several possible uses. One is to understand the relationship between the two variables. Note that understanding a relationship between two variables in regression does not imply that one variable *causes* the other. Casuality is a much more complicated issue and cannot be determined by a simple regression analysis.

A more common use of a regression analysis is **prediction**: providing estimates of values of the dependent variable by using the prediction equation: $\hat{Y} = b_0 + b_1 X$. It is important that prediction be done in the region of the data used in the estimation process. *You should be aware that using a regression for extrapolating outside the estimation range is risky, as the estimated relationship may not be appropriate outside this range.* This is demonstrated in Figure 9.27.

Point Predictions

It is very easy to produce point predictions using the estimated regression equation. All we need to do is substitute the value of X for which we want to predict Y into the prediction equation. In Example 9.1, suppose that American Express wants to predict charges on the card for a member who traveled 4,000 miles during a period equal to

FIGURE 9.27   The Danger of Extrapolation

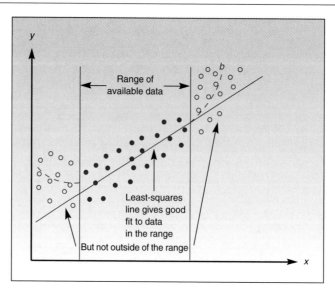

Prediction in American Express Study

FIGURE 9.28

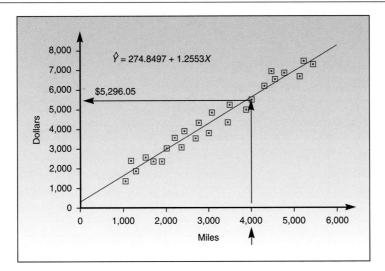

the one studied. (Note that $x = 4,000$ is in the range of X values used in the estimation.) We use the prediction equation, but with higher accuracy for b_1:

$$\hat{y} = 274.85 + 1.2553x = 274.85 + 1.2553(4,000) = 5,296.05 \text{ (dollars)}$$

The process of prediction in this example is demonstrated in Figure 9.28.

In Example 9.2, suppose we want to predict the hearing loss that would result from a person's prolonged exposure to 100 decibels. Using the prediction equation, substituting the estimates for the slope and intercept and using $x = 100$, we find:

$$y = -20.4 + 0.504(100) = 30.0.$$

Thus, the expected hearing loss at this level of exposure to noise is 30 percent.

Prediction Intervals

▲ Point predictions are not perfect and are subject to error. The error is due to the uncertainty in estimation as well as the natural variation of points about the regression line.

optional

A $(1 - \alpha)100\%$ prediction interval for Y:

$$\hat{y} \pm t_{\alpha/2}s \sqrt{1 + \frac{1}{n} + \frac{(x - \bar{x})^2}{\text{Sum of } x^2 - (\text{Sum of } x)^2/n}}$$

with degrees of freedom for t equal to $n - 2$.

As can be seen from the formula, the width of the interval depends on the distance of our value x (for which we wish to predict Y) from the mean, $\bar{x}$. This is shown in Figure 9.29.

FIGURE 9.29   Prediction Band and Its Width

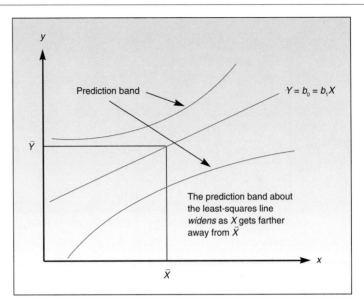

We will now use the equation to compute a 95 percent prediction interval for the amount charged on the American Express card by a member who traveled 4,000 miles:

$$5,296.05 \pm (2.069)(318.16)\sqrt{1 + 1/25 + (4,000 - 3,177.92)^2/40,947,557.84}$$

$$= 5,296.05 \pm 676.62 = [4,619,43, 5,972.67]$$

Based on the validity of the study, we are 95 percent confident that a cardholder who traveled 4,000 miles during a period of the given length will have charges on his or her card totaling anywhere from $4,619.43 to $5,972.67. What about the *average* total charge of all card holders who traveled 4,000 miles? The point estimate is equal to $\hat{y}$, but the confidence interval is different.

A Confidence Interval for the Average Y, Given a Particular Value of X

optional

205

▲ We may compute a *confidence interval* for the expected value of Y for a given X. Here the variation is smaller because we are dealing with the average Y for a given X, rather than a particular Y. Thus, the confidence interval is narrower than a prediction interval of the same confidence level.

A $(1 - \alpha)100\%$ confidence interval for the expected value of Y given X:

$$\hat{y} \pm t_{\alpha/2}s\sqrt{\frac{1}{n} + \frac{(x - \bar{x})^2}{\text{Sum of } x^2 - (\text{Sum of } x)^2/n}}$$

$$\text{df} = n - 2$$

The confidence band for the expected value of Y given X around the regression line looks like Figure 9.29 except that the band is narrower. The standard error of the estimator of the conditional mean is smaller than the standard error of the predicted Y. Therefore, the 1 is missing from the square root quantity as compared with the previous equation.

For the American Express example, let's now compute a 95 percent confidence interval for expected value of Y when $X = 4,000$ miles traveled:

$$5{,}296.05 \pm (2.069)(318.16)\sqrt{1/25 + (4{,}000 - 3{,}177.92)^2/40{,}947{,}557.84}$$
$$= 5{,}296.05 \pm 156.48 = [5{,}139.57, 5{,}452.53]$$

Being a confidence interval for a conditional mean, the interval is much narrower than the prediction interval, which has the same confidence level for covering *any given* observation at the level of X.

Computers are indispensable in constructing confidence intervals for mean response and prediction intervals. In MINITAB, the subcommand PREDICT followed by a value of the independent variable will produce a point prediction as well as prediction and confidence bounds. The default is 95 percent. Below we show how this is done for predicting charges for 4,000 miles traveled in Example 9.1.

```
MTB > REGRESS 'DOLLARS' on 1 predictor 'MILES';
SUBC> PREDICT 4000.
```

The regression equation is
dollars = 275 + 1.26 miles

Predictor	Coef	Stdev	t-ratio	p
Constant	274.8	170.3	1.61	0.120
miles	1.25533	0.04972	25.25	0.000

s = 318.2   R-sq = 96.5%   R-sq(adj) = 96.4%

Analysis of Variance

SOURCE	DF	SS	MS	F	p
Regression	1	65427736	64527736	634.60	0.000
Error	23	2328161	101224		
Total	24	66855896			

Fit	Stdev.Fit	95% C.I.	95% P.I.
5296.2	75.6	(5139.7, 5452.7)	(4619.5, 5972.8)

PROBLEMS

45. For the American Express example, give a 95 percent prediction interval for the amount charged by a member who traveled 5,000 miles. Compare the result with the one for $x = 4,000$ miles.
46. Give a 95 percent confidence interval for the expected value of Y given $X = 5,000$ miles traveled in the American Express example. Compare with your answer to Problem 45.
47. Produce some predictions for the regression in Problem 12.
48. Produce prediction and 95 percent prediction and confidence intervals for the percentage of coffee bean shrinkage for coffee roasted for 18 minutes using the regression in Problem 13.
49. Predict the number of suitors for an insect with 85 percent symmetry in the regression in Problem 14.
50. Explain why prediction intervals are useful.
51. What is the difference between a prediction interval and a confidence interval, of the same level (say 95 percent), for the mean level of the predicted variable?

52. In Problem 21, if course interest is rated 4.00, what will be the expected rating of the instructor? Also give a 95 percent prediction interval.
53. In Problem 21, if course difficulty is rated 3.5, what will be the best estimate for the instructor's rating? Also give a 95 percent prediction interval. Do you trust this prediction? Why?
54. In Example 9.2, predict hearing loss for noise exposure of 90 decibels. Provide prediction and confidence bands.

9.9 CORRELATION

In this section we finally come to the important topic of statistical correlation, to which we have made allusions throughout this chapter.

Recall that one of the assumptions of the regression model is that the independent variable, X, is fixed rather than random and that the only randomness in the values of Y comes from the error term, ϵ. Let's now relax this assumption and *assume that both X and Y are random variables*. In this new context, the study of the relationship between two variables is called **correlation analysis.**

In correlation analysis, we adopt a symmetric approach: We make no distinction between an independent variable and a dependent one. The correlation between two variables is a measure of the linear relationship between them. The correlation gives an indication of how well the two variables move together in a straight-line fashion. The correlation between X and Y is the same as the correlation between Y and X.

> The **correlation** between two variables, X and Y, is a measure of the *degree of linear association* between the two variables.

Two variables are highly correlated if they move well together. Correlation is indicated by the **correlation coefficient.**

> The population correlation coefficient is denoted by ρ (the Greek letter rho). The coefficient ρ can take on any value from -1, through 0, to 1.

The possible values of ρ and their interpretations are as follows:

1. When ρ is equal to zero, there is no correlation. That is, there is no linear relationship between the two random variables.
2. When $\rho = 1$, there is a perfect positive linear relationship between the two variables. That is, whenever one of the variables, X or Y, increases, the other variable also increases; and whenever one of the variables decreases, the other one must also decrease.
3. When $\rho = -1$, there is a perfect negative linear relationship between X and Y. When X or Y increases, the other variable decreases; and when one decreases, the other one must increase.
4. When the value of ρ is between 0 and 1 in absolute value, it reflects the relative strength of the linear relationship between the two variables. For example, a correlation of 0.90 implies a relatively strong positive relationship between the two variables. A correlation of -0.70 implies a weaker negative (as indicated by the negative sign) linear relationship. A correlation $\rho = 0.30$ implies a relatively weak positive linear relationship between X and Y.

A few sets of data on two variables, and their corresponding population correlation coefficients, are shown in Figure 9.30. The figure gives an idea as to what data from populations with different values of ρ may look like.

Several Possible Correlations between Two Variables   **FIGURE 9.30**

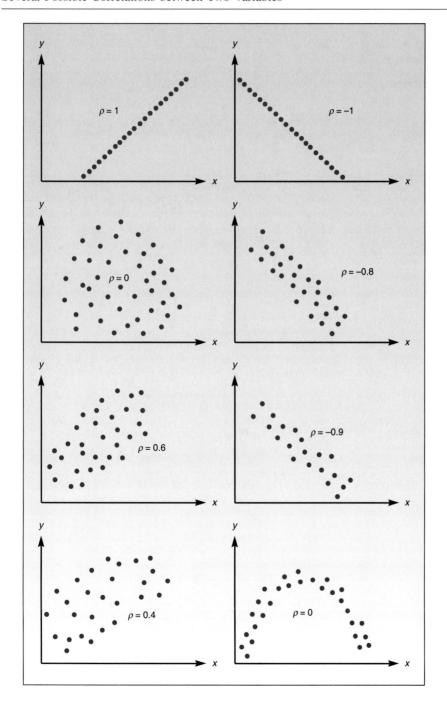

Like all population parameters, the value of ρ is not known to us, and we need to estimate it from our random sample of (X,Y) observation pairs. The estimate of the population correlation, ρ, is the *sample correlation coefficient,* denoted by r. This estimate, also referred to as the *Pearson product-moment correlation coefficient,* is given below.

The **sample correlation coefficient:**

$$r = \frac{\text{Sum of } xy - (\text{Sum of } x)(\text{Sum of } y)/n}{\sqrt{(\text{Sum of } x^2 - (\text{Sum of } x)^2/n)(\text{Sum of } y^2 - (\text{Sum of } y)^2/n)}}$$

Recall that in regression analysis, the square of the sample correlation coefficient, r^2, has a special meaning and importance. This statistic is the coefficient of determination associated with the regression.

Example 9.3 illustrates the use of the sample correlation coefficient.

EXAMPLE 9.3

A researcher is interested in the correlation between age and the maximum exercise pulse rate for women. In the sample of five pairs of measurements in Table 9.6, $X =$ age and $Y =$ maximum pulse. The table also gives the squares of the X and Y values, as well as the products of X and Y. The sums of all these quantities are used to find the sample correlation.

SOLUTION

Using the formula, first we subtract from the sum of XY the sum of X times the sum of Y divided by n:

$$29{,}714 - (176)(902)/5 = -2{,}036.4$$

The denominator is the square root of the sum of X-squared minus the (Sum of X) squared divided by n and similarly for Y:

$$\sqrt{(6{,}948 - (176)^2/5)(168{,}750 - (902)^2/5)} = 2{,}130.4$$

The sample correlation is therefore:

$$r = -2{,}036.4/2{,}130.4 = -0.956.$$

Now let's use MINITAB, to avoid lengthy hand calculations. Incidentally, many

---

TABLE 9.6   Data for Study on Age and Exercise Pulse Rate

X (Age)	Y (Pulse)	XY	X^2	Y^2
23	210	4,830	529	44,100
39	185	7,215	1,521	34,225
19	220	4,180	361	48,400
44	164	7,216	1,936	26,896
51	123	6,273	2,601	15,129
176	902	29,714	6,948	168,750

calculators will compute a correlation coefficient. Look for "r" or "corr" on your calculator. In MINITAB, the command is simply CORRELATION. The input and the output are shown below. This includes a scatterplot of the data on these two variables.

```
MTB   > SET C1
DATA  > 23 39 19 44 51
DATA  > END
MTB   > SET C2
DATA  > 210 185 220 164 123
DATA  > END
MTB   > CORRELATION C1 C2

Correlation of C1 and C2 = -0.956

MTB  > PLOT C2*C1


        -
        -   *
      210+       *
        -
   C2   -

        -
        -              *
      175+
        -
        -                    *
        -
        -
      140+
        -
        -                          *
        -
        -
        +---------+---------+---------+---------+---------+------C1
        18.0 24.0   30.0    36.0    42.0    48.0
```

We see that the maximum exercise pulse rate decreases with age—as evidenced by the *negative* correlation—at least for the five women in our sample.

We often use the sample correlation coefficient for descriptive purposes as a point estimator of the population correlation coefficient, ρ. When r is large and positive (close to +1), we say that the two variables are highly correlated in a positive way; when r is large and negative (toward −1), we say that the two variables are highly inversely correlated, and so on. That is, we view r as if it were the parameter ρ, which r estimates. It is possible, however, to use r as an estimator in testing hypotheses about the true correlation coefficient, ρ. When testing such hypotheses, the assumption of normal distributions of the two variables is required.

The most common test is a test of whether or not two random variables, X and Y, are correlated.

$$H_0: \rho = 0$$

$$H_1: \rho \neq 0$$

The test statistic for this particular test is as follows.

$$t_{(n-2)} = \frac{r}{\sqrt{\dfrac{1 - r^2}{n - 2}}}$$

This test statistic may also be used for carrying out tests for the existence of a positive only, or a negative only, correlation between X and Y. These would be one-tailed tests instead of the two-tailed test above, and the only difference is that the critical points for t would be the appropriate one-tailed values for a given α. The test statistic, however, is good *only* for tests where the null hypothesis assumes a zero correlation. When the true correlation between the two variables is anything but zero, the t distribution above does not apply; in such cases the distribution is more complicated.[4] The test above is the most common hypothesis test about the population correlation coefficient because it is a test for the existence of a linear relationship between two variables. Example 9.4 demonstrates this test.

EXAMPLE 9.4

Suppose that in the area described in the accompanying article a random sample of 27 lakes revealed a sample correlation of $r = 0.424$ between acidity (on an increasing scale) and fish mercury levels. Is the correlation statistically significant?

SOLUTION

We want to conduct the hypothesis test H_0: $\rho = 0$ versus H_1: $\rho \neq 0$. Using the test statistic, we get the following:

$$t_{(25)} = \frac{r}{\sqrt{(1 - r^2)/(n - 2)}} = \frac{0.424}{\sqrt{(1 - 0.424^2)/25}} = 2.34$$

From Appendix B, Table B.3, we find that the critical points for a t distribution with 25 degrees of freedom and $\alpha = 0.05$ are ± 2.060. Therefore, we reject the null hypothesis of no correlation in favor of the alternative that the two variables are linearly related. Since the critical points for $\alpha = 0.01$ are ± 2.787, and $2.787 > 2.34$, we may not reject the null hypothesis of no correlation between the two variables using the 0.01 level of significance. If we wanted to test (before looking at our data) only for the existence of a positive correlation between the two variables, our test would have been H_0: $\rho \leq 0$ versus H_1: $\rho > 0$, and we would have used only the right-hand tail of the t distribution. At $\alpha = 0.05$, the critical point of t with 25 degrees of freedom is 1.708, and at $\alpha = 0.01$

---

[4]In cases where we want to test H_0: $\rho = a$ versus H_1: $\rho \neq a$, where a is some number other than zero, we may do so by using the Fisher transformation: $z' = (1/2)\log[(1 + r)/(1 - r)]$, where z' is approximately normally distributed with mean $\mu' = (1/2)\log[(1 + \rho)/(1 - \rho)]$ and standard deviation $\sigma' = 1/\sqrt{n-3}$. (Here $\log$ is taken to mean *natural logarithm*.) Such tests are less common, and a more complete description may be found in advanced texts. As an exercise, the interested reader may try this test on some data. (You need to transform z' to an approximate standard normal: $z = (z' - \mu')/\sigma'$; use the null-hypothesized value of ρ in the formula for μ'.)

State Warns of Mercury in Its Lakes

High Levels Found in Freshwater Fish

By JON NORDHEIMER

BATSTO, N.J.—Deep in the heart of the Pine Barrens, surrounded by the protected forests of Wharton State Park, Batsto Lake presents a picture of New Jersey that never emerges when late-night talk show hosts make jokes about the state.

Tall pines and cedars are mirrored on its smooth, broad surface. Bald eagles wheel high overhead. All that breaks the midday quiet are the calls of birds and the whoosh of water sliding over a spillway near a century-old sawmill that has been preserved as a museum.

Yet something unnerving is going on. Notices posted on trees at the water's edge read: "Do not eat pickerel, bass or yellow bullhead from this body of water. Preliminary data from a study funded by the New Jersey Department of Environmental Protection and Energy have found elevated levels of mercury in these fish species."

Role of Acidity

With one organism consuming another, the mercury is accumulated and passed up the food chain. As smaller fish are eaten by bigger fish, it becomes "bio-magnified" into dangerous concentrations in large fish.

Fish with one part per million of methyl mercury may be found in water that has a mercury content of only one part per trillion, an amplification of one million, said Dr. William F. Fitzgerald, professor of marine sciences at the University of Connecticut in Groton.

The yearly introduction from the atmosphere of just one gram of mercury, the amount found in a single hearing aid battery, would be all that would be needed in a lake the size of 20 football fields to reach levels harmful to humans in the biggest fish there, Dr. Fitzgerald said.

Scientists are also struck by evidence that problems of methyl mercury tend to be found in lakes with higher acidity levels—a common trait of lakes in the Pine Barrens, the Adirondacks, Minnesota and Michigan—and ones like Batsto and New Brooklyn that are tea-colored and drain bogs and peat land.

Reprinted by permission of *The New York Times,* May 18, 1994, p. B5.

it is 2.485. The null hypothesis would, again, be rejected at the 0.05 level but not at the 0.01 level of significance.

In regression analysis, the test for the existence of a linear relationship between X and Y is a test of whether or not the regression slope, β_1, is equal to zero. The regression slope parameter is related to the correlation coefficient. (As an exercise, compare the equations of the estimates r and b_1.) When two random variables are uncorrelated, the population regression slope is zero.

We end this section with two words of caution. First, the existence of a correlation between two variables does not necessarily mean that one of the variables *causes* the other one. The determination of **causality** is a difficult question that cannot be directly answered in the context of correlation analysis or regression analysis. Second, the statistical determination that two variables are correlated may not always mean that they are correlated in any direct, meaningful way. For example, if we study any two population-related variables and find that both variables increase "together," this may merely be a reflection of the general increase in population rather than any direct correlation between the two variables. We should look for outside variables that may affect both variables under study.

PROBLEMS

55. What is the main difference between correlation analysis and regression analysis?
56. Compute the sample correlation coefficient for the data of Problem 12.
57. Compute the sample correlation coefficient for the data of Problem 13.
58. Using the data in Problem 14, conduct the hypothesis test for the existence of a correlation between the two variables. Use $\alpha = 0.01$.
59. Is it possible that a sample correlation of 0.51 between two variables will not indicate that the two variables are really correlated, while a sample correlation of 0.04 between another pair of variables will be statistically significant? Explain.
60. The following data, from the September 1993 issue of *International Financial Statistics*, are indexed prices of gold and copper over the last 10 years. Assuming these indexed values constitute a random sample from the population of possible values, test for the existence of a correlation between the indexed prices of the two metals.

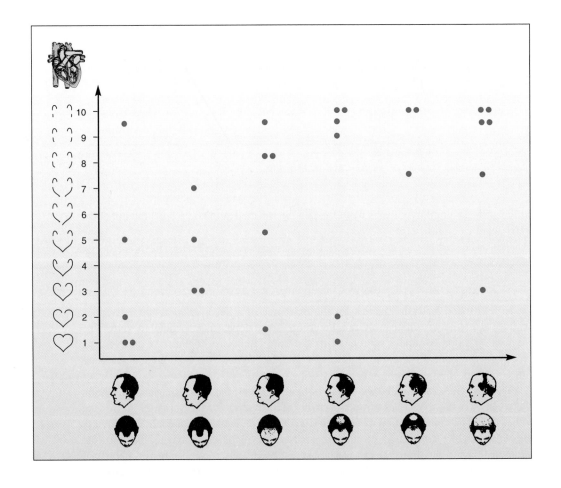

Gold: 76 62 70 59 52 53 53 56 57 56

Copper: 80 68 73 63 65 68 65 63 65 66

Also, state one limitation of the data set.

61. Follow daily stock price quotations in *The Wall Street Journal* for a pair of stocks of your choice, and compute the sample correlation coefficient. Also, test for the existence of a nonzero correlation in the "population" of prices of the two stocks. For your sample, use as many daily prices as you can.

62. Again using *The Wall Street Journal* as a source of data, determine whether there is a correlation between morning and afternoon price quotations in London for an ounce of gold (for the same day). Any ideas?

63. The Framingham Heart Study, which ended in April 1993 and was reported widely in the media, inferred that a relationship existed between male-pattern baldness and heart disease. Researchers visited male hospital patients recovering from heart attacks or open-heart surgery and showed each of them a chart of various degrees of baldness. Each patient was asked to evaluate his own level of baldness based on the chart. The exhibit on the previous page shows the baldness chart and the severity of the heart trouble, on a scale of 1 (mild heart attack) to 10 (triple bypass surgery with 90 percent artery blockage) for a sample of patients. Do you believe that the two variables are correlated? Explain. Also, could we state that baldness *causes* heart trouble? (Think about this!) Explain carefully.

64. The two charts on the bottom of the page show results of a study on sleep patterns. Are the correlations statistically significant? Explain. Comment on the scatterplots. Also comment on the *p*-values.

65. The display on page 456 is from a study of factors in learning. The sample size used was $n = 223$ teachers. The teachers responded on the variables listed in the table related to the use of concept maps like the one shown in the figure. Are the correlations statistically significant? Explain.

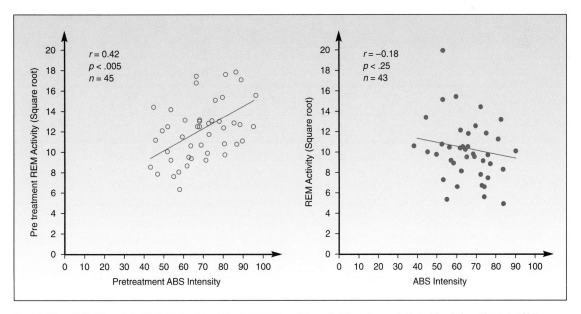

Reprinted from E. Nofziger et al., "Affect Intensity and Phasic REM Sleep," *Journal of Consulting and Clinical Psychology* 62, no. 1 (1994), pp. 83–91.

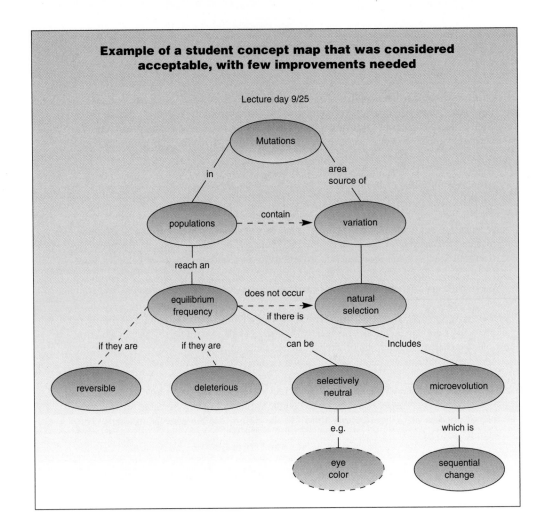

Example of a student concept map that was considered acceptable, with few improvements needed

Possible Predictors of Course Grades

Correlations	r
Course grade with number of maps made	.050
Course grade with number of revisions per map	.093
Final exam grade with map index	.099
Course grade with total map points earned	.198
Course grade with time spent on map construction	.214

Reprinted from J. Trowbridge and J. Wandersee, "Identifying Critical Junctures in Learning," *Journal of Research in Science Teaching* 31, no. 5 (1994), pp. 459–73.

SUMMARY

In this chapter we learned how to model a **straight-line relationship** between two quantitative variables. The technique we learned is called **regression analysis,** and we saw how the **least-squares method** gives us estimates of a **slope parameter** and an **intercept parameter** for the straight line we desire. A test for the existence of a **linear relationship** between the two variables was presented. We learned how to measure the strength of the linear relationship between the two variables using the estimated **coefficient of determination.** We saw how the examination of **residual plots** can help us detect violations of model assumptions, such as curvature of unequal variance. The use of the estimated regression model in **predicting** values of the dependent variable based on values of the independent variable was demonstrated. We discussed a different approach to studying the linear relationship between two variables: **correlation analysis.** Here we saw how the square root of the coefficient of determination, **the correlation coefficient,** is used both in a descriptive analysis and in the context of inference about the population relationship.

KEY FORMULAS

The coefficient of correlation:

$$r = \frac{\text{Sum of } xy - (\text{Sum of } x)(\text{Sum of } y)/n}{\sqrt{[\text{Sum of } x^2 - (\text{Sum of } x)^2/n][\text{Sum of } y^2 - (\text{Sum of } y)^2/n]}}$$

The coefficient of determination:

$$r^2 = \frac{\text{SSR}}{\text{SST}} = 1 - \frac{\text{SSE}}{\text{SST}} = \text{square of correlation coefficient}$$

ADDITIONAL PROBLEMS

66. Skin cancer rates have been steadily increasing in recent years. A study was undertaken to assess the effect of atmospheric ozone depletion on these rates. The following data are ozone depletion rates over various geographical regions and the rates of melanoma (a form of skin cancer) in these regions:

    Ozone dep (%): 5, 7, 13, 14, 17, 20, 26, 30, 34, 39, 44

    Melanoma (%): 1, 1, 3, 4, 6, 5, 6, 8, 7, 10, 9

    Run a regression of melanoma rates versus ozone depletion rates.

67. Conduct a residual analysis for the regression in Problem 66.

68. Refer again to Problem 66. What percentage of the variation in rates of melanoma is explained by the relationship between this form of cancer and the level of atmospheric ozone depletion? Explain.

69. Refer again to Problem 66. In the spring of 1993, scientists discovered that 40 percent of the ozone was depleted in the region of Hamburg in northern Germany. What would you expect to be the rate of melanoma in this area in the near future? Explain the limitations of such predictions. What other factors may play a role? Could you account for such factors?

70. Look at the display on page 458 titled "Vacations as Americans Take Them." Take New York as the residence of all respondents, and find the distance in miles of all destinations from New York. Do you believe that greater distance from home makes a vacation destination more desirable for Americans (or at least for New Yorkers)?

71. Explain in detail the uses of regression analysis. Does regression automatically imply causality? Explain.

72. Refer to the graph showing data on page 459 about meteoroids.
    a. Explain the graph in the context of regression.
    b. What is the *actual relationship* between number of years to occurrence and intensity of impact?

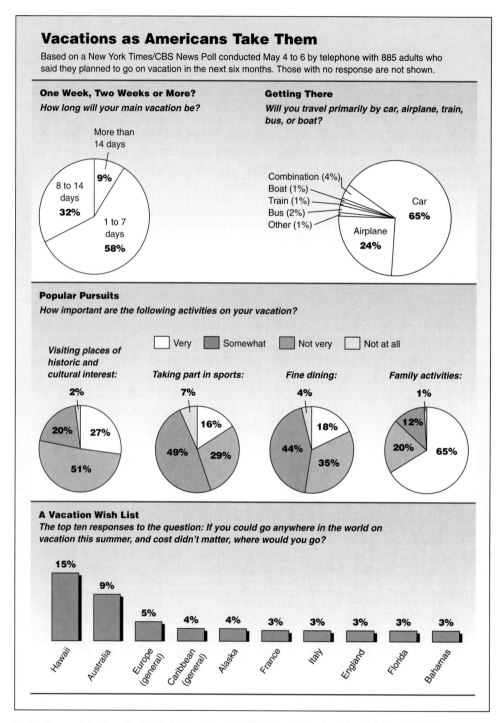

Vacations as Americans Take Them

Based on a New York Times/CBS News Poll conducted May 4 to 6 by telephone with 885 adults who said they planned to go on vacation in the next six months. Those with no response are not shown.

One Week, Two Weeks or More?

How long will your main vacation be?

More than 14 days 9%

8 to 14 days 32%

1 to 7 days 58%

Getting There

Will you travel primarily by car, airplane, train, bus, or boat?

Combination (4%)
Boat (1%)
Train (1%)
Bus (2%)
Other (1%)

Car 65%

Airplane 24%

Popular Pursuits

How important are the following activities on your vacation?

☐ Very   ◼ Somewhat   ◼ Not very   ☐ Not at all

Visiting places of historic and cultural interest:

2%   20%   27%   51%

Taking part in sports:

7%   16%   49%   29%

Fine dining:

4%   18%   44%   35%

Family activities:

1%   12%   20%   65%

A Vacation Wish List

The top ten responses to the question: If you could go anywhere in the world on vacation this summer, and cost didn't matter, where would you go?

Hawaii	Australia	Europe (general)	Caribbean (general)	Alaska	France	Italy	England	Florida	Bahamas
15%	9%	5%	4%	4%	3%	3%	3%	3%	3%

Reprinted by permission from *The New York Times,* May 23, 1993, p. 23, with data from New York Times/CBS News Poll conducted May 4 to 6 by telephone with 1,233 adults.

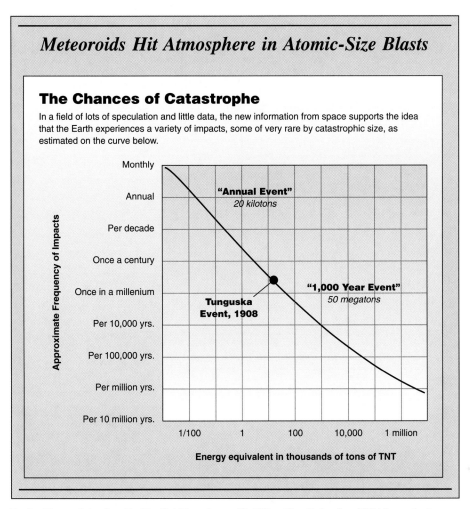

Meteoroids Hit Atmosphere in Atomic-Size Blasts

The Chances of Catastrophe

In a field of lots of speculation and little data, the new information from space supports the idea that the Earth experiences a variety of impacts, some of very rare by catastrophic size, as estimated on the curve below.

Reprinted by permission from *The New York Times,* January 25, 1994, p. C1, with data from NASA International Near-Earth-Object Detection Workshop.

73. Look at the table showing data from a study on attitudes toward physical exercise. These are regressions of attitude versus intention to exercise. Interpret the findings.

Behaviors Studied	r^2	Attitude
Jogging	.49	$b = .41$**
Mountain climbing	.37	$b = .08$
Biking	.45	$b = .53$**

** $p < .01$.

Reprinted from G. Godin, "The Theories of Reasoned Action and Planned Behavior: Exercise Promotion," *Journal of Applied Sport Psychology* 5 (1993), pp. 141–57.

74. Use Table A.3 in Appendix A. Run a simple linear regression of output per hour (X) versus real compensation (Y). Estimate the parameters of the regression model, test for the existence of a linear relationship, and report the r^2. Use residual plots to test for model inadequacies.

75. Using the regression model you developed in Problem 74, predict real compensation for output per hour equalling 215.

INTERVIEW WITH
Mary Guinan

"The accuracy of projections varies with the different methodologics. What is the least well defined is the number of new cases of HIV infection occurring each year."

Mary E. Guinan is Special Assistant for Evaluation in the HIV Deputy Director's Office at the Centers for Disease Control in Atlanta. She received a PhD from the University of Texas and an MD from Johns Hopkins and is credited with being the first to identify AIDS as epidemic.

Aczel: Where did you go to school and what type of statistics courses did you take? What did you think of them?

Guinan: I received my PhD from the University of Texas and MD from Johns Hopkins. I took an introductory course to statistics in graduate school. I later took statistics in an epidemiology training program here at [the Centers for Disease Control and Prevention] CDC called the Epidemic Intelligence Service.

Aczel: What special interest or focus brought you to the CDC?

Guinan: When I first graduated from college I worked in a university laboratory as a research technician, then I was employed as a flavor chemist for chewing gum by a large pharmaceutical company.

I came to the Centers for Disease Control and Prevention because I was interested in disease control

and prevention from a population perspective rather than the individual perspective. In epidemiology, statistics are of paramount importance in tracking disease and injury in the community. Statistical data are used to determine health priorities in communities and to convince Congress, and state and local legislatures to invest in prevention and control of various health problems including AIDS.

Aczel: What type of research do you do at the CDC?

Guinan: I do epidemiologic research, which involves studying the risk factors and determinants of disease spread in the community for AIDS, which is caused by the human immunodeficiency virus.

Aczel: Could you talk about the current spread of AIDS, the population it is infecting, and the uncertainty of this epidemic.

Guinan: AIDS is a unique syndrome. HIV impairs the immune system and the manifestation of disease is very varied, from infections to tumors. AIDS is defined as one of these infections, tumors, or conditions, such as wasting syndrome. The time between HIV infection and disease manifestation averages 10 years, during which time the person may not know he/she is infected, but is infectious and can spread disease either sexually or by sharing injection drugs. So, a completely healthy appearing individual may be spreading infection. AIDS cases are indicators of long-term HIV infection. About 350,000 cases have been reported in the United States as of 1994. The proportion of cases in homosexual men has leveled off and the proportion of drug users is only rising slightly, but the proportion of cases transmitted during heterosexual intercourse is rapidly rising and of great concern. The heterosexual population is much greater than the populations of gay men or drug users, and the potential for spread in this population is well recognized by health authorities. However, political and religious concerns have kept public policy makers from confronting this issue.

Aczel: What caused the CDC to perceive that the AIDS virus was an epidemic?

Guinan: The CDC as part of its mission to prevent and control disease and injury often receives reports of unusual occurrences of disease. A group of physicians in San Francisco first contacted CDC when they noticed the occurrence of a rare pneumonia in previously healthy young gay men. The CDC has a weekly newsletter that it sends to public health workers, in which it reported these cases. Subsequent to that publication, CDC received hundreds of calls about similar cases occurring in New York, Florida, Georgia, Texas, and other states. An investigation was begun to determine the cause/causes of the unusual pattern of this pneumonia.

Aczel: So, now what are the issues you are working on?

Guinan: I am now focused on trying to emphasize prevention of transmission of HIV through encouraging healthy sexual behavior. Because we have no vaccine to prevent disease or no cure for HIV, we have to focus on behavior. This is very difficult and controversial. Many people want the message to be "just say no" to sex for adolescents and young people. I think that this is a good message, but not one that all young people will accept. CDC advocates the use of condoms in those persons who choose to have sex.

Aczel: How about the forecasting, how does the CDC gather the data for projections?

Guinan: CDC has a surveillance system for AIDS. States determine which diseases or conditions are reportable to health departments. In all states, AIDS is a reportable disease. We gather the data electronically from the state health departments and analyze it for case numbers, geographic distribution, risk factors, and trends.

Aczel: Are there any unique statistical methods employed in tracking the AIDS virus? Or any data-gathering issues you had to confront with AIDS?

Guinan: What is unique about CDC's statistics is that we employ strict surveillance case definitions. What is a case of AIDS is very difficult to determine, because it is not one disease. So we establish criteria that determine what we will count as a case. This means, for example, that many true cases may not be accepted in our counting system because the criteria have not been met. It is important to have a standard case definition because we must be counting the same thing in order to have accurate statistics. We have changed the case definition several times as we have learned more about HIV infection. This complicates the analysis of trends because we are counting different things. We have methods to determine trends using the old case definition and for analyzing how many of the increased annual cases are due to true incidence changes or to changes in the case definition.

Aczel: The press tends to report on "new AIDS cases"; is that reliable?

Guinan: The press is rarely reliable in reporting stories on AIDS.

Aczel: Back to the projections—with different spread rates for different populations, how do you forecast? Do you model each?

Guinan: We project the number of future cases of AIDS using a back calculation method which uses the slope of the curves of trends in annual cases for each transmission subpopulation. For example, the slope of the curve in gay men has flattened out while the slope of the heterosexually transmitted cases is rising acutely. The spread of the virus among various subpopulations has been estimated by doing surveys of samples of special populations. For example, injection drug users who come into treatment will be tested for HIV. The proportion of positives in various drug populations is different. For example, in New York up to 60 percent of some drug user populations are infected with HIV, but in California less than 10 percent of drug user populations tested are HIV positive. The spread of HIV into the heterosexual population has occurred primarily through injection drug users. This has been determined by the surveillance data in which the risk factors for HIV are recorded. Having a sex partner who uses injection drugs is the most common reported risk factor for heterosexual cases.

The accuracy of projections varies with the different methodologies. What is the least well defined is the number of new cases of HIV infection occurring each year. We estimated in 1991 that approximately 1 million persons in the United States had been infected with HIV. This was estimated from a large group of HIV seroprevalence studies done on college campuses, in health clinics, in women delivering babies, and in various high risk populations, such as gay men and injection drug users.

Aczel: As an MD, what are your particular concerns or frustrations with forecasting for disease control?

Guinan: Most modelers of the AIDS epidemic do not have enough information on human behavior in order to make accurate projections. For example, we do not know what proportion of the U.S. population is engaging in unsafe sexual behavior. Any attempts to study sexual behavior are so controversial that they rarely get done properly. Public sentiment about sexual behavior studies is very polarized. Also, prevention efforts are very controversial such as promoting condom use among sexually active teens. Political and social issues have been the biggest barrier to

trying to determine the extent of the HIV epidemic and the most effective measures to control it. Because of the discrimination against HIV-infected persons in housing, jobs, schools, and social settings, large groups of people at risk refuse to be tested. Unrealistic fears of HIV positive persons has complicated the approach to recognizing the infected, which is necessary in order to offer early treatment.

Aczel: Do you have any advice or suggestions to students, if this is their first, possibly only, statistics course?

Guinan: Students today should look at how the media use statistics, how politicians use statistics, and how we have become a nation "hooked on numbers." Statistics can be used to lure you into investing money, buying one particular item over another, or even where to go to college. Everyone should have a healthy skepticism about statistics if they are presented in order to convince you to buy something, or to vote a certain way. One should always ask, "What is it that they are not telling me?" For automobiles, for example, one particular model may be a best seller but its repair record may be poor. You can bet that the auto salesman doesn't give you the statistics on repair rates. Mark Twain once said there are lies, great lies, and statistics. The general public usually doesn't understand the use of statistics. Even one of our past presidents of the United States was shown lacking in this area. President Eisenhower was very surprised to find out that half of the American population had an IQ below average. Do you think he should have been surprised?

CHAPTER

10

The Quest for Quality

Courtesy of AIDO MAURO.

10.1 Control Charts   466

10.2 The $\bar{x}$ Chart   468

10.3 The R Chart and the s Chart   472

10.4 The p Chart   474

10.5 The c Chart   476

I n 1950, Japan was trying to revive from the devastation of World War II. Japanese industry was all but destroyed, and its leaders knew that industry must be rebuilt well if the nation was to survive. But how? By an ironic twist of fate, Japanese industrialists decided to hire an American statistician as their consultant. The man they chose was the late W. Edwards Deming, at the time a virtually unknown government statistician. No one in America had paid much attention to Deming's theories on how statistics could be used to improve industrial quality. The Japanese wanted to listen. They brought Deming to Japan in 1950, and in July of that year he met with the top management of Japan's leading companies. He then gave the first of many lecture series to Japanese management. The title of the course was "Elementary Principles of the Statistical Control of Quality," and it was attended by 230 Japanese managers of industrial firms, engineers, and scientists.

The Japanese listened closely to Deming's message. In fact, they listened so well that in a few short decades, Japan became one of the most successful industrial nations on earth. Whereas "Made in Japan" once meant low quality, the phrase has now come to denote the highest quality. In 1960, Emperor Hirohito awarded Dr. Deming the Medal of the Sacred Treasure. The citation with the medal stated that the Japanese people attribute the rebirth of Japanese industry to W. Edwards Deming. In addition, the Deming Award was instituted in Japan to recognize outstanding developments and innovations in the field of quality improvement. On the walls of the main lobby of Toyota's headquarters in Tokyo hang three portraits. One portrait is of the company's founder, another is of the current chairman, and the largest portrait is of Ed Deming.

Ironically, Dr. Deming's ideas did get recognized back in the United States—alas, when he was 80 years old. For years, American manufacturing firms had been feeling the pressure to improve quality, but not much was actually being done while the Japanese were conquering the world markets. In June 1980, Dr. Deming appeared in a network television documentary entitled "If Japan Can, Why Can't We?" Starting the next morning, Dr. Deming's mail quadrupled, and the phone was constantly ringing. Offers came from Ford, General Motors, Xerox, and many others.

While in his 90s, Dr. Ed Deming was one of the most sought-after consultants to American industry. His appointment book was filled years in advance, and companies were willing to pay very high fees for an hour of his time. He traveled around the country lecturing on quality and how to achieve it.

In all fairness, Dr. Deming did not invent the idea of using statistics to control and improve quality; that honor goes to his colleague, Walter Shewhart. What Deming did was to expand the theory and demonstrate how it could be used very successfully in industry. After that, Deming's theories went beyond statistics and quality control, to encompass the entire firm. His tenets of management are the well-known "14 Points" he advocated, which deal with the desired corporate approach to costs, prices, profits, labor, and other factors. In the years prior to his passing, Deming even liked to expound about antitrust laws and capitalism, and to have fun with his audience. At a lecture attended by the author around Deming's 90th birthday, Dr. Deming opened by writing on a transparency: "Deming's Second Theorem: 'Nobody gives a hoot about profits.'" He then stopped and addressed the audience, "Ask me what is Deming's First Theorem." He looked expectantly at his listeners and answered, "I haven't thought of it yet!" The philosophical approach to quality and the whole firm, and how it relates to profits and costs, is described in the ever-growing literature on this subject. It is sometimes referred to as **total quality management (TQM).** Joseph Juran and Armand Feigenbaum, author of an early book on total quality management, should also be mentioned in this context. Here we want to concentrate on Deming's, and others', *statistical* ideas.

10.1 CONTROL CHARTS

The first modern ideas on how statistics could be used in quality control came in the mid-1920s from Walter Shewhart of Bell Laboratories. Shewhart invented the **control chart** for industrial processes. A control chart is a graphical display of measurements (usually aggregated in the form of means or other statistics) of an industrial process through time. By carefully scrutinizing the chart, a quality-control engineer can identify any potential problems with the production process. The idea is that when a process is in control, the variable being measured—the mean of every four observations, for example—should remain stable through time. The mean should stay somewhere around the middle line (the grand mean for the process) and not wander off too much. By now you understand what "too much" means in statistics: more than several standard deviations of the process. The required number of standard deviations is chosen so that there will be a small probability of exceeding them when the process is in control. Addition and subtraction of the required number of standard deviations (generally three) gives us the **upper control limit** (UCL) and the **lower control limit** (LCL) of the control chart. The UCL and LCL are similar to the "tolerance" limits in the story of the pyx (Chapter 6). When the bounds are breached, the process is deemed **out of control** and must be corrected. Control charts are illustrated in Figure 10.1. We assume throughout that the variable being charted is at least approximately normally distributed.

In addition to looking for the process exceeding the bounds, quality-control workers also look for patterns and trends in the charted variable. For example, if the mean of four observations at a time keeps increasing or decreasing, or it stays too long above or below the center line (even if the UCL and LCL are not breached), the process may be out of control.

A *control chart* is a time plot of a statistic, such as a sample mean, range, standard deviation, or proportion, with a center line and *upper and lower control limits.* The limits give the desired range of values of the statistic. When the statistic is outside the bounds, or when its time plot reveals certain patterns, the process may be *out of control.*

Central to the idea of a control chart—and, in general, to the use of statistics in quality control—is the concept of *variance.* If we were to summarize the entire field of statistical quality control (also called *statistical process control,* or SPC) in one word, that word would have to be *variance.* Shewhart, Deming, and others wanted to bring the statistical concept of variance down to the shop floor. If foremen and production line workers could understand the existence of variance in the production process, then this awareness itself could be used to help minimize the variance. Furthermore, the variance in the production process could be partitioned into two kinds: the natural, random variation of the process, and variation due to *assignable causes.* Examples of assignable causes are fatigue of workers and breakdown of components. Variation due to assignable causes is especially undesirable because it is due to something being wrong with the production process, and may result in low quality of the produced items. The chart helps us detect an assignable cause by allowing us to ask what has happened at a particular point where the process looks unusual.

51

A process is considered in **statistical control** when it has no assignable causes, only natural variation.

Actually, *any* kind of variance is undesirable in a production process. Even the natural variance of a process due to purely random causes rather than to assignable causes can be detrimental. The control chart, however, will detect only assignable causes. As the following story shows, one could do very well by removing all variance:

An American car manufacturer was having problems with transmissions made at one of its domestic plants, and warranty costs were enormous. The identical type of transmission made in Japan was not causing any problems at all. Engineers carefully examined 12 transmissions made at the company's American plant. They found that variations existed among the 12 transmissions, but there were no assignable causes and a control chart revealed nothing unusual. All transmissions were well within specifications. Then they looked at 12

Courtesy of Toyota USA.

FIGURE 10.1   A Production Process in, and out of, Statistical Control

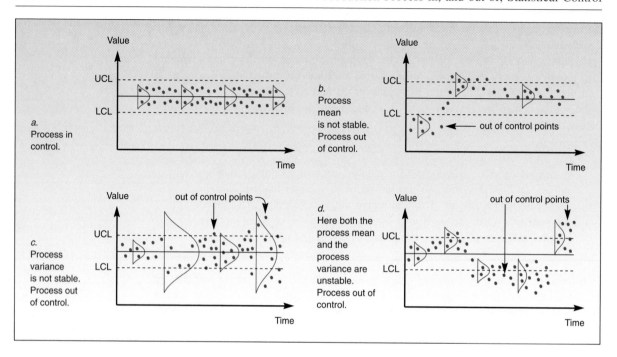

transmissions made at the Japanese plant. The engineer who made the measurements reported that the measuring equipment was broken: in testing one transmission after the other, the needle did not move at all. A closer investigation revealed that the measuring equipment was perfectly fine: the transmissions simply had no variation. *They did not just satify specifications; for all practical purposes, the 12 transmissions were identical!*[1]

Such perfection may be difficult to achieve, but the use of control charts can go a long way toward improving quality.

10.2 THE $\bar{x}$ CHART

We want to compute the center line and the upper and lower control limits for a process believed to be in control. Then future observations can be checked against these bounds to make sure the process remains in control. To do this, we first conduct an *initial run.* We determine trial control limits to test for control of past data and then we remove out-of-control observations and recompute the control limits. We then apply these improved control limits to future data. This is the philosophy behind all of the control charts discussed in this chapter. Although we present the $\bar{x}$ chart first, in an actual quality-control program we would first want to test that the process variation is under control. This is done using the R (range) or the s (standard deviation) chart.

---

[1]From L. Dobyns, "Ed Deming Wants Big Changes and He Wants Them Fast," *Smithsonian,* August 1990, p. 80.

Unless the process variability is under statistical control, there is no stable distribution of values with a fixed mean.

An **x̄ chart** can help us to detect shifts in the process mean. One reason for a control chart for the process mean (rather than for a single observation) has to do with the *central limit theorem.* We want to be able to use the known properties of the normal curve in designing the control limits. By the central limit theorem, the distribution of the sample mean tends toward a *normal distribution* as the sample size increases. Thus, when we aggregate data from a process, the aggregated statistic, the sample mean, becomes closer to a normal random variable than the original, unaggregated quantity. Typically, a set number of observations will be aggregated and averaged. For example, a set of four measurements of rod diameter will be made every hour of production. The four rods will be chosen randomly from all rods made during that hour. If the distribution of rod diameters is roughly mound-shaped, then the sample means of the groups of four diameters will have a distribution closer to normal.

The mean of the random variable $\overline{X}$ is the population mean μ, and the standard deviation of $\overline{X}$ is $\sigma/\sqrt{4}$, where σ is the population standard deviation. We know all this from the theory in Chapter 5. We also know from the theory that the probability that a normal random variable will exceed three of its standard deviations on either side of the mean is 0.0026. (Check this using the normal table.) Thus, the interval: $\mu \pm 3\sigma/\sqrt{n}$ should contain about 99.74 percent of the sample means. This is, in fact, the logic of the control chart for the process mean. The idea is the same as that of a hypothesis test (conducted in a form similar to a confidence interval). We try to select the bounds so that they will be as close as possible to the above equation. Then we chart the bounds, an estimate of μ in the center (the center line), and the upper and lower bounds (UCL and LCL) as close as possible to the bounds of the interval specified by the equation. Out of 1,000 x̄s, fewer than 3 are expected to be out of bounds. Therefore, with a limited number of x̄s on the control chart, observing even one of them out of bounds is cause to reject the null hypothesis that the process is in control, in favor of the alternative that it is out of control. (One could also compute a *p*-value here, although it is more complicated since we have *several* x̄s on the chart, and in general this is not done.)

Note that the assumption of random sampling is important here as well. If somehow the process is such that successively produced items have values that are correlated—thus violating the independence assumption of random sampling—the interpretation of the chart may be misleading. Various new techniques have been devised to solve this problem.

To construct the control chart for the sample mean, we need estimates of the parameters in the equation. The grand mean of the process, that is, the mean of all the sample means (the mean of all the observations of the process), is our estimate of μ. This is our center line. To estimate σ, we use s, the standard deviation of all the process observations. However, this estimate is only good for large samples (here, $n > 10$). For smaller sample sizes we use an alternative procedure. When sample sizes are small, we use the *range* of the values in each sample used to compute an x̄. Then we average these ranges, giving us a mean range, $\overline{R}$. When the mean range, $\overline{R}$, is multiplied by a constant, which we call A_2, the result is a good estimate for 3σ. Values of A_2 for all sample sizes up to 25 are found in Appendix B, Table B.6, at the end of the book. The table also contains the values for all other constants required for the quality-control charts discussed in this chapter.

The following box shows how we compute the center line and the upper and lower control limits when constructing a control chart for the process mean.

158
157

Elements of a control chart for the process mean:

Center line:
$$\bar{\bar{x}} = \frac{\text{Sum of } \bar{x}_i}{k}$$

UCL: $\bar{\bar{x}} + A_2\bar{R}$ LCL: $\bar{\bar{x}} - A_2\bar{R}$

where: k = Number of samples, each of size n
           $\bar{x}_i$ = Sample mean for the ith sample
           R_i = Range of the ith sample $\bar{R} = \dfrac{\text{Sum of } R_i}{k}$

If the sample size in each group is over 10, then:

$$\text{UCL} = \bar{\bar{x}} + 3\frac{\bar{s}/c_4}{\sqrt{n}} \qquad\qquad \text{LCL} = \bar{\bar{x}} - 3\frac{\bar{s}/c_4}{\sqrt{n}}$$

where $\bar{s}$ is the average of the standard deviations of all groups, and c_4 is a constant found in Appendix B, Table B.6.

In addition to a sample mean being outside the bounds given by the UCL and LCL, other occurrences on the chart may lead us to conclude that there is evidence that the process is out of control. Several such sets of rules have been developed, and the idea behind them is that they represent occurrences that have a very low probability when the process is indeed in control. The set of rules we use is given in Table 10.1.[2]

EXAMPLE 10.1

A pharmaceutical manufacturer needs to control the concentration of the active ingredient in a formula used to restore hair to bald people. The concentration should be around 10 percent, and a control chart is desired to check the sample means of 30 observations, aggregated in groups of 3. The data and the MINITAB program, as well as the control chart it produced, are given in Figure 10.2. As can be seen from the control chart, there is no evidence here that the process is out of control.

The grand mean is $\bar{x} = 10.253$. The ranges of the groups of three observations each are: 0.15, 0.53, 0.69, 0.45, 0.55, 0.71, 0.90, 0.68, 0.11, and 0.24. Thus, $\bar{R} = 0.501$. From the table in the appendix we find for $n = 3$, $A_2 = 1.023$. Thus, UCL = 10.253 + 1.023(0.501) = 10.766, and LCL = 10.253 − 1.023(0.501) = 9.74. The MINITAB program estimates using the sample standard deviation instead of $\bar{R}$ (the use of $\bar{R}$ is a special option), hence the slight differences.

PROBLEMS

1. What is the logic behind the control chart for the sample mean, and how is the chart constructed?
2. Legal Seafoods in Boston prides itself on having instituted an advanced quality-control system that includes the control of both the food quality and the service quality. The

[2]This particular set of rules was provided courtesy of Dr. Lloyd S. Nelson of the Nashua Corporation, one of the pioneers in the area of quality control. See L. S. Nelson, "The Shewhart Control Chart—Tests for Special Causes," *Journal of Quality Technology* 16 (1984), pp. 237–39. The MINITAB package tests for special causes using Nelson's criteria.

Tests for Assignable Causes   **TABLE 10.1**

Test 1: One point beyond 3σ ($3s$)
Test 2: Nine points in a row on one side of the center line
Test 3: Six points in a row steadily increasing or decreasing
Test 4: Fourteen points in a row alternating up and down
Test 5: Two out of three points in a row beyond 2σ ($2s$)
Test 6: Four out of five points in a row beyond 1σ ($1s$)
Test 7: Fifteen points in a row within 1σ ($1s$) of the center line
Test 8: Eight points in a row on both sides of the center line, all beyond 1σ ($1s$)

MINITAB-Produced Output for Example 10.1   **FIGURE 10.2**

```
MTB > SET C1

DATA< 10.22 10.25 10.37 10.46 10.06 10.59 10.82 10.52 10.13 9.88

DATA> 10.31 10.33 9.92 9.94 9.39 10.15 10.85 10.14 10.69 10.32

DATA> 9.79 10.12 10.80 10.26 10.31 10.23 10.20 10.07 10.15 10.31

DATA> END
MTB > XBARCHART C1 3
```

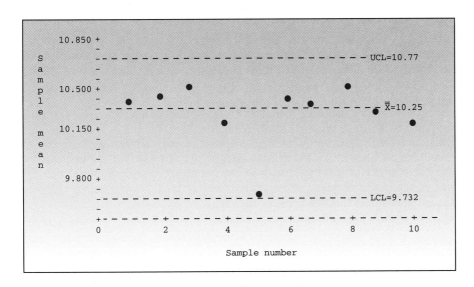

following are successive service times at one of the chain's restaurants on a Saturday night
in May (in minutes from customer entry to appearance of waitperson):

5, 6, 5, 5.5, 7, 4, 12, 4.5, 2, 5, 5.5, 6, 6, 13, 2, 5, 4, 4.5, 6.5, 4, 1,
2, 3, 5.5, 4, 4, 8, 12, 3, 4.5, 6.5, 6, 7, 10, 6, 6.5, 5, 3, 6.5, 7

Aggregate the data into groups of four, and construct a control chart for the process mean.
Is the waiting time at the restaurant under control?

3. What assumptions are necessary for constructing an $\bar{x}$ chart?

4. The manufacturer of jet engines needs to control the maximum power delivered by
engines. The following are readings related to power for successive engines produced:

121, 122, 121, 125, 123, 121, 129, 123, 122, 122, 120, 121, 119, 118, 121,
125, 139, 150, 121, 122, 120, 123, 127, 123, 128, 129, 122, 120, 128, 120

Aggregate the data in groups of three, and create a control chart for the process mean. Use the chart to test the assumption that the production process is under control.

5. The following data are tensile strengths, in pounds, for a sample of string for industrial use made at a plant. Construct a control chart for the mean, using groups of five observations each. Test for statistical control of the process mean.

5, 6, 4, 6, 5, 7, 7, 7, 6, 5, 3, 5, 5, 5, 6, 5, 5, 5, 6, 7, 7, 7, 7, 6, 7, 5,
5, 5, 6, 7, 7, 7, 7, 7, 5, 5, 6, 4, 6, 6, 6, 7, 6, 6, 6, 6, 6, 7, 5, 7, 6

10.3 THE R CHART AND THE s CHART

In addition to the process mean, we want to control the process variance as well. When the variation in the production process is high, it means that produced items will have a wider range of values, and this jeopardizes the product's quality. Recall also that in general we want as small a variance as possible. As noted earlier, it is advisable to check the process variance first and then check its mean. Two charts are commonly used to achieve this aim. The more common of the two is a control chart for the process range, called the **R chart.** The other is a control chart for the process standard deviation, the **s chart.** A third chart, called the s^2 chart, is for the actual variance, but we will not discuss it since it is the least commonly used of the three.

The R Chart

Like the $\bar{x}$ chart, the R chart also contains a center line and upper and lower control limits. The lower limit is bounded by zero, and sometimes turns out to be zero. Similar to what was done in the previous section for the process mean, we want to specify the upper and lower control limits in the following form:

$$\bar{R} \pm 3\sigma_{\bar{R}}$$

Although the distribution of the sample range is not normal, it is still a common practice to use symmetric bounds such as the ones given above. These bounds still assure us that the probability of breaching them is small. Note that we are most interested in any breach of the upper control limit.

Quality-control experts have calculated that appropriate bounds for small samples (as used in practice) in accordance with the bounds of the equation are given using constants D_3 and D_4 multiplied by $\bar{R}$. The first gives us the LCL; the second gives us the UCL. The constants D_3 and D_4 are given in Appendix B, Table B.6. The center line, UCL, and LCL for the control chart for the process range are given in the following box.

The elements of an R chart:

Center Line: $\bar{R}$
LCL: $D_3\bar{R}$
UCL: $D_4\bar{R}$

where $\bar{R}$ = Sum of group ranges/Number of groups.

Returning to Example 10.1, we find that $\bar{R} = 0.501$, and from Table B.6, $D_3 = 0$ and $D_4 = 2.574$. Thus, the center line is 0.501, the lower control limit is 0, and the upper control limit is $(0.501)(2.574) = 1.29$. Figure 10.3 gives the MINITAB-produced control chart for the process range for this example.

There is a small difference in the UCL because of the way MINITAB computes it. The test for control in the case of the process range is just to look for at least one observation outside the bounds. Based on the *R* chart for Example 10.1, we conclude that the process range seems to be in control.

The *s* Chart

The *R* chart is in common use because it is easier (by *hand*) to compute ranges than to compute standard deviations. Today (as compared with the 1920s, when these charts were invented), computers are usually used to create control charts, and an *s* chart should be at least as good as an *R* chart. Note, however, that the standard deviation suffers from the same nonnormality (skewness) as does the range. Again, symmetric bounds as suggested by the equation, with *s* replacing *R*, are still used. The control chart for the process standard deviation is similar to that for the range. Here we use constants B_3 and B_4, also found in Appendix B, Table B.6. The bounds and the center line are given in the following box.

Elements of the *s* chart:

Center line: $\bar{s}$

LCL: $B_3\bar{s}$

UCL: $B_4\bar{s}$

where $\bar{s}$ is the sum of group standard deviations divided by the number of groups.

MINITAB-Produced *R* Chart for Example 10.1   **FIGURE 10.3**

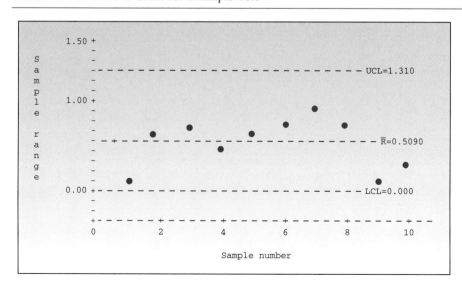

The *s* chart for Example 10.1 is given in Figure 10.4.

Again, we note that the process standard deviation seems to be in control. Since *s* charts are done by computer, we will not carry out the computations of the standard deviations of all the groups.

PROBLEMS

6. Why do we need a control chart for the process range?
7. Compare and contrast the control charts for the process range and the process standard deviation.
8. What are the limitations of symmetric LCL and UCL? Under what conditions are symmetric bounds impossible in practice?
9. Create an *R* chart for the process in Problem 2.
10. Create an *R* chart for the process in Problem 4.
11. Create an *R* chart for the process in Problem 5.
12. Create an *s* chart for the process in Problem 2.
13. Create an *s* chart for the data in Problem 4.
14. Is the standard deviation of the process in Problem 5 under control?

10.4 THE *p* CHART

The number of defective items in a random sample chosen from a population has a binomial distribution: the number of successes, *x*, out of a number of trials, *n*, with a constant probability of success, *p*, in each trial. The parameter *p* is the proportion of defective items in the population. If the sample size, *n*, is fixed in repeated samplings, then the sample proportion, $\hat{P}$, has a distribution related to the binomial. Recall that the binomial distribution is symmetric when $p = 0.5$, and it is skewed for other values of *p*. By the central limit theorem, as *n* increases, the distribution of $\hat{P}$ approaches a normal distribution. Thus, a normal approximation to the binomial should work well with large sample sizes; a relatively small sample size would suffice if $p = 0.5$ because of the symmetry of the binomial in this case.

FIGURE 10.4   An *s* Chart for the Process of Example 10.1

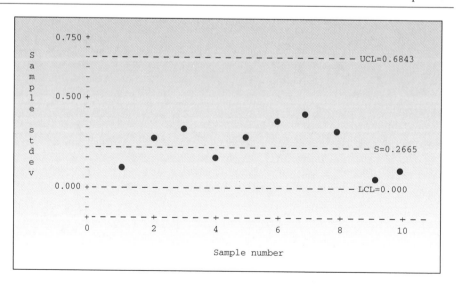

The central limit theorem is the rationale for the control chart for *p*, called the **p chart.** Using the normal approximation, we want bounds of the form:

$$\hat{p} \pm 3\sigma_{\hat{p}}$$

The idea is, again, that the probability of a sample proportion falling outside the bounds is small when the process is under control. When the process is not under control, the proportion of defective or nonconforming items will tend to exceed the upper bound of the control chart. The lower bound is sometimes zero, which happens when $\hat{p}$ is sufficiently small. Being at the lower bound of zero defectives is, of course, a very good occurrence.

Recall that the sample proportion, $\hat{p}$, is given by the number of defectives, *x*, divided by the sample size, *n*. We estimate the population proportion, *p*, by the total number of defectives in all the samples of size *n* we have obtained, divided by the entire sample size (all the items in all our samples). This is denoted by $\bar{p}$, and serves as the center line of the chart. Also recall that the standard deviation of this statistic is given by the following:

$$\sqrt{\frac{\bar{p}(1 - \bar{p})}{n}}$$

Thus, the control chart for the proportion of defective items is given in the following box. The process is believed to be out of control when at least one sample proportion falls outside the bounds.

The elements of a control chart for the process proportion:

Center line: $\qquad\qquad\qquad\qquad\quad \bar{p}$

LCL:

$$\bar{p} - 3\sqrt{\frac{\bar{p}(1 - \bar{p})}{n}}$$

UCL:

$$\bar{p} + 3\sqrt{\frac{\bar{p}(1 - \bar{p})}{n}}$$

where *n* is the number of items in each sample, and $\bar{p}$ is the proportion of defectives in the combined, overall sample.

A tire manufacturer randomly samples 40 tires at the end of each shift to test for tires that are defective. The number of defectives in 12 shifts are as follows: 4, 2, 0, 5, 2, 3, 14, 2, 3, 4, 12, 3. Construct a control chart for this process. Is the production process under control?

EXAMPLE 10.2

We use MINITAB. We enter the *number* of defective items for all shifts into column c1 and then state the constant sample size (40 for each shift). The results are shown in Figure 10.5. Our estimate of *p*, the center line, is the sum of all the defective tires divided by 40×12. It is $\bar{p} = 0.1125$. Our estimated standard error of $\bar{p}$ is

SOLUTION

FIGURE 10.5   MINITAB-Produced p Chart for Example 10.2

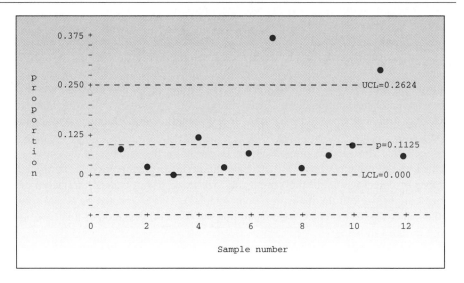

Courtesy of The Goodyear Tire
& Rubber Company.

$\sqrt{\bar{p}(1-\bar{p})/n}$ = 0.05; thus, LCL = 0.1125 − 3(0.05) = −0.0375, which means that the LCL should be *zero*. Similarly, UCL = 0.1125 + 3(0.05) = 0.2625.

As we can see from the figure, two sample proportions are outside the UCL. These correspond to the samples with 14 and 12 defective tires, respectively. There is ample evidence that the production process is out of control.

PROBLEMS

15. The manufacturer of steel rods looks at random samples of 20 items from each production shift and notes the number of nonconforming rods in these samples. The results of 10 shifts are 8, 7, 8, 9, 6, 7, 8, 6, 6, 8. Is there evidence that the process is out of control? Explain.

16. A battery manufacturer looks at samples of 30 batteries at the end of every day of production and notes the number of defective batteries. Results (numbers of defectives in samples of 30) are 1, 1, 0, 0, 1, 2, 0, 1, 0, 0, 2, 5, 0, 1. Is the production process under control?

17. BASF Inc. makes 3.5-inch two-sided double-density disks for use in microcomputers. A quality-control engineer at the plant tests batches of 50 disks at a time and plots the proportions of defective disks on a control chart. The first 10 batches used to create the chart had the following numbers of defective disks: 8, 7, 6, 7, 8, 4, 3, 5, 5, 8. Construct the chart and interpret the results.

18. If the proportion of defective items in a production process is very small, and few items are tested in each batch, what problems do you foresee? Explain.

10.5 THE c CHART

It often happens in production activities that we want to control the *number of defects or imperfections per item*. When fabric is woven, for example, it is of interest to keep a record of the number of blemishes per yard, and take corrective action when this number is out of control.

The random variable representing the count of the number of errors occurring in a fixed period of time or space is often modeled using the *Poisson distribution*. For the

Poisson distribution, the mean and the variance are both equal to the same parameter. Here we call that parameter *c,* and our chart for the number of defects per item (or yard, etc.) is the **c chart.** In this chart we plot a random variable, the number of defects per item. We estimate *c* by $\bar{c}$, which is the average number of defects per item, the total number averaged over all the items we have. The standard deviation of the random variable is thus the square root of *c*. The Poisson distribution can be approximated by a limiting normal distribution, which again suggests the form:

$$\bar{c} \pm 3\sqrt{\bar{c}}$$

This equation leads to the control bounds and center line given in the box that follows.

Elements of the *c* chart:

Center line: $\bar{c}$

LCL: $\bar{c} - 3\sqrt{\bar{c}}$

UCL: $\bar{c} + 3\sqrt{\bar{c}}$

where $\bar{c}$ is the average number of defects or imperfections per item (or area, volume, etc.).

The following data are the numbers of nonconformities in bolts for use in cars made by the Ford Motor Company:[3] 9, 15, 11, 8, 17, 11, 5, 11, 13, 7, 10, 12, 4, 3, 7, 2, 3, 3, 6, 2, 7, 9, 1, 5, 8. Is there evidence that the process is out of control?

EXAMPLE 10.3

MINITAB-Produced Control Chart for Example 10.3

FIGURE 10.6

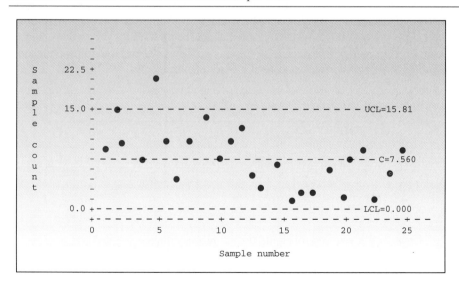

[3]From T. P. Ryan, *Statistical Methods for Quality Improvement* (New York: Wiley, 1989), p. 198.

SOLUTION

We need to find the mean number of nonconformities per item. This is the sum of the numbers divided by 25, or 7.56. The standard deviation of the statistic is the square root of this number, or 2.75, and the control limits are obtained as shown in the box. Figure 10.6 gives the MINITAB solution.

From the figure we see that one observation is outside the upper control limit, indicating that the production process may be out of control. We also note a general downward trend, which should be investigated (maybe the process is improving).

PROBLEMS

19. The following are the numbers of imperfections per yard of yarn: 5, 3, 4, 8, 2, 3, 1, 2, 5, 9, 2, 2, 2, 3, 4, 2, 1. Is there evidence that the process is out of control?

20. The following are the numbers of blemishes in the coat of paint of new automobiles: 12, 25, 13, 20, 5, 22, 8, 17, 31, 40, 9, 62, 14, 16, 9, 28. Is there evidence that the painting process is out of control?

21. The following are the numbers of imperfections in rolls of wallpaper: 5, 6, 3, 4, 5, 2, 7, 4, 5, 3, 5, 5, 3, 2, 0, 5, 5, 6, 7, 6, 9, 3, 3, 4, 2, 6. Construct a *c* chart for the process, and determine whether there is evidence that the process is out of control.

22. What are the assumptions underlying the use of the *c* chart?

SUMMARY

In this chapter we introduced the basic concepts of **quality control and improvement.** We learned how **control charts** can be used in checking whether a production process is under **statistical control.** We saw how a control chart contains two elements: an **upper control limit,** and a **lower control limit.** We developed some rules for determining when a process is out of control, based on the control limits being breached. We studied the $\bar{x}$ chart for checking whether the mean of a process is under control. We defined the **R chart** and the **s chart** for testing whether the variability of a process is under control. We also learned about the **p chart** for proportions, and the **c chart** for the number of defects per item. We discussed the important philosophy behind quality control efforts and total quality management.

KEY FORMULAS

Upper and lower control limits for the process mean:

$$\text{LCL} = \bar{x} - A_2\overline{R} \qquad \text{UCL} = \bar{x} + A_2\overline{R} \qquad \text{Center Line} = \bar{x} = \text{Sum of sample means/number of samples}$$

Upper and lower control limits for the process range:

$$\text{LCL} = D_3\overline{R} \qquad \text{UCL} = D_4\overline{R} \qquad \text{Center Line} = \overline{R} = \text{Sum of group ranges/number of groups}$$

Upper and lower control limits for the process standard deviation:

$$\text{LCL} = B_3\bar{s} \qquad \text{UCL } B_4\bar{s} \qquad \text{Center Line} = \begin{array}{l}\text{Sum of group} \\ \text{standard deviations/} \\ \text{number of groups}\end{array}$$

Upper and lower control limits for the process proportion.

$$\text{LCL} = \bar{p} - 3\sqrt{\bar{p}(1-\bar{p})/n} \quad \text{UCL} = \bar{p} + 3\sqrt{\bar{p}(1-\bar{p})/n}$$

Center Line $= \bar{p} =$ Proportion of defectives in combined, overall sample.

Upper and lower control limits for the number of defects:

$$\text{LCL} = \bar{c} - 3\sqrt{\bar{c}} \quad \text{UCL} = \bar{c} + 3\sqrt{\bar{c}} \quad \text{Center Line} = \begin{array}{l}\text{Average number} \\ \text{of imperfections} \\ \text{per item}\end{array}$$

ADDITIONAL
PROBLEMS

23. Discuss and compare the various control charts discussed in this chapter.
24. The numbers of blemishes in rolls of tape coming out of a production process are as follows: 17, 12, 13 18, 12, 13, 14, 11, 18, 29, 13, 13, 15, 16. Is there evidence that the production process is out of control?
25. The number of defective items out of random samples of 100 windshield wipers selected at the end of each production shift at a factory are as follows: 4, 4, 5, 4, 4, 6, 6, 3, 3, 3, 3, 2, 2, 4, 5, 3, 4, 6, 4, 12, 2, 2, 0, 1, 1, 1, 2, 3, 1. Is there evidence that the production process is out of control?
26. Weights of pieces of tile (in ounces) are as follows: 2.5, 2.66, 2.8, 2.3, 2.5, 2.33, 2.41, 2.88, 2.54, 2.11, 2.26, 2.3, 2.41, 2.44, 2.17, 2.52, 2.55, 2.38, 2.89, 2.9, 2.11, 2.12, 2.13, 2.16. Create an R chart for these data, using subgroups of four. Is the process variation under control?
27. Use the data in Problem 26 to create an $\bar{x}$ chart to test whether the process mean is under control.
28. Create an s chart for the data in Problem 26.
29. In 1979, the Nashua Corporation, with an increasing awareness of the importance of always maintaining and improving quality, invited Dr. W. Edwards Deming for a visit and a consultation. Following many suggestions by Deming, Nashua hired Dr. Lloyd S. Nelson the following year as director of statistical methods. The idea was to teach everyone at the company about quality and how it can be maintained and improved using statistics.

    Dr. Nelson instituted various courses and workshops lasting 4 to 10 weeks for all the employees. Workers on the shop floor are now familiar with statistical process control (SPC) charts and their use in maintaining and improving quality. Nashua uses $\bar{x}$, R, and p charts.

    Among the many products Nashua makes is thermally responsive paper, which is used in printers and recording instruments. The paper is coated with a chemical mixture that is sensitive to heat, thus producing marks in a printer or instrument when heat is applied by a print head or stylus. The variable of interest is the amount of material coated on the paper (the "weight coat"). Large rolls, some as long as 35,000 feet, are coated, and samples are taken from the ends of the rolls. A template 12 × 18 inches is used in cutting through four layers of the paper—first from an area that was coated and second from an uncoated area. A gravimetric comparison of the coated and uncoated samples gives four measurements of the weight coat. The average of these is the individual x value for that roll.

Assume that 12 rolls are coated per shift and that each roll is tested as described above. For two shifts, the 24 values of weight coat, in pounds per 3,000 square feet, were as follows:

3.46, 3.56, 3.58 3.49, 3.45, 3.51, 3.54, 3.48, 3.54, 3.49, 3.55, 3.60,
3.62, 3.60, 3.53, 3.60, 3.51, 3.54, 3.60, 3.61, 3.49, 3.60, 3.60, 3.49

Construct $\bar{x}$ and R charts for these data. Is the production process in statistical control? Explain. Discuss any possible actions or solutions.[4]

---

[4]I am indebted to Dr. Lloyd S. Nelson of the Nashua Corporation for providing me with this interesting and instructive problem.

"Those who think of quality as strictly manufacturing oriented have not thought it out. All of my writings are based on people working, not on an industry. All work is a process and all processes can be designed, measured, and improved."

Phil Crosby was a Navy Medic in the Korean war after studying podiatric medicine at Ohio College. He subsequently worked for Bendix-Mishawaka, Martin-Marrietta, and ITT where he formulated his "zero defects" concept. He is the founder of Philip Crosby Associates, Inc., and author of the well-known Quality Is Free.

Aczel: Where did you take your first statistics courses? What did you think of them?

Crosby: I studied at the Ohio College of Podiatric Medicine, graduating in 1950. I was recalled to the Navy at that time and never practiced medicine. When first working in industry, I took several statistical quality control courses and found them to be committed to the inevitability of error. I learned to understand the statistics used to determine status. Darrel Huff's book *How to Lie with Statistics* was very helpful to me in that regard.

Aczel: What did you do when you first graduated?

Crosby: My first job after college (and the Korean War) was as a junior technician at the Crosley Corp. in Richmond, Ind. There I realized that my ideas were of *prevention*, and were picked up in medical studies and experience. Two years later I went to Bendix-

Mishawaka Indiana as a reliability engineer. There I began writing and talking about defect prevention and found myself in conflict with the conventional wisdom of quality. Five years after leaving the Navy I went to Martin-Marrietta in Orlando as a senior quality engineer. The next year I became a project quality manager.

Aczel: How did you first get interested in statistics? How did it lead or not lead to your quality efforts?

Crosby: I never really became interested in statistics theory as such. I regarded them as a useful tool. They were often used in manufacturing, but only to control a process here and there. The literature on SQC and SPC was abstract and complex but never seemed to be put into actual practice. All of this motivated my search for a concept of quality management that would actually produce products and services in conformance to the agreed requirements. Statistics was used often, in fact, to show that, theoretically, this could not be accomplished.

Aczel: How did you come up with the title (and the subject material) for your book *Quality Is Free?*

Crosby: The title of *Quality Is Free*, as explained on the dedication page of the book, came from Harold Geneen, who was CEO of ITT corporation where I worked 14 years. He was my boss and I learned a great deal from him. I defined quality as "conformance to agreed requirements." That meant doing what we said we would do. It didn't cost more to do things right. Expense comes from doing things wrong. When people use acceptable quality levels (AQL) and probability of error, they commit to spending more money and time. Zero defects means doing things right the first time.

Aczel: When you were writing it, did you have any expectations about the interest it would generate?

Crosby: I thought *QIF* would generate interest because it was written specifically for executives in their language. Having served in that role at ITT for 14 years, I knew how to do that. However, I did not think that it would become a classic. Fifteen years later it still sells 15,000 paperbacks a month in a dozen languages and is used in most colleges.

Aczel: What do you think about the great attention given recently to Deming and his ideas about quality?

Crosby: Dr. Deming was a remarkable man and a great statistician. However he never actually worked in quality management, or ran anything. We will know whether his ideas were useful, or whether people reacted to his personality and style, as time goes on. That which deserves to live, lives. I never met him.

Aczel: What do you think about the state of the teaching of quality and statistics today compared to your own education. Is it better or worse and how so?

Crosby: I would like to see texts relate more to practical cases rather than to the concepts. The work-

ings of math are not that important anymore given the technology of calculators and computers. The real message we need to get out is "What use can management make of these data in meeting the goals of the organization?"

Aczel: Do you follow all the recent conversations/ writings about the quality movement in nonmanufacturing settings such as health care, service industries, education, and government?

Crosby: I always considered that every individual was already in the service business. Each of us works to procedures and other agreements whether we are running a lathe or checking a patient into a health care unit. Those who think of quality as strictly manufacturing oriented have not thought it out. All of my writings are based on people working, not on an industry. All work is a process and all processes can be designed, measured, and improved.

Aczel: Is applying Quality Control in those industries likely to lead to significant improvements?

Crosby: Quality Control per se will not help any industry since it is dedicated to the inevitability of error and the adjustment for it. *Quality management,* on the other hand, will be a tremendous help because it is based on the prevention of error. Medical people relate to prevention.

Aczel: Is there any advice you would give to students who will be facing the job market in the 21st century?

Crosby: Learn to be useful.

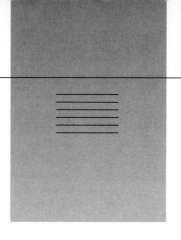

Glossary

Alpha   The probability of a Type I error in hypothesis testing.

Alternative hypothesis   The opposite of the null hypothesis; usually the hypothesis the statistician is trying to prove (also called the research hypothesis).

Analysis of variance (ANOVA)   A statistical technique for testing for the existence of differences among several population means.

ANOVA table   A tabular display of the results of an analysis of variance (ANOVA).

Bar graph   A graphical technique that displays values or frequencies of categorical data.

Bayes theorem   A theorem in probability theory that allows reversing of a conditional probability.

Beta   The probability of a Type II error in hypothesis testing.

Binomial distribution   The probability distribution of a binomial random variable.

Binomial random variable   The random variable that counts the number of successes in a sequence of n trials where the probability of success in each trial is independent of other trials and equal to p.

Blocking   An experimental design where observations are grouped into blocks of several observations that are similar in some way.

Box plot   A graphical method for displaying the distribution of quantitative observations in the form of a box, featuring the median and pointing out outliers.

***c*-chart**   A quality control chart for the number of defects per item.

Categorical variable   A variable that denotes a category rather than a quantity that can be measured on a scale.

Central limit theorem   A mathematical theorem stating that the sampling distribution of many statistics approaches a normal distribution.

Chi-square distribution   An asymmetrical probability distribution used in statistics, including contingency table analysis.

Coefficient of determination (r^2)   A measure of the strength of a linear regression relationship.

Complement of a set or event   The points of the sample space not belonging to the set in question; the opposite of an event.

Conditional probability   The probability of one event, given that another event occurs.

Confidence coefficient   The level of confidence of a confidence interval.

Confidence interval   An interval of numbers constructed in such a way that we may have a given confidence (say, 95 percent) that an unknown parameter of interest (such as the population mean) is within the interval.

Confidence level   The level of confidence of a confidence interval.

Contingency table   A table displaying counts by levels of two variables (which may be categorical).

Control chart   A chart designed to detect when a production process is out of statistical control.

Control limit   A line in a quality control chart, that, when breached, may indicate a process is out of statistical control.

Correlation   A measure of how well two variables move together in a linear way.

Critical point   A point defining a beginning of the rejection region in a statistical hypothesis test.

Data   A set of observations.

Data collection   The collection of observations, often using some experimental design.

Data distribution   The distribution of the frequencies of occurrence of values in a data set.

Decision rule   An objective, theory-derived rule that indicates when to reject the null hypothesis in a statistical test, so that the probability of a Type I error will be at most a given alpha.

Density   A smooth curve that is the probability rule for a continuous random variable; probabilities are defined as areas under this curve.

Distribution   The frequencies of occurrence of data values or the probability rule of a random variable.

Empirical rule   A normal-theory rule for mound-shaped data distributions that says about 68 percent of the data will lie within one standard deviation of the mean, 95 percent within two standard deviations of the mean, and 99.7 percent within three standard deviations of the mean.

Event   Something that may or may not happen and whose probability is of interest.

Expectation   The expected value, the mean, of a random variable.

Expected value   The mean of the distribution of a random variable.

Experiment   A process that leads to one of several outcomes.

Experimental design   Designing an experiment for the purpose of studying particular factors of interest while minimizing the effects of extraneous factors. Randomization plays an important role in experimental design.

F-distribution   An asymmetrical probability distribution useful in statistics, notably in analysis of variance and regression analysis.

Frequency polygon   A graphical display of data frequencies that joins the midpoints of data intervals.

Gaussian distribution   Another name for the normal distribution.

Histogram   A very useful frequency chart for data where the sides of bars of different heights are joined.

Hypothesis testing   A statistical method for testing which of two assertions about some population parameter (such as the population mean) is correct.

Independence   A very important concept in statistics. Two events are independent of each other if knowledge about the occurrence of one of them does not change the other event's probability. Two random variables are independent of each other if they do not affect one another.

Intercept   The point on the y axis where a regression line crosses.

Interquartile range (IQR)   The distance from the first quartile to the third.

Intersection of events   The joint occurrence of two or more events.

Interval scale   A scale of measurement where distances have meaning, but not necessarily their ratios.

Joint probability   The probability of the intersection of events.

Left-tailed test   A statistical hypothesis test where the rejection region lies to the left side (also called a left-sided test).

Level of significance   The probability of a Type I error.

Linear relationship   A statistical relationship that may be generally described by a straight line.

Law of unions   The rule that allows the computation of the union of two (or more) events.

Lower control limit   A line on the bottom of a quality control chart that, when breached, may indicate that a process is out of statistical control.

Measurement   A numerical piece of information based on an observation from a sample or a population.

Mean   The (arithmetic) average of a data set; the expected value of a random variable.

Mean squares   Average-like quantities computed in the analysis of variance and reported in the ANOVA table.

Median   A point in the middle of a data distribution or a probability distribution that splits the distribution in two halves.

Mode   A most frequent value in a distribution.

Mutually exclusive events   Events such that the occurrence of one event precludes the possibility that the other event(s) will occur.

Nominal scale   The weakest scale of measurement, which merely describes category memberships.

Nonrejection region   The region of values of a test statistic where rejection of a null hypothesis may not occur.

Normal approximation   An approximation of a probability distribution by a normal distribution.

Normal data   A data set whose distribution is similar to a normal distribution.

Normal Distribution   A commonly occurring probability distribution. By the central limit theorem, many random variables have distributions that approach it.

Null hypothesis   The hypothesis enjoying the assumption of truth until it can be rejected.

Ogive   A cumulative-frequency graph.

One-tailed test   A statistical hypothesis test where the rejection region stretches to one side of the distribution only (also called a one-sided test).

Ordinal scale   A scale of measurement where data may be ordered from smallest to largest but where actual distances may not have meaning.

Outlier   A measurement that lies away from the rest of the data.

p-chart   A quality-control chart for the proportion of defective items in a production process.

p-value   The attained level of significance of a test; the smallest significance level (alpha) at which a null hypothesis may be rejected.

Paired observations   Data from two groups that may be paired in some meaningful way.

Parameter   A quantity that describes a population.

Percentile   A value below which lie P% of the values in the data set or population.

Pie chart   A graphical display of categorical data (resembling a pie) where the proportion of the total belonging to each category is represented by the size of the wedge of pie for the category.

Population   The collection of all the elements of interest.

Power of a test   The probability of rejecting a false null hypothesis.

Practical significance   The meaningfulness of rejecting a null hypothesis in practical—as contrasted with statistical—terms.

Prediction   Forecasting values of a dependent variable in regression based on the values of the independent variable (or variables).

Probability   A quantitative measure of uncertainty.

Probability distribution   The distribution of the probabilities of the values of a discrete random variable, or the density of a continuous random variable.

Qualitative variable   A variable that denotes membership in categories (also called a categorical variable), rather than a meaningful numerical value.

Quality control   A statistical theory aimed at improving quality of production and other processes.

Quantitative variable   A variable that measures a quantity, rather than a quality.

Quartile   The 25th, the 50th, or the 75th percentile of a distribution.

r   The coefficient of correlation between two variables.

r^2   The coefficient of determination in regression analysis.

R-chart   A quality control chart for the range of the observations.

Random sample   A sample drawn randomly from a population of interest.

Random sampling   The process of drawing a random sample.

Random variable   A quantity determined by chance.

Range   The distance from the smallest observation to the largest observation.

Ratio scale   The strongest scale of measurement, where ratios of distances between measurements may be meaningful.

Regression analysis   The statistical method of estimating a straight line to explain the relationship between two (or more) variables.

Rejection region   The region in the distribution of a test statistic that will lead to rejection of the null hypothesis at a given level of the probability of a Type I error.

Resistant measure   A measure, such as the median, that is not strongly affected by outliers (as contrasted with the mean, which is affected by outliers).

Right-tailed test   A statistical test where the rejection region stretches to the right side of the distribution (also called a right-sided test).

s-chart   A quality-control chart for the standard deviation of a production process.

Sample space   The set of all possible outcomes of an experiment.

Sampling   The process of obtaining a sample.

Sampling distribution   The probability distribution of a sample statistic (for example, the sample mean) as a random variable.

Scatterplot   A bivariate display of data.

Significance level   The probability of a Type I error.

Significant result   A result that leads to the rejection of a null hypothesis at some level of significance.

Skewness   Asymmetry of a distribution (to the left or the right).

Simple linear regression   A straight-line regression of a dependent variable (y) on an independent variable (x).

Simple random sample (SRS)   A random sample obtained by assigning a number to each element in the population and then randomly choosing the numbers of the items to be chosen.

Slope of regression   The slope of the regression line: the number of units the line moves up or down per horizontal step of one unit.

Standard deviation   The square root of the variance (of a sample, a population, or a random variable).

Standard normal distribution   The probability distribution of a normal random variable with zero mean and unit variance.

Statistic   A quantity computed from a random sample, for example, the sample mean.

Statistical significance   The quality of a result that leads to the rejection of a null hypothesis at some level.

Stemplot (stem-and-leaf display)   An easily constructed graphical display of a data distribution using leftmost digits as a stem and the ones on the right as leaves.

Student's distribution   Another name for the t-distribution, in honor of 'student' (W. Gossett, its discoverer).

Sum of squares principle   An important principle in analysis of variance and regression, assuring us that sums of squares are additive.

t distribution   A probability distribution commonly used in statistics when population standard deviations are unknown. See *Student's distribution.*

Test statistic   A quantity computed from a random sample using the assumption in the null hypothesis and employed in determining whether the null hypothesis can be rejected.

Time plot   A plot of some variable versus time.

Total quality management (TQM)   A philosophy of quality and its continuing pursuit.

Transformation of normal random variables   A simple algebraic transformation of a normal random variable to the standard normal random variable so that a table may be used.

Treatment   The factor of interest in analysis of variance (as distinguished from factors not of interest to the researcher, lumped together as 'error').

Two-tailed test   A statistical hypothesis test where the rejection region stretches both to the left side and to the right side of the distribution (also called a two-sided test).

Type I error   In statistical hypothesis testing, the error of rejecting a true null hypothesis.

Type II error   In statistical hypothesis testing, the error of failing to reject a false null hypothesis.

Uniform distribution   A continuous probability distribution, where the density is a straight line parallel to the horizontal axis.

Union of events   The occurrence of either of two (or more) events.

Upper control limit   The top line in a quality control chart; when breached, it may indicate that a process is out of statistical control.

Variability   The spread of a set of observations or a random variable; usually measured by the variance, the standard deviation, or the range.

Variance   A measure of variation for a population, a sample, or the distribution of a random variable.

X-bar chart   A quality control chart for the process mean.

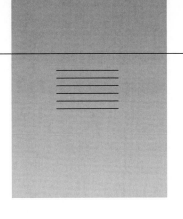

Bibliography

Barnett, V., and T. Lewis. *Outliers in Statistical Data.* 3rd ed. New York: Wiley, 1994.

Chambers, J.M.; W. S. Cleveland; B. Kleiner; and P. A. Tukey. *Graphical Methods for Data Analysis.* New York: Chapman & Hall, 1983. This book presents an interesting approach to graphical display of data.

Deming, W. E. *Out of the Crisis.* Cambridge Mass.: MIT, 1986. A treatise by the well-known quality guru.

Emory, C., and D. Cooper. *Business Research Methods.* 4th ed. Homewood, IL.: 1991. An excellent introduction to Business research.

Evans, M.; N. Hastings; and B. Peacock. *Statistical Distributions.* 2nd Ed. New York: Wiley, 1993.

Fairley, W., and F. Mosteller, eds. *Statistics and Public Policy.* Reading, MA: Addison-Wesley, 1977. Presents interesting articles on the use of statistics in public policy.

Fisher, R. A. *The Design of Experiments.* 7th ed. Edinburgh:

Oliver & Boyd, 1960. A classic treatise on statistics.

Friedman, D.; R. Pisani; R. Purves; and A. Adhikari. *Statistics.* 2d ed. New York: W. W. Norton, 1991. An introduction to statistics using a no-formula, approach.

Gitlow, H.; S. Gitow; A. Oppenheim; and R. Oppenheim. *Tools and Methods for the Improvement of Quality.* Homewood, IL: Richard D. Irwin, 1989. An accessible introduction to the field of quality control.

Hinkelmann, K., and O. Kempthorne. *Design and Analysis of Experiments: Introduction to Experimental Design.* New York: Wiley, 1994.

Hoaglin, D.; F. Mosteller; and J. Tukey, eds. *Understanding Robust and Exploratory Data Analysis.* New York: Wiley, 1983. An excellent introduction to data analysis.

Huff, D. *How to Lie with Statistics.* New York: Norton, 1954. This is an old favorite on statistics, graphs, and the truth; very readable.

Kahneman, D.; P. Slovic; and A. Tversky. *Judgement Under Uncertainty: Heuristics and Biases.* New York: Cambridge University Press, 1987.

Kotz, S. and N. Johnson, eds, *Encyclopedia of Statistical Sciences.* New York: Wiley, 1982–1989. The volumes of this encyclopedia are very useful for looking up statistical terms and their explanations by experts.

MINITAB Reference Manual. College Station, PA: MINITAB Inc., 1993. The user guide accompanying the computing package.

Neter, J.; W. Wasserman; and M. Kutner; *Applied Linear Regression Models.* 2nd ed. Homewood, IL: Richard D. Irwin, 1989. A comprehensive introduction to applied regression analysis.

Paulos, J. *Innumeracy.* New York: Hill and Wang, 1988. A popular guide to looking at probabilities and other numerical data with an open mind.

Ryan, B., and B. Joiner. *Minitab Handbook.* 3rd ed. Belmont, CA: Duxbury Press, 1994. An excellent, easy-to-understand guide to using MINITAB.

Schaeffer, R. L.; W. Mendenhall; and L. Ott. *Elementary Survey Sampling.* 4th ed. Boston: PWS-Kent, 1990. An introduction to the important area of sampling.

Stigler, S. *The History of Statistics.* Cambridge, MA: Harvard University Press, 1986. An enjoyable, exciting survey of the history of statistics.

Tanur, J., ed. *Statistics: A Guide to the Unknown.* Belmont, CA: Wadsworth, 1989. Excellent, readable, nontechnical articles on the use of statistics in a wide variety of areas.

Thompson, S. *Sampling.* New York: Wiley, 1992.

Tufte, E. R. *The Visual Display of Quantitative Information.* Cheshire, CT.: Graphics Press, 1983. A visually pleasing array of graphical techniques.

A

Data Tables

Rank Ordering of States, by Selected Demographic Indicators: November 1994 **TABLE A.1**

State	Population 65 Years and Over		Population 18 Years and Over					
			White		Black		Hispanic[1]	
	Percent	Rank	Percent	Rank	Percent	Rank	Percent	Rank
Alabama	17.8	22	76.0	44	23.0	7	0.6	46
Alaska	6.3	51	78.1	42	3.7	33	3.0	22
Arizona	18.9	11	90.4	23	2.7	35	17.6	4
Arkansas	20.2	5	85.1	32	13.6	14	1.0	40
California	14.6	44	80.2	39	7.3	24	24.7	3
Colorado	13.7	49	93.0	16	3.9	31	11.8	7
Connecticut	18.7	13	90.2	24	7.9	22	6.4	12
Delaware	17.0	29	81.5	38	16.5	10	2.1	29
District of Columbia	16.8	32	35.6	51	62.2	1	5.8	14
Florida	24.6	1	86.1	30	12.1	17	13.1	5
Georgia	13.9	48	73.3	45	25.0	6	1.8	31
Hawaii	15.8	39	40.7	50	2.4	38	7.1	11
Idaho	16.6	34	97.0	6	0.2	48	5.1	15
Illinois	17.1	27	82.7	36	14.0	13	7.7	10
Indiana	17.3	25	91.5	21	7.4	23	1.8	32
Iowa	20.7	3	97.1	5	1.8	41	1.2	37
Kansas	18.8	12	91.8	20	5.6	28	3.7	21
Kentucky	17.1	28	92.5	18	6.8	25	0.6	47
Louisiana	16.0	38	70.0	48	28.5	3	2.5	26
Maine	18.4	15	98.8	2	0.2	49	0.5	48
Maryland	15.0	42	71.0	47	25.1	5	2.8	23
Massachusetts	18.4	14	92.2	19	5.0	29	4.6	18

State	Population 65 Years and Over		Population 18 Years and Over					
			White		Black		Hispanic[1]	
	Percent	Rank	Percent	Rank	Percent	Rank	Percent	Rank
Michigan	16.9	31	84.6	33	13.5	15	2.1	28
Minnesota	17.1	26	95.3	10	1.9	39	1.2	39
Mississippi	17.5	24	67.1	49	32.0	2	0.7	44
Missouri	19.1	10	88.7	27	9.9	20	1.2	38
Montana	18.1	19	94.4	12	0.2	50	1.3	34
Nebraska	19.4	9	95.0	11	3.4	34	2.4	27
Nevada	14.6	45	87.7	29	6.1	27	11.2	8
New Hampshire	16.0	37	98.1	3	0.4	47	0.9	41
New Jersey	18.2	18	82.4	37	13.3	16	10.2	9
New Mexico	15.7	40	89.0	26	1.8	40	37.2	1
New York	17.7	23	78.7	41	16.5	11	11.9	6
North Carolina	16.8	33	77.5	43	20.3	8	1.2	36
North Dakota	20.1	7	95.5	8	0.4	45	0.4	50
Ohio	18.1	21	88.6	28	10.2	19	1.3	35
Oklahoma	18.3	16	84.6	34	6.7	26	2.6	25
Oregon	18.1	20	94.0	13	1.5	42	3.8	20
Pennsylvania	20.8	2	89.8	25	8.7	21	2.0	30
Rhode Island	20.2	6	93.7	14	3.8	32	4.8	16
South Carolina	16.1	36	71.6	46	27.5	4	0.9	42
South Dakota	20.3	4	92.9	17	0.4	46	0.4	51
Tennessee	17.0	30	84.5	35	14.5	12	0.7	43
Texas	14.3	47	85.8	31	11.4	18	24.9	2
Utah	13.6	50	95.4	9	0.7	43	4.8	17
Vermont	16.1	35	99.1	1	-	51	0.5	49
Virginia	14.8	43	78.8	40	17.9	9	2.7	24
Washington	15.7	41	90.4	22	2.7	37	4.3	19
West Virginia	20.0	8	96.5	7	2.7	36	0.6	45
Wisconsin	18.2	17	93.6	15	4.6	30	1.7	33
Wyoming	14.6	46	97.1	4	0.6	44	5.8	13

Note: Order of states determined by percentages derived from detailed data before rounding.

–Represents zero or rounds to zero.

[1]Hispanic persons may be of any race.

Source: U.S. Department of Commerce, Bureau of the Census: "Population Estimates and Projections," May 1994, p. 16

TABLE A.2 Percent distribution of occupations with the highest number of fatal occupational injuries and month the injury occurred, 1992

Occupation[1]	Total Number	Total Percent	Jan	Feb	Mar	Apr	May	Jun	Jul	Aug	Sep	Oct	Nov	Dec
Total[2]	6,083	100.0	8.0	7.0	7.2	7.6	7.9	9.6	9.4	8.8	9.4	9.6	8.3	7.2
Truck drivers	685	100.0	8.3	7.2	5.8	7.9	8.8	10.8	8.6	8.6	8.5	11.5	7.2	6.9
Farmers, except horticultural	257	100.0	5.1	3.5	5.8	12.1	8.2	10.9	8.6	16.0	7.4	11.7	7.0	3.9
Supervisors and proprietors, sales occupations	232	100.0	9.5	7.3	6.5	6.9	10.8	6.5	7.8	8.6	8.6	7.8	9.1	10.8
Construction laborers	226	100.0	9.3	5.8	5.8	8.4	6.6	11.9	7.5	9.3	13.3	11.1	5.3	5.8
Farm workers	204	100.0	3.4	2.9	3.9	6.9	9.3	13.2	10.8	11.3	10.3	15.7	6.4	5.9
Managers and administrators, n.e.c.	202	100.0	10.4	8.9	10.4	7.4	8.9	9.4	5.9	6.4	9.9	10.4	5.0	6.9
Sales workers, retail and personal services	183	100.0	8.2	8.2	7.7	12.0	11.5	9.3	8.2	9.3	6.0	6.0	4.9	8.7
Laborers, except construction	173	100.0	5.2	5.2	5.8	9.2	8.7	6.4	11.0	9.8	8.7	10.4	9.8	9.8
Material moving equipment operators	163	100.0	8.0	8.6	4.9	8.0	6.1	9.8	11.0	8.6	11.0	8.0	9.8	6.1
Technicians, except health, engineering, and science	140	100.0	11.4	13.6	5.7	7.1	4.3	17.1	6.4	5.0	9.3	2.1	7.1	10.7
Vehicle and mobile equipment mechanics and repairers	140	100.0	7.1	8.6	9.3	4.3	7.1	8.6	10.7	10.7	14.3	5.0	7.9	6.4
Police and detectives	139	100.0	9.4	5.8	5.8	7.2	9.4	7.9	5.0	10.8	10.1	10.1	8.6	10.1
Timber cutting and logging occupations	133	100.0	6.0	12.8	6.8	5.3	4.5	10.5	8.3	9.0	9.0	6.8	11.3	9.8
Taxicab drivers and chauffeurs	105	100.0	7.6	8.6	12.4	6.7	8.6	4.8	12.4	13.3	3.8	8.6	7.6	5.7
Cleaning and building service occupations, except households (includes janitors and cleaners)	102	100.0	9.8	2.9	9.8	5.9	8.8	7.8	12.7	8.8	7.8	5.9	9.8	9.8
Guards	100	100.0	15.0	4.0	9.0	7.0	8.0	9.0	14.0	6.0	6.0	11.0	5.0	6.0
Carpenters	100	100.0	16.0	-	9.0	8.0	6.0	11.0	7.0	12.0	7.0	11.0	8.0	-
Electricians	93	100.0	5.4	4.3	9.7	4.3	11.8	10.8	16.1	11.8	-	12.9	7.5	-
Food preparation and service occupations	84	100.0	-	7.1	14.3	8.3	16.7	-	6.0	10.7	4.8	7.1	11.9	6.0
Machine operators, assorted materials	83	100.0	9.6	4.8	3.6	7.2	9.6	8.4	13.3	6.0	16.9	10.8	6.0	3.6
Freight, stock, and material handlers	82	100.0	3.7	6.1	9.8	3.7	13.4	6.1	4.9	9.8	8.5	9.8	13.4	11.0
Managers, food serving and lodging establishments	78	100.0	-	7.7	7.7	9.0	7.7	14.1	5.1	11.5	10.3	12.8	7.7	-
Fishers	75	100.0	13.3	5.3	-	6.7	-	14.7	6.7	13.3	14.7	9.3	9.3	-
Managers, farms, except horticultural	68	100.0	-	-	4.4	7.4	7.4	14.7	11.8	4.4	17.6	14.7	4.4	7.4
Supervisors, farm workers	54	100.0	-	5.6	-	5.6	9.3	14.8	5.6	9.3	20.4	9.3	5.6	7.4
Structural metal workers	48	100.0	8.3	16.7	8.3	-	-	8.3	8.3	10.4	6.3	12.5	6.3	6.3
Roofers	47	100.0	-	-	-	8.5	6.4	-	14.9	8.5	21.3	10.6	6.4	10.6
Driver-sales workers	45	100.0	6.7	8.9	13.3	-	-	6.7	17.8	-	-	8.9	11.1	15.6
Painters, construction and maintenance	43	100.0	-	9.3	9.3	9.3	7.0	20.9	9.3	11.6	-	7.0	-	-
Military occupations	154	100.0	5.8	8.4	5.8	6.5	5.8	8.4	11.7	3.9	3.2	14.3	19.5	6.5

Header: **Month the Injury Occurred**

[1]Based on the 1990 Occupational Classification System developed by the Bureau of the Census. [2]Total includes all occupational categories. All categories are not shown. n.e.c. = not elsewhere classified. Percentages may not add to totals due to rounding. Dashes indicate no data reported or data that do not meet publication guidelines.

Source: "Fatal Workplace Injuries in 1992" Washington DC: U.S. Dept. of Labor, Bureau of Labor Statistics, April 1994, p. 78.

TABLE A.3   Output per hour of all persons and real compensation per hour in the business sector, 1950-92

Year	Output per Hour of All Persons	Real Compensation per Hour
1950	100.0	100.0
1951	103.6	101.8
1952	107.4	106.2
1953	110.8	112.6
1954	113.6	115.4
1955	117.5	118.9
1956	119.0	125.0
1957	122.3	128.9
1958	126.2	131.1
1959	129.4	135.9
1960	131.5	139.4
1961	136.5	143.4
1962	141.2	148.6
1963	147.0	152.2
1964	153.4	158.1
1965	157.5	161.6
1966	161.9	168.0
1967	166.0	172.3
1968	171.0	178.8
1969	172.0	181.9
1970	174.3	185.0
1971	180.1	188.6
1972	185.8	194.4
1973	190.5	198.8
1974	186.9	196.6
1975	191.4	198.2
1976	197.0	204.5
1977	200.3	207.4
1978	201.5	209.9
1979	199.3	206.8
1980	197.7	201.7
1981	200.2	200.0
1982	200.5	202.6
1983	205.1	203.9
1984	210.1	203.9
1985	213.1	205.7
1986	217.6	212.1
1987	219.8	211.9
1988	222.0	212.4
1989	220.3	209.7
1990	220.8	209.9
1991	221.0	210.7
1992	227.1	212.1

Source: "Productivity and Economy," Washington, DC: U.S. Dept. of Labor, Bureau of Labor Statistics September 1993, p. 70.

B

Statistical Tables

The Binomial Distribution
Entries in the table are the probabilities of x successes in n trials of a binomial experiment, where p is the probability of success in one trial. For example, with five trials and $p=.04$, the probability of two successes is 0.0142.

TABLE B.1

						p				
n	x	.01	.02	.03	.04	.05	.06	.07	.08	.09
2	0	.9801	.9604	.9409	.9216	.9025	.8836	.8649	.8464	.8281
	1	.0198	.0392	.0582	.0768	.0950	.1128	.1302	.1472	.1638
	2	.0001	.0004	.0009	.0016	.0025	.0036	.0049	.0064	.0081
3	0	.9703	.9412	.9127	.8847	.8574	.8306	.8044	.7787	.7536
	1	.0294	.0576	.0847	.1106	.1354	.1590	.1816	.2031	.2236
	2	.0003	.0012	.0026	.0046	.0071	.0102	.0137	.0177	.0221
	3	.0000	.0000	.0000	.0001	.0001	.0002	.0003	.0005	.0007
4	0	.9606	.9224	.8853	.8493	.8145	.7807	.7481	.7164	.6857
	1	.0388	.0753	.1095	.1416	.1715	.1993	.2252	.2492	.2713
	2	.0006	.0023	.0051	.0088	.0135	.0191	.0254	.0325	.0402
	3	.0000	.0000	.0001	.0002	.0005	.0008	.0013	.0019	.0027
	4	.0000	.0000	.0000	.0000	.0000	.0000	.0000	.0000	.0001
5	0	.9510	.9039	.8587	.8154	.7738	.7339	.6957	.6591	.6240
	1	.0480	.0922	.1328	.1699	.2036	.2342	.2618	.2866	.3086
	2	.0010	.0038	.0082	.0142	.0214	.0299	.0394	.0498	.0610
	3	.0000	.0001	.0003	.0006	.0011	.0019	.0030	.0043	.0060
	4	.0000	.0000	.0000	.0000	.0000	.0001	.0001	.0002	.0003
	5	.0000	.0000	.0000	.0000	.0000	.0000	.0000	.0000	.0000

					p					
n	x	.01	.02	.03	.04	.05	.06	.07	.08	.09
6	0	.9415	.8858	.8330	.7828	.7351	.6899	.6470	.6064	.5679
	1	.0571	.1085	.1546	.1957	.2321	.2642	.2922	.3164	.3370
	2	.0014	.0055	.0120	.0204	.0305	.0422	.0550	.0688	.0833
	3	.0000	.0002	.0005	.0011	.0021	.0036	.0055	.0080	.0110
	4	.0000	.0000	.0000	.0000	.0001	.0002	.0003	.0005	.0008
	5	.0000	.0000	.0000	.0000	.0000	.0000	.0000	.0000	.0000
	6	.0000	.0000	.0000	.0000	.0000	.0000	.0000	.0000	.0000
7	0	.9321	.8681	.8080	.7514	.6983	.6485	.6017	.5578	.5168
	1	.0659	.1240	.1749	.2192	.2573	.2897	.3170	.3396	.3578
	2	.0020	.0076	.0162	.0274	.0406	.0555	.0716	.0886	.1061
	3	.0000	.0003	.0008	.0019	.0036	.0059	.0090	.0128	.0175
	4	.0000	.0000	.0000	.0001	.0002	.0004	.0007	.0011	.0017
	5	.0000	.0000	.0000	.0000	.0000	.0000	.0000	.0001	.0001
	6	.0000	.0000	.0000	.0000	.0000	.0000	.0000	.0000	.0000
	7	.0000	.0000	.0000	.0000	.0000	.0000	.0000	.0000	.0000
8	0	.9227	.8508	.7837	.7214	.6634	.6096	.5596	.5132	.4703
	1	.0746	.1389	.1939	.2405	.2793	.3113	.3370	.3570	.3721
	2	.0026	.0099	.0210	.0351	.0515	.0695	.0888	.1087	.1288
	3	.0001	.0004	.0013	.0029	.0054	.0089	.0134	.0189	.0255
	4	.0000	.0000	.0001	.0002	.0004	.0007	.0013	.0021	.0031
	5	.0000	.0000	.0000	.0000	.0000	.0000	.0001	.0001	.0002
	6	.0000	.0000	.0000	.0000	.0000	.0000	.0000	.0000	.0000
	7	.0000	.0000	.0000	.0000	.0000	.0000	.0000	.0000	.0000
	8	.0000	.0000	.0000	.0000	.0000	.0000	.0000	.0000	.0000
9	0	.9135	.8337	.7602	.6925	.6302	.5730	.5204	.4722	.4279
	1	.0830	.1531	.2116	.2597	.2985	.3292	.3525	.3695	.3809
	2	.0034	.0125	.0262	.0433	.0629	.0840	.1061	.1285	.1507
	3	.0001	.0006	.0019	.0042	.0077	.0125	.0186	.0261	.0348
	4	.0000	.0000	.0001	.0003	.0006	.0012	.0021	.0034	.0052
	5	.0000	.0000	.0000	.0000	.0000	.0001	.0002	.0003	.0005
	6	.0000	.0000	.0000	.0000	.0000	.0000	.0000	.0000	.0000
	7	.0000	.0000	.0000	.0000	.0000	.0000	.0000	.0000	.0000
	8	.0000	.0000	.0000	.0000	.0000	.0000	.0000	.0000	.0000
	9	.0000	.0000	.0000	.0000	.0000	.0000	.0000	.0000	.0000
10	0	.9044	.8171	.7374	.6648	.5987	.5386	.4840	.4344	.3894
	1	.0914	.1667	.2281	.2770	.3151	.3438	.3643	.3777	.3851
	2	.0042	.0153	.0317	.0519	.0746	.0988	.1234	.1478	.1714
	3	.0001	.0008	.0026	.0058	.0105	.0168	.0248	.0343	.0452
	4	.0000	.0000	.0001	.0004	.0010	.0019	.0033	.0052	.0078
	5	.0000	.0000	.0000	.0000	.0001	.0001	.0003	.0005	.0009
	6	.0000	.0000	.0000	.0000	.0000	.0000	.0000	.0000	.0001
	7	.0000	.0000	.0000	.0000	.0000	.0000	.0000	.0000	.0000
	8	.0000	.0000	.0000	.0000	.0000	.0000	.0000	.0000	.0001
	9	.0000	.0000	.0000	.0000	.0000	.0000	.0000	.0000	.0001
	10	.0000	.0000	.0000	.0000	.0000	.0000	.0000	.0000	.0001

						p				
n	x	.01	.02	.03	.04	.05	.06	.07	.08	.09
12	0	.8864	.7847	.6938	.6127	.5404	.4759	.4186	.3677	.3225
	1	.1074	.1922	.2575	.3064	.3413	.3645	.3781	.3837	.3827
	2	.0060	.0216	.0438	.0702	.0988	.1280	.1565	.1835	.2082
	3	.0002	.0015	.0045	.0098	.0173	.0272	.0393	.0532	.0686
	4	.0000	.0001	.0003	.0009	.0021	.0039	.0067	.0104	.0153
	5	.0000	.0000	.0000	.0001	.0002	.0004	.0008	.0014	.0024
	6	.0000	.0000	.0000	.0000	.0000	.0000	.0001	.0001	.0003
	7	.0000	.0000	.0000	.0000	.0000	.0000	.0000	.0000	.0000
	8	.0000	.0000	.0000	.0000	.0000	.0000	.0000	.0000	.0000
	9	.0000	.0000	.0000	.0000	.0000	.0000	.0000	.0000	.0000
	10	.0000	.0000	.0000	.0000	.0000	.0000	.0000	.0000	.0000
	11	.0000	.0000	.0000	.0000	.0000	.0000	.0000	.0000	.0000
	12	.0000	.0000	.0000	.0000	.0000	.0000	.0000	.0000	.0000
15	0	.8601	.7386	.6333	.5421	.4633	.3953	.3367	.2863	.2430
	1	.1303	.2261	.2938	.3388	.3658	.3785	.3801	.3734	.3605
	2	.0092	.0323	.0636	.0988	.1348	.1691	.2003	.2273	.2496
	3	.0004	.0029	.0085	.0178	.0307	.0468	.0653	.0857	.1070
	4	.0000	.0002	.0008	.0022	.0049	.0090	.0148	.0223	.0317
	5	.0000	.0000	.0001	.0002	.0006	.0013	.0024	.0043	.0069
	6	.0000	.0000	.0000	.0000	.0000	.0001	.0003	.0006	.0011
	7	.0000	.0000	.0000	.0000	.0000	.0000	.0000	.0001	.0001
	8	.0000	.0000	.0000	.0000	.0000	.0000	.0000	.0000	.0000
	9	.0000	.0000	.0000	.0000	.0000	.0000	.0000	.0000	.0000
	10	.0000	.0000	.0000	.0000	.0000	.0000	.0000	.0000	.0000
	11	.0000	.0000	.0000	.0000	.0000	.0000	.0000	.0000	.0000
	12	.0000	.0000	.0000	.0000	.0000	.0000	.0000	.0000	.0000
	13	.0000	.0000	.0000	.0000	.0000	.0000	.0000	.0000	.0000
	14	.0000	.0000	.0000	.0000	.0000	.0000	.0000	.0000	.0000
	15	.0000	.0000	.0000	.0000	.0000	.0000	.0000	.0000	.0000
18	0	.8345	.6951	.5780	.4796	.3972	.3283	.2708	.2229	.1831
	1	.1517	.2554	.3217	.3597	.3763	.3772	.3669	.3489	.3260
	2	.0130	.0443	.0846	.1274	.1683	.2047	.2348	.2579	.2741
	3	.0007	.0048	.0140	.0283	.0473	.0697	.0942	.1196	.1446
	4	.0000	.0004	.0016	.0044	.0093	.0167	.0266	.0390	.0536
	5	.0000	.0000	.0001	.0005	.0014	.0030	.0056	.0095	.0148
	6	.0000	.0000	.0000	.0000	.0002	.0004	.0009	.0018	.0032
	7	.0000	.0000	.0000	.0000	.0000	.0000	.0001	.0003	.0005
	8	.0000	.0000	.0000	.0000	.0000	.0000	.0000	.0000	.0001
	9	.0000	.0000	.0000	.0000	.0000	.0000	.0000	.0000	.0000
	10	.0000	.0000	.0000	.0000	.0000	.0000	.0000	.0000	.0000
	11	.0000	.0000	.0000	.0000	.0000	.0000	.0000	.0000	.0000
	12	.0000	.0000	.0000	.0000	.0000	.0000	.0000	.0000	.0000
	13	.0000	.0000	.0000	.0000	.0000	.0000	.0000	.0000	.0000
	14	.0000	.0000	.0000	.0000	.0000	.0000	.0000	.0000	.0000
	15	.0000	.0000	.0000	.0000	.0000	.0000	.0000	.0000	.0000
	16	.0000	.0000	.0000	.0000	.0000	.0000	.0000	.0000	.0000
	17	.0000	.0000	.0000	.0000	.0000	.0000	.0000	.0000	.0000
	18	.0000	.0000	.0000	.0000	.0000	.0000	.0000	.0000	.0000

		p								
n	*x*	.01	.02	.03	.04	.05	.06	.07	.08	.09
20	0	.8179	.6676	.5438	.4420	.3585	.2901	.2342	.1887	.1516
	1	.1652	.2725	.3364	.3683	.3774	.3703	.3526	.3282	.3000
	2	.0159	.0528	.0988	.1458	.1887	.2246	.2521	.2711	.2818
	3	.0010	.0065	.0183	.0364	.0596	.0860	.1139	.1414	.1672
	4	.0000	.0006	.0024	.0065	.0133	.0233	.0364	.0523	.0703
	5	.0000	.0000	.0002	.0009	.0022	.0048	.0088	.0145	.0222
	6	.0000	.0000	.0000	.0001	.0003	.0008	.0017	.0032	.0055
	7	.0000	.0000	.0000	.0000	.0000	.0001	.0002	.0005	.0011
	8	.0000	.0000	.0000	.0000	.0000	.0000	.0000	.0001	.0002
	9	.0000	.0000	.0000	.0000	.0000	.0000	.0000	.0000	.0000
	10	.0000	.0000	.0000	.0000	.0000	.0000	.0000	.0000	.0000
	11	.0000	.0000	.0000	.0000	.0000	.0000	.0000	.0000	.0000
	12	.0000	.0000	.0000	.0000	.0000	.0000	.0000	.0000	.0000
	13	.0000	.0000	.0000	.0000	.0000	.0000	.0000	.0000	.0000
	14	.0000	.0000	.0000	.0000	.0000	.0000	.0000	.0000	.0000
	15	.0000	.0000	.0000	.0000	.0000	.0000	.0000	.0000	.0000
	16	.0000	.0000	.0000	.0000	.0000	.0000	.0000	.0000	.0000
	17	.0000	.0000	.0000	.0000	.0000	.0000	.0000	.0000	.0000
	18	.0000	.0000	.0000	.0000	.0000	.0000	.0000	.0000	.0000
	19	.0000	.0000	.0000	.0000	.0000	.0000	.0000	.0000	.0000
	20	.0000	.0000	.0000	.0000	.0000	.0000	.0000	.0000	.0000

		p								
n	*x*	.10	.15	.20	.25	.30	.35	.40	.45	.50
2	0	.8100	.7225	.6400	.5625	.4900	.4225	.3600	.3025	.2500
	1	.1800	.2550	.3200	.3750	.4200	.4550	.4800	.4950	.5000
	2	.0100	.0225	.0400	.0625	.0900	.1225	.1600	.2025	.2500
3	0	.7290	.6141	.5120	.4219	.3430	.2746	.2160	.1664	.1250
	1	.2430	.3251	.3840	.4219	.4410	.4436	.4320	.4084	.3750
	2	.0270	.0574	.0960	.1406	.1890	.2389	.2880	.3341	.3750
	3	.0010	.0034	.0080	.0156	.0270	.0429	.0640	.0911	.1250
4	0	.6561	.5220	.4096	.3164	.2401	.1785	.1296	.0915	.0625
	1	.2916	.3685	.4096	.4219	.4116	.3845	.3456	.2995	.2500
	2	.0486	.0975	.1536	.2109	.2646	.3105	.3456	.3675	.3750
	3	.0036	.0115	.0256	.0469	.0756	.1115	.1536	.2005	.2500
	4	.0001	.0005	.0016	.0039	.0081	.0150	.0256	.0410	.0625
5	0	.5905	.4437	.3277	.2373	.1681	.1160	.0778	.0503	.0312
	1	.3280	.3915	.4096	.3955	.3602	.3124	.2592	.2059	.1562
	2	.0729	.1382	.2048	.2637	.3087	.3364	.3456	.3369	.3125
	3	.0081	.0244	.0512	.0879	.1323	.1811	.2304	.2757	.3125
	4	.0004	.0022	.0064	.0146	.0284	.0488	.0768	.1128	.1562
	5	.0000	.0001	.0003	.0010	.0024	.0053	.0102	.0185	.0312

						p				
n	x	.10	.15	.20	.25	.30	.35	.40	.45	.50
6	0	.5314	.3771	.2621	.1780	.1176	.0754	.0467	.0277	.0156
	1	.3543	.3993	.3932	.3560	.3025	.2437	.1866	.1359	.0938
	2	.0984	.1762	.2458	.2966	.3241	.3280	.3110	.2780	.2344
	3	.0146	.0415	.0819	.1318	.1852	.2355	.2765	.3032	.3125
	4	.0012	.0055	.0154	.0330	.0595	.0951	.1382	.1861	.2344
	5	.0001	.0004	.0015	.0044	.0102	.0205	.0369	.0609	.0938
	6	.0000	.0000	.0001	.0002	.0007	.0018	.0041	.0083	.0156
7	0	.4783	.3206	.2097	.1335	.0824	.0490	.0280	.0152	.0078
	1	.3720	.3960	.3670	.3115	.2471	.1848	.1306	.0872	.0547
	2	.1240	.2097	.2753	.3115	.3177	.2985	.2613	.2140	.1641
	3	.0230	.0617	.1147	.1730	.2269	.2679	.2903	.2918	.2734
	4	.0026	.0109	.0287	.0577	.0972	.1442	.1935	.2388	.2734
	5	.0002	.0012	.0043	.0115	.0250	.0466	.0774	.1172	.1641
	6	.0000	.0001	.0004	.0013	.0036	.0084	.0172	.0320	.0547
	7	.0000	.0000	.0000	.0001	.0002	.0006	.0016	.0037	.0078
8	0	.4305	.2725	.1678	.1001	.0576	.0319	.0168	.0084	.0039
	1	.3826	.3847	.3355	.2670	.1977	.1373	.0896	.0548	.0312
	2	.1488	.2376	.2936	.3115	.2965	.2587	.2090	.1569	.1094
	3	.0331	.0839	.1468	.2076	.2541	.2786	.2787	.2568	.2188
	4	.0046	.0185	.0459	.0865	.1361	.1875	.2322	.2627	.2734
	5	.0004	.0026	.0092	.0231	.0467	.0808	.1239	.1719	.2188
	6	.0000	.0002	.0011	.0038	.0100	.0217	.0413	.0703	.1094
	7	.0000	.0000	.0001	.0004	.0012	.0033	.0079	.0164	.0312
	8	.0000	.0000	.0000	.0000	.0001	.0002	.0007	.0017	.0039
9	0	.3874	.2316	.1342	.0751	.0404	.0207	.0101	.0046	.0020
	1	.3874	.3679	.3020	.2253	.1556	.1004	.0605	.0339	.0176
	2	.1722	.2597	.3020	.3003	.2668	.2162	.1612	.1110	.0703
	3	.0446	.1069	.1762	.2336	.2668	.2716	.2508	.2119	.1641
	4	.0074	.0283	.0661	.1168	.1715	.2194	.2508	.2600	.2461
	5	.0008	.0050	.0165	.0389	.0735	.1181	.1672	.2128	.2461
	6	.0001	.0006	.0028	.0087	.0210	.0424	.0743	.1160	.1641
	7	.0000	.0000	.0003	.0012	.0039	.0098	.0212	.0407	.0703
	8	.0000	.0000	.0000	.0001	.0004	.0013	.0035	.0083	.0176
	9	.0000	.0000	.0000	.0000	.0000	.0001	.0003	.0008	.0020
10	0	.3487	.1969	.1074	.0563	.0282	.0135	.0060	.0025	.0010
	1	.3874	.3474	.2684	.1877	.1211	.0725	.0403	.0207	.0098
	2	.1937	.2759	.3020	.2816	.2335	.1757	.1209	.0763	.0439
	3	.0574	.1298	.2013	.2503	.2668	.2522	.2150	.1665	.1172
	4	.0112	.0401	.0881	.1460	.2001	.2377	.2508	.2384	.2051
	5	.0015	.0085	.0264	.0584	.1029	.1536	.2007	.2340	.2461
	6	.0001	.0012	.0055	.0162	.0368	.0689	.1115	.1596	.2051
	7	.0000	.0001	.0008	.0031	.0090	.0212	.0425	.0746	.1172
	8	.0000	.0000	.0001	.0004	.0014	.0043	.0106	.0229	.0439
	9	.0000	.0000	.0000	.0000	.0001	.0005	.0016	.0042	.0098
	10	.0000	.0000	.0000	.0000	.0000	.0000	.0001	.0003	.0010

						p				
n	x	.10	.15	.20	.25	.30	.35	.40	.45	.50
12	0	.2824	.1422	.0687	.0317	.0138	.0057	.0022	.0008	.0002
	1	.3766	.3012	.2062	.1267	.0712	.0368	.0174	.0075	.0029
	2	.2301	.2924	.2835	.2323	.1678	.1088	.0639	.0339	.0161
	3	.0853	.1720	.2362	.2581	.2397	.1954	.1419	.0923	.0537
	4	.0213	.0683	.1329	.1936	.2311	.2367	.2128	.1700	.1208
	5	.0038	.0193	.0532	.1032	.1585	.2039	.2270	.2225	.1934
	6	.0005	.0040	.0155	.0401	.0792	.1281	.1766	.2124	.2256
	7	.0000	.0006	.0033	.0115	.0291	.0591	.1009	.1489	.1934
	8	.0000	.0001	.0005	.0024	.0078	.0199	.0420	.0762	.1208
	9	.0000	.0000	.0001	.0004	.0015	.0048	.0125	.0277	.0537
	10	.0000	.0000	.0000	.0000	.0002	.0008	.0025	.0068	.0161
	11	.0000	.0000	.0000	.0000	.0000	.0001	.0003	.0010	.0029
	12	.0000	.0000	.0000	.0000	.0000	.0000	.0000	.0001	.0002
15	0	.2059	.0874	.0352	.0134	.0047	.0016	.0005	.0001	.0000
	1	.3432	.2312	.1319	.0668	.0305	.0126	.0047	.0016	.0005
	2	.2669	.2856	.2309	.1559	.0916	.0476	.0219	.0090	.0032
	3	.1285	.2184	.2501	.2252	.1700	.1110	.0634	.0318	.0139
	4	.0428	.1156	.1876	.2252	.2186	.1792	.1268	.0780	.0417
	5	.0105	.0449	.1032	.1651	.2061	.2123	.1859	.1404	.0916
	6	.0019	.0132	.0430	.0917	.1472	.1906	.2066	.1914	.1527
	7	.0003	.0030	.0138	.0393	.0811	.1319	.1771	.2013	.1964
	8	.0000	.0005	.0035	.0131	.0348	.0710	.1181	.1647	.1964
	9	.0000	.0001	.0007	.0034	.0116	.0298	.0612	.1048	.1527
	10	.0000	.0000	.0001	.0007	.0030	.0096	.0245	.0515	.0916
	11	.0000	.0000	.0000	.0001	.0006	.0024	.0074	.0191	.0417
	12	.0000	.0000	.0000	.0000	.0001	.0004	.0016	.0052	.0139
	13	.0000	.0000	.0000	.0000	.0000	.0001	.0003	.0010	.0032
	14	.0000	.0000	.0000	.0000	.0000	.0000	.0000	.0001	.0005
	15	.0000	.0000	.0000	.0000	.0000	.0000	.0000	.0000	.0000
18	0	.1501	.0536	.0180	.0056	.0016	.0004	.0001	.0000	.0000
	1	.3002	.1704	.0811	.0338	.0126	.0042	.0012	.0003	.0001
	2	.2835	.2556	.1723	.0958	.0458	.0190	.0069	.0022	.0006
	3	.1680	.2406	.2297	.1704	.1046	.0547	.0246	.0095	.0031
	4	.0700	.1592	.2153	.2130	.1681	.1104	.0614	.0291	.0117
	5	.0218	.0787	.1507	.1988	.2017	.1664	.1146	.0666	.0327
	6	.0052	.0301	.0816	.1436	.1873	.1941	.1655	.1181	.0708
	7	.0010	.0091	.0350	.0820	.1376	.1792	.1892	.1657	.1214
	8	.0002	.0022	.0120	.0376	.0811	.1327	.1734	.1864	.1669
	9	.0000	.0004	.0033	.0139	.0386	.0794	.1284	.1694	.1855
	10	.0000	.0001	.0008	.0042	.0149	.0385	.0771	.1248	.1669
	11	.0000	.0000	.0001	.0010	.0046	.0151	.0374	.0742	.1214
	12	.0000	.0000	.0000	.0002	.0012	.0047	.0145	.0354	.0708
	13	.0000	.0000	.0000	.0000	.0002	.0012	.0045	.0134	.0327
	14	.0000	.0000	.0000	.0000	.0000	.0002	.0011	.0039	.0117
	15	.0000	.0000	.0000	.0000	.0000	.0000	.0002	.0009	.0031
	16	.0000	.0000	.0000	.0000	.0000	.0000	.0000	.0001	.0006
	17	.0000	.0000	.0000	.0000	.0000	.0000	.0000	.0000	.0001
	18	.0000	.0000	.0000	.0000	.0000	.0000	.0000	.0000	.0000

					p				
n	x	.10	.15	.20	.25	.30	.35	.40	.45	.50
20	0	.1216	.0388	.0115	.0032	.0008	.0002	.0000	.0000	.0000
	1	.2702	.1368	.0576	.0211	.0068	.0020	.0005	.0001	.0000
	2	.2852	.2293	.1369	.0669	.0278	.0100	.0031	.0008	.0002
	3	.1901	.2428	.2054	.1339	.0716	.0323	.0123	.0040	.0011
	4	.0898	.1821	.2182	.1897	.1304	.0738	.0350	.0139	.0046
	5	.0319	.1028	.1746	.2023	.1789	.1272	.0746	.0365	.0148
	6	.0089	.0454	.1091	.1686	.1916	.1712	.1244	.0746	.0370
	7	.0020	.0160	.0545	.1124	.1643	.1844	.1659	.1221	.0739
	8	.0004	.0046	.0222	.0609	.1144	.1614	.1797	.1623	.1201
	9	.0001	.0011	.0074	.0271	.0654	.1158	.1597	.1771	.1602
	10	.0000	.0002	.0020	.0099	.0308	.0686	.1171	.1593	.1762
	11	.0000	.0000	.0005	.0030	.0120	.0336	.0710	.1185	.1602
	12	.0000	.0000	.0001	.0008	.0039	.0136	.0355	.0727	.1201
	13	.0000	.0000	.0000	.0002	.0010	.0045	.0146	.0366	.0739
	14	.0000	.0000	.0000	.0000	.0002	.0012	.0049	.0150	.0370
	15	.0000	.0000	.0000	.0000	.0000	.0003	.0013	.0049	.0148
	16	.0000	.0000	.0000	.0000	.0000	.0000	.0003	.0013	.0046
	17	.0000	.0000	.0000	.0000	.0000	.0000	.0000	.0002	.0011
	18	.0000	.0000	.0000	.0000	.0000	.0000	.0000	.0000	.0002
	19	.0000	.0000	.0000	.0000	.0000	.0000	.0000	.0000	.0000
	20	.0000	.0000	.0000	.0000	.0000	.0000	.0000	.0000	.0000

					p				
n	x	.55	.60	.65	.70	.75	.80	.85	.90	.95
2	0	.2025	.1600	.1225	.0900	.0625	.0400	.0225	.0100	.0025
	1	.4950	.4800	.4550	.4200	.3750	.3200	.2550	.1800	.0950
	2	.3025	.3600	.4225	.4900	.5625	.6400	.7225	.8100	.9025
3	0	.0911	.0640	.0429	.0270	.0156	.0080	.0034	.0010	.0001
	1	.3341	.2880	.2389	.1890	.1406	.0960	.0574	.0270	.0071
	2	.4084	.4320	.4436	.4410	.4219	.3840	.3251	.2430	.1354
	3	.1664	.2160	.2746	.3430	.4219	.5120	.6141	.7290	.8574
4	0	.0410	.0256	.0150	.0081	.0039	.0016	.0005	.0001	.0000
	1	.2005	.1536	.1115	.0756	.0469	.0256	.0115	.0036	.0005
	2	.3675	.3456	.3105	.2646	.2109	.1536	.0975	.0486	.0135
	3	.2995	.3456	.3845	.4116	.4219	.4096	.3685	.2916	.1715
	4	.0915	.1296	.1785	.2401	.3164	.4096	.5220	.6561	.8145
5	0	.0185	.0102	.0053	.0024	.0010	.0003	.0001	.0000	.0000
	1	.1128	.0768	.0488	.0284	.0146	.0064	.0022	.0005	.0000
	2	.2757	.2304	.1811	.1323	.0879	.0512	.0244	.0081	.0011
	3	.3369	.3456	.3364	.3087	.2637	.2048	.1382	.0729	.0214
	4	.2059	.2592	.3124	.3601	.3955	.4096	.3915	.3281	.2036
	5	.0503	.0778	.1160	.1681	.2373	.3277	.4437	.5905	.7738

						p				
n	x	.55	.60	.65	.70	.75	.80	.85	.90	.95
6	0	.0083	.0041	.0018	.0007	.0002	.0001	.0000	.0000	.0000
	1	.0609	.0369	.0205	.0102	.0044	.0015	.0004	.0001	.0000
	2	.1861	.1382	.0951	.0595	.0330	.0154	.0055	.0012	.0001
	3	.3032	.2765	.2355	.1852	.1318	.0819	.0415	.0146	.0021
	4	.2780	.3110	.3280	.3241	.2966	.2458	.1762	.0984	.0305
	5	.1359	.1866	.2437	.3025	.3560	.3932	.3993	.3543	.2321
	6	.0277	.0467	.0754	.1176	.1780	.2621	.3771	.5314	.7351
7	0	.0037	.0016	.0006	.0002	.0001	.0000	.0000	.0000	.0000
	1	.0320	.0172	.0084	.0036	.0013	.0004	.0001	.0000	.0000
	2	.1172	.0774	.0466	.0250	.0115	.0043	.0012	.0002	.0000
	3	.2388	.1935	.1442	.0972	.0577	.0287	.0109	.0026	.0002
	4	.2918	.2903	.2679	.2269	.1730	.1147	.0617	.0230	.0036
	5	.2140	.2613	.2985	.3177	.3115	.2753	.2097	.1240	.0406
	6	.0872	.1306	.1848	.2471	.3115	.3670	.3960	.3720	.2573
	7	.0152	.0280	.0490	.0824	.1335	.2097	.3206	.4783	.6983
8	0	.0017	.0007	.0002	.0001	.0000	.0000	.0000	.0000	.0000
	1	.0164	.0079	.0033	.0012	.0004	.0001	.0000	.0000	.0000
	2	.0703	.0413	.0217	.0100	.0038	.0011	.0002	.0000	.0000
	3	.1719	.1239	.0808	.0467	.0231	.0092	.0026	.0004	.0000
	4	.2627	.2322	.1875	.1361	.0865	.0459	.0185	.0046	.0004
	5	.2568	.2787	.2786	.2541	.2076	.1468	.0839	.0331	.0054
	6	.1569	.2090	.2587	.2965	.3115	.2936	.2376	.1488	.0515
	7	.0548	.0896	.1373	.1977	.2670	.3355	.3847	.3826	.2793
	8	.0084	.0168	.0319	.0576	.1001	.1678	.2725	.4305	.6634
9	0	.0008	.0003	.0001	.0000	.0000	.0000	.0000	.0000	.0000
	1	.0083	.0035	.0013	.0004	.0001	.0000	.0000	.0000	.0000
	2	.0407	.0212	.0098	.0039	.0012	.0003	.0000	.0000	.0000
	3	.1160	.0743	.0424	.0210	.0087	.0028	.0006	.0001	.0000
	4	.2128	.1672	.1181	.0735	.0389	.0165	.0050	.0008	.0000
	5	.2600	.2508	.2194	.1715	.1168	.0661	.0283	.0074	.0006
	6	.2119	.2508	.2716	.2668	.2336	.1762	.1069	.0446	.0077
	7	.1110	.1612	.2162	.2668	.3003	.3020	.2597	.1722	.0629
	8	.0339	.0605	.1004	.1556	.2253	.3020	.3679	.3874	.2985
	9	.0046	.0101	.0207	.0404	.0751	.1342	.2316	.3874	.6302
10	0	.0003	.0001	.0000	.0000	.0000	.0000	.0000	.0000	.0000
	1	.0042	.0016	.0005	.0001	.0000	.0000	.0000	.0000	.0000
	2	.0229	.0106	.0043	.0014	.0004	.0001	.0000	.0000	.0000
	3	.0746	.0425	.0212	.0090	.0031	.0008	.0001	.0000	.0000
	4	.1596	.1115	.0689	.0368	.0162	.0055	.0012	.0001	.0000
	5	.2340	.2007	.1536	.1029	.0584	.0264	.0085	.0015	.0001
	6	.2384	.2508	.2377	.2001	.1460	.0881	.0401	.0112	.0010
	7	.1665	.2150	.2522	.2668	.2503	.2013	.1298	.0574	.0105
	8	.0763	.1209	.1757	.2335	.2816	.3020	.2759	.1937	.0746
	9	.0207	.0403	.0725	.1211	.1877	.2684	.3474	.3874	.3151
	10	.0025	.0060	.0135	.0282	.0563	.1074	.1969	.3487	.5987

						p				
n	x	.55	.60	.65	.70	.75	.80	.85	.90	.95
12	0	.0001	.0000	.0000	.0000	.0000	.0000	.0000	.0000	.0000
	1	.0010	.0003	.0001	.0000	.0000	.0000	.0000	.0000	.0000
	2	.0068	.0025	.0008	.0002	.0000	.0000	.0000	.0000	.0000
	3	.0277	.0125	.0048	.0015	.0004	.0001	.0000	.0000	.0000
	4	.0762	.0420	.0199	.0078	.0024	.0005	.0001	.0000	.0000
	5	.1489	.1009	.0591	.0291	.0115	.0033	.0006	.0000	.0000
	6	.2124	.1766	.1281	.0792	.0401	.0155	.0040	.0005	.0000
	7	.2225	.2270	.2039	.1585	.1032	.0532	.0193	.0038	.0002
	8	.1700	.2128	.2367	.2311	.1936	.1329	.0683	.0213	.0021
	9	.0923	.1419	.1954	.2397	.2581	.2362	.1720	.0852	.0173
	10	.0339	.0639	.1088	.1678	.2323	.2835	.2924	.2301	.0988
	11	.0075	.0174	.0368	.0712	.1267	.2062	.3012	.3766	.3413
	12	.0008	.0022	.0057	.0138	.0317	.0687	.1422	.2824	.5404
15	0	.0000	.0000	.0000	.0000	.0000	.0000	.0000	.0000	.0000
	1	.0001	.0000	.0000	.0000	.0000	.0000	.0000	.0000	.0000
	2	.0010	.0003	.0001	.0000	.0000	.0000	.0000	.0000	.0000
	3	.0052	.0016	.0004	.0001	.0000	.0000	.0000	.0000	.0000
	4	.0191	.0074	.0024	.0006	.0001	.0000	.0000	.0000	.0000
	5	.0515	.0245	.0096	.0030	.0007	.0001	.0000	.0000	.0000
	6	.1048	.0612	.0298	.0116	.0034	.0007	.0001	.0000	.0000
	7	.1647	.1181	.0710	.0348	.0131	.0035	.0005	.0000	.0000
	8	.2013	.1771	.1319	.0811	.0393	.0138	.0030	.0003	.0000
	9	.1914	.2066	.1906	.1472	.0917	.0430	.0132	.0019	.0000
	10	.1404	.1859	.2123	.2061	.1651	.1032	.0449	.0105	.0006
	11	.0780	.1268	.1792	.2186	.2252	.1876	.1156	.0428	.0049
	12	.0318	.0634	.1110	.1700	.2252	.2501	.2184	.1285	.0307
	13	.0090	.0219	.0476	.0916	.1559	.2309	.2856	.2669	.1348
	14	.0016	.0047	.0126	.0305	.0668	.1319	.2312	.3432	.3658
	15	.0001	.0005	.0016	.0047	.0134	.0352	.0874	.2059	.4633
18	0	.0000	.0000	.0000	.0000	.0000	.0000	.0000	.0000	.0000
	1	.0000	.0000	.0000	.0000	.0000	.0000	.0000	.0000	.0000
	2	.0001	.0000	.0000	.0000	.0000	.0000	.0000	.0000	.0000
	3	.0009	.0002	.0000	.0000	.0000	.0000	.0000	.0000	.0000
	4	.0039	.0011	.0002	.0000	.0000	.0000	.0000	.0000	.0000
	5	.0134	.0045	.0012	.0002	.0000	.0000	.0000	.0000	.0000
	6	.0354	.0145	.0047	.0012	.0002	.0000	.0000	.0000	.0000
	7	.0742	.0374	.0151	.0046	.0010	.0001	.0000	.0000	.0000
	8	.1248	.0771	.0385	.0149	.0042	.0008	.0001	.0000	.0000
	9	.1694	.1284	.0794	.0386	.0139	.0033	.0004	.0000	.0000
	10	.1864	.1734	.1327	.0811	.0376	.0120	.0022	.0002	.0000
	11	.1657	.1892	.1792	.1376	.0820	.0350	.0091	.0010	.0000
	12	.1181	.1655	.1941	.1873	.1436	.0816	.0301	.0052	.0002
	13	.0666	.1146	.1664	.2017	.1988	.1507	.0787	.0218	.0014
	14	.0291	.0614	.1104	.1681	.2130	.2153	.1592	.0700	.0093
	15	.0095	.0246	.0547	.1046	.1704	.2297	.2406	.1680	.0473
	16	.0022	.0069	.0190	.0458	.0958	.1723	.2556	.2835	.1683
	17	.0003	.0012	.0042	.0126	.0338	.0811	.1704	.3002	.3763
	18	.0000	.0001	.0004	.0016	.0056	.0180	.0536	.1501	.3972

(concluded)

						p				
n	*x*	.55	.60	.65	.70	.75	.80	.85	.90	.95
20	0	.0000	.0000	.0000	.0000	.0000	.0000	.0000	.0000	.0000
	1	.0000	.0000	.0000	.0000	.0000	.0000	.0000	.0000	.0000
	2	.0000	.0000	.0000	.0000	.0000	.0000	.0000	.0000	.0000
	3	.0002	.0000	.0000	.0000	.0000	.0000	.0000	.0000	.0000
	4	.0013	.0003	.0000	.0000	.0000	.0000	.0000	.0000	.0000
	5	.0049	.0013	.0003	.0000	.0000	.0000	.0000	.0000	.0000
	6	.0150	.0049	.0012	.0002	.0000	.0000	.0000	.0000	.0000
	7	.0366	.0146	.0045	.0010	.0002	.0000	.0000	.0000	.0000
	8	.0727	.0355	.0136	.0039	.0008	.0001	.0000	.0000	.0000
	9	.1185	.0710	.0336	.0120	.0030	.0005	.0000	.0000	.0000
	10	.1593	.1171	.0686	.0308	.0099	.0020	.0002	.0000	.0000
	11	.1771	.1597	.1158	.0654	.0271	.0074	.0011	.0001	.0000
	12	.1623	.1797	.1614	.1144	.0609	.0222	.0046	.0004	.0000
	13	.1221	.1659	.1844	.1643	.1124	.0545	.0160	.0020	.0000
	14	.0746	.1244	.1712	.1916	.1686	.1091	.0454	.0089	.0003
	15	.0365	.0746	.1272	.1789	.2023	.1746	.1028	.0319	.0022
	16	.0139	.0350	.0738	.1304	.1897	.2182	.1821	.0898	.0133
	17	.0040	.0123	.0323	.0716	.1339	.2054	.2428	.1901	.0596
	18	.0008	.0031	.0100	.0278	.0669	.1369	.2293	.2852	.1887
	19	.0001	.0005	.0020	.0068	.0211	.0576	.1368	.2702	.3774
	20	.0000	.0000	.0002	.0008	.0032	.0115	.0388	.1216	.3585

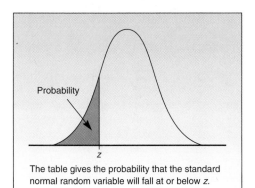

Probability

The table gives the probability that the standard normal random variable will fall at or below z.

The Standard Normal Probability Distribution   **TABLE B.2**

z	.00	.01	.02	.03	.04	.05	.06	.07	.08	.09
−3.4	.0003	.0003	.0003	.0003	.0003	.0003	.0003	.0003	.0003	.0002
−3.3	.0005	.0005	.0005	.0004	.0004	.0004	.0004	.0004	.0004	.0003
−3.2	.0007	.0007	.0006	.0006	.0006	.0006	.0006	.0005	.0005	.0005
−3.1	.0010	.0009	.0009	.0009	.0008	.0008	.0008	.0008	.0007	.0007
−3.0	.0013	.0013	.0013	.0012	.0012	.0011	.0011	.0011	.0010	.0010
−2.9	.0019	.0018	.0018	.0017	.0016	.0016	.0015	.0015	.0014	.0014
−2.8	.0026	.0025	.0024	.0023	.0023	.0022	.0021	.0021	.0020	.0019
−2.7	.0035	.0034	.0033	.0032	.0031	.0030	.0029	.0028	.0027	.0026
−2.6	.0047	.0045	.0044	.0043	.0041	.0040	.0039	.0038	.0037	.0036
−2.5	.0062	.0060	.0059	.0057	.0055	.0054	.0052	.0051	.0049	.0048
−2.4	.0082	.0080	.0078	.0075	.0073	.0071	.0069	.0068	.0066	.0064
−2.3	.0107	.0104	.0102	.0099	.0096	.0094	.0091	.0089	.0087	.0084
−2.2	.0139	.0136	.0132	.0129	.0125	.0122	.0119	.0116	.0113	.0110
−2.1	.0179	.0174	.0170	.0166	.0162	.0158	.0154	.0150	.0146	.0143
−2.0	.0228	.0222	.0217	.0212	.0207	.0202	.0197	.0192	.0188	.0183
−1.9	.0287	.0281	.0274	.0268	.0262	.0256	.0250	.0244	.0239	.0233
−1.8	.0359	.0351	.0344	.0336	.0329	.0322	.0314	.0307	.0301	.0294
−1.7	.0446	.0436	.0427	.0418	.0409	.0401	.0392	.0384	.0375	.0367
−1.6	.0548	.0537	.0526	.0516	.0505	.0495	.0485	.0475	.0465	.0455
−1.5	.0668	.0655	.0643	.0630	.0618	.0606	.0594	.0582	.0571	.0559
−1.4	.0808	.0793	.0778	.0764	.0749	.0735	.0721	.0708	.0694	.0681
−1.3	.0968	.0951	.0934	.0918	.0901	.0885	.0869	.0853	.0838	.0823
−1.2	.1151	.1131	.1112	.1093	.1075	.1056	.1038	.1020	.1003	.0985
−1.1	.1357	.1335	.1314	.1292	.1271	.1251	.1230	.1210	.1190	.1170
−1.0	.1587	.1562	.1539	.1515	.1492	.1469	.1446	.1423	.1401	.1379
−0.9	.1841	.1814	.1788	.1762	.1736	.1711	.1685	.1660	.1635	.1611
−0.8	.2119	.2090	.2061	.2033	.2005	.1977	.1949	.1922	.1894	.1867
−0.7	.2420	.2389	.2358	.2327	.2296	.2266	.2236	.2206	.2177	.2148
−0.6	.2743	.2709	.2676	.2643	.2611	.2578	.2546	.2514	.2483	.2451
−0.5	.3085	.3050	.3015	.2981	.2946	.2912	.2877	.2843	.2810	.2776
−0.4	.3446	.3409	.3372	.3336	.3300	.3264	.3228	.3192	.3156	.3121
−0.3	.3821	.3783	.3745	.3707	.3669	.3632	.3594	.3557	.3520	.3483
−0.2	.4207	.4168	.4129	.4090	.4052	.4013	.3974	.3936	.3897	.3859
−0.1	.4602	.4562	.4522	.4483	.4443	.4404	.4364	.4325	.4286	.4247
−0.0	.5000	.4960	.4920	.4880	.4840	.4801	.4761	.4721	.4681	.4641
−3.5	.0002									
−4.0	.00003									
−4.5	.000003									
−5.0	.0000003									
−6.0	.000000001									

z	.00	.01	.02	.03	.04	.05	.06	.07	.08	.09
0.0	.5000	.5040	.5080	.5120	.5160	.5199	.5239	.5279	.5319	.5359
0.1	.5398	.5438	.5478	.5517	.5557	.5596	.5636	.5675	.5714	.5753
0.2	.5793	.5832	.5871	.5910	.5948	.5987	.6026	.6064	.6103	.6141
0.3	.6179	.6217	.6255	.6293	.6331	.6368	.6406	.6443	.6480	.6517
0.4	.6554	.6591	.6628	.6664	.6700	.6736	.6772	.6808	.6844	.6879
0.5	.6915	.6950	.6985	.7019	.7054	.7088	.7123	.7157	.7190	.7224
0.6	.7257	.7291	.7324	.7357	.7389	.7422	.7454	.7486	.7517	.7549
0.7	.7580	.7611	.7642	.7673	.7704	.7734	.7764	.7794	.7823	.7852
0.8	.7881	.7910	.7939	.7967	.7995	.8023	.8051	.8078	.8106	.8133
0.9	.8159	.8186	.8212	.8238	.8264	.8289	.8315	.8340	.8365	.8389
1.0	.8413	.8438	.8461	.8485	.8508	.8531	.8554	.8577	.8599	.8621
1.1	.8643	.8665	.8686	.8708	.8729	.8749	.8770	.8790	.8810	.8830
1.2	.8849	.8869	.8888	.8907	.8925	.8944	.8962	.8980	.8997	.9015
1.3	.9032	.9049	.9066	.9082	.9099	.9115	.9131	.9147	.9162	.9177
1.4	.9192	.9207	.9222	.9236	.9251	.9265	.9279	.9292	.9306	.9319
1.5	.9332	.9345	.9357	.9370	.9382	.9394	.9406	.9418	.9429	.9441
1.6	.9452	.9463	.9474	.9484	.9495	.9505	.9515	.9525	.9535	.9545
1.7	.9554	.9564	.9573	.9582	.9591	.9599	.9608	.9616	.9625	.9633
1.8	.9641	.9649	.9656	.9664	.9671	.9678	.9686	.9693	.9699	.9706
1.9	.9713	.9719	.9726	.9732	.9738	.9744	.9750	.9756	.9761	.9767
2.0	.9772	.9778	.9783	.9788	.9793	.9798	.9803	.9808	.9812	.9817
2.1	.9821	.9826	.9830	.9834	.9838	.9842	.9846	.9850	.9854	.9857
2.2	.9861	.9864	.9868	.9871	.9875	.9878	.9881	.9884	.9887	.9890
2.3	.9893	.9896	.9898	.9901	.9904	.9906	.9909	.9911	.9913	.9916
2.4	.9918	.9920	.9922	.9925	.9927	.9929	.9931	.9932	.9934	.9936
2.5	.9938	.9940	.9941	.9943	.9945	.9946	.9948	.9949	.9951	.9952
2.6	.9953	.9955	.9956	.9957	.9959	.9960	.9961	.9962	.9963	.9964
2.7	.9965	.9966	.9967	.9968	.9969	.9970	.9971	.9972	.9973	.9974
2.8	.9974	.9975	.9976	.9977	.9977	.9978	.9979	.9979	.9980	.9981
2.9	.9981	.9982	.9982	.9983	.9984	.9984	.9985	.9985	.9986	.9986
3.0	.9987	.9987	.9987	.9988	.9988	.9989	.9989	.9989	.9990	.9990
3.1	.9990	.9991	.9991	.9991	.9992	.9992	.9992	.9992	.9993	.9993
3.2	.9993	.9993	.9994	.9994	.9994	.9994	.9994	.9995	.9995	.9995
3.3	.9995	.9995	.9995	.9996	.9996	.9996	.9996	.9996	.9996	.9997
3.4	.9997	.9997	.9997	.9997	.9997	.9997	.9997	.9997	.9997	.9998
3.5	.9998									
4.0	.99997									
4.5	.999997									
5.0	.9999997									
6.0	.999999999									

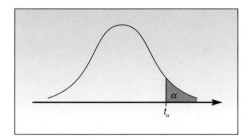

Critical Values of the t Distribution   **TABLE B.3**

Degrees of Freedom	$t_{.100}$	$t_{.050}$	$t_{.025}$	$t_{.010}$	$t_{.005}$
1	3.078	6.314	12.706	31.821	63.657
2	1.886	2.920	4.303	6.965	9.925
3	1.638	2.353	3.182	4.541	5.841
4	1.533	2.132	2.776	3.747	4.604
5	1.476	2.015	2.571	3.365	4.032
6	1.440	1.943	2.447	3.143	3.707
7	1.415	1.895	2.365	2.998	3.499
8	1.397	1.860	2.306	2.896	3.355
9	1.383	1.833	2.262	2.821	3.250
10	1.372	1.812	2.228	2.764	3.169
11	1.363	1.796	2.201	2.718	3.106
12	1.356	1.782	2.179	2.681	3.055
13	1.350	1.771	2.160	2.650	3.012
14	1.345	1.761	2.145	2.624	2.977
15	1.341	1.753	2.131	2.602	2.947
16	1.337	1.746	2.120	2.583	2.921
17	1.333	1.740	2.110	2.567	2.898
18	1.330	1.734	2.101	2.552	2.878
19	1.328	1.729	2.093	2.539	2.861
20	1.325	1.725	2.086	2.528	2.845
21	1.323	1.721	2.080	2.518	2.831
22	1.321	1.717	2.074	2.508	2.819
23	1.319	1.714	2.069	2.500	2.807
24	1.318	1.711	2.064	2.492	2.797
25	1.316	1.708	2.060	2.485	2.787
26	1.315	1.706	2.056	2.479	2.779
27	1.314	1.703	2.052	2.473	2.771
28	1.313	1.701	2.048	2.467	2.763
29	1.311	1.699	2.045	2.462	2.756
30	1.310	1.697	2.042	2.457	2.750
40	1.303	1.684	2.021	2.423	2.704
60	1.296	1.671	2.000	2.390	2.660
120	1.289	1.658	1.980	2.358	2.617
∞	1.282	1.645	1.960	2.326	2.576

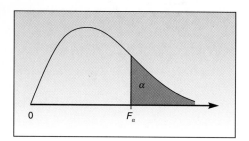

TABLE B.4

Critical Values of the F Distribution for $\alpha = 0.10$

Denominator Degrees of Freedom (k_2)	Numerator Degrees of Freedom (k_1)								
	1	**2**	**3**	**4**	**5**	**6**	**7**	**8**	**9**
1	39.86	49.50	53.59	55.83	57.24	58.20	58.91	59.44	59.86
2	8.53	9.00	9.16	9.24	9.29	9.33	9.35	9.37	9.38
3	5.54	5.46	5.39	5.34	5.31	5.28	5.27	5.25	5.24
4	4.54	4.32	4.19	4.11	4.05	4.01	3.98	3.95	3.94
5	4.06	3.78	3.62	3.52	3.45	3.40	3.37	3.34	3.32
6	3.78	3.46	3.29	3.18	3.11	3.05	3.01	2.98	2.96
7	3.59	3.26	3.07	2.96	2.88	2.83	2.78	2.75	2.72
8	3.46	3.11	2.92	2.81	2.73	2.67	2.62	2.59	2.56
9	3.36	3.01	2.81	2.69	2.61	2.55	2.51	2.47	2.44
10	3.29	2.92	2.73	2.61	2.52	2.46	2.41	2.38	2.35
11	3.23	2.86	2.66	2.54	2.45	2.39	2.34	2.30	2.27
12	3.18	2.81	2.61	2.48	2.39	2.33	2.28	2.24	2.21
13	3.14	2.76	2.56	2.43	2.35	2.28	2.23	2.20	2.16
14	3.10	2.73	2.52	2.39	2.31	2.24	2.19	2.15	2.12
15	3.07	2.70	2.49	2.36	2.27	2.21	2.16	2.12	2.09
16	3.05	2.67	2.46	2.33	2.24	2.18	2.13	2.09	2.06
17	3.03	2.64	2.44	2.31	2.22	2.15	2.10	2.06	2.03
18	3.01	2.62	2.42	2.29	2.20	2.13	2.08	2.04	2.00
19	2.99	2.61	2.40	2.27	2.18	2.11	2.06	2.02	1.98
20	2.97	2.59	2.38	2.25	2.16	2.09	2.04	2.00	1.96
21	2.96	2.57	2.36	2.23	2.14	2.08	2.02	1.98	1.95
22	2.95	2.56	2.35	2.22	2.13	2.06	2.01	1.97	1.93
23	2.94	2.55	2.34	2.21	2.11	2.05	1.99	1.95	1.92
24	2.93	2.54	2.33	2.19	2.10	2.04	1.98	1.94	1.91
25	2.92	2.53	2.32	2.18	2.09	2.02	1.97	1.93	1.89
26	2.91	2.52	2.31	2.17	2.08	2.01	1.96	1.92	1.88
27	2.90	2.51	2.30	2.17	2.07	2.00	1.95	1.91	1.87
28	2.89	2.50	2.29	2.16	2.06	2.00	1.94	1.90	1.87
29	2.89	2.50	2.28	2.15	2.06	1.99	1.93	1.89	1.86
30	2.88	2.49	2.28	2.14	2.05	1.98	1.93	1.88	1.85
40	2.84	2.44	2.23	2.09	2.00	1.93	1.87	1.83	1.79
60	2.79	2.39	2.18	2.04	1.95	1.87	1.82	1.77	1.74
120	2.75	2.35	2.13	1.99	1.90	1.82	1.77	1.72	1.68
∞	2.71	2.30	2.08	1.94	1.85	1.77	1.72	1.67	1.63

Denominator Degrees of Freedom (k_2)	Numerator Degrees of Freedom (k_1)									
	10	12	15	20	24	30	40	60	120	∞
1	60.19	60.71	61.22	61.74	62.00	62.26	62.53	62.79	63.06	63.33
2	9.39	9.41	9.42	9.44	9.45	9.46	9.47	9.47	9.48	9.49
3	5.23	5.22	5.20	5.18	5.18	5.17	5.16	5.15	5.14	5.13
4	3.92	3.90	3.87	3.84	3.83	3.82	3.80	3.79	3.78	3.76
5	3.30	3.27	3.24	3.21	3.19	3.17	3.16	3.14	3.12	3.10
6	2.94	2.90	2.87	2.84	2.82	2.80	2.78	2.76	2.74	2.72
7	2.70	2.67	2.63	2.59	2.58	2.56	2.54	2.51	2.49	2.47
8	2.54	2.50	2.46	2.42	2.40	2.38	2.36	2.34	2.32	2.29
9	2.42	2.38	2.34	2.30	2.28	2.25	2.23	2.21	2.18	2.16
10	2.32	2.28	2.24	2.20	2.18	2.16	2.13	2.11	2.08	2.06
11	2.25	2.21	2.17	2.12	2.10	2.08	2.05	2.03	2.00	1.97
12	2.19	2.15	2.10	2.06	2.04	2.01	1.99	1.96	1.93	1.90
13	2.14	2.10	2.05	2.01	1.98	1.96	1.93	1.90	1.88	1.85
14	2.10	2.05	2.01	1.96	1.94	1.91	1.89	1.86	1.83	1.80
15	2.06	2.02	1.97	1.92	1.90	1.87	1.85	1.82	1.79	1.76
16	2.03	1.99	1.94	1.89	1.87	1.84	1.81	1.78	1.75	1.72
17	2.00	1.96	1.91	1.86	1.84	1.81	1.78	1.75	1.72	1.69
18	1.98	1.93	1.89	1.84	1.81	1.78	1.75	1.72	1.69	1.66
19	1.96	1.91	1.86	1.81	1.79	1.76	1.73	1.70	1.67	1.63
20	1.94	1.89	1.84	1.79	1.77	1.74	1.71	1.68	1.64	1.61
21	1.92	1.87	1.83	1.78	1.75	1.72	1.69	1.66	1.62	1.59
22	1.90	1.86	1.81	1.76	1.73	1.70	1.67	1.64	1.60	1.57
23	1.89	1.84	1.80	1.74	1.72	1.69	1.66	1.62	1.59	1.55
24	1.88	1.83	1.78	1.73	1.70	1.67	1.64	1.61	1.57	1.53
25	1.87	1.82	1.77	1.72	1.69	1.66	1.63	1.59	1.56	1.52
26	1.86	1.81	1.76	1.71	1.68	1.65	1.61	1.58	1.54	1.50
27	1.85	1.80	1.75	1.70	1.67	1.64	1.60	1.57	1.53	1.49
28	1.84	1.79	1.74	1.69	1.66	1.63	1.59	1.56	1.52	1.48
29	1.83	1.78	1.73	1.68	1.65	1.62	1.58	1.55	1.51	1.47
30	1.82	1.77	1.72	1.67	1.64	1.61	1.57	1.54	1.50	1.46
40	1.76	1.71	1.66	1.61	1.57	1.54	1.51	1.47	1.42	1.38
60	1.71	1.66	1.60	1.54	1.51	1.48	1.44	1.40	1.35	1.29
120	1.65	1.60	1.55	1.48	1.45	1.41	1.37	1.32	1.26	1.19
∞	1.60	1.55	1.49	1.42	1.38	1.34	1.30	1.24	1.17	1.00

TABLE B.4 (*continued*)   Critical Values of the *F* Distribution for $\alpha = 0.05$

Denominator Degrees of Freedom (k_2)	Numerator Degrees of Freedom (k_1)								
	1	**2**	**3**	**4**	**5**	**6**	**7**	**8**	**9**
1	161.4	199.5	215.7	224.6	230.2	234.0	236.8	238.9	240.5
2	18.51	19.00	19.16	19.25	19.30	19.33	19.35	19.37	19.38
3	10.13	9.55	9.28	9.12	9.01	8.94	8.89	8.85	8.81
4	7.71	6.94	6.59	6.39	6.26	6.16	6.09	6.04	6.00
5	6.61	5.79	5.41	5.19	5.05	4.95	4.88	4.82	4.77
6	5.99	5.14	4.76	4.53	4.39	4.28	4.21	4.15	4.10
7	5.59	4.74	4.35	4.12	3.97	3.87	3.79	3.73	3.68
8	5.32	4.46	4.07	3.84	3.69	3.58	3.50	3.44	3.39
9	5.12	4.26	3.86	3.63	3.48	3.37	3.29	3.23	3.18
10	4.96	4.10	3.71	3.48	3.33	3.22	3.14	3.07	3.02
11	4.84	3.98	3.59	3.36	3.20	3.09	3.01	2.95	2.90
12	4.75	3.89	3.49	3.26	3.11	3.00	2.91	2.85	2.80
13	4.67	3.81	3.41	3.18	3.03	2.92	2.83	2.77	2.71
14	4.60	3.74	3.34	3.11	2.96	2.85	2.76	2.70	2.65
15	4.54	3.68	3.29	3.06	2.90	2.79	2.71	2.64	2.59
16	4.49	3.63	3.24	3.01	2.85	2.74	2.66	2.59	2.54
17	4.45	3.59	3.20	2.96	2.81	2.70	2.61	2.55	2.49
18	4.41	3.55	3.16	2.93	2.77	2.66	2.58	2.51	2.46
19	4.38	3.52	3.13	2.90	2.74	2.63	2.54	2.48	2.42
20	4.35	3.49	3.10	2.87	2.71	2.60	2.51	2.45	2.39
21	4.32	3.47	3.07	2.84	2.68	2.57	2.49	2.42	2.37
22	4.30	3.44	3.05	2.82	2.66	2.55	2.46	2.40	2.34
23	4.28	3.42	3.03	2.80	2.64	2.53	2.44	2.37	2.32
24	4.26	3.40	3.01	2.78	2.62	2.51	2.42	2.36	2.30
25	4.24	3.39	2.99	2.76	2.60	2.49	2.40	2.34	2.28
26	4.23	3.37	2.98	2.74	2.59	2.47	2.39	2.32	2.27
27	4.21	3.35	2.96	2.73	2.57	2.46	2.37	2.31	2.25
28	4.20	3.34	2.95	2.71	2.56	2.45	2.36	2.29	2.24
29	4.18	3.33	2.93	2.70	2.55	2.43	2.35	2.28	2.22
30	4.17	3.32	2.92	2.69	2.53	2.42	2.33	2.27	2.21
40	4.08	3.23	2.84	2.61	2.45	2.34	2.25	2.18	2.12
60	4.00	3.15	2.76	2.53	2.37	2.25	2.17	2.10	2.04
120	3.92	3.07	2.68	2.45	2.29	2.17	2.09	2.02	1.96
∞	3.84	3.00	2.60	2.37	2.21	2.10	2.01	1.94	1.88

Denominator Degrees of Freedom (k_2)	Numerator Degrees of Freedom (k_1)									
	10	12	15	20	24	30	40	60	120	∞
1	241.9	243.9	245.9	248.0	249.1	250.1	251.1	252.2	253.3	254.3
2	19.40	19.41	19.43	19.45	19.45	19.46	19.47	19.48	19.49	19.50
3	8.79	8.74	8.70	8.66	8.64	8.62	8.59	8.57	8.55	8.53
4	5.96	5.91	5.86	5.80	5.77	5.75	5.72	5.69	5.66	5.63
5	4.74	4.68	4.62	4.56	4.53	4.50	4.46	4.43	4.40	4.36
6	4.06	4.00	3.94	3.87	3.84	3.81	3.77	3.74	3.70	3.67
7	3.64	3.57	3.51	3.44	3.41	3.38	3.34	3.30	3.27	3.23
8	3.35	3.28	3.22	3.15	3.12	3.08	3.04	3.01	2.97	2.93
9	3.14	3.07	3.01	2.94	2.90	2.86	2.83	2.79	2.75	2.71
10	2.98	2.91	2.85	2.77	2.74	2.70	2.66	2.62	2.58	2.54
11	2.85	2.79	2.72	2.65	2.61	2.57	2.53	2.49	2.45	2.40
12	2.75	2.69	2.62	2.54	2.51	2.47	2.43	2.38	2.34	2.30
13	2.67	2.60	2.53	2.46	2.42	2.38	2.34	2.30	2.25	2.21
14	2.60	2.53	2.46	2.39	2.35	2.31	2.27	2.22	2.18	2.13
15	2.54	2.48	2.40	2.33	2.29	2.25	2.20	2.16	2.11	2.07
16	2.49	2.42	2.35	2.28	2.24	2.19	2.15	2.11	2.06	2.01
17	2.45	2.38	2.31	2.23	2.19	2.15	2.10	2.06	2.01	1.96
18	2.41	2.34	2.27	2.19	2.15	2.11	2.06	2.02	1.97	1.92
19	2.38	2.31	2.23	2.16	2.11	2.07	2.03	1.98	1.93	1.88
20	2.35	2.28	2.20	2.12	2.08	2.04	1.99	1.95	1.90	1.84
21	2.32	2.25	2.18	2.10	2.05	2.01	1.96	1.92	1.87	1.81
22	2.30	2.23	2.15	2.07	2.03	1.98	1.94	1.89	1.84	1.78
23	2.27	2.20	2.13	2.05	2.01	1.96	1.91	1.86	1.81	1.76
24	2.25	2.18	2.11	2.03	1.98	1.94	1.89	1.84	1.79	1.73
25	2.24	2.16	2.09	2.01	1.96	1.92	1.87	1.82	1.77	1.71
26	2.22	2.15	2.07	1.99	1.95	1.90	1.85	1.80	1.75	1.69
27	2.20	2.13	2.06	1.97	1.93	1.88	1.84	1.79	1.73	1.67
28	2.19	2.12	2.04	1.96	1.91	1.87	1.82	1.77	1.71	1.65
29	2.18	2.10	2.03	1.94	1.90	1.85	1.81	1.75	1.70	1.64
30	2.16	2.09	2.01	1.93	1.89	1.84	1.79	1.74	1.68	1.62
40	2.08	2.00	1.92	1.84	1.79	1.74	1.69	1.64	1.58	1.51
60	1.99	1.92	1.84	1.75	1.70	1.65	1.59	1.53	1.47	1.39
120	1.91	1.83	1.75	1.66	1.61	1.55	1.50	1.43	1.35	1.25
∞	1.83	1.75	1.67	1.57	1.52	1.46	1.39	1.32	1.22	1.00

TABLE B.4 (*continued*)   Critical Values of the *F* Distribution for α = 0.025

Denominator Degrees of Freedom (k_2)	Numerator Degrees of Freedom (k_1)								
	1	**2**	**3**	**4**	**5**	**6**	**7**	**8**	**9**
1	647.8	799.5	864.2	899.6	921.8	937.1	948.2	956.7	963.3
2	38.51	39.00	39.17	39.25	39.30	39.33	39.36	39.37	39.39
3	17.44	16.04	15.44	15.10	14.88	14.73	14.62	14.54	14.47
4	12.22	10.65	9.98	9.60	9.36	9.20	9.07	8.98	8.90
5	10.01	8.43	7.76	7.39	7.15	6.98	6.85	6.76	6.68
6	8.81	7.26	6.60	6.23	5.99	5.82	5.70	5.60	5.52
7	8.07	6.54	5.89	5.52	5.29	5.12	4.99	4.90	4.82
8	7.57	6.06	5.42	5.05	4.82	4.65	4.53	4.43	4.36
9	7.21	5.71	5.08	4.72	4.48	4.32	4.20	4.10	4.03
10	6.94	5.46	4.83	4.47	4.24	4.07	3.95	3.85	3.78
11	6.72	5.26	4.63	4.28	4.04	3.88	3.76	3.66	3.59
12	6.55	5.10	4.47	4.12	3.89	3.73	3.61	3.51	3.44
13	6.41	4.97	4.35	4.00	3.77	3.60	3.48	3.39	3.31
14	6.30	4.86	4.24	3.89	3.66	3.50	3.38	3.29	3.21
15	6.20	4.77	4.15	3.80	3.58	3.41	3.29	3.20	3.12
16	6.12	4.69	4.08	3.73	3.50	3.34	3.22	3.12	3.05
17	6.04	4.62	4.01	3.66	3.44	3.28	3.16	3.06	2.98
18	5.98	4.56	3.95	3.61	3.38	3.22	3.10	3.01	2.93
19	5.92	4.51	3.90	3.56	3.33	3.17	3.05	2.96	2.88
20	5.87	4.46	3.86	3.51	3.29	3.13	3.01	2.91	2.84
21	5.83	4.42	3.82	3.48	3.25	3.09	2.97	2.87	2.80
22	5.79	4.38	3.78	3.44	3.22	3.05	2.93	2.84	2.76
23	5.75	4.35	3.75	3.41	3.18	3.02	2.90	2.81	2.73
24	5.72	4.32	3.72	3.38	3.15	2.99	2.87	2.78	2.70
25	5.69	4.29	3.69	3.35	3.13	2.97	2.85	2.75	2.68
26	5.66	4.27	3.67	3.33	3.10	2.94	2.82	2.73	2.65
27	5.63	4.24	3.65	3.31	3.08	2.92	2.80	2.71	2.63
28	5.61	4.22	3.63	3.29	3.06	2.90	2.78	2.69	2.61
29	5.59	4.20	3.61	3.27	3.04	2.88	2.76	2.67	2.59
30	5.57	4.18	3.59	3.25	3.03	2.87	2.75	2.65	2.57
40	5.42	4.05	3.46	3.13	2.90	2.74	2.62	2.53	2.45
60	5.29	3.93	3.34	3.01	2.79	2.63	2.51	2.41	2.33
120	5.15	3.80	3.23	2.89	2.67	2.52	2.39	2.30	2.22
∞	5.02	3.69	3.12	2.79	2.57	2.41	2.29	2.19	2.11

Denominator Degrees of Freedom (k_2)	Numerator Degrees of Freedom (k_1)									
	10	12	15	20	24	30	40	60	120	∞
1	968.6	976.7	984.9	993.1	997.2	1001	1006	1010	1014	1018
2	39.40	39.41	39.43	39.45	39.46	39.46	39.47	39.48	39.49	39.50
3	14.42	14.34	14.25	14.17	14.12	14.08	14.04	13.99	13.95	13.90
4	8.84	8.75	8.66	8.56	8.51	8.46	8.41	8.36	8.31	8.26
5	6.62	6.52	6.43	6.33	6.28	6.23	6.18	6.12	6.07	6.02
6	5.46	5.37	5.27	5.17	5.12	5.07	5.01	4.96	4.90	4.85
7	4.76	4.67	4.57	4.47	4.42	4.36	4.31	4.25	4.20	4.14
8	4.30	4.20	4.10	4.00	3.95	3.89	3.84	3.78	3.73	3.67
9	3.96	3.87	3.77	3.67	3.61	3.56	3.51	3.45	3.39	3.33
10	3.72	3.62	3.52	3.42	3.37	3.31	3.26	3.20	3.14	3.08
11	3.53	3.43	3.33	3.23	3.17	3.12	3.06	3.00	2.94	2.88
12	3.37	3.28	3.18	3.07	3.02	2.96	2.91	2.85	2.79	2.72
13	3.25	3.15	3.05	2.95	2.89	2.84	2.78	2.72	2.66	2.60
14	3.15	3.05	2.95	2.84	2.79	2.73	2.67	2.61	2.55	2.49
15	3.06	2.96	2.86	2.76	2.70	2.64	2.59	2.52	2.46	2.40
16	2.99	2.89	2.79	2.68	2.63	2.57	2.51	2.45	2.38	2.32
17	2.92	2.82	2.72	2.62	2.56	2.50	2.44	2.38	2.32	2.25
18	2.87	2.77	2.67	2.56	2.50	2.44	2.38	2.32	2.26	2.19
19	2.82	2.72	2.62	2.51	2.45	2.39	2.33	2.27	2.20	2.13
20	2.77	2.68	2.57	2.46	2.41	2.35	2.29	2.22	2.16	2.09
21	2.73	2.64	2.53	2.42	2.37	2.31	2.25	2.18	2.11	2.04
22	2.70	2.60	2.50	2.39	2.33	2.27	2.21	2.14	2.08	2.00
23	2.67	2.57	2.47	2.36	2.30	2.24	2.18	2.11	2.04	1.97
24	2.64	2.54	2.44	2.33	2.27	2.21	2.15	2.08	2.01	1.94
25	2.61	2.51	2.41	2.30	2.24	2.18	2.12	2.05	1.98	1.91
26	2.59	2.49	2.39	2.28	2.22	2.16	2.09	2.03	1.95	1.88
27	2.57	2.47	2.36	2.25	2.19	2.13	2.07	2.00	1.93	1.85
28	2.55	2.45	2.34	2.23	2.17	2.11	2.05	1.98	1.91	1.83
29	2.53	2.43	2.32	2.21	2.15	2.09	2.03	1.96	1.89	1.81
30	2.51	2.41	2.31	2.20	2.14	2.07	2.01	1.94	1.87	1.79
40	2.39	2.29	2.18	2.07	2.01	1.94	1.88	1.80	1.72	1.64
60	2.27	2.17	2.06	1.94	1.88	1.82	1.74	1.67	1.58	1.48
120	2.16	2.05	1.94	1.82	1.76	1.69	1.61	1.53	1.43	1.31
∞	2.05	1.94	1.83	1.71	1.64	1.57	1.48	1.39	1.27	1.00

TABLE B.4 (*concluded*)   Critical Values of the *F* Distribution for $\alpha = 0.01$

Denominator Degrees of Freedom (k_2)	Numerator Degrees of Freedom (k_1)								
	1	2	3	4	5	6	7	8	9
1	4,052	4,999.5	5,403	5,625	5,764	5,859	5,928	5,982	6,022
2	98.50	99.00	99.17	99.25	99.30	99.33	99.36	99.37	99.39
3	34.12	30.82	29.46	28.71	28.24	27.91	27.67	27.49	27.35
4	21.20	18.00	16.69	15.98	15.52	15.21	14.98	14.80	14.66
5	16.26	13.27	12.06	11.39	10.97	10.67	10.46	10.29	10.16
6	13.75	10.92	9.78	9.15	8.75	8.47	8.26	8.10	7.98
7	12.25	9.55	8.45	7.85	7.46	7.19	6.99	6.84	6.72
8	11.26	8.65	7.59	7.01	6.63	6.37	6.18	6.03	5.91
9	10.56	8.02	6.99	6.42	6.06	5.80	5.61	5.47	5.35
10	10.04	7.56	6.55	5.99	5.64	5.39	5.20	5.06	4.94
11	9.65	7.21	6.22	5.67	5.32	5.07	4.89	4.74	4.63
12	9.33	6.93	5.95	5.41	5.06	4.82	4.64	4.50	4.39
13	9.07	6.70	5.74	5.21	4.86	4.62	4.44	4.30	4.19
14	8.86	6.51	5.56	5.04	4.69	4.46	4.28	4.14	4.03
15	8.68	6.36	5.42	4.89	4.56	4.32	4.14	4.00	3.89
16	8.53	6.23	5.29	4.77	4.44	4.20	4.03	3.89	3.78
17	8.40	6.11	5.18	4.67	4.34	4.10	3.93	3.79	3.68
18	8.29	6.01	5.09	4.58	4.25	4.01	3.84	3.71	3.60
19	8.18	5.93	5.01	4.50	4.17	3.94	3.77	3.63	3.52
20	8.10	5.85	4.94	4.43	4.10	3.87	3.70	3.56	3.46
21	8.02	5.78	4.87	4.37	4.04	3.81	3.64	3.51	3.40
22	7.95	5.72	4.82	4.31	3.99	3.76	3.59	3.45	3.35
23	7.88	5.66	4.76	4.26	3.94	3.71	3.54	3.41	3.30
24	7.82	5.61	4.72	4.22	3.90	3.67	3.50	3.36	3.26
25	7.77	5.57	4.68	4.18	3.85	3.63	3.46	3.32	3.22
26	7.72	5.53	4.64	4.14	3.82	3.59	3.42	3.29	3.18
27	7.68	5.49	4.60	4.11	3.78	3.56	3.39	3.26	3.15
28	7.64	5.45	4.57	4.07	3.75	3.53	3.36	3.23	3.12
29	7.60	5.42	4.54	4.04	3.73	3.50	3.33	3.20	3.09
30	7.56	5.39	4.51	4.02	3.70	3.47	3.30	3.17	3.07
40	7.31	5.18	4.31	3.83	3.51	3.29	3.12	2.99	2.89
60	7.08	4.98	4.13	3.65	3.34	3.12	2.95	2.82	2.72
120	6.85	4.79	3.95	3.48	3.17	2.96	2.79	2.66	2.56
∞	6.63	4.61	3.78	3.32	3.02	2.80	2.64	2.51	2.41

Denominator Degrees of Freedom (k_2)	Numerator of Degrees of Freedom (k_1)									
	10	12	15	20	24	30	40	60	120	∞
1	6,056	6,106	6,157	6,209	6,235	6,261	6,287	6,313	6,339	6,366
2	99.40	99.42	99.43	99.45	99.46	99.47	99.47	99.48	99.49	99.50
3	27.23	27.05	26.87	26.69	26.60	26.50	26.41	26.32	26.22	26.13
4	14.55	14.37	14.20	14.02	13.93	13.84	13.75	13.65	13.56	13.46
5	10.05	9.89	9.72	9.55	9.47	9.38	9.29	9.20	9.11	9.02
6	7.87	7.72	7.56	7.40	7.31	7.23	7.14	7.06	6.97	6.88
7	6.62	6.47	6.31	6.16	6.07	5.99	5.91	5.82	5.74	5.65
8	5.81	5.67	5.52	5.36	5.28	5.20	5.12	5.03	4.95	4.86
9	5.26	5.11	4.96	4.81	4.73	4.65	4.57	4.48	4.40	4.31
10	4.85	4.71	4.56	4.41	4.33	4.25	4.17	4.08	4.00	3.91
11	4.54	4.40	4.25	4.10	4.02	3.94	3.86	3.78	3.69	3.60
12	4.30	4.16	4.01	3.86	3.78	3.70	3.62	3.54	3.45	3.36
13	4.10	3.96	3.82	3.66	3.59	3.51	3.43	3.34	3.25	3.17
14	3.94	3.80	3.66	3.51	3.43	3.35	3.27	3.18	3.09	3.00
15	3.80	3.67	3.52	3.37	3.29	3.21	3.13	3.05	2.96	2.87
16	3.69	3.55	3.41	3.26	3.18	3.10	3.02	2.93	2.84	2.75
17	3.59	3.46	3.31	3.16	3.08	3.00	2.92	2.83	2.75	2.65
18	3.51	3.37	3.23	3.08	3.00	2.92	2.84	2.75	2.66	2.57
19	3.43	3.30	3.15	3.00	2.92	2.84	2.76	2.67	2.58	2.49
20	3.37	3.23	3.09	2.94	2.86	2.78	2.69	2.61	2.52	2.42
21	3.31	3.17	3.03	2.88	2.80	2.72	2.64	2.55	2.46	2.36
22	3.26	3.12	2.98	2.83	2.75	2.67	2.58	2.50	2.40	2.31
23	3.21	3.07	2.93	2.78	2.70	2.62	2.54	2.45	2.35	2.26
24	3.17	3.03	2.89	2.74	2.66	2.58	2.49	2.40	2.31	2.21
25	3.13	2.99	2.85	2.70	2.62	2.54	2.45	2.36	2.27	2.17
26	3.09	2.96	2.81	2.66	2.58	2.50	2.42	2.33	2.23	2.13
27	3.06	2.93	2.78	2.63	2.55	2.47	2.38	2.29	2.20	2.10
28	3.03	2.90	2.75	2.60	2.52	2.44	2.35	2.26	2.17	2.06
29	3.00	2.87	2.73	2.57	2.49	2.41	2.33	2.23	2.14	2.03
30	2.98	2.84	2.70	2.55	2.47	2.39	2.30	2.21	2.11	2.01
40	2.80	2.66	2.52	2.37	2.29	2.20	2.11	2.02	1.92	1.80
60	2.63	2.50	2.35	2.20	2.12	2.03	1.94	1.84	1.73	1.60
120	2.47	2.34	2.19	2.03	1.95	1.86	1.76	1.66	1.53	1.38
∞	2.32	2.18	2.04	1.88	1.79	1.70	1.59	1.47	1.32	1.00

Source: M. Merrington and C. M. Thompson, "Tables of Percentage Points of the Inverted Beta (F)-Distribution," *Biometrika* 33 (1943) pp. 73–88. Reproduced by permission of the *Biometrika* Trustees.

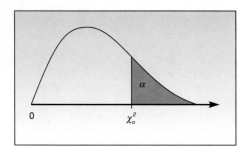

TABLE B.5   Critical Values of the Chi-Square Distribution

Degrees of Freedom	$X^2.995$	$X^2.990$	$X^2.975$	$X^2.950$	$X^2.900$
1	0.0000393	0.0001571	0.0009821	0.0039321	0.0157908
2.	0.0100251	0.0201007	0.0506356	0.102587	0.210720
3	0.0717212	0.114832	0.215795	0.351846	0.584375
4	0.206990	0.297110	0.484419	0.710721	1.063623
5	0.411740	0.554300	0.831211	1.145476	1.61031
6	0.675727	0.872085	1.237347	1.63539	2.20413
7	0.989265	1.239043	1.68987	2.16735	2.83311
8	1.344419	1.646482	2.17973	2.73264	3.48954
9	1.734926	2.087912	2.70039	3.32511	4.16816
10	2.15585	2.55821	3.24697	3.94030	4.86518
11	2.60321	3.05347	3.81575	4.57481	5.57779
12	3.07382	3.57056	4.40379	5.22603	6.30380
13	3.56503	4.10691	5.00874	5.89186	7.04150
14	4.07468	4.66043	5.62872	6.57063	7.78953
15	4.60094	5.22935	6.26214	7.26094	8.54675
16	5.14224	5.81221	6.90766	7.96164	9.31223
17	5.69724	6.40776	7.56418	8.67176	10.0852
18	6.26481	7.01491	8.23075	9.39046	10.8649
19	6.84398	7.63273	8.90655	10.1170	11.6509
20	7.43386	8.26040	9.59083	10.8508	12.4426
21	8.03366	8.89720	10.28293	11.5913	13.2396
22	8.64272	9.54249	10.9823	12.3380	14.0415
23	9.26042	10.19567	11.6885	13.0905	14.8479
24	9.88623	10.8564	12.4011	13.8484	15.6587
25	10.5197	11.5240	13.1197	14.6114	16.4734
26	11.1603	12.1981	13.8439	15.3791	17.2919
27	11.8076	12.8786	14.5733	16.1513	18.1138
28	12.4613	13.5648	15.3079	16.9279	18.9392
29	13.1211	14.2565	16.0471	17.7083	19.7677
30	13.7867	14.9535	16.7908	18.4926	20.5992
40	20.7065	22.1643	24.4331	26.5093	29.0505
50	27.9907	29.7067	32.3574	34.7642	37.6886
60	35.5346	37.4848	40.4817	43.1879	46.4589
70	43.2752	45.4418	48.7576	51.7393	55.3290
80	51.1720	53.5400	57.1532	60.3915	64.2778
90	59.1963	61.7541	65.6466	69.1260	73.2912
100	67.3276	70.0648	74.2219	77.9295	82.3581

Degrees of Freedom	$X^2.100$	$X^2.050$	$X^2.025$	$X^2.010$	$X^2.005$
1	2.70554	3.84146	5.02389	6.63490	7.87944
2	4.60517	5.99147	7.37776	9.21034	10.5966
3	6.25139	7.81473	9.34840	11.3449	12.8381
4	7.77944	9.48773	11.1433	13.2767	14.8602
5	9.23635	11.0705	12.8325	15.0863	16.7496
6	10.6446	12.5916	14.4494	16.8119	18.5476
7	12.0170	14.0671	16.0128	18.4753	20.2777
8	13.3616	15.5073	17.5346	20.0902	21.9550
9	14.6837	16.9190	19.0228	21.6660	23.5893
10	15.9871	18.3070	20.4831	23.2093	25.1882
11	17.2750	19.6751	21.9200	24.7250	26.7569
12	18.5494	21.0261	23.3367	26.2170	28.2995
13	19.8119	22.3621	24.7356	27.6883	29.8194
14	21.0642	23.6848	26.1190	29.1413	31.3193
15	22.3072	24.9958	27.4884	30.5779	32.8013
16	23.5418	26.2962	28.8454	31.9999	34.2672
17	24.7690	27.5871	30.1910	33.4087	35.7185
18	25.9894	28.8693	31.5264	34.8053	37.1564
19	27.2036	30.1435	32.8523	36.1908	38.5822
20	28.4120	31.4104	34.1696	37.5662	39.9968
21	29.6151	32.6705	35.4789	38.9321	41.4010
22	30.8133	33.9244	36.7807	40.2894	42.7956
23	32.0069	35.1725	38.0757	41.6384	44.1813
24	33.1963	36.4151	39.3641	42.9798	45.5585
25	34.3816	37.6525	40.6465	44.3141	46.9278
26	35.5631	38.8852	41.9232	45.6417	48.2899
27	36.7412	40.1133	43.1944	46.9630	49.6449
28	37.9159	41.3372	44.4607	48.2782	50.9933
29	39.0875	42.5569	45.7222	49.5879	52.3356
30	40.2560	43.7729	46.9792	50.8922	53.6720
40	51.8050	55.7585	59.3417	63.6907	66.7659
50	63.1671	67.5048	71.4202	76.1539	79.4900
60	74.3970	79.0819	83.2976	88.3794	91.9517
70	85.5271	90.5312	95.0231	100.425	104.215
80	96.5782	101.879	106.629	112.329	116.321
90	107.565	113.145	118.136	124.116	128.299
100	118.498	124.342	129.561	135.807	140.169

Source: C. M. Thompson, "Tables of the Percentage Points of the X^2-Distribution," *Biometrika* 32 (1941), pp. 188–89. Reproduced by permission of the *Biometrika* Trustees.

Control Chart Constants

n	For Estimating Sigma		For X̄ Chart		For X̄ Chart (Standard Given)	For R Chart		For R Chart (Standard Given)		For s Chart (Standard Given)			
	c_4	d_2	A_2	A_3	A	D_3	D_4	D_1	D_2	B_3	B_4	B_5	B_6
2	0.7979	1.128	1.880	2.659	2.121	0	3.267	0	3.686	0	3.267	0	2.606
3	0.8862	1.693	1.023	1.954	1.732	0	2.575	0	4.358	0	2.568	0	2.276
4	0.9213	2.059	0.729	1.628	1.500	0	2.282	0	4.698	0	2.266	0	2.088
5	0.9400	2.326	0.577	1.427	1.342	0	2.115	0	4.918	0	2.089	0	1.964
6	0.9515	2.534	0.483	1.287	1.225	0	2.004	0	5.078	0.030	1.970	0.029	1.874
7	0.9594	2.704	0.419	1.182	1.134	0.076	1.924	0.205	5.203	0.118	1.882	0.113	1.806
8	0.9650	2.847	0.373	1.099	1.061	0.136	1.864	0.387	5.307	0.185	1.815	0.179	1.751
9	0.9693	2.970	0.337	1.032	1.000	0.184	1.816	0.546	5.394	0.239	1.761	0.232	1.707
10	0.9727	3.078	0.308	0.975	0.949	0.223	1.777	0.687	5.469	0.284	1.716	0.276	1.669
15	0.9823	3.472	0.223	0.789	0.775	0.348	1.652	1.207	5.737	0.428	1.572	0.421	1.544
20	0.9869	3.735	0.180	0.680	0.671	0.414	1.586	1.548	5.922	0.510	1.490	0.504	1.470
25	0.9896	3.931	0.153	0.606	0.600	0.459	1.541	1.804	6.058	0.565	1.435	0.559	1.420

Source: Reprinted by permission from T. P. Ryan, *Statistical Methods for Quality Improvement* (New York: John Wiley and Sons, 1989).

Random Numbers

1559	9068	9290	8303	8508	8954	1051	6677	6415	0342
5550	6245	7313	0117	7652	5069	6354	7668	1096	5780
4735	6214	8037	1385	1882	0828	2957	0530	9210	0177
5333	1313	3063	1134	8676	6241	9960	5304	1582	6198
8495	2956	1121	8484	2920	7934	0670	5263	0968	0069
1947	3353	1197	7363	9003	9313	3434	4261	0066	2714
4785	6325	1868	5020	9100	0823	7379	7391	1250	5501
9972	9163	5833	0100	5758	3696	6496	6297	5653	7782
0472	4629	2007	4464	3312	8728	1193	2497	4219	5339
4727	6994	1175	5622	2341	8562	5192	1471	7206	2027
3658	3226	5981	9025	1080	1437	6721	7331	0792	5383
6906	9758	0244	0259	4609	1269	5957	7556	1975	7898
3793	6916	0132	8873	8987	4975	4814	2098	6683	0901
3376	5966	1614	4025	0721	1537	6695	6090	8083	5450
6126	0224	7169	3596	1593	5097	7286	2686	1796	1150
0466	7566	1320	8777	8470	5448	9575	4669	1402	3905
9908	9832	8185	8835	0384	3699	1272	1181	8627	1968
7594	3636	1224	6808	1184	3404	6752	4391	2016	6167
5715	9301	5847	3524	0077	6674	8061	5438	6508	9673
7932	4739	4567	6797	4540	8488	3639	9777	1621	7244
6311	2025	5250	6099	6718	7539	9681	3204	9637	1091
0476	1624	3470	1600	0675	3261	7749	4195	2660	2150
5317	3903	6098	9438	3482	5505	5167	9993	8191	8488
7474	8876	1918	9828	2061	6664	0391	9170	2776	4025
7460	6800	1987	2758	0737	6880	1500	5763	2061	9373
1002	1494	9972	3877	6104	4006	0477	0669	8557	0513
5449	6891	9047	6297	1075	7762	8091	7153	8881	3367
9453	0809	7151	9982	0411	1120	6129	5090	2053	7570
0471	2725	7588	6573	0546	0110	6132	1224	3124	6563
5469	2668	1996	2249	3857	6637	8010	1701	3141	6147
2782	9603	1877	4159	9809	2570	4544	0544	2660	6737
3129	7217	5020	3788	0853	9465	2186	3945	1696	2286
7092	9885	3714	8557	7804	9524	6228	7774	6674	2775
9566	0501	8352	1062	0634	2401	0379	1697	7153	6208
5863	7000	1714	9276	7218	6922	1032	4838	1954	1680
5881	9151	2321	3147	6755	2510	5759	6947	7102	0097
6416	9939	9569	0439	1705	4680	9881	7071	9596	8758
9568	3012	6316	9065	0710	2158	1639	9149	4848	8634
0452	9538	5730	1893	1186	9245	6558	9562	8534	9321
8762	5920	8989	4777	2169	7073	7082	9495	1594	8600
0194	0270	7601	0342	3897	4133	7650	9228	5558	3597
3306	5478	2797	1605	4996	0023	9780	9429	3937	7573
7198	3079	2171	6972	0928	6599	9328	0597	5948	5753
8350	4846	1309	0612	4584	4988	4642	4430	9481	9048
7449	4279	4224	1018	2496	2091	9750	6086	1955	9860
6126	5399	0852	5491	6557	4946	9918	1541	7894	1843
1851	7940	9908	3860	1536	8011	4314	7269	7047	0382
7698	4218	2726	5130	3132	1722	8592	9662	4795	7718
0810	0118	4979	0458	1059	5739	7919	4557	0245	4861
6647	7149	1409	6809	3313	0082	9024	7477	7320	5822
3867	7111	5549	9439	3427	9793	3071	6651	4267	8099
1172	7278	7527	2492	6211	9457	5120	4903	1023	5745
6701	1668	5067	0413	7961	7825	9261	8572	0634	1140
8244	0620	8736	2649	1429	6253	4181	8120	6500	8127
8009	4031	7884	2215	2382	1931	1252	8088	2490	9122
1947	8315	9755	7187	4074	4743	6669	6060	2319	0635
9562	4821	8050	0106	2782	4665	9436	4973	4879	8900
0729	9026	9631	8096	8906	5713	3212	8854	3435	4206
6904	2569	3251	0079	8838	8738	8503	6333	0952	1641

C Answers to Most Odd-Numbered Problems

CHAPTER 1

1. Age = quantitative, ratio; Sex = qualitative, nominal; Number of people = quantitative, ratio; Use electric heat? = qualitative; Number of appliances = quantitative, ratio; Thermostat setting = quantitative, interval; Average number of heating hours = quantitative, ratio; Average number of heating days = quantitative, ratio; Household income = quantitative, ratio; Average monthly electric bill = quantitative, ratio; Ranking of electric company = quantitative, ordinal.

3. In terms of their relative meaning, scales may be ordered from weakest to strongest: Nominal < Ordinal < Interval < Ratio.

5. Ordinal.

7. Neighborhood = a sample from town, but definitely not random. One hundred people from all neighborhoods, randomly chosen using a frame = a random sample.

9. A sample drawn from a population in a random way, so that every member of the population has an equal chance of being selected for the sample.

11. Ordinal (because the color of the belt denotes ability level).

15. Heart and blood vessel diseases are the leading cause of death, followed by all forms of cancer as the second leading cause (only somewhat over one-half as prevalent as heart diseases); other causes are far behind.

17. There seems to be some improvement in increase to higher percentage of on-time arrivals from first to third quarter.

19. Departures look better in terms of higher percentage on-time. The reason may lie in the fact that the airports are in control of departure times (usually airplanes wait till they are allowed to take off), while airports cannot control the arrival time (they can only, possibly, shorten circling time overhead once the plane has arrived at the destination).

21. There are over twice as many cellular phones per 100 people in the U.S. as there are in Europe or Japan (both of which have roughly the same number per 100 people). The percentage of households with cable TV in the U.S. is roughly 4 times that in Europe or Japan; for computers per 100 people, the U.S. figure is roughly 3 times that for Europe and 3.5 times that for Japan.

25. Increase in the male-parent-only group from 1980 to 1992.

27. Seventh is centered, the rest spread out; first five and ninth are right-skewed.

31. Based on the displayed information, it seems the least

bad tax increase (except for the Reagan tax cut of 1981).

33. LQ = 121, Median = 128, UQ = 133.5, 10th percentile = 114.8, 15th percentile = 118.1, 65th percentile = 131.1, IQR = 12.5

35. President: LQ = 51, UQ = 60.5, Median = 55, 10th percentile = 42.8, 90th percentile = 65, President Clinton is at 17th percentile of the presidents' age distribution. Vice president: LQ = 49.5, UQ = 60.0, Median = 53, 10th percentile = 41.3, 90th percentile = 68.7, Vice President Gore is at 15th percentile of the vice presidents' age distribution.

37. Median = 70.

39. The greatest rank is 225. It goes to the place with the smallest population (Saint Pierre and Miquelon, population 7,000). By definition, the median is the population with rank (n + 1)/2 = 226/2 = 113. That median population is 5,924,000 and belongs to Papua New Guinea. The interesting thing about this problem was that having had the ranks for a large data set made it very easy to find the median.

41. A graphical display of a data distribution consisting of five measures: the median, the quartiles, the smallest and largest observations.

43. Set 1 is by far smallest in values, tightly distributed, right-skewed, no outliers. Set 2 larger than 1 in values, roughly similar to Set 3, but with a lower center. Set 3 has a low-valued outlier. Set 4 is left-skewed, largest in

values, and with a larger spread than the other three sets.

47. Median = average of 2 center observations = average of 2.45 and 3.09 = 2.77.

49. Median age of ships is 14 years. The distribution is right-skewed. There is one clear outlier: 33, the average age of U.S. government-owned ships.

51. Presidents: age distribution is slightly to right of VP distribution, symmetric, no outliers. Vice Presidents: age distribution slightly to left of Presidents, right-skewed, no outliers.

55. Mean = 64.

57. British pound: Mean = 0.65279, Median = 0.65150; French franc: Mean = 5.4733, Median = 5.4823; German mark: Mean = 1.6168, Median = 1.6240; Italian lira: Mean = 1489.1, Median = 1491.4; Singapore dollar: Mean = 1.6516, Median = 1.6556.

59. The mean for 1982 is 14.04375, for 1987 it is 14.06875, a very small increase.

61. Presidents: Mean age = 54.59, Median is 55. Vice presidents: Mean = 54.90, Median is 53. Mean being greater than median for VPs indicates a right-skewed distribution (as we concluded in Problem 51; although the boxplot for presidents did not reveal a left-skew).

63. Mean is largest, then comes median, and then mode.

67. Population: denominator in formula is the population size, N. Sample: denominator in formula is n − 1.

69. Largest standard deviation is Italian Lira: 22.04, smallest is British Pound: 0.0076. But the reason is the units of measurement of the data. To truly measure volatility we need a standardized measure.

71. Variance = 22.75, std. dev = 4.77, mean = 13.25. Fifty-eight percent lie within one standard deviation of the mean, and 100 percent within 2 standard deviations. Empirical rule doesn't seem to apply (probably due to the very 'discrete' nature of the data, and small size).

75. Mean = 74.7, Variance = 194.4, Standard deviation = 13.94.

77. Mean = 74.36, Standard deviation = 19.51.

79. Median = 51, LQ = 30.5, UQ = 194.25, IQR = 163.75, 45th percentile = 42.2.

81. Clearly, in the three states and the United States as a whole, single-parent families led by a mother have lowest incomes, then single-father families, and then two-parent households.

85. The moving-average evens out the random variability from year to year, allowing us to see a smoother trend. The graph seems to show a warming trend.

87. Calories: Mean = 136.7, Std. dev = 51.8; Sodium: Mean = 555.8, Std. dev = 208.2.

89. Team is categorical, nominal scale. Points is quantitative, ratio scale (40 points is twice as good as 20).

91. Boxplot shows a tight, almost symmetric distribution around a center of roughly 17 percent, with two outliers: a low-valued

outlier (6.3%)—Alaska, and a high-valued outlier (24.6%)—Florida. (The two outliers make sense, based on states' known demographics!)

CHAPTER 2

1. Objective and subjective.
3. The set of all possible elementary outcomes of an experiment.
5. Union = either a girl or a baby of over 5 pounds (or both); intersection = a baby girl of over 5 pounds.
7. Union = acceptance to grad school or getting a job offer; intersection = getting accepted to grad school and getting a job offer.
9. Union = purchase stock or bonds (or both); intersection = purchase both stock and bonds.
11. Probability = 1 in $2 \times 2 \times 2 \times 200 = 1/1600$.
13. Probability = 1/1000.
15. Probability = 3/16.
17. Probability = 0.80 (or 0.90, or thereabouts).
19. Quite good.
21. *a.* Mutually exclusive. *b.* 0.035. *c)* 0.985.
23. P(TUR) = P(T) + P(R) − P (T ∩ R) = .25 + .34 − .10 = 0.49.
25. P(CUW) = 0.45 + 0.82 − 0.38 = 0.89.
27. *a.* 4/7. *b.* 5/7. *c.* 3/7 + 4/7 − 1/7 = 6/7. *d.* Yes.
29. (0.8)(0.4) = 0.32.
31. (0.25)(0.10) = 0.025; 2.5 percent.
33. (0.94)(0.65) = 0.611; 61.1 percent.
35. *a.* 3/10. *b.* 3/10. *c.* 2/3. *d.* 1/3.
37. P (Die | Age) is not necessarily increasing with Age (according to article).

39. 1 − (2/3) (2/3) (2/3) (5/6) (5/6) (19/20) (19/20) = 0.814.
41. 20 flights: 0.00004; 50 flights: 0.0001; 100 flights: 0.0002.
43. P(D and S) = 18/100 = 0.18; P(D) = 50/100 = 0.5; P(S) = 18/30 = 0.6; now: (0.6)(0.5) = 0.3, which does not equal 0.18. Therefore, not independent.
45. (112/246) (119/246) = 0.22, not equal to 34/246 = 0.138; therefore not independent.
47. P(at least one disease) = 1 − P (not Malaria) P (not Shis.) P (not Sleep.sckness) = 1 − (1 − 110/2100) (1 − 200/600)(1 − 25/50000) = 0.369.
49. P(work) = 1 − P (all fail) = 1 − (0.04)(0.09)(0.20) = 0.99928.
51. P(All 4 from top quartile) = (1/4) (1/4) (1/4) (1/4) = 0.0039; P (at least one) = 1 − (3/4) (3/4) (3/4) (3/4) = 0.6836.
53. 0.1602.
55. 0.85.
57. Surprisingly still 1/3.
59. 0.0248.
61. P (at least 1) = 1 − P (all fail) = 1 − (0.7)(0.8)(0.85) = 0.524.
63. P (at least one) = 1 − P (none) = 1 − (0.08)(0.08)(0.08) = 0.999488.
65. To minimize all probabilities of red lights: From office go East, East again, then S, S again, E, S, E, and E.
67. Roughly: 2.1/250 = 0.0084 Limitations: frequency of illness may change with time.
69. 1/2 (The two could be both above the median, prob =

1/4; both below the median, prob = 1/4; one above and one below: 1/4 + 1/4 = 1/2.)
71. Yes (see introduction to Chapter 3 for solution).
72. 39/51 = 0.765; 11/51 = 0.216; 9/51 = 0.176.

CHAPTER 3

1. The probabilities sum to 1.00. P (more than 2 calls) = P(3) + P(4) + P(5) = 0.1 + 0.1 + 0.1 = 0.3.
3. Continuous.
5. Continuous.
7. P(x) add to 1.00. P(At least 400,000) = sum of probs from 400,000 and onward: 0.3 + 0.2 = 0.5.
9. Probs add to 1.00. Prob of at least 12: 0.2 + 0.15 + 0.1 +0.05 = 0.5.
11.

x	P(x)
2	1/48
3	2/48
4	3/48
5	4/48
6	5/48
7	6/48
8	6/48
9	6/48
10	5/48
11	4/48
12	3/48
13	2/48
14	1/48

13.

x	P(x)
0	0.05
1	0.15
2	0.22
3	0.38
4	0.20

P (from 1 to 3 colonies) = 0.15 + 0.22 + 0.38 = 0.75.

15.

x	$P(x)$
2	1/13
3	1/13
4	1/13
5	1/13
6	1/13
7	1/13
8	1/13
9	1/13
10	4/13
11	1/13

17. Mean = 3.19; variance = 2.214.

19. Mean = 2.2; Standard deviation = 1.4.

21. Std. dev. = 2.415.

23. Mean = 11.75 Std. dev. = 1.512.

25. Mean = 8.

27. Mean = 4.44.

29. Mean = 2E(x) = 2(5) = 10; Std. dev. = $\sqrt{(4 \times 1)} = 2$.

31. If the probability of scoring a point is the same for all trials and trials are independent.

33. cdf (7) = 0.9992.

35. cdf (7) − cdf (2) = 0.9832 − 0.0498 = 0.9334.

37. 0.033.

39. No! Members of a family are related: independence of trials assumption is violated.

41. 0.0037; mean = 4.5, std. dev. = 1.77.

43. 1 − cdf(2) = 0.0381.

45. $p = 0.13$, $n = 8$, P (X ≥ 3) = 1 − cdf (2) = 1 − 0.9257 = 0.0743. Not too likely, *but:* TV show crowd is definitely not a random sample of the population of women.

47. $p = 0.03$, $n = 15$, P (X ≥ 2) = 1 − cdf (1) = 1 − 0.927 = 0.073.

49. Probability = Area under density between two points.

51. *a.* Density is a right-triangle, base = 0 to 2, height at 2 is 1.00.
b. Total probability = Total area under density = base x height/2 = 2 × 1/2 = 1.00.
c. Prob between 0 and 1 is area of smaller triangle from 0 to 1 = 1 × (1/2)/2 = 1/4.

53. 2(1/7) = 2/7.

55. 5.5/7 = 0.7857.

57. *a.* sum of probabilities = 1.00, so yes.
b. P(X ≥ 6) = 0.01 + 0.01 + 0.01 + 0.01 = 0.04.
c. E(X) = 0.73 visits per year.

59. 1.00 (so that base x height/2 will equal 1.00).

61. 1 − cdf(1) = 1 − 0.3758 = 0.6242.

63. *a.* Time in mall is random variable; Average time = mean of r.v.
b. Mall of America: 87/3 = 29 dollars per hour. Y = 29X Average US Mall: Y = 49X.
c. No. See (b) above.

65. 1 − cdf(1) = 1 − 0.1951 = 0.8049.

67. 1.5 percent.

69. Continuous.

CHAPTER 4

1. 0.6826, 0.95, 0.9802, 0.9951, 0.9974.

3. 0.1828.

5. 0.0215.

7. 0.9901.

9. A number close to zero.

11. 0.9544.

13. Not likely; probability = 0.00003.

15. $z = 0.19$.

17. $z = 0.05$.

19. $z = -1.96$ and $z = 1.96$.

21. $z = -1.645$ and $z = 1.645$.

23. $z = -0.6745$.

25. 0.927.

27. 0.003.

29. 0.86, 0.21, 0.63.

31. 0.0931.

33. 0.121, 0.0037, less than 0.00003—do not believe claim.

35. 0.0808.

37. By independence: P (both exceed 160) = P(X > 160) P(Y > 160) = (0.1587)(0.0918) = 0.0146 P (at least one exceeds 160) = 0.1587 + 0.0918 − 0.0146 = 0.2359.

39. 126.6.

41. 1500 ± 1.645(500) = 677.5 to 2322.5 mhz.

43. 13.94 and 50.06.

45. 832.6, 435.5.

47. 18,130.2, 35,887.8.

49. 31 ± 1.96(8) = 15.32 to 46.68.

51. 273.73.

53. Assume independence of arriving parties. Using binomial, prob. = 0.2375. Using normal approximation: prob. = 0.2321.

55. 0.9922.

57. Assume independence of students. 0.0738.

59. For men in North America, Oceania, and Europe, MINITAB does not give a straight-line plot. Data not close to normal.

61. MINITAB plot of nscores for all men does look close to a straight-line. Data set is close to normal.

65. 0.2743.

67. 58 + 2.33(18) = 99.94, close to 100 hours.

69. 0.9332.

71. 7.02 to 8.98.

73. 1,556; 1,373 to 3,323.

75. 6015.6.

77. 46.93 minutes.

79. 0.50 (Prob. that normal r.v. will fall below its mean).

CHAPTER 5

3. For all presidential races, poll results seem to be randomly distributed around the actual election results.

5. Polls provide statistical estimates of population parameters.

7. The probability distribution of a sample statistic.

9. Mean = 0.5.

11. For large n, normal. Mean = population mean; standard deviation = population standard deviation/square root of n.

13. 50/10 = 5.

15. 0.923.

17. 0.0912.

19. A range of values with a given confidence of containing an actual population parameter.

23. [9.05, 9.55] percent.

25. [32.92, 35.48] hours.

27. [5.747, 6.253] microgram/deciliter.

29. With strangers: $87 \pm 1.645(7/9.48) = [85.79, 88.21]$ mmHg.
Alone: $84 \pm 1.645 (8/9.48) = [82.61, 85.39]$ mmHg.
With family: $83 \pm 1.645(7/9.48) = [81.79, 84.21]$ mmHg.

31. MTB > describe c1

	N	MEAN	MEDIAN	TRMEAN	STDEV	SE MEAN
C1	36	8.361	8.000	8.156	4.189	0.698
	MIN	MAX	Q1	Q3		
C1	2.000	20.000	5.000	11.000		

MTB > zinterval 95 4.189 c1
THE ASSUMED SIGMA = 4.19

	N	MEAN	STDEV	SE MEAN	95.0 PERCENT C.I.
C1	36	8.361	4.189	0.698	(6.991, 9.731)

33. Four times as large (because the width goes down with the square root of the sample size).

34. $10(1.645/1.96) = 8.39$.

35. Estimating the mean when the population standard deviation is not known (and sample size is small).

37. It approaches the standard normal distribution, Z.

39. $5.91 \pm 2.228 (2.7/3.31) = [4.095, 7.723]$ days.

41. $3.6 \pm 3.499 (1.51/2.83) = [1.73, 5.47]$ meters.

43. $346.5 \pm 2.262 (53.44/3.16) = [308.27, 384.73]$ mg/d1.

45. (Using correct data mean and standard deviation: 35.6, 9.42, respectively): $35.6 \pm 1.833 (9.42/3.16) = [30.14, 41.06]$.

47. $10.0 \pm 2.110 (8.02/4.24) = [6.01, 13.99]$ years.

49. For large samples, normal with mean p and standard deviation $\sqrt{pq/n}$.

51. [0.158, 0.309].

53. $\hat{p} = 0.41$, C.I. is $0.41 \pm 1.96 \sqrt{0.41 \times 0.59 /776} = [0.375, 0.445]$.

55. $\hat{p} = 0.22$, C.I. is $0.22 \pm 1.96 \sqrt{0.22 \times 0.78/776} = [0.191, 0.249]$.

57. [0.7857, 0.8143].

59. 0.5.

61. 0.019.

63. $\hat{p} = 121/220 = 0.55$; $0.55 \pm 1.645 \sqrt{0.55 \times 0.45/220} = [0.495, 0.605]$.

65. $\hat{p} = 72/400 = 0.18$; $0.18 \pm 1.96 \sqrt{.18 \times .82/400} = [0.142, 0.218]$.

67. $n = (1.96)^2 (.5)(.5)/0.01 = 96.04$, so sample $n = 97$ people.

69. $n = 125$.

71. $n = (1.645)^2 (0.7)(0.3)/(0.05)^2 = 227.3$, so sample 228.

73. 865 accounts

75. $\hat{p} = 1900/3349 = 0.567$. C.I. is $0.567 \pm 1.645 \sqrt{.567 \times .433/3349} = [0.5529, 0.5811]$.

77. $n = 1769$. Increase risk of sex crime: $0.63 \pm 1.96 \sqrt{.63 \times .37/1769} = [0.608, 0.652]$
Interpret as unfair: $0.85 \pm 1.96 \sqrt{0.85 \times 0.15/1769}$

= [0.833, 0.867].
Parents should consider:
0.88 ± 1.96
$\sqrt{0.88 \times 0.12/1769}$
= [0.865, 0.895].

79. $\hat{p} = 48/240 = 0.2$. C.I. is $0.2 \pm 2.576 \sqrt{0.2 \times 0.8/240}$ = [0.133, 0.267].

81. 95 percent C.I. $212 \pm 2.160 (38/3.74) = [190.1, 233.9]$ calories.
98 percent C.I. $212 \pm 2.650 (38/3.74) = [185.1, 238.9]$ calories.

83. $\hat{p} = 34/70 = 0.486$. C.I. is 0.486 ± 1.282
$\sqrt{0.486 \times 0.514/70}$
= [0.409, 0.563].

85. $n = 271$.

87. $n = 38.4$, so sample 39.

89. Using Z: [52.86, 54.34], a slightly narrower interval.

91. $136,080 \pm 1.282 (29,100/11.31) = [132,783, 139,377]$

93. $1.75 \pm 1.96\ 0.5/6.16 = [1.59, 1.91]$ ppm.
Data do not look like a random sample randomly scattered in space—seem more concentrated, and on a straight line.

95. Generalizing the procedure of problem 94: The probability that median falls between lowest and highest data points is $1 - P$ (all data are on same size of median). Whatever first data point does, to miss the median the remaining $n - 1$ points must all be on same side of median. This has probability $1/2$ to the power $n - 1$. So confidence level is $1 - (1/2)^{n-1}$. Limitation: such an interval will get very wide as sample size gets large. Other methods give more reasonable intervals.

CHAPTER 6

1. Random selection is needed for correct inference and fairness of the trial.

3. Lower conviction rate of masters of the mint.

5. ± 1.645.

7. To determine whether its value is likely or not likely under the assumption that the null hypothesis is true.

9. Type I: Reject a true null hypothesis; probability. α Type II: Fail to reject a false null hypothesis; probability β.

11. Because there is a known, small probability of doing so when the null hypothesis is actually true.

13. A test of whether a population parameter is equal to a given number. Rejection may occur on either of two tails of the probability distribution.

15. p-value is the actual significance level of the test.

17. $z = -2.575$; reject null hypothesis; p-value close to 0.01; average mpg probably different from 31.5.

19. $z = -19.72$; strongly reject the null hypothesis; changes in service are necessary as average miles traveled per day are probably less than 5. p-value is very small (close to zero).

21. $z = -1.456$; cannot reject the null hypothesis; p-value = 0.1454.

23. $z = -9.5$; p-value is very small (close to zero). Reject the null hypothesis that the average vehicle occupancy rate is 6.8 seats and conclude it is probably smaller.

25. p-value = 0.26; cannot reject the null hypothesis.

27. $z = (1.5 - 3.8)/\ 1/31.6 = -72.73$. Very strong rejection of the null hypothesis; p-value extremely small (near zero).

29. These are p-values from clinical trials, reported as proof that treatment with Vantin is effective.

31. Yes, there is cause for action; null hypothesis of stated cholesterol level can be strongly rejected in favor of more cholesterol present. Test statistic $z = 9.6$; p-value is very small.

33. $z = (27 - 30)/\ 6/10 = -5$. Test statistic is on left tail of the distribution, in rejection region. p-value = 0.0000003. Competitor may reject Toshiba's claim.

35. $z = (65 - 50)/\ 20/10 = 7.5$, in rejection region. p-value very small. Reject the null hypothesis. Probably more than 50 percent fat may be lost, on average.

37. $z = -5.39$ (Sample mean = 82.6 and std. dev = 8.68); p-value is very small; average purity is probably less than 90 percent.

39. Sample mean = 3.5 days and std. dev = 2.13; test statistic $t = 0.25$; p-value = 0.81. Can't reject the null hypothesis.

41. $t = (9560 - 10000)/\ 2380/2.24 = -0.4$. Cannot reject the null hypothesis. There is no statistical evidence based on the five measurements that average (population) concentration of autocrine for this patient is below 10,000.

43. $t = (115-60)/\ 32/3.87 = 6.66$; statistic $t(14)$ is far in rejection region; p-value is very small (less than 0.005);

it is likely that average radiation exceeded danger levels.

45. $t(2) = (735.3 - 710)/19.47/1.73 = 2.22$. This value is less than the critical point for $t(2)$ for $\alpha = 0.05$ in a right-tailed test, which is 2.92. There is no statistical evidence that air traffic is more dangerous. Major limitation: extremely small sample (n = 3), and the assumption of a random sample (which it is not).

47. Sample mean = 152, std. dev = 90.6; test statistic value $t = 1.49$; p-value = 0.15. No evidence to reject the null hypothesis.

49. $z = (0.51 - 0.50)/\sqrt{0.5 \times 0.5/774} = 0.556$, in nonrejection region at any conventional level of significance, so we cannot reject the null hypothesis that his approval rating was about 50 percent; p-value = 0.578. Reported margin of error of 4 percent is too high: the actual margin of error is closer to 3.5 percent.

51. $p_0 = 5/1000 = 0.005$, $\hat{p} = 78/10000 = 0.0078$; $z = (0.0078 - 0.005)/\sqrt{0.005 \times 0.995/10000} = 3.97$. Reject the null hypothesis. p-value = 0.00003. There is strong statistical evidence that the success rate in implanting the human gene in pigs is higher than 5 in 1000 attempts.

53. $z = (0 - 0.1)/\sqrt{.1 \times .9/300} = -5.77$; p-value is very small (close to zero). Reject the null hypothesis that nudists' infection rate for Lyme disease is the same as for non-nudists in favor of the alternative that nudists are less susceptible to the disease.

55. $\hat{p} = 601/835 = 0.72$; $z = (0.72 - 0.68)/\sqrt{0.68 \times 0.32/835} = 2.46$. Reject the null hypothesis that the percentage is 68 percent in favor of the alternative that the percentage is higher. p-value = 0.0069.

57. $z = (0.02 - 0.05)/\sqrt{0.05 \times 0.95/155} = -1.71$. Some (but not very strong) evidence against the null hypothesis, in favor of the alternative that the proportion is smaller than 0.05. p-value = 0.0436.

59. $\hat{p} = 20/220 = 0.0909$ $z = (0.0909 - .010)/\sqrt{.1 \times .9/220} = -0.45$. No evidence against the null hypothesis; do not reject claim. p-value = 0.3264.

61. $z - (.5 - .666)/\sqrt{.66 \times .33/450} = -7.5$; strongly reject the null hypothesis; proportion is probably less than 2/3. p-value is very small.

63. No! The small p-value indicates strong *statistical* significance, not necessarily practical significance. The actual population mean may be close to the null-hypothesized one and still give us a small p-value (when n is large and/or std. dev. is small).

65. Neither is true.

67. Yes. Since the sample mean is smaller than the null-hypothesized population mean, the test statistic will fall to the left of zero—in the nonrejection region (regardless of n and s).

69. 0.909.

71. Power at 60 is 0.984.

73. Power is a measure of our ability to reject the null hypothesis when it should be rejected.

75. Low power means lower ability to reject a wrong null hypothesis.

77. $\hat{p} = 210/849 = 0.247$; $z = (0.247 - 0.5)/\sqrt{.5 \times .5/849} = -14.7$; p-value is extremely small. Strong statistical evidence that the proportion of injured roller bladers is less than 50 percent.

79. $z = -4.86$. Reject the null hypothesis; p-value is very small. Power at 4 percent is 0.521.

81. $z = (20.5 - 25)/8/10.24 = -5.76$. Reject the null hypothesis; there is strong evidence for the competitor's claim of lower average number of recharges; p-value is very small.

83. There is strong statistical evidence against the null hypothesis in the first case, weak in the second. The first result is highly statistically significant, the second barely significant.

85. The probability that I committed a Type II error is identically zero because I rejected the null hypothesis and a Type II error occurs only when I fail to reject the null (and the null is false).

87. $z = (.87 - .6)/\sqrt{.6 \times .4/62} = 4.35$ Strong statistical evidence that the success rate is over 60 percent. p-value is very small (less than 0.00003).

89. $z = 12.3$; highly significant; p-value very small.

CHAPTER 7

1. Reduction of extraneous errors.

3. $t(11) = -2.03$, cannot reject the null hypothesis using $\alpha = 0.05$.

5. Differences, D-I: -1, -2, -1, 1, $-.5$, 1, -1, 2, -2, -6, -2, -4. Mean $= -1.29$, sd $= 2.2$. 95% CI: -1.29 ± 2.201 $(2.2/3.46) = [-2.69, 0.106]$. Note that the interval contains zero, which agrees with the nonrejection of the null hypothesis of zero difference in Problem 3.

7. The eleven differences between Category 1 and Category 2 (ignoring the "Total" columns) sites are: -24, -5, -22, 2, -7, 12, -11, -14, -12, -1, -4. Mean $= -7.818$, std. dev $= 10.43$. A 99% CI for the average difference is: $-7.818 \pm 3.169 (10.43/3.316) = [-17.8, 2.15]$. The interval contains zero.

9. The population average difference is 3.8 years. Depending on your actual sample, you may or may not reject the null hypothesis (most samples should lead to rejection).

11. Mean difference $= -0.1$, std. dev $= 4.58$. 90% CI for difference is; $-.1 \pm 1.833$ $(4.58/3.16) = [-2.75, 2.55]$. Interval contains zero.

13. Methods of designing meaningful statistical studies that reduce the influence of outside factors.

15. The paired-sample tests are more precise, as they reduce extraneous variation.

17. Positive verbal test for difference between FH+ and FH− men. For Male interaction: $z = (17.9 - 18.55)/$
$$\sqrt{4.17^2/20 + 5.48^2/20} =$$
-0.65, in nonrejection region for any level of significance. For Female interaction: $z = (18.6 - 19.25)/$
$$\sqrt{4.02^2/20 + 5.99^2/20} =$$
-0.48. Again, not significant at any conventional α.

19. $z = (5.8 - 3.5)/$
$$\sqrt{1.2^2/294 + 1.4^2/294} =$$
21.4; extremely significant, p-value very close to zero. Based on these numbers, sunscreen is strongly proven effective in reducing average number of keratoses.

21. Using a computer: $t = 0.89$, df $= 10$, p-value $= 0.39$. No evidence to reject the null hypothesis that both fertilizers are equally effective.

23. $z = 3.3$, p-value $= 0.001$. Reject the null hypothesis of equal population averages. On average, Bel Air properties are probably higher priced.

25. Using a computer: $[-1.93, 4.5]$ interval contains zero, agreeing with finding of no significant difference in Problem 21.

27. 90 percent confidence interval (using z): $56,210 \pm 28,023 = [28,186, 84,233]$ dollars difference. In agreement with rejection of null hypothesis of zero average difference in Problem 23.

29. Using a computer: Mean for early period is 116.5 inches, mean for the later period is 132.6 inches. $t = -0.97$ (for early period − later period). So even in the samples, there is more average snow in the 1984-5 to present period. Of course this leads to nonrejection of the null hypothesis in a right-tailed test for more snow in the earlier period. There is no evidence for the old Vermonters' belief.

31. $z = 1.71$ (for: New engine − Old engine), p-value $= 0.04$. There is some statistical evidence that the new engine is more economical.

33. $z = 1.273$, p-value $= 0.102$. Cannot reject the null hypothesis. No evidence of increased customer satisfaction.

35. $z = 2.29$, p-value $= 0.011$. There is evidence of an increase of over 2 pounds on average.

37. $t(26) = 2.479$, p-value $= 0.02$. Reject the null hypothesis of equal average computing time.

39. $t(29) = 1.08$, cannot reject the null hypothesis of no increase in average occupancy rate.

41. $z = 4.68$, p-value is very small. Reject the null hypothesis. There is strong evidence that the percentage of popular recommendations is higher in small towns.

43. $z = .68 - .26/$
$$\sqrt{.45 \times .55 \times 1/25 + 1/29}$$
$= 3.09$; p-value $= 0.001$ (one-tailed). There is statistical evidence for the benefit of broccoli. Limitation: normal distribution assumption with small data sets. (Note also that 26 percent of 29 rats is

7.54, apparently the data in article are rounded up. Ignore this fact when computing combined sample proportion.)

45. p(traditional) = 17/358 = 0.047; p(Medit.) = 4/412 = 0.0097; difference between the two minus one percent is: 0.0273. z = 0.0273/

$$\sqrt{(.047 \times .953/358 + .0097 \times .9903/412)} = 2.24; p\text{-value}$$

(one-tail) = 0.0125. There is statistical evidence that the Mediterranean diet reduces the percentage of second heart attacks by at least one percent. A 95 percent confidence interval for the difference in proportions of second heart attack between the two diets: 0.047 − 0.0097 ± 1.96

$$\sqrt{(.047 \times .953/358 + .0097 \times .9903/412)} = [0.013, 0.061].$$

47. p(Cheetahs) = 67/100000 = 0.00067; p (Humans) = 0.10. z = 0.1/

$$\sqrt{(.1 \times .9) \times (2/100000)} = 74.5.$$ Overwhelming evidence (the sample sizes are huge). CI(90%) for the difference: 0.1 ± 1.645

$$\sqrt{(.1 \times .9/100000 + .00067 \times .99933/100000)} = [0.098, 0.101].$$

49. Combined proportion = 266/377 = 0.71. 1/268 + 1/109 = 0.013 p (main) = 190/268 = 0.71; p (client) = 76/109 = 0.70 z = 0.01/

$$\sqrt{0.71 \times 0.29 \times 0.013} = 0.23,$$ not significant at any α.

51. Combined proportion = 0.467 Diff. between props = 0.014 z = 0.014/

$$\sqrt{0.467 \times 0.533 \times 0.013} = 0.25.$$ Not significant at any α.

53. 975 − 899/

$$\sqrt{300^2/3500 + 280^2/2800} = 10.37;$$ very small p-value. Difference is highly statistically significant.

55. z = 1.99; p-value (one-tail) = 0.0228; some statistical evidence of increase in proportion.

57. z = 0.11/

$$\sqrt{0.67 \times 0.33 \times 0.0018} = 5.46.$$ p-value is very small. There is strong statistical evidence for a difference.

59. z = 0.3/

$$\sqrt{1.2^2/10000 + 1.1^2/6000} = 16.14.$$ Highly significant.

61. z = 1.09, result is not statistically significant.

63. There is a statistically significant difference between the two group means. With 95 percent confidence, that average difference is anywhere from 8 to 11.61.

65. p(share) = 0.41 p(never) = 0.288 p(combined) = 0.34 z = 0.125/

$$\sqrt{0.34 \times 0.66 \times (1/46 + 1/59)} = 1.34,$$ not significant. p-value (two-tailed) = 0.18. There is no statistical evidence for unethical physician behavior (but note the small samples and their limitation, especially use of the normal approximation).

CHAPTER 8

1. Under null hypothesis: the four means are equal. Under alternative: two equal and different from other equal two; or one different from three equal; all four different.

3. No control over the overall level of significance of the separate tests.

5. $F(3,176)$ = 12.53 > 3.78, which is the critical point from the F table for α = 0.01. So reject the null hypothesis of equal population means; p-value < 0.01.

7. $F(2,1675)$ = 380.04 > 4.61, the critical point for the F distribution at the 0.01 level. In fact, the statistic value is much larger than the critical point. The p-value is extremely small, so there is overwhelming statistical evidence that averge respect levels are not equal for all three categories.

9. The average error is zero. Same for average treatment deviation.

11. SSTO = SSTR + SSE.

13. MSTR and MSE are sample statistics given to natural variation. Whatever is obtained from samples has to be determined statistically significant—that is, different beyond natural variation and implying a population difference before conclusions can be made.

15. When data vary little around their means but the means themselves have a large variance then the population means are very likely to be different from one another.

17. No; because MSTR and MSE are obtained from SSTR and SSE by division by different numbers (the appropriate degrees of freedom).

19. $F(7, 792)$ = 108.5; p-value is extremely small. There is strong evidence that not all eight brands have equal average consumer rating.

21. MS (Factor) = 12,467; MSE = 7,075; $F(3, 44) = 1.76$. This result is not significant (p-value is greater than 0.10); no evidence that there are differences in average rent in the four cities.

23. Using a computer: $F(2, 63) = 69.42$; p-value extremely small; strong evidence that not all methods are equally effective. A plot of separate confidence intervals for the population means shows great separation between method B and the other two, both of which are also separated from each other, but less so.

25. Using a computer: $F(3, 96) = 66.41$; p-value is extremely small; there is strong evidence that not all four are equally liked, on average. Confidence interval for Jingle population mean is largest, separated widely from Celebrity, which follows with some overlap with Prod. Package (so, possibly, no difference between these two), and last, widely separated from rest, is Cast.

27. Using a computer: $F(2, 37) = 13.79$; p-value is very small. Reject null hypothesis that all three accomodation types are equally liked. Highest confidence interval is for House, widely separated from other two. Next is Apartment, which overlaps with Dorms, which follows it in third place. Probably no significant difference between latter two.

29. Blocking reduces the variation due to outside factors and allows for more precise testing of treatment effects.

31. Chi square (1 df) = 84.841; p-value is extremely small; reject null hypothesis of equal fatality rate. Seatbelt/airbag combination is probably effective in reducing fatalities.

33. When testing all three groups using a computer, Chi square (8 df) = 536.8, extremely significant. Strong evidence that the proportions of skiers in each self-assessed ability level are not equal across the three classes: Alpine, Snowboard, Both. When testing only the non-intersection groups (Alpine vs. Snowboard) the results are still extremely significant: Chi square (4 df) = 157.8.

35. Number of people with children: 905 + 67 = 972, Without children: 358 + 208 = 566. Forty percent of 972 is 389, sixty percent is 583. Twenty-five percent of 566 is 142, seventy-five percent is 424. Enter these data into the computer. We get: Chi square (1 df) = 35.28, highly statistically significant. There is strong evidence that people without children ski more days in this resort. (Note that percentages given in the survey are rounded.)

37. Enter the numbers: 90, 33, 10 for Rogaine and 71, 26, 21 for Placebo. The computer-derived statistic value is Chi square (2 df) = 6.101. The p-value is slightly less than 0.05 (0.0473, using the MINITAB cdf command). There is slight statistical evidence of a difference in the proportions of the three effects for Rogaine as compared with the Placebo. The Upjohn Company would do well to test the side effects further.

39. Doing the analysis the same way as in Problem 37, we get: Chi square (2 df) = 3.007; not significant (p-value > 0.10). No statistical difference between the Placebo and Rogaine groups with respect to proportions of patients showing symptoms in these three areas.

41. Assumptions: independent random sampling; sample sizes large enough for normality leading to a Chi square distribution under the equal-proportion null hypothesis assumption; large-enough expected cell counts—this assumption is violated in the overall test in Problem 40, where two cells have expected counts less than 5 (cells should be combined). Rogaine should probably be tested further, since there are statistically significant differences between the Rogaine and Placebo groups in a number of symptom areas.

43. ANOVA: When both sample means are identical, the MSTR will be zero, and therefore so will the value of the F statistic, making it impossible to reject the null hypothesis of equal population means. Using a two-sample t-test: We get the same result since the numerator of the t-statistic is the difference between the two sample means. This gives a t statistic value of zero, again making it

impossible to reject the equal-population-mean null hypothesis.

45. Summer: 1000 by car, 500 by air. Winter: 1000 by air, 500 by car (by multiplying 1500 by the given percentages). Chi square (1 df) = 333.3; extremely significant difference.

47. Summer: 753 grads, 747 not; Winter: 1103 grads, 397 not (percentage rounded in problem). Chi square (1 df) = 173; result is extremely statistically significant. Strong evidence that proportion of college graduates is higher for the winter versus summer visitors. For white collar workers: Chi square (1 df) = 433.9, same conclusion as for college grads.

49. SS (Factor) = 2245; MS (Factor) = 748.3; MSE = 9.2; $F(3, 63)$ = 81.34; highly significant, p-value close to zero.

51. The CI for the Nothing group is high and separated from the other two. The Placebo and Drug CIs overlap (Drug slightly below Placebo).

53. ANOVA tests for differences among means (quantitative variable); Chi square analysis tests for differences among proportions (qualitative variable).

55. Yes, results are highly significant, p-value is close to zero.

CHAPTER 9

1. A formula and set of assumptions that capture the essential elements in a real-world problem.

3. Straight-line relationship between X and Y; Y is random but X not; Errors are uncorrelated, normally distributed, with mean zero and constant variance.

5. A point exactly on the regression line (intercept + slope times x).

7. The error accounts for the effects of extraneous factors; it is used in testing the model.

9. Slope = 8; Y–intercept = 5.

11. Least squares finds the line that best fits the data in the sense that it minimizes the sum of the squared errors that result. This tends to be a line that lies right in the "middle" of the data points moving up or down with X.

13. Y = 1.61 + 1.01X.

15. $t(13)$ = 16.01; p-value is extremely small; there is strong evidence for the existence of a linear relationship between the two rates.

17. $t(5)$ = 14.77; very small p-value; strong evidence for a linear regression relationship.

19. $t(10)$ = 1.554; p-value > 0.10; no statistical evidence for a linear relationship.

21. Regression of Instructor rating on Course Difficulty: $t(42)$ = −1.48, p-value = 0.146. Not statistically significant. Surprisingly, there is not enough evidence that perceived difficulty lowers an instructor's student evaluations (these are real data). For Course Interest as a predictor of Instructor Rating: $t(42)$ = 10.49, p-value is extremely small. It seems that Course Interest is highly statistically significant as a predictor of the instructor's rating. (Again, an interesting finding.)

23. *a.* No. *b.* No. *c.* No. The regression explains almost nothing.

25. r-square = 96.3%, a very high value.

27. r-square = 1 − 2850/7965 = 0.642, or 64.2%.

29. $F(1, 7)$ = 177.25 (the square of the t statistic, which is 13.31). The p-value is very small, same as for the t statistic.

31. r-square = 72.4%. This means that 72.4 percent of the variation in overall instructor ratings can be explained by how students rate interest of the course.

33. This means that all the data are right on the regression line with no variation at all. This will not likely occur in most real-life situations, where there is at least some variation due to outside factors.

35. ANOVA in regression gives us another test for the existence of a linear relationship. This is especially useful in multiple regression, where there are several explanatory variables.

37. *a.* no trend with time; *b.* increasing trend with time (and increasing variance); time should be included in the model, variance should be stabilized (both require advanced work).

39. No model inadequacy.

41. No apparent model inadequacy.

43. No model inadequacy. A plot of the residuals against the X variable looks random;

normal-probability plot looks close to a straight line.

45. [5,854.4, 7,248.3] dollars.

47. For example, for Fed. Funds Rate = 6.5 percent, the predicted Prime Rate is 7.87 percent.

49. Predicted number of suitors when symmetry is 85 is: 3.63. A 95 percent prediction interval is from 3.00 to 4.25.

51. The prediction interval is for a single point, so it has more variability—it is a wider interval—than a confidence interval for the average value for the given X.

53. Prediction is 4.1707, 95 percent P.I. is [3.2929, 5.0485]. Note: We would never use this prediction since we have determined early on that there is no evidence of a linear relationship between the two variables.

55. Correlation: Both variables, assumed normally distributed, are symmetrically considered as random variables in their own right. Regression: There is a distinction between X, not considered random, and Y—the variable of interest to us—which is considered a random variable.

57. $r = 0.981$.

59. Yes! A correlation computed from the data is a sample estimate of the population correlation coefficient. We use the sample correlation in testing whether or not the population correlation is zero (testing for significance). For example, if our sample correlation of 0.51 was obtained from a sample of size $n = 3$, our t-statistic value would be: 0.59, which is not significant at any usual α level. On the other hand, if our sample correlation of 0.04 was obtained from a sample of 5,000 then our t statistic (close to Z now) would be equal to 2.83, which is definitely significant (two-tailed p-value = 0.0046).

63. From the figure, it looks like there is a correlation. To draw a conclusion, one would have to compute the sample correlation from the actual data, and test for significance. Correlation does not imply causality. Both variables (hair loss, heart trouble) could be caused jointly by a third variable—such as hormone levels.

65. Significance of variables, at $\alpha = 0.05$, (in table order): No, no, no, yes (two-tailed p-value = 0.0026), yes (two-tailed p-value = 0.0012).

67. Residuals seem normally distributed; a plot of the residuals versus X could possibly indicate some curvature. More data may be necessary.

69. Predicted Melanoma is 9.286 percent; the 95 percent prediction interval is from 6.75 to 11.82 percent. Other factors, such as the percentage of people with light skin in the area, use of sunscreens, percentage of sunny days, may all affect the dependent variable.

71. Regression does not imply causality (other, detailed analysis may be required).

73. Relation with Jogging is significant, with highest r-square; relation with Biking is also statistically significant; lower r-square. The relation with mountain climbing was not found statistically significant.

75. Predicted value = 214.6.

CHAPTER 10

3. Random sampling.

5. Process passes all tests, seems in control.

15. Process in control.

17. Process in control.

19. Process may be out of control (one point on UCL).

21. Process in control.

25. Out of control: sample point valued 12 is well above the UCL.

27. Last group of four points have mean below the LCL, process mean out of control.

29. Process seems in control.

Index

A

Alpha, 251
Alternative hypothesis, 249
Analysis of variance (ANOVA), 358–84
ANOVA table, 374–78
  in regression, 435–37
Antarctica ozone hole, 57
Arbuthnott, John, 107

B

Bar graph, 12
Bayes' Theorem, 97–100
  extended, 99
Beta, 251, 297
Binomial distribution, 130–37
Binomial random variable, 130–37
Blackcap, 223
Blocks, 316, 379–80
Blocking design, 316
Box plot, 42

C

c–Chart, 476–78
Categorical variable, 4
Causality, 444, 453
Cellular phones, 146
Central limit theorem, 158, 200, 222
Chi–square distribution, 386–92
Chi–square test for equality of proportions, 389–92
Chi–square test for independence, 384–88
Club Med, 376–78
Coefficient of determination (r^2), 432–35

Complement of a set or event, 75
  rule of, 82
Conditional probability, 87–89
Confidence coefficient, 208
Confidence interval, 205
  for population mean, 207
    with known sigma, 207–12
    with unknown sigma, 213–19
Confidence level, 208
Contingency table, 88, 384–86
Control chart, 466–78
Control limit, 466
Correlation, 41n, 448–54
  coefficient, 448, 450
    tests about, 451–54
  and causality, 453
Critical point, 252
Critical region, 252
Crosby, Phil (interview), 481–82

D

Data, 5
Data collection, 5
Data distribution, 7
Decision rule, 250
Degrees of freedom, 372
Deming, W. Edwards, 465–66
Density, 141
  uniform, 141
Dependent variable, 408
Distribution, 115
  binomial, 130–37
    mean, 135
    standard deviation, 135
    variance, 135
  chi–square, 386–92

Distribution, *cont'd*
  mean, 386
  variance, 386
  Gaussian, 155–82
  F, 361–62, 374
  normal, 155–82
    as an approximation, 176–77
  standard normal, 159–66
  Student's, 214
  t, 214
  Uniform, 141
Drake, Sir Francis, 357

E

Efron, Bradley (interview), 239–42
Empirical rule, 55
Error deviation, 367
Ethico, 58–59
Evans, Sir Arthur, 3
Event, 77
Expectation, 124
Expected value, 124
Experiment, 77
Experimental design, 316, 379–80

F

Falvo, Toni, 351–55
F–distribution, 361–62, 374
Fences (of box plot), 42
Fermat, Pierre de, 73
Frame, 9
Frequency polygon, 19

G

Galton, Sir Francis, 407
Gaussian distribution, 157n

Goode, Bud (interview), 185–88
Grand mean (ANOVA), 366
Guinan, Mary (interview), 461–63

H

Heteroscedasticity, 438
Hinges (of box plot), 42
Histogram, 13
Hypothesis testing, 249
    ANOVA, 358–83
    Chi square, 384–92
    about difference in means, 322–42
    about difference in proportions,
        336–44
    left–tailed, 263–65
    about linear regression relationship,
        425–30
    about population mean, 249–80
        with known sigma, 257–77
        with unknown sigma, 274–84
    about population proportion,
        284–87
    one–tailed, 260–65
    right–tailed, 260–63
    two tailed, 256–58

I

Independence, 92
    of events, 92
        product rules for, 93–94
Independent variable, 408
Inference, statistical, 19, 196
Intercept (regression), 416–17, 416
Interquartile range (IQR), 36, 51
Intersection of events, 76
Interval scale, 5

J

Joint probability, 83

K

Kirkendall, Nancy (interview),
    401–4
Knossos, 3, 156

L

Least–squares method, 414–20
Left–tailed–test, 263–65
Level of statistical significance, 252
Literary Digest, 192–96
Lower control limit, 472

M

Malta, 71
Marinatos, Spyridon, 155
Mean, 45
    of population, 47
    of sample, 46
Mean squares, 373
Measurement, 4
Median, 36, 42, 43
MINITAB, 7
Mode, 45
Model, statistical, 409–10
    simple linear regression, 410–14
Moivre, Abraham de, 73
Monte Carlo casino, 73
Mozart, 213
Mutually exclusive events, 83

N

Nominal scale, 4
Nonrejection region, 252
Nonresponse bias, 6
Normal approximation, 176–77
Normal data, 178–80
Normal distribution, 155–82; *see*
    Distribution, normal
Null hypothesis, 249

O

Odds, 82
Ogive, 20
One–tailed–test, 260–65; *see*
    Hypothesis testing, one–tailed
Ordinal scale, 4
Outlier, 8
Ozone depletion, 57, 281

P

p–chart, 474–76
p–value, 265–69
Paired observations, 316
Parameter, 46
Pascal, Blaise, 73
Percentile, 35
Phaestos, 3
    disk, 3
Pie chart, 9
Polls, presidential election, 191–96,
    233–34
Population, 5
Power of a test, 297–302

Power function, 300
Practical significance, 295–96
Prediction (in regression), 444–47
Probability, 72–111
    conditional, 87–89
    density function, 141
    distribution, 115
    histogram, 116
    joint, 83
    objective, 74
    posterior, 98
    prior, 98
    subjective, 74
Probability distribution;
    see Distribution
Pyx, 245–50
    chamber, 245
    trial, 245–50
    today, 247

Q

Qualitative variable, 4
Quality control, 466–79
Quantitative variable, 4
Quartile, 35

R

r, 448–54
r^2, 432–35
r–chart, 472–74
Random sample, 196–98;
    see Sampling, random
Random sampling, 196–98;
    see Sampling, random
Random variable, 115–49
    continuous, 121, 140–44
    discrete, 120
        expected values, 124–30
        standard deviation, 128
        variance, 127
Randomized complete block design,
    380–81
Range, 51
Ratio scale, 5
Regression analysis, 407–59
Rejection region, 252
Remedy (pyx trial), 248
Residual, 410
    analysis (in regression), 437–41
Resistant measure, 36, 47
Right–tailed test, 260–63;
    see Hypothesis testing,
    right–tailed